THE INVENTION OF IMPROVEMENT

The Invention of Improvement

Information and Material Progress in Seventeenth-Century England

PAUL SLACK

OXFORD

UNIVERSITY PRESS

Great Clarendon Street, Oxford, OX2 6DP,
United Kingdom

Oxford University Press is a department of the University of Oxford.
It furthers the University's objective of excellence in research, scholarship,
and education by publishing worldwide. Oxford is a registered trade mark of
Oxford University Press in the UK and in certain other countries

© Paul Slack 2015

The moral rights of the authors have been asserted

First Edition published in 2015

All rights reserved. No part of this publication may be reproduced, stored in
a retrieval system, or transmitted, in any form or by any means, without the
prior permission in writing of Oxford University Press, or as expressly permitted
by law, by licence or under terms agreed with the appropriate reprographics
rights organization. Enquiries concerning reproduction outside the scope of the
above should be sent to the Rights Department, Oxford University Press, at the
address above

You must not circulate this work in any other form
and you must impose this same condition on any acquirer

Published in the United States of America by Oxford University Press
198 Madison Avenue, New York, NY 10016, United States of America

British Library Cataloguing in Publication Data
Data available

Library of Congress Control Number: 2014937426

ISBN 978-0-19-964591-6

Links to third party websites are provided by Oxford in good faith and
for information only. Oxford disclaims any responsibility for the materials
contained in any third party website referenced in this work.

For Alison and Kate

Preface

I have tried in this book to describe and explain how the notion of improvement—that is to say, of gradual, piecemeal, and cumulative betterment—came to be a familiar item in English public discourse in the seventeenth century. The importance of the topic first attracted my attention when I was completing my Ford Lectures, published in *From Reformation to Improvement: Public Welfare in Early Modern England* (Oxford, 1999). I was able then to say something about how the word improvement had first become common in discussion of economic and social well-being in the 1640s and 1650s, and I was intrigued by the fact that it was being applied after 1650 to public and private enterprises of every kind. The history and application of the concept, and what it meant for attitudes and behaviour, seemed to be worth further investigation.

Pursuit of the theme has led me in multiple directions. I have looked back before 1650 to examine the conditions which created what became a culture of improvement, and forward into the early eighteenth century in order to explore the consequences which it had. Central to the culture, as the subtitle of this book indicates, was a new appreciation of material progress as a process which could be investigated and measured, and the creation and collection of new kinds of knowledge and information for the purpose. Only the English, moreover, invented a word of their own for material progress, and that gave improvement a distinct identity and particular rhetorical resonance. Improvement was one of the things which made the English different from everyone else.

I am conscious that there are aspects of improvement which I have neglected, chiefly through lack of space. The most obvious is the theme of self-improvement, the betterment of the intellectual and moral capacities of every individual. Some of the authors important in what follows, Bacon and Hartlib among them, were at least as interested in that as in material improvement; and Locke, to whom I will refer in various contexts, revolutionized ways of thinking about it. That topic, however, was still being developed in novel ways in the mid-eighteenth century, and it merits a book of its own. Aspirations towards material improvement, on the other hand, along with the conviction that it was an indispensable foundation for intellectual and moral progress, were fully formed by 1740.

My story stops before what has generally been thought, with some justice, to have been the great age of an improvement ideology in England, in the later eighteenth and early nineteenth centuries. That is deliberate. My aim is to explain how it was that the pursuit of improvement came to enjoy that privileged position, so entrenched in English mentality in the middle of the eighteenth century that it was possible for a contemporary of the great improver Capability Brown to remark that he hoped to die before him, in order to see Heaven before even that had been improved.[1]

In the course of working on this book I have accumulated many debts. Some of them will be evident from the authors and works cited in my footnotes, and I have drawn very often on information in the *Oxford Dictionary of National Biography*, while only citing its entries when they were especially valuable for my purpose. I am grateful for the help of librarians, not only and often in the Bodleian and British Libraries, but also in the Royal Society and the Beinecke Library at Yale University, when I was working there. The staff of Oxford University Press, especially Christopher Wheeler, Robert Faber, Stephanie Ireland, and Cathryn Steele, have been unfailingly helpful and encouraging at every stage, and its anonymous readers were generous and constructively critical.

I owe most of all to the friends and colleagues with whom I have discussed part or the whole of my theme, especially Toby Barnard, Faramerz Dabhoiwala, John Elliott, Janet Godden, Julian Hoppit, Martin Ingram, Joanna Innes, Richard Ovenden, Margaret Pelling, Steve Pincus, Richard Smith, Keith Thomas, Keith Wrightson, and Tony Wrigley. My daughters have known that this book was in the making for much longer than anyone else, and may be as tired of hearing about improvement as the neighbours of Capability Brown, but here it is, and I dedicate it to them.

Paul Slack

Oxford
August 2014

[1] Brenda Colvin, *Land and Landscape: Evolution, Design and Control* (1970), p. 65.

Contents

List of Illustrations	xi
Abbreviations and Conventions	xii

1. **Introduction: Varieties of Improvement** — 1
   - Words and culture — 4
   - Economic circumstances — 8

2. **The Discovery of England** — 15
   - The face of the kingdom — 16
   - Wider perspectives — 31
   - Information and its uses — 43

3. **Elizabethan Foundations 1570–1640** — 53
   - Reforming the commonwealth — 54
   - Private enterprise for the public good — 66
   - Economic thinking and the crisis of the 1620s — 76

4. **Revolutions 1640–1670** — 91
   - Great expectations — 92
   - Infinite improvement — 102
   - Political economy — 116

5. **Wealth and Happiness 1670–1690** — 129
   - Decay or prosperity? — 129
   - Consumers and cities — 142
   - High living — 153

6. **Challenges to Affluence 1690–1730** — 170
   - Public politics — 170
   - War and empire — 179
   - A stationary state? — 191
   - Luxury — 202

7. **England's Improvement** — 215
   - Comfort and progress — 215
   - A knowledge economy — 229
   - The European context — 242

8. **Conclusion** — 257

Bibliography 265
   Manuscript sources 265
   Electronic resource 265
   Printed primary sources 265
   Secondary works 278
Index 307

List of Illustrations

1. *A Table of the cheiffest Citties, and Townes in England, as they ly from London* (*c.*1600). Provided by the Society of Antiquaries of London.   24
2. Title page of John Fitzherbert, *Here begynneth a ryght frutefull mater: and hath to name the boke of surueyeng and improumētes* (1523). Provided by the Bodleian Library.   27
3. Frontispiece of Walter Blith, *The English Improver Improved* (1652). Provided by the Bodleian Library.   107
4. Title page of William Potter, *The Key of Wealth, or, A new Way, for Improving of Trade* (1650). Provided by the Bodleian Library.   110
5. Title page of Andrew Yarranton, *England's Improvement by Sea and Land* (1677). Provided by the Bodleian Library.   162

Abbreviations and Conventions

BIHR	*Bulletin of the Institute of Historical Research*
BL	British Library, London
Davenant, *PCW*	Charles Davenant, *The Political and Commercial Works*, ed. Sir Charles Whitworth (5 vols, 1771)
EcHR	*Economic History Review*
EHR	*English Historical Review*
ESTC	English Short Title Catalogue (<http://estc.bl.uk>)
HMC	Historical Manuscripts Commission
HP	*The Hartlib Papers: A Complete Text and Image Database of the Papers of Samuel Hartlib (c.1600–1662)* (2nd edn on CDROM, HROnline, University of Sheffield, 2002)
NS	New Series
ODNB	*Oxford Dictionary of National Biography* (<http://www.oxforddnb.com>)
OED	*Oxford English Dictionary* (<http://dictionary.oed.com>)
OFB	*The Oxford Francis Bacon*
P&P	*Past and Present*
Petty, *EW*	*The Economic Writings of Sir William Petty, together with the Observations upon the Bills of Mortality*, ed. Charles Henry Hull (Cambridge, 1899)
SCED	Joan Thirsk and J. P. Cooper, eds, *Seventeenth-Century Economic Documents* (Oxford, 1972)
SP	State Papers
TNA	The National Archives, London
TRHS	*Transactions of the Royal Historical Society*
VCH	*Victoria County History*
Vickers, *Bacon*	*Francis Bacon: A Critical Edition of the Major Works*, ed. Brian Vickers (Oxford, 1996)

Dates are given in Old Style, except that the year is reckoned to begin on 1 January. In quotations from contemporary sources the spelling has generally been modernized. In cases where the surname of an author has changed its usual spelling since the seventeenth century (e.g. Ralegh as opposed to the original Raleigh), I have left the latter in my citations, but not in my text. In references to printed books, the place of publication is London, unless otherwise stated.

1
Introduction
Varieties of Improvement

Improvement means gradual, piecemeal, but cumulative betterment. It can refer to mental capacities as much as material circumstances, and to the capabilities of individuals as well as to the resources of whole societies and countries. Nowadays we take the pursuit of all kinds of improvement for granted, perhaps regard it as fundamental to human and social appetites for change, and enquire only into ways of achieving it and whether they are effective. Such aspirations have a history, however. They have not been equally prominent in all past societies, and they have sometimes been impeded and sometimes promoted by different cultural and economic circumstances.

The purpose of this book is to show that there was something wholly novel about the way in which they evolved in England during the seventeenth century. There, as in other countries, there had for centuries been innovations which would now be called improvements, including new technologies and more efficient modes of communication which created wealth and better standards of living. In seventeenth-century England, however, the word improvement was applied to all of these things for the first time, and the word and the notion were extended to the country as a whole, so that improvement became a fundamental part of the national culture, governing how the English saw themselves and the condition of the nation to which they belonged, and their expectations of how it might alter in the future.

Improvement came to rival and eventually replace alternative roads to better things, such as 'reformation' or 'revolution', which implied a total restoration or the sudden creation of some ideal state of affairs. Instead, the condition of England would be bettered by gradual and piecemeal change, and essential to that was wealth and the well-being it engendered. The idea of national improvement carried with it an appreciation of material progress as a wholly beneficent process, whose causes and consequences needed to be understood if it was to be prolonged. It also stimulated the collection of information about them, and the creation of new kinds of knowledge. By the beginning of the eighteenth century the quest for improvement distinguished England from other countries. It had become part of the collective mentality, and it made England distinctive.

It might also be said to have been one of the things that made England modern, or more accurately early modern. Historians and sociologists trying to specify the differences between modern and traditional societies, or the qualities which gave the West competitive advantages relatively early in modern global history, have compiled a long list of defining characteristics. Among them are the use of calculation and measurement as instruments of understanding and control in 'information-rich societies', a conviction that moral and material progress are to be encouraged, a belief in the virtues and promise of the future, and a readiness to break with tradition, custom, and the past.[1] All of these were aspects of English improvement, but most of them were also part of a general European culture which came into being between the fifteenth and the eighteenth centuries, in what has consequently been termed 'early modern' Europe.[2] First in Italy and then in other countries, Europeans began to distinguish themselves from the ancients, to engage deliberately in sustained scientific and technological innovation, and to use the term 'modern' about their achievements in something like its present sense.[3]

By the middle of the sixteenth century, in France as well as England, writers were also stressing the virtues of diligence and industry which enabled agriculture to flourish and the need to encourage commerce, provided it did not lead to an outflow of treasure.[4] In Spain in particular, they directed attention towards the role of markets in determining value, questioned medieval concepts of the 'just price', and explored the relationship between money and wealth.[5] In a century of rapid price inflation and political and economic competition between European states, all of them were beginning to examine the conditions which made one nation richer and more powerful than another, and to conceive of them in new ways.[6] We shall see at various points in this book that the English read and learnt from foreign publications and adopted their insights, just as they developed foreign technologies, for their own purposes.[7]

If some of the elements in English improvement had been borrowed from elsewhere, however, in England they were welded together in the course of the seventeenth century into a particularly powerful and coherent narrative about the country's betterment and how it could be prolonged. The Renaissance appetite for new inventions and discoveries, for example, which had led Italians even in the fourteenth century to think that such advances would 'never come to an end', was channelled by the disciples of Francis Bacon into the empirical investigations of the

---

[1] See, for example, C. A. Bayly, *The Birth of the Modern World, 1780–1914* (Oxford, 2004), pp. 55, 292–3; John Darwin, *After Tamerlane: The Rise and Fall of Global Empires, 1400–2000* (2008), pp. 187–8, 339–40; Mark Harrison, *Disease and the Modern World: 1500 to the Present Day* (Cambridge, 2004), pp. 1–2; Daniel Roche, *A History of Everyday Things: The Birth of Consumption in France, 1600–1800* (Cambridge, 2000), p. 8. For an introduction to modernization theory, see Agnes Heller, *A Theory of Modernity* (Oxford, 1999).
[2] Phil Withington, *Society in Early Modern England: The Vernacular Origins of Some Powerful Ideas* (Cambridge, 2010), pp. 45–8, 57–68.
[3] Withington, *Society in Early Modern England*, pp. 73–8; below, pp. 39–42; Robert Friedel, *A Culture of Improvement: Technology and the Western Millennium* (Cambridge, Mass., 2007), pp. 6–7.
[4] e.g. Claude de Seyssel, *La Monarchie de France et deux autres fragments politiques*, ed. Jacques Poujol (Paris, 1961), pp. 125, 160–4; below, pp. 18, 84.
[5] Marjorie Grice-Hutchinson, *Economic Thought in Spain: Selected Essays*, ed. Laurence S. Moss and Christopher K. Ryan (Aldershot, 1993), pp. xvi–xvii, 12–13; Marjorie Grice-Hutchinson, *The School of Salamanca: Readings in Spanish Monetary Theory, 1544–1605* (Oxford, 1952), *passim*.
[6] See, for example, Antonio Serra, *A Short Treatise on the Wealth and Poverty of Nations (1613)*, ed. Sophus A. Reinert, trans. Jonathan Hunt (2011).
[7] Below, pp. 15, 45–6, 152–3, 230–1.

Royal Society 'for the improving of natural knowledge'.[8] The systematic search for hard facts, 'matters of fact', 'certain knowledge' about peoples and kingdoms, which began with Italian and French political theorists in the sixteenth century, similarly flourished in England, influencing intellectual enterprise and shaping what eventually became the social sciences.[9]

One of its most striking manifestations was the deliberate creation of political economy, a new science with its own theoretical models, and using what William Petty christened 'political arithmetic' to measure the resources of a nation, how they might be increased, and—crucially for the notion of improvement—how they had changed over time.[10] In 1600 no one knew what the size, national income, or population of England were. By 1700 all these had been calculated within acceptable margins of error and were widely known; they could be related to one another, so that average incomes per head and the distribution of population and taxable wealth could be determined; and they could be compared with data from other countries and from the past, where they were available. New information enabled England's improvement, its material progress, to be measured.

Material progress was itself a subject of European-wide debate and argument which similarly pushed English thinking about improvement in new directions. The whole character of economic activity altered in the more prosperous parts of Europe in the seventeenth and eighteenth centuries as what have sometimes been termed 'consumer revolutions' followed changes in international and internal trade, and an 'industrious revolution' gave labourers the rising incomes which made them consumers as well as producers. Much has been written also about developments in moral and political philosophy which tried to accommodate the 'passions and interests' now visibly at work in new, competitive market societies.[11] When it came, material progress everywhere raised questions about the morality of conspicuous wealth, private indulgence, and the pursuit of profit, and about their compatibility with public happiness and the welfare of whole societies. But they posed particular problems for English advocates of continuing and cumulative betterment, and their success in evading, though not resolving, these issues was one reason why Bernard Mandeville's *Fable of the Bees*, the savage satire which ignited debate about luxury across the Continent in the 1720s, was written first for an English audience.[12]

---

[8] Friedel, *Culture of Improvement*, p. 91; Thomas Sprat, *History of the Royal Society*, ed. Jackson I. Cope and Harold Whitmore Jones (1959), p. 1.

[9] Mary Poovey, *A History of the Modern Fact: Problems of Knowledge in the Sciences of Wealth and Society* (Chicago, 1998); Barbara J. Shapiro, *A Culture of Fact: England, 1550–1720* (Ithaca, NY, 2002).

[10] Terence W. Hutchison, *Before Adam Smith: The Emergence of Political Economy, 1662–1776* (Oxford, 1988); Poovey, *History of the Modern Fact*, pp. 92–143; Andrea Finkelstein, *Harmony and the Balance: An Intellectual History of Seventeenth-Century English Economic Thought* (Ann Arbor, 2000).

[11] Maxine Berg and Elizabeth Eger, eds, *Luxury in the Eighteenth Century: Debates, Desires, and Delectable Goods* (Basingstoke, 2003); Jan de Vries, *The Industrious Revolution: Consumer Behaviour and the Household Economy, 1650 to the Present* (Cambridge, 2008); Albert O. Hirschman, *The Passions and the Interests: Political Arguments for Capitalism Before its Triumph* (Princeton, 1976).

[12] Christopher J. Berry, *The Idea of Luxury: A Conceptual and Historical Investigation* (Cambridge, 1994).

Improvement could not so easily have repelled Mandeville's challenge, however, if England had not been the chief beneficiary from the rapid changes in economic circumstances which had raised concern about material wealth and its contribution to well-being in the first place. Shifts in the distribution of wealth within and between countries presented the issues of luxury, consumption, and industriousness in different forms in different places, and left winners and losers with different responses to economic betterment. The fact that the English enjoyed high and rising real incomes per head for a century after 1650, and had overtaken the Dutch in command of global trade by 1720, was conspicuous proof that English improvement worked, and powerful propaganda for the cultural attitudes which seemed to have delivered it. Fifty years later, when Britain was becoming the first industrial nation, and distributing its culture with its empire across the world, improvement was no longer a cause that needed defending at home, but a conviction fortified by complacency.

Whether the culture contributed to changes in economic circumstances, rather than being simply a reflection of them, is a question which has been often debated, particularly by historians of the industrial revolution, and the final chapters of this book will argue that it did, that new kinds of information and aspirations to material betterment had a 'positive-feedback' effect on economic performance. The ways in which people think about their economic status, future prospects, and general well-being affect their behaviour as economic and political actors. Economies and cultures evolve together. In the case of improvement, however, cultural change came first; and its evolution over a 'long' seventeenth century between the end of the sixteenth and beginning of the eighteenth is my central theme. Improvement made England different, not only in its public frame of mind but in the kinds of economic behaviour that frame of mind encouraged.

WORDS AND CULTURE

A large part of the peculiarity of English improvement lay in the word itself. Although it came to have profound cultural significance, its origin and evolution were the product of linguistic accident not design. It was invented in the early sixteenth century, and used first about land and its uses, before being extended far beyond that in the seventeenth century to apply to every aspect of human and social endeavour. It was also a word for which there was, and still is, no equivalent in other languages. From the start, and almost by definition, English improvement could not be replicated elsewhere.

For most of the sixteenth century to improve was to make a profit from land. The verb had its roots in an Anglo-Norman term 'emprower' and in the medieval Latin 'approare', with the consequence that in the fifteenth century 'enprowment' or 'emprowment', and 'approvement', were both nouns used about land and profit.[13]

---

[13] *OED sub* 'improve, *v.* 2', 'improvement, *n.*'

Introduction 5

There was some early uncertainty about spelling and usage. In 1510 the citizens of Coventry were discussing what 'enprowement... opprowment or profit' might be gained from enclosure of the town commons; and parliament was still debating 'approwement' in 1549 when it referred to medieval legislation about the use of commons and wastes. The new statute which emerged, however, was explicitly called an Act 'concerning the improvement of commons and waste grounds'.[14] The spelling and the word had been popularized by John Fitzherbert's *Boke of surveyeng and improumentes*, published in 1523, and they were used after that about all kinds of profitable agrarian innovation. By 1613, according to one dictionary, 'improve' simply meant 'to raise rents'.[15]

The word was already being applied by analogy to other kinds of betterment, chiefly it would seem through the influence of Bacon. In 1605 his *Advancement of Learning* taught that learning might be improved, 'improved and converted by the industry of man' in order to 'correct ill husbandry'; and his essays and letters referred to the improvement of the king's lands and revenues and to 'improvements of things invented'. In the 1640s Baconian reformers of agriculture and education found it natural to transfer ideas of enclosing, nurturing, and improving from one sphere to the other, and once 'what is good in children' and their 'intellectual abilities' could be improved, so, by 1650, could the whole of nature and the whole nation.[16] The process is a good example of the ways in which words often change their colour and meaning through metaphorical extension, so that initially neutral or descriptive terms of narrow application come to confer approval on disparate things, and in this case give coherence to a host of apparently unrelated activities.[17] Dictionaries caught up with usage only slowly. In 1658, however, one of them included in its definition of improvement 'a thriving, a benefitting in any kind of profession'. In 1721 another said simply 'bettering, progress'; and progress, as we shall see, was another word extending its meaning in the same way, at the same time, and with similar effect.[18]

Unlike progress, however, improvement was a new English coinage, and its singularity was reinforced by the fact that no other European language had a synonym for it. Compilers of bilingual dictionaries and translators of English tracts

---

[14] Paul Slack, *From Reformation to Improvement: Public Welfare in Early Modern England* (Oxford, 1999), p. 70 n. 65; *Journals of the House of Commons*, ed. T. Vardon and T. E. May, vol. i (1852), p. 16; M. Luders et al., eds, *Statutes of the Realm* (11 vols, 1820–8), 3 and 4 Ed. VI, *c*. 3. The 1549 statute was called the 'Statute of Improvement' in the Grand Remonstrance in 1641, clause 32.

[15] Bill Shannon, 'Approvement and Improvement in the Lowland Wastes of Early Modern Lancashire', in Richard W. Hoyle, ed., *Custom, Improvement and the Landscape in Early Modern Britain* (Farnham, 2011), pp. 175–7; Paul Warde, 'The Idea of Improvement *c*.1520–1700', in Hoyle, ed., *Custom, Improvement and the Landscape in Early Modern Britain*, pp. 129–30; Robert Cawdrey, *A Table Alphabeticall* (3rd edn, 1613), *sub* 'improve'.

[16] Slack, *Reformation to Improvement*, pp. 69, 80–1; Vickers, *Bacon*, p. 34.

[17] For a lucid account of the evolution of verbal usage in the ways suggested here, see Quentin Skinner, 'The Idea of a Cultural Lexicon', in *Visions of Politics* (3 vols, Cambridge, 2002), i. 158–74. For related discussion of 'key words' in the early modern period, see Mark Knights, 'Towards a Social and Cultural History of Keywords and Concepts by the Early Modern Research Group', *History of Political Thought*, 31 (2010), pp. 427–48.

[18] Slack, *Reformation to Improvement*, p. 81; below, pp. 217–18.

had to use numerous words to cover its several applications. As might be expected, the most common equivalents were versions of the English words betterment, increase, or advance. To improve is still rendered as *améliorer, mejorar, migliorare, verbessern*, or *augmenter, aumentar, aumentare, erweitern*, in French, Spanish, Italian, and German, respectively; and improvement becomes the noun based on such verbs, as with *verbetering* and *vooruitgang* in Dutch. James Howell's *Lexicon tetraglotton: an English–French–Italian–Spanish dictionary* in 1660 used several of these alternatives, but had to resort to translating whole phrases like 'a lord that improves his rents' in order to convey the word's full flavour, and Guy Miège's French–English dictionary in 1677 had to distinguish between improving land, knowledge, learning, and sincerity for the same reason. In 1736 a *Dictionary English, German, and French*, published in Leipzig and Frankfurt, had more than twenty German equivalents of 'to improve' and gave separate translations of 'capable of improvement' and 'improver'.[19] No single word would do.

One obstacle to easy translation seems to have been a quirk of linguistic development. In some languages, to improve meant to disapprove, disprove, or reject, from the Latin *improbo* or *improbare*. It was occasionally used in that sense in English in the sixteenth century, chiefly in theological texts,[20] but it persisted longer in Italian and in French. The result could be some lexicographical confusion, as with John Florio's Italian–English Dictionary which rendered the Italian *improvare* and *improbare* as 'to improve' even in a 1659 edition, by which time it meant the opposite in English.[21] In France, dictionaries at the end of the seventeenth century show that *improuver* meant *condamner, désapprouver, ne pas approuver*, and the word was still being used there, in that dismissive sense, in the early eighteenth century.[22] English improvement can have had even less appeal in countries with a Romance language than in those like Germany and Holland which had no word as capacious in its range as improve at all.

Translating English texts about improvement into any other language was therefore a difficult business, requiring linguistic dexterity and considerable circumlocution. Separate references to improvement in a single text had to be replaced by different words, depending on the context. When translated into French in 1674, William Temple's remarks about improvements of land and improvements of manufactures in the Dutch Republic became descriptions of the 'incomparable advantage' gained from land and the 'industry' applied to manufacturing there,

---

[19] James Howell, *Lexicon tetraglotton: an English–French–Italian–Spanish dictionary* (1660) *sub* 'to improve', 'improved'; Guy Miège, *A New Dictionary French and English* (1677), *sub* 'improve', 'improved', 'an improving or improvement'; Christian Ludovici, *A Dictionary English, German and French* (2nd edn, Leipzig and Frankfurt, 1736), p. 337.

[20] *OED sub* 'improve, *v*. 1', cf. 'improve 2'. Christopher Wase, *Dictionarium minus* (1662) translated 'improbo' as 'disapprove' or 'disprove'.

[21] John Florio, *Vocabolario italiano & inglese* (1659), p. 171. Cf. John Florio, *A Worlde of Wordes* (1598), p. 171. 'Improverare' in Italian still meant 'to reproach' in 1777: Ferdinando Bottarelli, *The New Italian, English and French Pocket-Dictionary* (3 vols, 1777), *sub* 'improverare'.

[22] Pierre Richelet, *Dictionnaire françois* (Geneva, 1680), *sub* 'improuver'; Antoine Furetière, *Dictionnaire universel* (The Hague, 1690), *sub* 'improuver'. For later French usage, see Émile Littré, *Dictionnaire de la langue française* (2 vols, Paris, 1863–73), *sub* 'improuver'.

without the repetition. What Josiah Child had written about the improvement of trade and the improvement of riches in England in 1690 became, in a much later French translation, an account of the 'extension' of commerce and an 'increase' in wealth.[23] There are similar examples in translations of English authors into German as well as French, and in English texts which translated several words meaning economic 'increase' or 'advance' in foreign books as simply 'improvement'.[24]

Improvement lost in translation the whole of the rhetorical force which the word's constant repetition in multiple contexts had given it in seventeenth-century England. As we will see in later chapters, it was first commonly used about all kinds of betterment in what one historian calls its 'revolutionary moment' of the 1640s and 1650s, when 'inventions and improvements' became a catchphrase summarizing what useful knowledge and economic advance could be expected to achieve.[25] By 1700 the titles of tracts about agriculture, navigation, commerce, manufactures, poor relief, the rebuilding of London after the Great Fire, and even the use of time, demonstrated that everything could be improved. 'England's Improvement' had been part of the running head of Walter Blith's agricultural tract of 1649, *The English Improver*, which was itself further *Improved* in 1652. After that other authors picked up the baton, in *England's Improvement Revived* (1670), for example, *England's Improvements* (1675), *England's Improvement by Sea and Land* (1677), and *England's Improvements Justified* (1680) yet again.[26] Improvement had become a commonplace, a national slogan, and slogans and catchphrases denote and promote changes in culture.

The number of English publications with titles referring to improvement was still tiny in relation to total published output, scarcely reaching 1 per cent of the whole in the last two decades of the seventeenth century, when there were 360 of them. In the last two decades of the eighteenth century, in what is commonly regarded as the great age of improvement, there were nearly 3,000 of them, but they still amounted to less than 3 per cent of the total number of publications.[27] Yet England's improvement had begun and been seen to have begun long before then. In 1700 improvement was more than an everyday idiom of expression; it was a word essential to political discussion of national affairs and an integral part of English culture. It privileged certain kinds of public and private behaviour above others, encouraging innovative, industrious, and in every sense profitable activities,

---

[23] Le Chevalier Temple, *Remarques sur l'estat des Provinces Unies des Païs-bas* (The Hague, 1674), pp. 272, 277; Sir William Temple, *Observations upon the United Provinces of the Netherlands*, ed. Sir George Clark (Oxford, 1972), pp. 108, 110–11; Josiah Child, *Traités sur le commerce* (Amsterdam and Berlin, 1754), pp. 36, 120; Josiah Child, *A New Discourse of Trade* (1751 edn), pp. xxviii, 32. See also below, p. 248.

[24] For an instance of an English translation of a Spanish economic text, compare Gerónimo de Uztáriz, *Theorica y practica de commercio y de marina* (Madrid, 1757), pp. 1, 12, with the English rendering of an earlier edition in *The Theory and Practice of Commerce and Maritime Affairs*, trans. John Kippax (2 vols, 1751), i. 1, 25.

[25] Warde, 'Idea of Improvement', p. 128; Slack, *Reformation to Improvement*, pp. 80–1.

[26] Cf. below, p. 115.

[27] Calculations based on *ESTC*, accessed 21 Mar. 2011. Cf. Sarah Tarlow, *The Archaeology of Improvement in Britain, 1750–1850* (Cambridge, 2007), pp. 11–19.

while discouraging their opposites. It sustained a story about England's progress and helped to bring it into being. Still novel and circumscribed in its uses in the sixteenth century, in the seventeenth century improvement became a particular way of thinking, a turn of mind, which distinguished the English from everyone else.

ECONOMIC CIRCUMSTANCES

Cultures are shaped by more than words, however. As suggested earlier, economic circumstances had also made England distinctive and helped to form the character of English improvement by the early eighteenth century, and something more needs to be said by way of introduction to indicate why that was so. It involves looking at economic and cultural change over the very long run, and asking the kinds of question which thinkers of the eighteenth-century Enlightenment were already posing, not only about what made one country in Europe different from another, but what differentiated western Europe in general from other civilizations.

China was already the great puzzle for early modern Europeans who were learning about its early invention of printing, the compass, and gunpowder, and who wondered, as Voltaire did, why a civilization which had cultivated sciences and arts in past centuries had recently made 'so little progress'.[28] In the later seventeenth and eighteenth centuries both China and Japan displayed many of the features of conspicuously successful economies. They were information-rich societies, and undergoing their own educational, industrious, and consumer revolutions.[29] China might no longer have had so strong a culture of invention as in the twelfth century, but in the later seventeenth century it was publishing books on how to improve health and on mathematics and the physical and historical sciences just as England was,[30] and there were the same complaints about a new luxury corroding traditional values and social distinctions as there were in Europe at the same time.[31] Chinese cities were still larger than any in Europe, even if they were no longer growing, while a burst of urbanization in Japan was producing cities bigger than

---

[28] Voltaire, *Essai sur les mœurs et l'esprit des nations*, ch. I, pp. 42–3, in Bruno Bernard et al., eds *Les Œuvres complètes de Voltaire*, vol. xxii (Oxford, 2009).

[29] Bayly, *Birth of the Modern World*, pp. 51–9; Geoffrey Parker, *Global Crisis: War, Climate Change and Catastrophe in the Seventeenth Century* (New Haven, 2013), p. 502; Mary Elizabeth Berry, *Japan in Print: Information and Nation in the Early Modern Period* (Berkeley, 2006), pp. 170–2; Jiang Wu, *Enlightenment in Dispute: The Reinvention of Chan Buddhism in Seventeenth-Century China* (Oxford, 2008), pp. 106–8; Alan Macfarlane, *The Savage Wars of Peace: England, Japan and the Malthusian Trap* (Oxford, 1997), p. 83; Timothy Brook, *The Confusions of Pleasure: Commerce and Culture in Ming China* (Berkeley, 1998), pp. 167–79.

[30] Mark Elvin, *The Pattern of the Chinese Past* (Stanford, Calif., 1973), pp. 178, 199, 203; Peter Burke, 'History, Myth, and Fiction; Doubts and Debates', in José Rabasa et al., eds, *The Oxford History of Historical Writing*, iii: *1400–1800* (Oxford, 2012), pp. 278–9; Kenneth Pomeranz, *The Great Divergence: China, Europe, and the Making of the Modern World Economy* (Princeton, 2000), p. 44; Parker, *Global Crisis*, pp. 650–1.

[31] Craig Clunas, 'Things in Between: Splendour and Excess in Ming China', in Frank Trentmann, ed., *The Oxford Handbook of the History of Consumption* (Oxford, 2012), pp. 52–4; Brook, *Confusions of Pleasure*, p. xvii; Timothy Brook, *Vermeer's Hat: The Seventeenth Century and the Dawn of the Global World* (2008), pp. 25, 173.

any in England or the Dutch Republic. By 1700 Edo, the Japanese capital, with a population of about a million people, was the world's largest city and twice the size of London; and Kyoto and Osaka, with over 300,000, were bigger than Amsterdam and ten times larger than any English provincial town.[32]

Neither empire, however, was deliberately investing in the large-scale maritime enterprises which stimulated economic exchange and competition in north-western Europe, perhaps because they had no obvious need for them; and since both had large populations working for low wages, they lacked the incentives to labour-saving which encouraged early industrialization. In the fifteenth century China had deliberately turned away from maritime exploration, and by the eighteenth century its manufacturing industries and transport networks were sufficiently well developed to meet existing demand, while improvements in agriculture in the later seventeenth century had pushed productivity per acre to the limit of what was possible. It has been argued that it had reached a 'high-level equilibrium trap'.[33] Eighteenth-century Japan was similar in many respects, resistant to commercial contacts with the West, and, since wages were low, inventing new technologies which increased rather than reduced the employment of labour, so that even the wheel, though known, was little used.[34]

The capacity of parts of Asia to sustain large cities and dense populations was a sign of their long-run economic success over centuries, but it was cheap labour which gave China and India the largest cotton industries in the world in the seventeenth century.[35] In such conditions there was arguably little incentive to disturb traditional cultures which for the most part looked to a past golden age and past precedent for examples of how things might be better, and did not envisage economic growth as something to be welcomed and encouraged in its own right, as many Europeans did.[36] It was only in the eighteenth century that writers in Japan began to investigate the economic importance of markets, commerce, and the circulation of wealth, inspired partly by new developments in Confucian thought and by observation of the Dutch.[37]

---

[32] Elvin, *Pattern*, pp. 175–8; Darwin, *After Tamerlane*, p. 133. Nanjing had c.750,000 people in the early seventeenth century: R. Po-chia Hsia, *A Jesuit in the Forbidden City: Matteo Ricci 1552–1610* (Oxford, 2010), p. 175.

[33] Darwin, *After Tamerlane*, pp. 42–5; Parker, *Global Crisis*, p. 619; Elvin, *Pattern*, pp. 203–34, 298–9.

[34] Parker, *Global Crisis*, pp. 487, 492; Macfarlane, *Savage Wars of Peace*, p. 45.

[35] Ronald Findlay and Kevin H. O'Rourke, *Power and Plenty: Trade, War, and the World Economy in the Second Millennium* (Princeton, 2007), pp. 353–5; Robert C. Allen, *Global Economic History: A Very Short Introduction* (Oxford, 2011), pp. 116–18; Robert C. Allen, *The British Industrial Revolution in Global Perspective* (Cambridge, 2009), pp. 212–13.

[36] Bayly, *Birth of the Modern World*, pp. 292–3; Rana Mitter, *Modern China: A Very Short Introduction* (Oxford, 2008), pp. 14–15.

[37] Mark Metzler and Gregory Smits, 'Introduction: The Autonomy of Market Activity and the Emergence of *Keizai* Thought', in Bettina Gramlich-Oka and Gregory Smits, eds, *Economic Thought in Early Modern Japan* (Leiden, 2010), pp. 5–6, 12–15. I am aware that the broad-brush generalizations in this and the previous paragraph require considerable qualification. For a more nuanced view of trade, see Findlay and O'Rourke, *Power and Plenty*, pp. 353–64, and of cultural attitudes, Pomeranz, *Great Divergence*, pp. 44–8. A more radical argument about culture's irrelevance in China is presented in

In some respects the countries of early modern Europe, exploring the oceans for new markets, and seeking to produce cheap textiles to rival those they imported from the East, had been trying to catch up with Asia, until coal and colonies gave some of them the resources for early industrialization and global empires. But their earlier histories had also given them what turned out to be political and institutional advantages in any contest for power and plenty. Warfare between the several city-states and kingdoms of western Europe had taught them how to build states and market economies in which property was taxed but protected. Competition between them had also led them to borrow skills and technologies from one another and create a common culture.[38] Through accidents of history or geography, however, some of them were better placed than others for long-term economic success, and by the end of the seventeenth century the countries of north-western Europe had markedly different economies from those elsewhere. Europe was different from Asia, but it was also diverse.

We can see some of the reasons for the diversity in what was happening to occupational structures and standards of living in different countries. In 1500 the most urbanized parts of Europe were northern Italy and the Low Countries, where less than half of the population was engaged in agriculture, compared with three-quarters in England, France, and Germany. By 1670, however, the agricultural proportion had shrunk to no more than 60 per cent in England and by 1750 to 46 per cent. The agrarian component fell much more slowly elsewhere, except in the Netherlands where urbanization increased until 1700 and then ground to a halt.[39] Trends in the relative value of wages tell a similar story about the divergence between north-western Europe and the rest of the Continent. The best data come from cities, but they are probably broadly illustrative of living standards in their hinterlands. In the early sixteenth century the purchasing power of wages was much the same in London and Amsterdam as it was in Florence and Vienna. After that it rose rapidly only in London and Amsterdam, with the result that by 1700 wages were worth twice as much there as in the other cities. After 1700 their value continued to rise in London while stagnating in Amsterdam.[40]

R. Bin Wong, *China Transformed: Historical Change and the Limits of European Experience* (Ithaca, NY, 1997), pp. 54–5.

[38] Darwin, *After Tamerlane*, pp. 190–1; E. L. Jones, *The European Miracle: Environments, Economies and Geopolitics in the History of Europe and Asia* (3rd edn, Cambridge, 2003); Findlay and O'Rourke, *Power and Plenty*, p. 359; below, pp. 242–5.

[39] Allen, *British Industrial Revolution*, p. 17; E. A. Wrigley, *People, Cities and Wealth: The Transformation of Traditional Society* (Oxford, 1987), pp. 170, 180–1; E. A. Wrigley, 'The Transition to an Advanced Organic Economy: Half a Millennium of English Agriculture', *EcHR* 59 (2006), pp. 466–70; Stephen Broadberry and Bishnupriya Gupta, 'The Early Modern Great Divergence: Wages, Prices and Economic Development in Europe and Asia, 1500–1800', *EcHR* 59 (2006), p. 27. On the early 'modernization' of the Dutch economy by 1500, see Jan Luiten Van Zanden, 'Taking the Measure of the Early Modern Economy: Historical National Accounts for Holland in 1510/14', *European Review of Economic History*, 6 (2002), pp. 131–63.

[40] Allen, *British Industrial Revolution*, p. 34. In 1725 real wages in Beijing were similar to those in Florence and Vienna; in Delhi they were much lower from at least 1575 onwards: Allen, *British Industrial Revolution*, p. 34.

The advantages of the north-west sprang partly from new patterns of Atlantic trade, and partly from demographic changes since the fifteenth century which had also accelerated economic growth. The Black Death of 1347–52 had reduced European populations by a third, sometimes more, and it raised living standards for the survivors, but only for as long as it took for population fully to recover. In some parts of Europe, in England as compared with southern France and Italy, for example, recovery came exceptionally slowly; and in north-west Europe generally the Black Death seems to have established a distinctive marriage pattern, which kept fertility low by encouraging celibacy and delaying family-formation until viable independent households could be afforded. In consequence, populations grew more quickly and standards of living declined earlier in other parts of Europe than in England, and left them closer once again to a potential Malthusian ceiling by 1600.[41]

England and Holland were therefore in a much stronger economic position than most other countries at the end of the seventeenth century. They could feed highly urbanized populations, either by a combination of intensive agriculture and imports of foodstuffs funded by a flourishing commerce, as the Dutch did, or, as in England, solely from the agrarian improvements which were being stimulated by rising urban demand. When bad harvests occurred, they did not mean famine there, as they did in large parts of France and in Scotland in the 1690s. One of their causes may have been a climate colder and wetter between 1670 and 1700 than it had been for centuries, but even that 'Little Ice Age' proved to be a positive incentive to improvement in England and Holland. It prompted the adoption of new crops like turnips, and in the case of England boosted demand for coal to keep the fires of London and other European cities burning when timber resources no longer met growing demand.[42] Both countries were enjoying the prosperity which came with what has been termed an 'advanced organic economy'; and while that meant a stationary economic equilibrium in Holland, England's inorganic and accessible coal resources provided one of the foundations in the eighteenth century for its continuing and increasingly rapid economic growth.[43]

The other foundation was English and then British command of international trade, delivered by naval and military power sufficient to defeat first Dutch and later

---

[41] Findlay and O'Rourke, *Power and Plenty*, pp. 116–20, 361–2; Richard Smith, 'Plagues and Peoples: The Long Demographic Cycle, 1250–1670', in Paul Slack and Ryk Ward, eds, *The Peopling of Britain: The Shaping of a Human Landscape* (Oxford, 2002), pp. 199–200, 203–4; Allen, *British Industrial Revolution*, pp. 13–14.

[42] Parker, *Global Crisis*, p. 619; Mark Overton, *Agricultural Revolution in England: The Transformation of the Agrarian Economy, 1500–1850* (Cambridge, 1996), pp. 202–3. On the Little Ice Age, see Parker, *Global Crisis*, pp. 3–25; T. C. Smout, *Nature Contested: Environmental History in Scotland and Northern England since 1600* (Edinburgh, 2000), pp. 52–3. For scepticism about whether it occurred at all, see Morgan Kelly and Cormac Ó Gráda, 'The Waning of the Little Ice Age: Climate Change in Early Modern Europe', *Journal of Interdisciplinary History*, 44.3 (2014), pp. 301–25.

[43] Wrigley, 'Transition to an Advanced Organic Economy', pp. 435–80; Wrigley, *People, Cities and Wealth*, pp. 181–3. On the coal industry, see John Hatcher, *The History of the British Coal Industry*, i: *Before 1700* (Oxford, 1993), pp. 40–5.

French competitors.⁴⁴ In 1700 England was poised to outpace a Dutch Republic which had for a century been the miracle of economic growth which the rest of Europe sought to copy. Agricultural and industrial output in England had probably more than doubled since 1500, and both were accelerating with regional specialization. The value of the country's foreign trade had risen fivefold since 1600, and particularly quickly after 1630.⁴⁵ The pace of urbanization was also remarkable, even by comparison with the Netherlands. Between 1500 and 1700 the proportion of the English population living in towns multiplied threefold, and most of the increase occurred in London, which grew from perhaps 55,000 people in 1520 to 200,000 in 1600 and 575,000 in 1700. Though scarcely comparable to Edo or Beijing, by 1700 it had overtaken Paris to become the largest city in Europe apart from Constantinople, and it was the engine and exemplar of the country's economic growth.⁴⁶

The exceptional character of the English economy in the second half of the seventeenth century emerges clearly from recent calculations of long-term changes in the national income over the centuries. After rising only slowly in the sixteenth century, annual national income (GDP) appears to have doubled in the seventeenth. In 1700 the country was twice as rich as it had been in 1600. What that meant for general living standards can be measured by calculating national income per head. Average incomes scarcely increased at all between 1500 and 1650, since the population was rising, and in some decades they declined. When the population was stable, as it was for most of the century after 1650, however, income per head increased. It rose by as much as a half between the 1640s and 1700, and continued to grow with the gross national income for another half-century, though less rapidly. For the first time in more than a century, standards of living were rising. The only comparable period of such sustained improvement in material living standards in England since the early Middle Ages was the half-century after the Black Death. That had raised national income per head to a level in the early fifteenth century which was only achieved again in 1700.⁴⁷

Independent evidence from calculations of changes in the value of wages between 1300 and 1750 corroborates that general picture, while sharpening the contrasts within it. The real value of the wages of unskilled labourers in the building industry appears to have increased even more than GDP per head between 1350 and 1450, and to have declined much more sharply between 1500 and 1650. After 1650 wages rose more or less in line with per capita GDP. Given different sets of

---

⁴⁴ Findlay and O'Rourke, *Power and Plenty*, pp. 338–9.
⁴⁵ Stephen Broadberry et al., 'British Economic Growth, 1270–1870', working paper, as at 10 Jan. 2011, pp. 13–17, 44 (fig. 4), 47 (fig. 5); D. W. Jones, 'The Workings and Measurement of Pre-Industrial "Organic" Economies: Conjectures on English Agrarian Growth, 1660–1820', *Journal of European Economic History*, 35 (2006), pp. 205–7; Ralph Davis, *English Overseas Trade 1500–1700* (1973), pp. 7, 9, and *passim*.
⁴⁶ E. A. Wrigley, *Energy and the English Industrial Revolution* (Cambridge, 2010), pp. 58–61; E. A. Wrigley, 'A Simple Model of London's Importance in Changing English Society and Economy, 1650–1750', in *People, Cities and Wealth*, pp. 133–56.
⁴⁷ N. J. Mayhew, 'Prices in England, 1170–1750', *P&P* 219 (May 2013), table 8, p. 37; Broadberry et al., 'British Economic Growth', pp. 22–4, 53 (fig. 9), 55 (fig. 10).

data and the different ways in which they have been manipulated, it is difficult to judge which is the more reliable.[48] The wage evidence certainly seems a better reflection of what was happening in the sixteenth century, when inequalities in the distribution of wealth were becoming more pronounced and the numbers suffering real poverty increasing;[49] and if it is reliable across the whole period, it suggests that even in the eighteenth century English living standards had not yet reached the level they had attained at the end of the fifteenth.[50] The evidence from wages is important to the history of material improvement because it prevents any easy assumption that English labourers were better off in the early eighteenth century than they had ever been.

The evidence for GDP is equally important for the history of improvement, however, because it points clearly to one simple and crucial difference between conditions after the Black Death and conditions after 1650. GDP had fallen sharply in the earlier period but increased markedly in the later seventeenth century. In 1700 English wage-earners may not yet have enjoyed the living standards they had had in the later Middle Ages, but the total national cake had grown much bigger and they had a share in the benefits. Moreover, that greater national income had funded improvements in agriculture, transport, and industry, and in the military power which captured a rising share of international commerce. Labourers and their families may well have had to work longer hours and engage in an industrious revolution in order to boost household incomes and improve their diet along with other material comforts, but the landed and commercial elite and a better-funded state had acquired the resources to invest in innovation for the future. Economic growth and a more integrated national economy had enabled London to grow to ten times its size in the later Middle Ages,[51] and created a much more literate population with an appetite for information, new consumer goods, and novelties of all kinds.

These were the circumstances which accompanied the evolution of an improvement culture in the later seventeenth and early eighteenth centuries, encouraged its formation, and fortified it against moralists who condemned luxury and the pursuit of profit and material comfort. In different conditions the culture would certainly have been less popular and less resilient than it was, and it is possible that notions of improvement would not have grown into something that can be called a culture at all. It is striking also that the culture and the economic circumstances which sustained it were both coming into existence in the 1650s, after a political revolution,

---

[48] Broadberry et al., 'British Economic Growth', pp. 30–1, 59 (fig. 13); Gregory Clark, 'The Condition of the Working Class in England 1209–2004', *Journal of Political Economy*, 113 (2005), pp. 1307–40; Robert C. Allen, 'The Great Divergence in European Wages and Prices from the Middle Ages to the First World War', *Explorations in Economic History*, 38 (2001), pp. 411–47.

[49] See for example Alexandra Shepard and Judith Spicksley, 'Worth, Age, and Social Status in Early Modern England', *EcHR* 64 (2011), pp. 516–24; Paul Slack, *Poverty and Policy in Tudor and Stuart England* (1988), pp. 40–7.

[50] Broadberry et al., 'British Economic Growth', p. 24. See also Allen, *British Industrial Revolution*, p. 40, fig. 2.3.

[51] Caroline Barron, *London in the Later Middle Ages: Government and People, 1200–1500* (Oxford, 2004), p. 241, suggests a population for London in 1500 of *c*.50,000.

and we will need to consider in the Conclusion to this book how far that was more than a coincidence. Nonetheless, the word and the concept had been introduced much earlier, in the very different economic circumstances of the century before 1650, and we should begin the history of improvement there if we are to understand the conditions—intellectual and political as well as social and economic—which explain why it was able first to take root, and then to blossom and flourish in the way that it did.

The next chapter is concerned with the intellectual background to the story which follows. It describes how the English thought about England and about their place in time and space in the later sixteenth century, and how those perceptions changed afterwards. It is intended to establish the framework of perceptions and assumptions which determined contemporary understanding of innovation and change for the better, and to survey the terrain which improvement first occupied and ultimately transformed. The chapters which follow have a narrower chronological focus, beginning with Chapter 3, which explores the enterprises, projects, and government policies, from the 1570s to the 1630s, which conditioned and constrained what improvement came to mean.

Chapters 4 and 5 are the heart of the book. The first is designed to show how, in a context of political and intellectual revolution, an ideology of improvement was manufactured which brought together multiple aspirations for betterment, explored their potential consequences, and publicized the results. Chapter 5 argues that when the reality of material progress became visible, and as widely recognized as it was by the 1680s, that too was incorporated, though rather less comfortably, into what was becoming a national culture. The culture had internal contradictions, and it faced formidable new challenges after 1689, when wars against France had to be paid for and won; and Chapter 6 describes how they were accommodated, though never wholly overcome, so that improvement, like the nation, not only survived but thrived. Chapter 7 summarizes what England's improvement had achieved by 1730, and widens the focus once more by placing it in its European context and trying to show what was special about it.

The headings to most of the chapters have been given dates as a guide to readers, since they present a chronological narrative about change over time. The chapters overlap with one another, however, each concentrating on the different topics indicated by the titles of their sub-sections. I hope that readers will be able easily to identify the sections of particular interest to them. But I hope also that, taken together and in sequence, they show how improvement came to be central to English public discourse and the historical impact which that had.

2
The Discovery of England

English improvement depended on knowledge that it was happening. It could only be perceived, thought about, and articulated as a concept, if people knew something about the present state of England, how it was changing, and how it compared with other countries and other times. In the period covered by this book, between the 1570s and 1730s, information about all those things increased in volume, became more explicitly comparative, and was used in wholly new ways. In an age of discoveries of many kinds, the land and people of England were also looked at afresh, and they were found to have improved. This chapter therefore sets the scene, in more senses than one, for the chapters which follow.

It begins in Elizabeth's reign with what W. G. Hoskins called 'the rediscovery of England', the first collective effort to document a country which seemed to be 'waiting to be explored and described'. In the 1570s William Lambarde's *Perambulation of Kent*, Christopher Saxton's maps of every county, and William Harrison's 'Description' of England inaugurated a series of similar publications which continued for more than a century, and presented images of the kingdom to a wide audience.[1] They had earlier models, in manuscript itineraries, chronicles, prospects, and plans,[2] but publication made all the difference, encouraging emulation and disseminating knowledge. It was not a kind of discovery unique to England, however. Other countries were simultaneously engaged in compiling their own national and local histories, comparing them with one another, and publishing their findings in far greater quantity.[3] Thanks to translations and borrowings from them, English readers were able to set their own rediscovery in a context increasingly better informed by new discoveries about Europe, its history, and its position in the wider world.

By 1700 therefore, there was plentiful published information available in England about the country's improvement for those who wished to acquire it. It was also being used and discussed there in ways which prompted European authors to look to

---

[1] W. G. Hoskins, *Provincial England: Essays in Social and Economic History* (1963), p. 222; D. M. Palliser, *The Age of Elizabeth: England under the Later Tudors, 1547–1603* (1983), pp. 8–9, 390.
[2] Richard Helgerson, *Forms of Nationhood: The Elizabethan Writing of England* (Chicago, 1992), p. 132; P. D. A. Harvey, *Maps in Tudor England* (1993), pp. 7–9.
[3] José Rabasa et al., eds, *The Oxford History of Historical Writing*, iii: *1400–1800* (Oxford, 2012), pp. 244–5, 332–4, 389. On the quantity of published output in different countries, see Ian Maclean, *Scholarship, Commerce, Religion: The Learned Book in the Age of Confessions, 1560–1630* (Cambridge, Mass., 2012), p. 5; Andrew Pettegree, 'Centre and Periphery in the European Book World', *TRHS* 18 (2008), p. 127.

English publications, as the English had previously looked to theirs, in order to learn what made one state richer and more powerful than another.[4] But that process of collecting and manipulating information and so creating new knowledge had itself begun in Elizabeth's reign. In 1603 England was already a society rich in information, even if it did not yet know what it wanted to do with it all. It had an increasing appetite for its accumulation; it was looking at other countries and at civilizations in the past in order to identify England's distinctiveness; and it was learning something in the process about prospects for alternative and perhaps better futures.

THE FACE OF THE KINGDOM

'The space of one hundred years... hath much altered the face of the kingdom', declared John Adams when justifying his large gazetteer of towns and villages and his new map of England and Wales in 1680. Older publications, most of them resting on the pioneering work of Elizabethan cartographers, were no longer fit for purpose. The same point was being made about the need to revise William Camden's monumental *Britannia*, first published in 1586 and much enlarged in 1610. 'The face of the kingdom' was 'so much changed to what it then was', according to Richard Blome in 1673;[5] and Edmund Gibson, who completed a revised edition in 1695, drew attention to the 'strange alteration in the face of things' since 1610. It had been brought about by 'the growth of trade, the increase of buildings [and] the number of inhabitants', in short by 'all these improvements'.[6] Still treated with respect, much copied, and plagiarized, the publications of the Elizabethan discoverers of England had become benchmarks against which recent change and improvement could be measured.

The Elizabethan pioneers had themselves been far from insensitive to change, but most of them were more interested in the distant than the recent past, and in what had been lost rather than what might have been gained. Many of them were antiquaries in the tradition of John Leland, whose manuscript collections about local antiquities had been compiled in Henry VIII's reign, and some of them, like Camden and Lambarde, were members of the Society of Antiquaries founded in 1586.[7] Lambarde's *Perambulation* was the first in a succession of what were called 'chorographies', local histories generally of a single county, describing its ancient monuments and its leading gentry, complete with their coats of arms and pedigrees

---

[4] Below, pp. 246–9.
[5] John Adams, *Index Villaris* (1680), sig. π4ᵛ; Richard Blome, *Britannia* (1673), sig. A2ᵛ.
[6] Edmund Gibson, *Camden's Britannia, Newly Translated into English with Large Additions and Improvements* (1695), sig. A1ʳ.
[7] Daniel Woolf 'Historical Writing in Britain from the Late Middle Ages to the Eve of the Enlightenment', in *Oxford History of Historical Writing*, iii. 489–91; Angus Vine, *In Defiance of Time: Antiquarian Writing in Early Modern England* (Oxford, 2010), pp. 7–8, 17–19; Helgerson, *Forms of Nationhood*, pp. 127–8, 137–9.

so as to demonstrate that they too had antiquity.[8] By 1700 there were published chorographies of twelve of the forty English counties, and those of another twenty were being worked on.[9] Towns also had new histories, the first of them John Stow's *Survey of London* which 'attempted the discovery' of his own 'native soil and country' in 1598,[10] and even some small parishes received close attention, particularly if they contained monuments to notable families in need of 'rescue... from the teeth of time and oblivion'.[11]

In much of the antiquarian and chorographical literature, especially before 1640, there was a pronounced nostalgia for what a place had once been. It can be found in Camden, and still more in Stow, who lamented the present 'declining time' when charity, hospitality, old pastimes, and ceremonies were in decay and London's churches and monuments neglected; and it echoed in much late sixteenth-century complaint about the damage done by the dissolution of the monasteries, inflation, and the social mobility and increase in poverty which came with them.[12] Even at this date there were other voices, however. No one praised the achievements of the Protestant Reformation more than Lambarde, Stow's 'loving friend', and another member of the Society of Antiquaries, Richard Carew, celebrated the flourishing tin mines as much as the old genealogies of Cornwall, and insisted (somewhat prematurely) that improvements in housing were at last bringing the 'civility' of eastern counties down to the south-west.[13]

There were equally mixed opinions about change in the first attempts to describe the present state of the kingdom as a whole. William Harrison's *Description of England* (1577 and 1587) welcomed many of its new civilities, including the 'amendment of lodging' over the past half-century which had brought chimneys, comfortable bedding, and pewter plates to the houses of Essex, though not yet to other counties 'further off from our southern parts'. But Harrison deplored the 'daily oppression' of small farmers by rack-renting landlords, and the growth in the number of lawyers, merchants, and tradesmen, whose greed for profit disrupted social hierarchies and corrupted the whole commonwealth.[14] In his 'State of England',

---

[8] e.g. Diarmaid MacCulloch, *Suffolk and the Tudors: Politics and Religion in an English County 1500–1600* (Oxford, 1986), pp. 118–20; Ann Hughes, *Politics, Society and Civil War in Warwickshire, 1620–1660* (Cambridge, 1987), p. 47.

[9] Calculations from C. R. J. Currie and C. P. Lewis, eds, *A Guide to English County Histories* (Stroud, 1994). Cf. Graham Parry, *The Trophies of Time: English Antiquarians of the Seventeenth Century* (Oxford, 1995), p. 241n.

[10] John Stow, *A Survey of London*, ed. Charles Lethbridge Kingsford (2 vols, Oxford, 1971), i. iii.

[11] Hoskins, *Provincial England*, p. 213.

[12] Patrick Collinson, 'John Stow and Nostalgic Antiquarianism', in J. F. Merritt, ed., *Imagining Early Modern London: Perceptions and Portrayals of the City from Stow to Strype, 1598–1720* (Cambridge, 2001), pp. 27–51; Ian Archer, 'The Nostalgia of John Stow', in David L. Smith et al., eds, *The Theatrical City: Culture, Theatre and Politics in London, 1576–1649* (Cambridge, 1995), pp. 17–34. Cf. Alexandra Walsham, *The Reformation of the Landscape: Religion, Identity, and Memory in Early Modern Britain and Ireland* (Oxford, 2011), p. 276.

[13] Collinson, 'John Stow', p. 28; Hoskins, *Provincial England*, p. 132. On standards of living in Cornwall, see below, p. 158.

[14] Hoskins, *Provincial England*, p. 138; William Harrison, *The Description of England*, ed. Georges Edelen (Ithaca, NY, 1968), pp. 115–20, 201–3; Ralph Houlbrooke, 'England', in Paulina Kewes et al., eds, *The Oxford Handbook of Holinshed's Chronicles* (Oxford, 2013), pp. 641–2, 644.

written for a foreign patron in 1601, Thomas Wilson similarly combined praise for the country's unparalleled natural resources and manufactured commodities with sharp criticism of the profit motives which, while making the 'common people... very rich', had eroded status distinctions between gentlemen, yeomen, and mere farmers.[15] Michael Drayton's very different 'chorographical description' of England, *Poly-Olbion* (1612 and 1622), was equally ambivalent. It evoked a historic landscape of copious rivers, forests, and fields, threatened only by the 'base avarice' of landowners, but claimed in its prefatory verses to display the 'power and plenty of Albion' and everything 'this spacious land contains', for 'profit' as well as for 'pleasure'.[16]

Drayton's idealization of an apparently unchanging countryside was a reflection of literary models with a long history behind them, and the texts from which they derived were themselves far from hostile to the pursuit of profit by farmers exploiting the fertility and productivity of the land. A popular translation of *Xenophons Treatise of Housholde*, which first appeared in 1532, commended the honourable, diligent, and healthy practice of husbandry, and the 'profit' to be obtained from close attention to the 'goodness and fertility of the ground', and Thomas Tusser's even more popular verses on the many *Points of Good Husbandry*, first published in 1557, underlined the importance of 'thrift' (derived from 'thriving') as a road to riches.[17] The land was already being regarded as a location of profitable and productive activity when the revival of interest in 'Georgic' literature at the end of the sixteenth century popularized such notions among the elite and made them fashionable.[18]

John Smyth of Nibley, for example, writing a history of his own part of Gloucestershire in 1639, might well have read Virgil's *Georgics*, with its image of the poet himself as an industrious ploughman. Smyth praised the 'sturdy, long-settled yeomanry' of the hundred of Berkeley, and drew particular attention to the 'noble' labour of its ploughmen, 'the only vocation wherein innocence remaineth'. He claimed to have once been a ploughman himself, but he owned his own farm, invested in colonial ventures, and lived in a county famous for its rural textile industry.[19] George Chapman's *The Georgicks*, a translation of Hesiod published in 1618, described the profits flowing from the labour of husbandmen, and may have caught the attention of Francis Bacon, to whom it was dedicated. When revising his essay 'Of Riches' in 1625, Bacon added a passage about 'the improvement of the

---

[15] Thomas Wilson, 'The State of England Anno Dom. 1600', ed. F. J. Fisher, in *Camden Miscellany XVI* (Camden Society, 3rd ser. LII, 1936), pp. vii, 10, 18–20, 24. Cf. below, p. 58.
[16] Vine, *Defiance of Time*, pp. 169–74, 177–8; Andrew McRae, *God Speed the Plough: The Representation of Agrarian England, 1500–1660* (Cambridge, 1996), pp. 255, 259.
[17] *Xenophons Treatise of Housholde* (1544 edn), fos 46–7, 59ᵛ; McRae, *God Speed the Plough*, pp. 143–51, 206, 211–17; Craig Muldrew, *The Economy of Obligation: The Culture of Credit and Social Relations in Early Modern England* (Basingstoke, 1998), pp. 161–7.
[18] On Georgic influence, see Anthony Low, *The Georgic Revolution* (Princeton, 1985), pp. 13–98. There has been some dispute about when classical models gave English Georgic poetry a more didactic and economic dimension. I have followed McRae, *God Speed the Plough*, pp. 198–228, in thinking the influence can be detected at the beginning of the seventeenth century if not earlier.
[19] David Rollison, *The Local Origins of Modern Society, Gloucestershire 1500–1800* (1992), pp. 76, 259, 263.

ground' being 'the most natural obtaining of riches', and observed that 'where men of great wealth do stoop to husbandry, it multiplieth riches exceedingly'.[20] Bacon's patron, James I, had already penned his own Horatian elegy, reminding the gentry that 'the country' provided their 'revenues', and that they should cherish 'the thrifty plough', 'your sheep, your corn, your cow'.[21]

The commonplace distinction which privileged 'the sweet country' with its 'innocent pleasures' above the avarice of 'unsavoury' towns and industries was therefore more qualified than it might at first appear;[22] and it was further complicated by other binary oppositions, between civility and barbarity, the tame and the wild, and the cultivated and the barren. The kind of countryside that was most admired was the one Camden found in Oxfordshire, 'a fertile county and plentiful', 'garnished with cornfields and meadows', 'the hills beset with woods'. It was a landscape ordered, exploited, and improved by man.[23] In 1620 Gervase Markham, another author who claimed once to have been a ploughman, thought farmers on hard and barren ground should be congratulated for 'having conquered nature by altering nature, and yet made nature better that she was before'.[24] Although that Baconian aspiration to improve nature by changing it was to resonate more strongly after 1640 than it had done before, and although agrarian improvement remained controversial in particular circumstances, there was no dispute that, other things being equal, the land was there to be exploited not simply conserved.

The members of Samuel Hartlib's circle during the Interregnum who set about changing public attitudes towards innovation in the ways described in later chapters were fully conscious that they were completing what had already been begun.[25] 'In Queen Elizabeth's days', one of them remarked, 'ingenuities, curiosities and good husbandry began to take place.'[26] Now, however, the time had come to realize the whole of Bacon's programme for the 'endowment and benefit of man's life' by means of a comprehensive 'natural and experimental history'. It would extend far beyond the few existing accounts of the progress of agriculture to embrace the 'history of arts' or 'history mechanical', and give an account of the whole of 'nature altered or wrought' by man.[27] Had it ever been completed, it would have brought

---

[20] McRae, *God Speed the Plough*, p. 212; Francis Bacon, *The Essayes or Counsels, Civill and Morall*, ed. Michael Kiernan (*OFB* xv, Oxford, 1985), p. 110.
[21] McRae, *God Speed the Plough*, p. 281.
[22] Walter Cary, *The Present State of England* (1626), p. 13; *The Diaries of Lady Anne Clifford*, ed. D. J. H. Clifford (Stroud, 1990), p. 112; Keith Thomas, *Man and the Natural World: Changing Attitudes in England 1500–1800* (1983), pp. 242–53.
[23] Thomas, *Man and the Natural World*, pp. 205, 254–8; William Camden, *Britain, or a chorographicall description of the most flourishing kingdoms, England, Scotland and Ireland, and the ilands adjoining*, trans. Philemon Holland (1610), p. 373.
[24] Paul Warde, 'The Idea of Improvement *c.*1520–1700', in Richard W. Hoyle, ed., *Custom, Improvement and the Landscape in Early Modern Britain* (Farnham, 2011), pp. 134–5; Gervase Markham, *Markhams farwell to Husbandry* (1620), pp. 3, 9. Cf. below, p. 40.
[25] Below, pp. 58–60, 106–8.
[26] Robert Child in *Samuel Hartlib his Legacie* (2nd edn, 1652), p. 40.
[27] Vickers, *Bacon*, pp. 176–8; Francis Bacon, *The Instauratio magna Part II: Novum organum*, ed. Graham Rees (*OFB* xi, Oxford, 2004), pp. 453, 473; *Part III*, ed. Rees (*OFB* xii, Oxford, 2007), pp. xxxi–xxxiii.

about a rediscovery of England as notable as that embarked upon by the Elizabethans. In practice it made new demands on writers of local chorographies and descriptions of the kingdom which it proved impossible for them ever to meet, but it altered the character of what they set out to describe and the images they presented of the face of the kingdom.

The first venture down this road, Joshua Childrey's *Britannia Baconica: Or, The Natural Rarities of England, Scotland and Wales* (1661), had only its title to recommend it. It consisted of apparently random references to the climate, flora and fauna, and antiquities of each county, and especially to their 'curiosities' which he tried, and largely failed, to account for in 'natural' terms.[28] In order to do better, Fellows of the Royal Society undertook to compile histories of trades and agricultural practices in every part of the kingdom, and that required a time-consuming collaborative exercise in data collection. In 1670 a group of Fellows and *virtuosi* circulated the first printed questionnaires for the purpose, partly to assist John Ogilby, the cartographer, with his own *Britannia*, a projected series of volumes which would include county surveys. His famous road-book was the only volume finally brought to publication.[29]

When publications did appear, they were rarely wholly satisfactory. Robert Plot's *Natural Histories* of Oxfordshire (1677) and Staffordshire (1686) were intended to cover 'all curiosities both of art and nature', not only the 'animals, plants and the universal furniture of the world' but also the 'inventions and improvements' in 'mechanic or liberal arts' which promoted trade. Plot's brief descriptions of things like Witney blankets and the special kind of cart used in Banbury scarcely amounted to what was promised.[30] Plot's pupil, the Welsh scholar Edward Lhuyd, spent a lifetime collecting materials for his *Archaeologia Britannica: An Account of the Languages, Histories, and Customs of Great Britain*, and he had 4,000 questionnaires printed for circulation; but his great 'Glossography', the first comparative study of the Celtic languages published in 1707, was a fragment of what he originally intended.[31]

Experimental natural history turned out to be more effectively pursued piecemeal, in contributions to periodicals like John Houghton's *Collection of Letters for the Improvement of Husbandry and Trade* begun in 1681, if they were ever published at all, and in private collections of information and correspondence.[32] They reflected the emergence of a nation-wide intellectual community, a network of gentlemen and clergy more widely diffused than that of Elizabethan antiquaries, and as keen to compile and exchange data about plants, minerals, and weather

---

[28] Stan A. E. Mendyk, *'Speculum Britanniae': Regional Study, Antiquarianism, and Science in Britain to 1700* (Toronto, 1989), pp. 166–9; *ODNB sub* Childrey; Walsham, *Reformation of the Landscape*, pp. 357–8, 362. Cf. Vine, *Defiance of Time*, pp. 202–3; below, p. 98.

[29] Mendyk, *Speculum Britanniae*, pp. 163–4; Adam Fox, 'Printed Questionnaires, Research Networks, and the Discovery of the British Isles 1650–1800', *Historical Journal*, 53 (2010), pp. 597–8.

[30] Fox, 'Printed Questionnaires', pp. 598–9; Mendyk, *Speculum Britanniae*, pp. 196–7; Robert Plot, *The Natural History of Oxford-shire* (2nd edn, Oxford, 1705), pp. 219–20, 262–3, 283–4.

[31] Mendyk, *Speculum Britanniae*, pp. 206–12; Fox, 'Printed Questionnaires', p. 603.

[32] Below, pp. 144–5, 165, 172–3.

patterns, as about genealogies and antiquities.[33] They had also inherited from the chorographic tradition a particular interest in what made one place different from another, not what they had in common but local 'curiosities' and idiosyncrasies in 'the manners of the people' which had to be rescued from oblivion because they might now be disappearing.[34] There was the same impulse behind the collections of dialect stories and vocabularies in the 1670s and 1680s which led John Evelyn to want a full 'dialect survey' of every county,[35] and it inspired collections of folk stories and superstitions which in the end produced Henry Bourne's *Antiquitates vulgares* (1725), the 'antiquities of the common people'.[36]

There was a related interest in what we might call environmental determinism, the view espoused by Aubrey that climates and soils determined the different characters of people in the cheese and chalk country of Wiltshire, and Thomas Fuller's supposition that each county had a 'particular genius' which inclined 'the natives... to be dexterous, some in one profession, some in another'.[37] Like the merchants and gentry from different counties who were meeting in separate dining clubs in London in the 1650s,[38] these authors must have been conscious that local differences were being eroded by the increasing cultural and economic dominance of London; but their primary purpose was the identification of difference, and that had been the purpose of chorographies from their beginning.[39]

Integration, on the other hand, the demonstration of connections between places, was the fundamental business of cartography, the other discipline which contributed to the discovery of England in the later sixteenth century, and which was similarly refashioned in the later seventeenth, but with greater success. The publications of the Elizabethan cartographers, Saxton, John Norden, and John Speed, were designed as much for pleasure as for profit, just as chorographies

---

[33] Fox, 'Printed Questionnaires', p. 616. See for example *A Seventeenth-Century Flora of Cumbria: William Nicolson's Catalogue of Plants, 1690*, ed. E. Jean Whittaker (Surtees Society 193, 1981); White Kennett, *Parochial Antiquities attempted in the History of Ambrosden, Burcester and other adjacent parts...* (Oxford, 1695); and on weather observations, Joshua Childrey, *Britannia Baconica* (1661), 'Preface'; Geoffrey Parker, *Global Crisis* (New Haven, 2013), p. 661; *The Great Diurnal of Nicholas Blundell of Little Crosby, Lancashire*, i: *1702–1711*, ed. J. J. Bagley (Lancashire and Cheshire Record Society 110, 1968), *passim*. Cf. Joanna Innes, *Inferior Politics: Social Problems and Social Policies in Eighteenth-Century Britain* (Oxford, 2009), pp. 130–1.

[34] Adam Fox, 'Vernacular Culture and Popular Customs in Early Modern England: Evidence from Thomas Machell's Westmorland', *Cultural and Social History*, 9 (2012), pp. 330–3.

[35] Roger Lass, ed., *The Cambridge History of the English Language*, iii: *1476–1776* (Cambridge, 1999), pp. 497, 501–2.

[36] Walsham, *Reformation of the Landscape*, pp. 540–1.

[37] Michael Hunter, *John Aubrey and the Realm of Learning* (1975), pp. 114–15; Thomas Fuller, *The History of the Worthies of England* (1662), p. 53. On differences between chalk and cheese countries, see David Underdown, *Revel, Riot and Rebellion: Popular Politics and Culture in England, 1603–1660* (Oxford, 1985), pp. 3–7 and *passim*. Cf. Childrey's remark that Cornishmen were bigger and stronger than others because 'the western people of most countries are the tallest and stoutest': *Britannia Baconica*, p. 2.

[38] Peter Clark, *British Clubs and Societies 1580–1800: The Origins of an Associational World* (Oxford, 2000), p. 50.

[39] Helgerson, *Forms of Nationhood*, pp. 131–9.

were.[40] Their maps of the counties of England and Wales made attractive illustrations for the backs of playing cards (since there were conveniently 52 of them), and they were hung on the walls of inns across the kingdom for centuries, even after the boundaries of these historic units of local government were redrawn in 1974.[41] Like maps of the whole kingdom and the British Isles, however, they were political statements as well as cultural icons, a means of conceptualizing space indispensable to the exercise of power and the imposition of control.

That was why William Cecil commissioned cartographers and drew sketch maps himself when he wanted to see how Ireland related to England, where coastal and inland fortifications should be situated, and how Catholic and loyal gentry were distributed across the country.[42] It was said that he carried in his pocket for easy reference Laurence Nowell's small atlas of Britain and Ireland, a volume which displayed the great contrast between the well-defined shires of England, with their towns and ports, and Ireland beyond the Pale which was a wilderness of forest and bog, still to be cultivated by English planters. Maps proved to be crucial to English strategy for subjugating and colonizing both Ireland and North America by means of planted towns, ordered landscapes, and hence civility.[43]

This is not to deny that they had more prosaic practical utility. The four thousand place-names on Saxton's great wall map of England situated each of them in relation to the others, so that people like the MPs in 1571 who had not seen Berwick or St Michael's Mount knew they could at least find them 'on the maps'.[44] Maps like Saxton's were said to be used by 'all noblemen and gentlemen' for 'their better instruction', and they could be instructive in more ways than one. When the young Bulstrode Whitelocke accompanied the Assize judges on their circuits and always 'carried with him... Camden and Speed, with the maps, and in every place compared the books with the information of the inhabitants', this aspiring young member of the governing class was doing more than simply finding his way.[45] A short handbook like Norden's *England: An Intended Guyde for English Travailers* (1625), with its tables of distances between major towns in each country,

---

[40] Helgerson, *Forms of Nationhood*, pp. 107–8; Harvey, *Maps in Tudor England*, pp. 54–60. On later maps see Bernhard Klein, *Maps and the Writing of Space in Early Modern England and Ireland* (Basingstoke, 2001); J. H. Andrews, *Shapes of Ireland: Maps and their Makers 1564–1839* (Dublin, 1997); William Ravenhill, 'John Adams, his Map of England, its Projection, and his *Index Villaris* of 1680', *Geographical Journal*, 144 (1978), pp. 424–37; Charles W. J. Withers, 'How Scotland came to Know Itself: Geography, National Identity and the Making of a Nation 1680–1790', *Journal of Historical Geography*, 21 (1995), pp. 376–9. Cf. below, p. 161.

[41] Victor Morgan, 'The Cartographical Image of "The Country"', *TRHS* 5th ser. 29 (1979), pp. 149–54; Harvey, *Maps in Tudor England*, pp. 64–5.

[42] Harvey, *Maps in Tudor England*, p. 47.

[43] William. J. Smyth, *Map-Making, Landscapes and Memory: A Geography of Colonial and Early Modern Ireland, c.1530–1750* (Cork, 2006), pp. 31–2, 421–50.

[44] Helgerson, *Forms of Nationhood*, p. 133; G. R. Elton, *The Parliament of England, 1559–1581* (Cambridge, 1986), pp. 227–8.

[45] J. B. Harley, 'Meaning and Ambiguity in Tudor Cartography', in Sarah Tyacke, ed., *English Map-Making 1500–1650: Historical Essays* (1983), p. 27; *The Diary of Bulstrode Whitelocke 1605–1675*, ed. Ruth Spalding (British Academy Records of Social and Economic History NS xiii, Oxford, 1990), p. 51.

said something about local urban hierarchies;[46] and there could be no more eye-catching demonstration of London's supremacy than a printed broadside of 1600 which showed provincial cities arrayed along points of the compass and in concentric circles, with a miniature prospect of the metropolis sitting in the centre (Figure 1, overleaf).[47] That was an abstract model which abandoned any pretence at accurate representation of space in order to convey a message.

It perhaps says something about the relative insignificance of more practical considerations that many early maps showed no roads.[48] When English cartography revived in the 1670s under the stimulus of the Royal Society, Ogilby's *Britannia* (1676) amply filled the gap, with its strip-maps of all the main roads leading out from 'our prime and great metropolis', most of them precisely measured, and a summary map of the kingdom on which roads were more prominent than county boundaries for the first time.[49] *Britannia* was too large and expensive to have been regularly used by travellers, like those in Bunyan's *Pilgrim's Progress* (1678) who looked at a 'book or map' to guide them to the Celestial City, but there were summaries of maps on broadsheets and in pocket books which might serve that purpose. Like Ogilby's volume, whose intention was to 'to improve our commerce and correspondency', they were responding to a growth of internal trade and communication which was already well under way.[50]

One of its symptoms was increasing traffic along rivers and roads, and especially the main roads used by the royal posts. The posts had been reorganized and opened up to private use in 1635, again explicitly 'for the advancement of all his Majesty's subjects in...trade and correspondence', and according to Hartlib they were already achieving that by 1647.[51] A more obvious consequence was to put pressure on a transport network still largely unchanged since the Middle Ages and produce calls for improvements, despite the local and private interests which had to be overridden to deliver them.[52] There had been occasional bills in parliament

---

[46] John Norden, *England: An Intended Guyde for English Travailers* (1625). Cf. Jacob van Langeren, *A Direction for the English Traviller* (1635).

[47] *A Table of the cheiffest Citties, and Townes in England, as they ly from London* (c.1600); Lena Cowen Orlin, ed., *Material London, ca. 1600* (Philadelphia, 2000), fig. 1.1, pp. 1–3.

[48] Andrew McRae, *Literature and Domestic Travel in Early Modern England* (Cambridge, 2009), p. 31. Main roads were described, however, in literary texts: e.g. McRae, *Literature and Domestic Travel*, pp. 76–7; William Smith, *The Particular Description of England, 1588*, ed. H. B. Wheatley and E. W. Ashbee (1879), pp. 69–72.

[49] McRae, *Literature and Domestic Travel*, pp. 78–81; Robert J. Mayhew, *Enlightenment Geography: The Political Languages of British Geography, 1650–1850* (Basingstoke, 2000), pp. 75–9. John Adams's rival new map of 1677, however, still showed no roads at all: McRae, *Literature and Domestic Travel*, p. 85.

[50] McRae, *Literature and Domestic Travel*, pp. 79, 120.

[51] James F. Larkin and Paul L. Hughes, eds, *Stuart Royal Proclamations* (2 vols, Oxford, 1973, 1983), ii. no. 202, pp. 468–9; *Samuel Hartlib and the Advancement of Learning*, ed. Charles Webster (Cambridge, 1970), p. 128; Adam Fox, *Oral and Literate Culture in England, 1500–1700* (Oxford, 2000), pp. 370–1. Cf. Kevin Sharpe, 'Sir Thomas Witherings and the Reform of the Foreign Posts, 1632–40', *BIHR* 57 (1984), pp. 149–64; Mark Brayshay, Philip Harrison, and Brian Chalkley, 'Knowledge, Nationhood and Governance: The Speed of the Royal Posts in Early Modern England', *Journal of Historical Geography*, 24 (1998), pp. 265–88.

[52] David Harrison, *The Bridges of Medieval England: Transport and Society 400–1800* (Oxford, 2004), pp. 1–7.

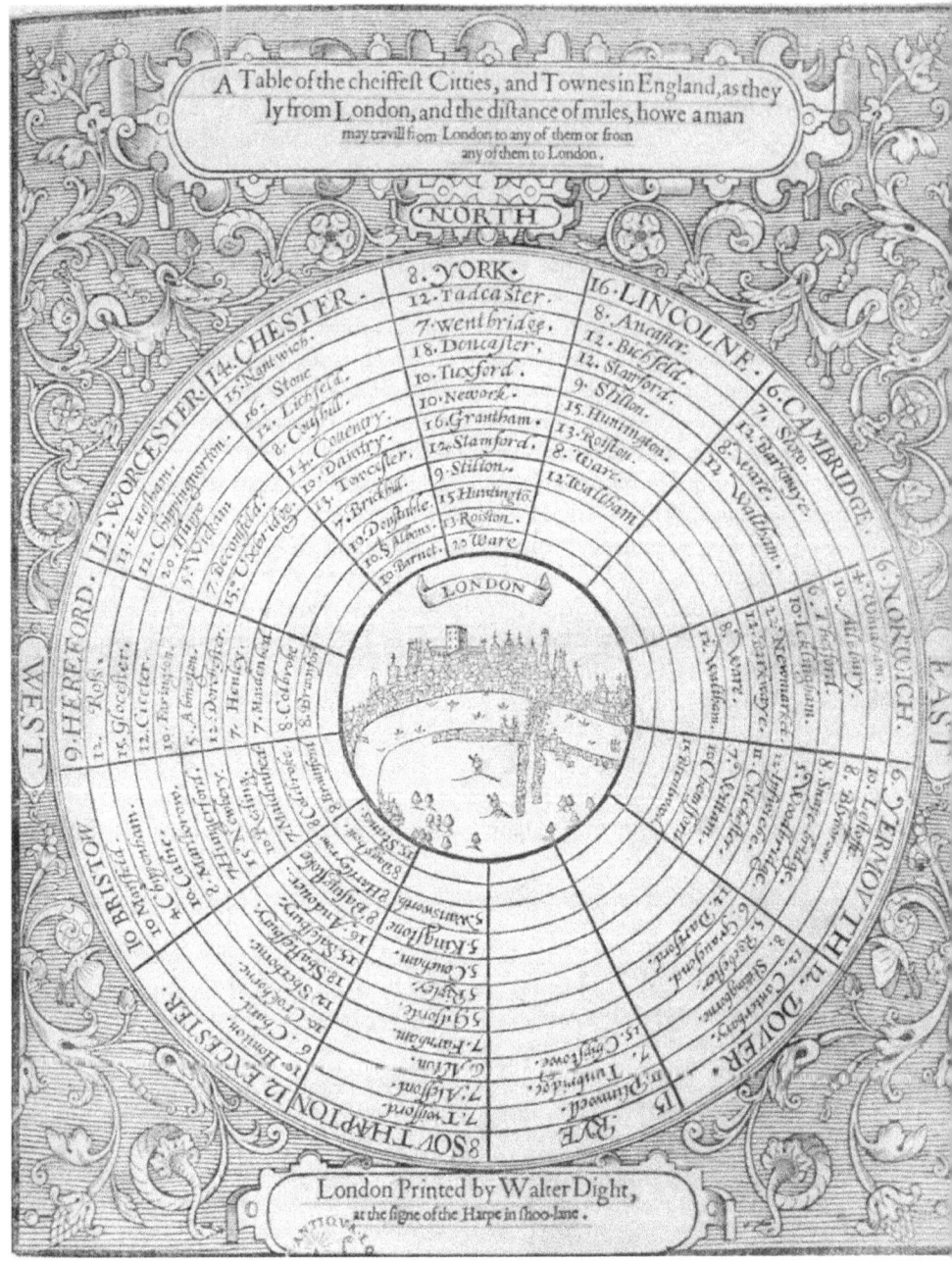

Figure 1. London at the centre: *A Table of the cheiffest Citties, and Townes in England, as they ly from London* (c.1600).

since at least 1571 for local improvements to river navigation, and an increase in their number in 1665 prompted the Speaker of the Commons to support them in terms which used the language of Drayton to make a point about political economy. 'Cosmographers', he said, were agreed that 'this island is incomparably furnished with pleasant rivers', and they were like veins in the natural body, conveying blood to every part, 'whereby the whole is nourished, and made useful'.[53] The physiological metaphor of circulation was a popular image often employed to justify commercial activity, and it was being applied to road as well as river improvements which were equally for 'the advancement of trade' and 'increase of wealth'.[54] But parliament was only able to pass Turnpike Acts in any number after 1689; and not until 1697, and only after some public pressure for it, was there legislation for roadside signposts to help people find their way.[55]

The expansion of trade and traffic had come first, driven by market forces in a more integrated national economy, and facilitated by new and quicker means of transport for people and goods, like the coaches introduced in the 1630s or the improved wagons with swivel axles which appeared in the 1650s. It had taken at least two days to get by road from Oxford to London until the advent of the 'flying coach' in 1671, which covered the ground in thirteen hours.[56] The benefits lay less in speed of communication, however, than in the frequency and reliability of transport services and the fact that they were available to the public. While the coasting trade in bulk goods grew by two-thirds between 1660 and 1702, the capacity of the regular carrying services by road in and out of London more than doubled between 1637 and 1715. Coach services numbered nearly a thousand a week by 1715 and reached most provincial centres. Weekly postal services from London expanded even more rapidly after the 1630s, and the penny post within ten miles of the General Post Office in Lombard Street was carrying nearly a million letters a year by 1700.[57] Public communication, 'correspondence' in the contemporary sense of the word, made distant places more accessible and familiar, brought them closer to London, and made the country seem smaller.

Although no one in England seems to have remarked on the fact, more accurate maps were having the same effect, making England and Wales and other parts of western Europe look smaller than they had before. When Louis XIV saw the new maps of France he had sponsored in 1693 he supposedly complained that his cartographers had cost his kingdom more land in a year than foreign armies had done in a century. English cartography had not yet reached the standards of the French, but the maps of the 1670s and 1680s made England and Wales narrower by

---

[53] T. S. Willan, *The Inland Trade* (Manchester, 1976), p. 23; McRae, *Literature and Domestic Travel*, pp. 44–54.

[54] McRae, *Literature and Domestic Travel*, pp. 86–7. Cf. below, pp. 71, 73, 78.

[55] Below, pp. 177–8; David Hey, *Packmen, Carriers and Packhorse Roads: Trade and Communications in North Derbyshire and South Yorkshire* (Leicester, 1980), p. 29.

[56] Hey, *Packmen, Carriers and Packhorse Roads*, p. 99; Alan Crossley, ed., *VCH Oxfordshire*, iv: *The City of Oxford* (Oxford, 1979), p. 290.

[57] John A. Chartres, 'The Marketing of Agricultural Produce', in Joan Thirsk, ed., *The Agrarian History of England and Wales*, v.ii: *1640–1750* (Cambridge, 1985), pp. 465–7; Michael Reed, *The Age of Exuberance 1500–1700* (1987), p. 318.

moving their western boundary eastward.[58] People who thought they knew their place precisely had been deluded. In 1701, a yeoman farmer in Shropshire, Richard Gough, had learnt from some now out-of-date 'computation of geographers' that his parish of Myddle was 'distant northwards from the earth's equator 52 deg. 53 min., and... in longitude... 21 deg. 37 min.' east of the Azores. Those coordinates, to which he attached such importance, would have placed him in the middle of Wales.[59]

Illusory or not, the apparent certainty of numbers had an obvious appeal; and accurate measurement had real significance for the distribution of property and power when it came to the measurement of area, whether that of an estate or a kingdom. Surveyors, who were emerging as a profession in the active land market of the sixteenth century, attracted controversy and hostility, especially in the colonial context of Ireland where they were almost unknown. The Irish decapitated a surveyor trying to draw a 'true and perfect' map of Ulster at the end of Elizabeth's reign, because 'they would not have their country discovered', and eight of the men working on William Petty's survey there in 1655 suffered a similar fate.[60] In England the tenants of improving landlords who did not want 'the quantities' of their land 'to be known by measuring' could only grumble that 'the world was merrier before measurings were used than it hath been since' and had to accept them. As early as 1523, in the first book on 'surveying and improvements', John Fitzherbert had insisted that landowners must have 'perfect knowledge' of the size of their estates in order to manage them effectively, and by 1551 Robert Recorde was using surveying as a prime example of the practical utility of mathematics, and reassuring his readers that 'proportion geometrical' offered no threat to anyone (Figure 2).[61] In 1610 another advocate of 'true information' from 'true surveys' added in their defence the fashionable argument that 'justice' could only be created out of confusion by the use of 'number, weight and measure'—the quantitative criteria of proportion sanctioned by Scripture for the understanding of God's creation.[62]

Measurement of a large area was nonetheless easier said than done. It involved some complicated geometry, and the larger the territory being measured, and the more irregular its boundaries, the more complicated the task became.[63] Writers

---

[58] Paul Glennie and Nigel Thrift, *Shaping the Day: A History of Timekeeping in England and Wales 1300–1800* (Oxford, 2009), pp. 337–9; Josef W. Konwitz, *Cartography in France 1660–1848: Science, Engineering and Statecraft* (Chicago, 1987), pp. 4–8.

[59] Richard Gough, *The History of Myddle*, ed. David Hey (Harmondsworth, 1981), p. 29. Adams's Gazetteer was more accurate (*Index Villaris*, p. 240), but Gough did not know it, and I have been unable to discover his own source.

[60] Smyth, *Map-Making*, pp. 54–5; below, pp. 97–8. For surveying in other colonial contexts and in reclaimed fenlands, see Roger J. P. Kain and Elizabeth Baigent, *The Cadastral Map in the Service of the State: A History of Property Mapping* (Chicago, 1992), pp. 255, 285–7.

[61] McRae, *God Speed the Plough*, pp. 172–3, 182–3, 186.

[62] McRae, *God Speed the Plough*, p. 188. On number, weight, and measure, see Wisdom of Solomon, 11: 20, and for uses of the text, John Dee, *The Elements of Geometrie of... Euclide* (1570), sig. biv; Jean Bodin, *The Six Bookes of a Common-weale* (1606), Book 4, p. 457; Charles Webster, *The Great Instauration: Science, Medicine and Reform, 1626–1660* (1975), p. 351; below, pp. 47, 80, 117, 188.

[63] For a later description of how to do it, see John Harris, *Lexicon technicum* (2 vols, 1704–10), *sub* 'area'; and for use of the methodology in calculating the acreage of England and Wales, *Nehemiah Grew and England's Economic Development: the means of a most ample increase of the wealth and strength of*

Figure 2. Advice to a farmer: John Fitzherbert's *Boke of surueyeng and improumētes* (1523). The first appearance of improvement on a title page.

wanting to make precise comparisons between the size of one county or country and another had no easy means of translating the spatial images on maps into square measures of acres or square miles, and often simply described their length and breadth and sometimes their circumference. Their areas could only be guessed at. In 1670 MPs spent 'a long time' disputing the size of England, some saying it

England, 1706–7, ed. Julian Hoppit (British Academy Records of Social and Economic History NS 47, Oxford, 2012), pp. 8–9.

was '50, some 36, some 30, some but 24 millions of acres'. One member thought it contained as many as 76 million, that having been judged from a map in the 1630s, presumably one by Speed or Norden.[64] The question could only be resolved quickly in the 1680s when Adams produced his much better map and Edmond Halley undertook the remarkable exercise of cutting it up, and comparing the weight of a measured circle of known area with the weight of the different counties and then the whole country. England and Wales were shown to contain just under 40 million acres.[65] (The modern figure is 37.3 million.) England, like France, was smaller than some contemporaries supposed.

The historical importance of Halley's calculation lies less in its accuracy, however, since that was sometimes questioned, than in the ambition which lay behind it and in the uses to which it was put by his contemporaries. In this as in other cases, their utility gave new numbers publicity and credibility. As we shall see in later chapters, precise knowledge of the area of every county was employed, along with other data, to plot the distribution of taxation, population, and even political representation across the kingdom, and show where they were concentrated. It prompted questions about inequalities and about how they might be remedied. As one of those who used Halley's figures appreciated, number, weight, and measure were tools for the discovery of the kingdom and for its improvement. 'The more we know of these islands', John Houghton observed in 1692, 'the better, I presume, may they be managed.'[66] England was not only becoming smaller and more integrated, it was being examined and inspected as a connected whole.

At the end of the seventeenth century, therefore, the face of the kingdom, actual and imagined, was very different from what it had been at the beginning. It was visibly marked now by agricultural improvement and the growth of internal commerce and communication; and it could be described, measured, and anatomized in new ways. Robert Morden's *New Description and State of England* of 1701 summarized much that was new while retaining the old chorographical framework. It printed up-to-date maps of the counties, had notes on their chief families and monuments, their natural history, soils, minerals, and products, and it could now add the size of every county in acres, its contribution to the land tax, the number of houses it contained (derived from the hearth tax), and an estimate of its population.[67] The numbers probably came from Halley and Houghton, or from the most recent edition of Edward Chamberlayne's *Angliae notitia: or, The Present State of England* which cited both authors and printed similar figures for England and

---

[64] *SCED*, pp. 672, 677.
[65] Andrew Browning, ed., *English Historical Documents 1660–1714* (1953), pp. 517–19 (Halley gave two estimates: 38.7 and 39.9 million acres). For a larger estimate in 1707, see *Nehemiah Grew*, p. xxiii, and for an earlier effort, below, p. 49. For Halley's method, which Petty may have suggested to him (*Nehemiah Grew*, p. xxiii), see Richard Stone, *Some British Empiricists in the Social Sciences 1650–1900* (Cambridge, 1997), p. 239. It had been used earlier: A. W. Richeson, *English Land Measuring to 1800: Instruments and Practices* (Cambridge, Mass., 1966), p. 123.
[66] John Houghton, *A Collection for the Improvement of Husbandry and Trade*, ed. Richard Bradley (4 vols, 1727–8), i. 9 (1692). For uses of Halley's acreages, see below, pp. 181–2.
[67] Robert Morden, *The New Description and State of England* (1701), *passim*.

Wales as a whole. According to Chamberlayne, the country was ten times the size of the Dutch Republic, less than half the size of France, and probably half as populous; and London with its 'vast traffic and commerce' was now 'the largest and most populous, the fairest and most opulent city... in all Europe, perhaps in the whole world'.[68]

Some account of the characteristic 'manners of the people' was still an essential item in such descriptions, and perceptions of them were changing too. Morden thought that the English had always been reckoned 'brave, valiant, beautiful, skilful and industrious', but he now found industrious artisans as well as industrious husbandmen in several counties.[69] Chamberlayne reported with regret that the English were generally thought 'wanting in industry', but manufacturing—'mechanics'—was the exception. There they were 'of all nations the greatest improvers', and successful 'most of all in improving' the inventions of foreigners.[70] Defoe was later to remark that it was becoming a proverb as well as a fact that the English were better at improvement than invention, and the observation seems to have originated, perhaps significantly, in comments about Londoners. According to James Howell in 1657, it was already agreed by 'all nations that, though the Londoners be not so apt to invent, yet when they have got the invention, they use always to improve it, and bring it to greater perfection'.[71]

Some determined antiquaries continued to ignore such things, and concentrated on what they took to be a still more perfect past. Ralph Thoresby, one of the contributors to the 1695 edition of *Britannia* and son of a wool merchant, was more interested in the Roman remains of Yorkshire than the flourishing industry of its West Riding.[72] Yet past history was becoming more distant and of less relevance to a literate public informed now about the kingdom's changing economy. As the councillors of York remarked, when coolly receiving Sir Thomas Widdrington's history of their city in 1660, it was no longer a splendid list of 'ancestors and predecessors, but wealth and estate which set a value upon men and places'.[73] Celia Fiennes, writing a narrative of her recent travels in 1702, encouraged her readers to look beyond 'pleasant prospects' and 'good buildings' and note, as she did, the 'different produces and manufactures of each place' which explained the

---

[68] Edward Chamberlayne, *Angliae notitia* (1700 edn), pp. 2, 46, 328, 332.

[69] Morden, *New Description*, pp. 1, 5, 82, 129, 180. It was customary earlier, and perhaps particularly after Cromwell's exploits, for the English to pride themselves on being of all nations the 'most eminent in arms': e.g. Richard Hawkins, *A Discourse of the Nationall Excellencies of England* (1658), Preface, p. 133; Edward Leigh, *England Described* (1659), pp. 1, 15–16.

[70] Chamberlayne, *Angliae notitia* (1700), pp. 47–8.

[71] James Howell, *Londinopolis: An Historical Discourse or Perlustration of the City of London* (1657), p. 396; Christine MacLeod, *Inventing the Industrial Revolution: The English Patent System, 1660–1800* (Cambridge, 1988), p. 208. Cf. below, pp. 172–3, 230; and Christopher Wren, quoted to the same effect, in Anthony Geraghty, *The Architectural Drawings of Sir Christopher Wren at All Souls College, Oxford* (Aldershot, 2007), p. 8.

[72] Parry, *Trophies of Time*, pp. 342–3.

[73] Sir Thomas Widdrington, *Analecta Eboracensia: Some Remaynes of the Ancient City of York*, ed. Caesar Caine (1897), pp. x–xi.

improvement and prosperity of towns like Colchester, Newcastle, Leeds, and Manchester. Only then would they get 'an idea of England' in the present.[74]

Agrarian improvement had marked the landscape even more profoundly, as a scholar of antiquarian bent less blinkered than Thoresby readily acknowledged. Writing to Edward Lhuyd from North Wales in the 1690s, John Lloyd reported that, thanks to good husbandry, 'most of our neighbourhood has been improved since Camden's days'. Although she was more interested in industries than farming and took enclosures for granted, even Celia Fiennes admired new crops with obvious commercial potential like the liquorice which she found in every garden around Pontefract; and she welcomed the investment going into draining marshes and fens in order to make them more 'useful'.[75] Looking back over the century in 1699, Charles Davenant, the political economist, was more eloquent on the point. In 1600 there had been 'a great deal more barren land; of that which was cultivated, very much was capable of melioration; and there were more forests, woods, coppices, commons, and waste ground, than there is now, which our wealth did enable us from time to time to inclose, cultivate, and improve'.[76]

When Gibson's contributors made their additions to Camden's classic in 1695 they necessarily reflected the same perceptions, if sometimes indirectly and—until a second edition appeared in 1722—with little evident enthusiasm. One of them updated the entries on ancient British and Roman coins, and noted, with an obvious eye to present conditions, that the Roman Conquest had proved 'exceedingly beneficial to the nation' by first introducing 'civility ... good husbandry too, and improvement of wealth and trade both by sea and land'. In a contribution unusual for its hyperbole, William Nicolson remarked that the 'wealth and commerce' of Newcastle had 'wonderfully increased since Camden's time' and its coal trade was now 'incredible'. By contrast, data on the fivefold growth of the navy 'since Mr Camden's time', communicated by Samuel Pepys, were simply tabulated in the 1695 *Britannia*. In the 1722 edition, however, they were used as evidence for the 'vast growth and improvement' in the navy 'in this and the last age'; and the rich soils of Suffolk mentioned in 1695 were similarly embellished by references to turnips and the 'vast improvement' they had brought to the county.[77]

These were the parts of Gibson's *Britannia* which must have appealed to Defoe in the 1720s, as he wrote up his own travels in his *Tour*. In an essay of 1697 he had already taken the roads of the Romans as symbols of the 'improvement and increase of arts and learning' and the 'civilizing and methodizing' which came with their conquests. Now, in the preface to his *Tour*, he paid Gibson the ultimate compliment of plagiarizing what he had said in his. 'If novelty pleases', Defoe told his readers, 'here is the present state of the country described, the improvement, as well

---

[74] McRae, *Literature and Domestic Travel*, pp. 204–9; *The Journeys of Celia Fiennes*, ed. Christopher Morris (1947), pp. 1–2, 142–3, 208–12, 219–20, 223–4.

[75] Joan Thirsk, ed., *The Agrarian History of England and Wales*, v.i: *1640–1750* (Cambridge, 1984), p. 395; McRae, *Literature and Domestic Travel*, p. 205; *Journeys of Celia Fiennes*, pp. 31, 158–9, 184–5.

[76] Davenant, *PCW*, ii. 221. Cf. below, pp. 192–3.

[77] Edmund Gibson, *Camden's Britannia* (1695), pp. xciii, 229–30, 871; (2nd edn, 1722), pp. 233–4, 437. Cf. Parry, *Trophies of Time*, pp. 344–5.

in culture as in commerce, the increase of people and employment for them . . . the increase of buildings . . . the increase of wealth, in many eminent particulars.'[78] The face of the kingdom had changed and what Fiennes called the 'idea of England' had changed along with it.

WIDER PERSPECTIVES

Ideas about England were given some context and set in perspective by what was known or being learnt about other places and about the past. Relevant information came from many sources. They included the works of ancient and modern historians, geographers, and travellers, and the large European compendia published in translation throughout the seventeenth century, from Pedro Mexía's *Treasury of Ancient and Modern Times* (1613) to Samuel Pufendorf's *History of the Principal Kingdoms and States of Europe* (1695).[79] The framework within which the English saw themselves in space and time was largely determined at the start of the period, however, by the discovery and exploration of America, and by the ways in which they were interpreted by Richard Hakluyt and Samuel Purchas. Their collections of navigations, voyages, and discoveries were as instrumental in the rediscovery of England as the works of Camden, Lambarde, and the early chorographers, and they had more lasting impact. As late as 1703, John Locke thought 'Hakluyt and Purchas' still in many ways 'very good', and fit to stand alongside the revised edition of Camden in any gentleman's library.[80]

Hakluyt's *Principal Navigations*, first published in 1589 and expanded into three large volumes between 1598 and 1600, set out to use 'geography and chronology . . . the sun and the moon of all history' in order to rescue the 'memorable exploits of late years by our English nation' from 'the devouring jaws of oblivion'.[81] The four folio volumes of *Purchas his Pilgrims* (1624–5) presented the same events in a 'theological and geographical' history of the whole world whose course had been divinely ordained 'from the Creation until this present'.[82] Hakluyt had more to say about England's need for colonies in order to relieve overpopulation at home and gain access to essential commodities. Purchas pictured them as fulfilling biblical injunctions to mankind to go forth and multiply and replenish the earth, and referred to the natural right of colonizers to exercise dominion over what he called

---

[78] McRae, *Literature and Domestic Travel*, p. 89; Defoe, *A Tour Through the Whole Island of Great Britain*, ed. G. D. H. Cole and D. C. Browning (2 vols, 1962), i. 1. Cf. above, p. 16, below, pp. 218–19.

[79] Pedro de Mexía, *The Treasurie of Auncient and Moderne Times* (1613); Samuel Pufendorf, *An Introduction to the History of the Principal Kingdoms and States of Europe* (1695). Another often cited translation was Pierre d'Avity, *The Estates, Empires & Principalities of the World* (1615).

[80] *Locke: Political Essays*, ed. Mark Goldie (Cambridge, 1997), p. 353.

[81] Andrew Hadfield, *Amazons, Savages, and Machiavels: Travel and Colonial Writing in English, 1550–1630: An Anthology* (Oxford, 2001), p. 27. For earlier literary ventures of this kind, see Hadfield, *Amazons*, pp. 15–19, 117; Roger Barlow, *A Brief Summe of Geographie* (Hakluyt Society, 2nd ser. 69, 1932), pp. 180–2.

[82] Samuel Purchas, *Hakluytus Posthumus, or Purchas his Pilgrimes* (4 vols, 1625), title page.

'vacant places'.[83] But both of them saw English navigation and colonization as necessary, justified by natural law and national self-interest, and providential in delivering civility and Christianity to 'the savage and the pagan'. Theirs was a vision of an English or (in Purchas's case) British empire which had begun with the 'opening of a new world' in the sixteenth century, and blossomed under what Purchas called 'the glorious sunshine of Queen Elizabeth'.[84]

Neither Purchas nor Hakluyt ever ventured far from home, but other kinds of travel writing were also beginning to flourish in the sun. They offered fresh perspectives on the old world as well as the new and could generally be fitted within a similar framework of a superior European civility which England exemplified, perhaps to perfection. Despite warnings of the temptations and dangers of foreign travel, especially south of Alps where William Cecil thought the young would learn only 'pride, blasphemy and atheism',[85] there were printed guides for those who went, telling them what to look for and note down.[86] In 1606 Thomas Palmer wanted information about 'the nature of the people', whether they were civil or barbarous, free or servile, religious or profane, warlike or effeminate, and about the size of the country, its population, cities, commodities, trades, and husbandry;[87] and the kinds of useful knowledge to be accumulated expanded like the contents of local chorographies across the seventeenth century.

Bacon advised travellers to collect information about antiquities, libraries, fortifications, warehouses, and merchants' exchanges, William Petty suggested currency, wages and prices, rents and interest rates, weights and measures, and Robert Boyle added aspects of natural history, climate, and prevalent diseases to the list. In 1696 John Woodward introduced an ethnographical strain, asking travellers in 'remote and uncivilised or pagan countries' to note their religious practices, notions of the supernatural, and distinctive arts and sciences.[88] As Palmer explained, the traveller should 'get knowledge for the bettering of himself and his country'. The aim was national and self-improvement, and James Howell claimed in 1642 that the English had indeed 'improved themselves infinitely by voyaging both by land and sea'.[89]

Some of the conclusions in the published travel literature must have amply reassured those who were sceptical about its value. For many authors, home was

---

[83] David Armitage, *The Ideological Origins of the British Empire* (Cambridge, 2000), p. 97. Cf. below, pp. 67–9.
[84] Armitage, *Ideological Origins*, pp. 68–87. Cf. Andrew Fitzmaurice, *Humanism and America: An Intellectual History of English Colonisation, 1500–1625* (Cambridge, 2003), pp. 90–1.
[85] Louis B. Wright, ed., *Advice to a Son: Precepts of Lord Burghley, Sir Walter Raleigh, and Francis Osborne* (Ithaca, NY, 1962), p. 11.
[86] Vine, *In Defiance of Time*, pp. 139–68; Alison Games, *The Web of Empire: English Cosmopolitans in an Age of Expansion, 1560–1660* (Oxford, 2008), pp. 18–46.
[87] Sir Thomas Palmer, *An Essay of the Meanes how to make our Trauailes, into forraine Countries, the more profitable and honourable* (1606), pp. 35, 53, 81–2, 87–91.
[88] Hadfield, *Amazons*, pp. 33–5; *The Petty Papers*, ed. Marquis of Lansdowne (2 vols, 1927), i. 175–8; Robert Boyle, *General Heads for the Natural History of a Country, Great or Small* (1692), pp. 2–7; John Woodward, *Brief Instructions For Making Observations in all Parts of the World* (1696), pp. 8–10.
[89] Palmer, *Essay*, p. 53; James Howell, *Instructions for Forreine Travell* (1642), p. 14.

best. In his popular *Glory of England* in 1618 Thomas Gainsford made his case by means of short chapters describing other countries and empires near and far, and finding them wanting: 'France compared, with a discovery of her defects', for example, and 'China compared, and her deficiency manifested'. England trumped 'all the nations of the world' in 'sufficiency and fullness of happiness'.[90] Twenty years later, having seen Persia and the East, the great traveller Thomas Herbert still concluded that 'this Island of Isles, Great Britain' contained 'the sum and abridge[ment] of all sorts of excellencies' to be found elsewhere.[91] Yet information about other places also heightened perceptions of difference, and the more information there was, the more it inhibited simple assumptions and contradicted chauvinistic prejudices.

Comparisons with other, richer, and more powerful, states in Europe were of most immediate consequence: with France throughout the period, with Spain until its long decline became evident in the mid-seventeenth century, and with the Dutch Republic from the 1620s as its share of European and global commerce grew ever larger. Competition between them created a demand for more precise information about them, and for works like those of Giovanni Botero in the new genre of political writing about 'reason of state'.[92] Part of Botero's work was published in English in 1601 in *The Travellers Breviat* which gave instructions for collecting information about the revenues and resources of different countries; and in 1652 Peter Heylyn looked back to Botero's *Greatness of Cities* when listing the criteria for judging the strength of a state in the first large-scale English *Cosmographie* worthy to stand beside its foreign competitors.[93]

'Surveys' and descriptions of the 'present state' of single countries also furnished comparative material. At first rudimentary in the information they presented, by the end of the seventeenth century guides to major countries were being regularly reissued and had more quantitative information.[94] The first successful English venture of this kind, Chamberlayne's *Angliae notitia*, first published in 1669, was modelled on *L'Estat Nouveau de la France dans sa perfection*, published in Paris in 1661, and it was itself quickly translated into French and published in Amsterdam and Paris so that foreigners could also have their 'understanding informed' about 'the present state of this considerable monarchy'.[95]

---

[90] Thomas Gainsford, *The Glory of England* (1618), title page, pp. 166, 236.
[91] Thomas Herbert, *A Relation of some yeares travaile, begunne Anno 1626* (1634), p. 2.
[92] Andrew Hadfield, *Literature, Travel, and Colonial Writing in the English Renaissance, 1545–1625* (Oxford, 1998), pp. 33–4; Maclean, *Scholarship, Commerce, Religion*, pp. 70, 280; below, p. 45. On reason of state, see Noel Malcolm, *Reason of State, Propaganda, and the Thirty Years' War: An Unknown Translation by Thomas Hobbes* (Oxford, 2007), pp. 92–4; Richard Tuck, *Philosophy and Government, 1572–1651* (Cambridge, 1993), pp. 65–113.
[93] Giovanni Botero, *The Travellers Breviat, Or An historicall description of the most famous kingdomes in the World* (1601); Peter Heylyn, *Cosmographie in Four Bookes* (1652), p. 5.
[94] Shapiro, *Culture of Fact*, pp. 77–9. For early examples, see Richard Sergier, *The Present State of Spaine* (1594); John Eliot, *The Survay or Topographical Description of France* (1592). It is striking how little numerical information about other countries, in comparison with England, George Abbot, Purchas's patron, was able to gather early in the century for his popular geography, *A Briefe Description of the whole World* (1634 edn, first published in 1599), pp. 12–13, 332.
[95] *ODNB sub* Edward Chamberlayne; Chamberlayne, *Angliae notitia* (1669 edn), sig. A3; (1679), sig. A2^r.

The comparisons were not always in England's favour. The Dutch were universally admired for their public virtues, especially their frugality, which accounted for a broad distribution of wealth. As an English traveller pointedly remarked early in the century, they lived in an 'uncorrupted' commonwealth where no one was 'exceeding rich, and few very poor'.[96] The rigid social distinctions between nobles and peasants in France, as compared with more fortunate England, on the other hand, were equally often commented on, and thought by John Evelyn in 1652 to explain why the French lacked the incentives to 'future industry' which followed from the social mobility of the English. Yet Evelyn found France superior in many other respects, and especially in the environment of its capital city, much better built and much healthier than London: each country had something to learn from the other.[97] London was to catch up later, but not until it had been rebuilt after the Great Fire could it be hailed as the European epitome of 'urbanity and civility'.[98]

Perceived contrasts between European countries paled into insignificance, however, when travellers reported on more distant parts of the world, India and China and even Japan, which their readers could find on a globe or on the maps which hung on the walls of taverns.[99] Europe was visibly the smallest continent, and Britain, in John Donne's phrase, merely 'the suburbs of the old world', even if also 'a bridge, a gallery to the new';[100] and England was only 'the thousandth part of the whole globe', according to a 'geometrical description' in a popular handbook for merchants of 1622.[101] It was difficult to claim, as one writer did in 1659, that European states were 'the most populous and eminent for arts and arms' in the world when writers such as Herbert were describing populous cities, states, and empires in the Levant, India, and China, of far greater wealth and power, and even perhaps civility.[102]

Like English chorographies, but on a global scale, travellers' accounts of other civilizations stimulated the collection of specimens of plants, minerals, dialects, customs, and manners, all far more exotic than those to be found in distant corners of England.[103] They emphasized difference, and uncovered distinct and distinctive cultures each in its own geographical and historical context with as much validity as

---

[96] Thomas Overbury, *Sir Thomas Overbury, his Observations in his travailes* (1626), pp. 4, 15–16. This work, published after his death, may not have been written by Overbury: *ODNB*, *sub*. Overbury.

[97] John Evelyn, *The State of France* (1652), pp. 38, 72, 75, 77, 107–8, 110–13; [John Evelyn], *A Character of England* (1659), 'To the Reader'. Cf Overbury, *Observations*, pp. 15–16. On the social contrasts of status between the two countries, see R. B. Grassby, 'Social Status and Commercial Enterprise under Louis XIV', *EcHR* 2nd ser. 13 (1960), pp. 19–38; below, pp. 72, 247.

[98] Howell, *Londinopolis*, p. 382; below, pp. 147–8.

[99] On the East India Company in Japan at the beginning of the seventeenth century, see Games, *Web of Empire*, pp. 102–9, 235; and for world maps on the walls of inns in Bristol c.1700, Carl B. Estabrook, *Urbane and Rustic England: Cultural Ties and Social Spheres in the Provinces, 1660–1780* (Manchester, 1998), pp. 56–7.

[100] John Donne, *A Sermon . . . To the Honourable Company of the Virginian Plantation* (1622), p. 44.

[101] Gerard Malynes, *Consuetudo vel lex mercatoria* (1622), pp. 66–7.

[102] Leigh, *England Described*, p. 1; Herbert, *Relation*, *passim*.

[103] Shapiro, *Culture of Fact*, pp. 72–6; Alix Cooper, *Inventing the Indigenous: Local Knowledge and Natural History in Early Modern Europe* (Cambridge, 2007); Richard Drayton, *Nature's Government: Science, Imperial Britain, and the 'Improvement' of the World* (New Haven, 2000), pp. 19–25.

any other. Even in Europe, as Palmer recognized, what was civility in one nation was 'unaccustomed and rejected in other states'.[104] Differences in manners and even morals were still greater across continents. In 1636 Henry Blount not only admired the 'incredible civility' of the Ottomans, but also, as Botero might have done, their 'numerous people . . . the foundation of all great empires'. He attributed it to the practice of polygamy (by men), a topic much discussed in western Europe after 1650 when the population there was no longer increasing.[105] Travel literature challenged conventional assumptions by opening eyes to alternatives. In 1609 it had already prompted Joseph Hall to think there might be people still 'more civil than we are' in the supposed southern continent of 'Terra Australis'. After all, he added, 'whoever expected such wit, such government in China? . . . We thought learning had dwelt in our corner of the world: they laugh at us for it, and well may.'[106]

The English could retain more than a little of their self-esteem nonetheless by looking to the New World, by considering what they had found and were now creating there, and by speculating that civility and civilization were marching across the globe from the eastern hemisphere to the western. That was what Nathaniel Carpenter supposed in his *Geography Delineated* (1625), and he predicted that the next movement of civilization would be south, to Africa, where there were new regions—'most barbarous, without laws, sciences, or civility'—only waiting to be improved by 'the industry of Europeans'.[107] Forty years later, writing his own *Geographical Description* of the world, Richard Blome had no doubt that England had caught up with China. China was still to be admired 'for its riches, for the great number and politeness of its inhabitants, for the beauty of its cities, for its manufactures', and for having had 'the inventions of silk, printing, paper, artillery etc. before us'. But England was now superior to all other nations in 'knowledge of arts and sciences, the secrets of nature' and everything necessary 'to the completing of a gentleman'; and it was communicating its civility to the Irish, who might still retain some of the 'absurd and ridiculous customs' of the Scots but had been even more 'barbarous till civilised by the English'.[108]

It was inevitable that the natives of North America should be seen in the same light, as targets for a civilizing mission, just as the Irish or the 'rude and unruly' Scottish highlanders or the 'wild' Welsh were, and as the ancient inhabitants of

---

[104] Palmer, *Essay*, p. 67.
[105] Anna Suranyi, *The Genius of the English Nation: Travel Writing and National Identity in Early Modern England* (Newark, Del., 2008), p. 58; Sir Henry Blount, *Voyage into the Levant* (2nd edn, 1636), p. 82; for debates on polygamy in England, see below, pp. 126, 210.
[106] Joseph Hall, *The Discovery of A New World or A Description of the South Indies, Hetherto Vnknowne* (1609), sig. A4ʳ.
[107] Nathaniel Carpenter, *Geography Delineated* (Oxford, 1625), pp. 221–2.
[108] Richard Blome, *A Geographical Description of the four parts of the World* (1670), pp. 76, 100, 126–7, 130. On respect for the politeness of India and China in the later seventeenth century, see Peter Burke, 'A Civil Tongue: Language and Politeness in Early Modern Europe', in Peter Burke, Brian Harrison, and Paul Slack, eds, *Civil Histories: Essays Presented to Sir Keith Thomas* (Oxford, 2000), p. 34; for an earlier reference in 1622, see below, p. 38.

England itself had once been.[109] In 1590 an edition of Thomas Harriot's report on the Roanoke colony published drawings of the first 'Virginians' ever seen by Europeans, together with five prints showing 'Picts' in order to demonstrate that 'the inhabitants of Great Britain have been in time past as savage' as they were;[110] and the point was commonly made by later propagandists for plantations that the English would have remained 'brutish, poor and naked Britons', if Julius Caesar had not begun to make them 'tame and civil'.[111] They were about to do for Virginia, what they were doing for Ireland, and what the Romans had done for them. They were still trying in 1750, according to a historian of the 'progressive improvements, and present state' of the American colonies. Compared with the people of China, 'the elder brother of all the nations of mankind' in government and improvement, the natives of America were 'brutish' and backward in husbandry and remained 'the youngest brother and meanest of mankind'.[112]

These persistent assumptions about the relative civility of peoples involved perceptions of history as well as geography, and about the progressive movement of civilization across time as well as across space. They carried with them also the implication that the civilizing process might not only move forward but grind to a halt, as Blome thought it had in China, and as it had done long ago in Rome itself. In the later seventeenth century, travellers' accounts of India were already picturing that too as a land of accumulated riches and once imperial splendour now sunk in corruption, indolence, and excess, badly cultivated, 'unmanured', and ready therefore for European improvers.[113] 'Geography without history seemeth a carcass without motion,' a governor of Bermuda remarked when writing its history in the 1620s,[114] but the more that was learnt about history, the more complicated and open to dispute the direction of that motion seemed to be.

The task of historical interpretation was not helped by the persistence of myths and legends which linked the distant past to the present, bridged the gap across

---

[109] Prys Morgan, 'Wild Wales: Civilizing the Welsh from the Sixteenth to the Nineteenth Centuries', in Burke et al., *Civil Histories*, pp. 266–8; John Darwin, 'Civility and Empire', in Burke et al., *Civil Histories*, pp. 322–4; Nicholas Canny, *Making Ireland British 1580–1650* (Oxford, 2001), pp. 121, 188; Camden, *Britain* (1610), 'Scotland', p. 5.

[110] Thomas Harriot, *A Briefe and True Report of the New Found Land of Virginia*, ed. Theodore de Bry (Frankfurt, 1590); Hadfield, *Literature, Travel, and Colonial Writing*, pp. 113–22. As Hadfield points out, the Virginians might even be thought to look less savage than the Picts.

[111] Robert Johnson, *Nova Britannia* (1609), sigs B1ᵛ, C1–2ʳ. Cf. Armitage, *Ideological Origins*, pp. 49–51, 68; Tristan Marshall, *Theatre and Empire: Great Britain on the London Stages under James VI and I* (Manchester, 2000), pp. 12, 23. For similar arguments later, see Jack P. Greene, 'Changing Identity in the British Caribbean: Barbados as a Case Study', in Nicholas Canny and Anthony Pagden, eds, *Colonial Identity in the Atlantic World 1500–1800* (Princeton, 1987), p. 240; Richard Hingley, *The Recovery of Roman Britain 1586–1906* (Oxford, 2008), pp. 64–5.

[112] William Douglass, *A Summary, Historical and Political, Of the first Planting, progressive Improvements, and present State of the British Settlements in North-America* (2 vols, Boston, 1749–51), i. 153. Cf. Kariann Akemi Yokota, *Unbecoming British: How Revolutionary America became a Postcolonial Nation* (Oxford, 2011), pp. 19–21 for images of native Virginians in the early eighteenth century.

[113] Pramod K. Nayar, *English Writing and India, 1600–1920: Colonizing Aesthetics* (2008), pp. 28–31.

[114] Games, *Web of Empire*, p. 11.

millennia, and foreshortened perceptions of historical time. The legend that civilization had been brought to Britain by Brutus coming from Troy, long before the Romans, remained popular in accounts of 'British' empires well into the seventeenth century, and scholars as familiar with new critical approaches to evidence as Camden and Sir Robert Cotton were reluctant wholly to dismiss it.[115] Like the chorographers, English historians were conscious of the need to reject mere rumour and fiction, and acknowledged that some historical 'facts' were more certain than others;[116] but it took time for the results of the new critical history espoused by sixteenth-century historians like Jean Bodin to work through into popular English publications.[117] Not until the later seventeenth century, for example, was the BC/AD distinction or the subdivision of history into centuries in common use.[118]

Different kinds of historical writing in early modern England nevertheless conveyed some sense of the chronological distance between antiquity and the present, and of the uneven flow of events across the measured spaces of time which lay in between. Sixteenth-century 'annals' of monarchs or towns or parishes set out successive events in sequence, and like medieval chronicles presented a linear view of historical change.[119] The simplified 'chronologies' in seventeenth-century almanacs, which were much the most widely read vehicles of historical information, had the same effect. They often began with the Creation, usually placed at some point between 3,900 and 4,000 BC, and they might sometimes indicate that reputable biblical scholars were divided over the precise date. They also provided dates for notable events where they could, not only for the Norman Conquest or the defeat of the Armada, but for the introduction of printing and gunpowder, the first glass windows in England, the first stone bridge across the Thames, and the first use of coal or tobacco.[120]

A reader of more substantial historical works in the 1660s, John Ward, was similarly interested in 'firsts', 'our first physician' around 1230, for instance; and his commonplace book leaves the impression that for many people history was an accumulation of discrete if curious and sometimes datable facts, without any particular rhyme or reason to it. The facts were not only about England, however, and some of them suggested that history might have some shape. Ward was as

---

[115] Armitage, *Ideological Origins*, pp. 38–9, 53–4; Kevin Sharpe, *Sir Robert Cotton 1586–1631: History and Politics in Early Modern England* (Oxford, 1979), p. 25; Parry, *Trophies of Time*, pp. 215, 310.

[116] Shapiro, *Culture of Fact*, pp. 63–4, 198–9.

[117] On Bodin, see Anthony Grafton, *What Was History? The Art of History in Early Modern Europe* (Cambridge, 2007), pp. 165–8; and for examples of his influence, see Daniel R. Woolf, *The Idea of History in Early Stuart England* (Toronto, 1990), pp. 67, 245. Carpenter had carefully read Bodin, and as a Devonian himself was sharply critical of the great historian's judgements about the 'blockishness and incivility' of the natives of his county: Carpenter, *Geography Delineated*, pp. 260–3.

[118] Glennie and Thrift, *Shaping the Day*, p. 59.

[119] On the slow decline of the chronicle format, see Daniel R. Woolf, *Reading History in Early Modern England* (Cambridge, 2000), pp. 11–78.

[120] Bernard Capp, *Astrology and the Popular Press: English Almanacs 1500–1800* (1979), pp. 23, 220–2. Cf. Keith Thomas, *The Perception of the Past in Early Modern England* (Creighton Trust Lecture, University of London, 1983), p. 6.

intrigued by the ancient customs of Byzantium as by those of London, and he noted that the value of money had declined since the Middle Ages.[121] As early as 1622, one of the almanacs noted (as Blome was to do much later) that the Chinese had been familiar with printing long before Gutenberg; and some of them organized history around a model taken from the prophet Daniel of four great world-empires, and saw the present age as the final phase of the fourth, the Roman Empire. If Byzantium and three of the great empires had disappeared, however, perhaps—despite all those 'firsts'—the pattern of history was not linear but cyclical. One almanac in 1682 recorded that in the past wars had often created poverty, poverty led to peace, peace to riches and pride, and wealth to wars again: 'the world goes round.'[122]

That cyclical view of historical change challenged any confident assumption that civilization was necessarily moving forwards through successive stages of a journey which had led from ancient to modern Britons, and which the Irish or the savages of Virginia were now embarking upon. Some of those first images of Virginians in the literature about America reinforced anxieties that the civilizing process might bring with it the corruption of pristine innocence and an erosion of martial and civic virtues, and one of the Jacobean promoters of plantations was conscious that precisely that had happened to the Romans. 'Their valour made them quiet and quiet wealthy', Robert Johnson observed; 'but according to the revolution of all things, with a swift and violent return, their wealth effeminated their valour with idleness, idleness occasioned disorder, disorder made ruin.'[123]

In 1690 William Temple described the 'revolutions' of other empires as similarly a matter of rotation. Parts of Egypt and Greece, once the site of great civilizations, were now 'as rude and barbarous' as France, Germany, and Britain had once been: 'science and arts have run their circles.'[124] Historians of imperial Rome were discovering that it had been far larger that it was now;[125] perhaps the same might one day be the fate of London and England. Even in the eighteenth century the Scottish historians who reformulated the successive stages of civilization into a story of linear progress, from the savage state of nature to the urbanized and commercial societies of modern western Europe, had to face the possibility that it must soon end.[126]

Enlightenment historians worried much less about another linear view of history, which had been more powerful in the seventeenth century, and which visualized the historical process as divinely preordained and moving ineluctably, and perhaps

---

[121] Woolf, *Reading History*, pp. 109–11.
[122] Capp, *Astrology*, pp. 222–4. For similar views, see below, p. 194.
[123] Fitzmaurice, *Humanism and America*, pp. 161–4; Andrew Fitzmaurice, 'American Corruption', in John F. McDiarmid, ed., *The Monarchical Republic of Early Modern England* (Aldershot, 2007), p. 222. Cf. below, pp. 67–8.
[124] Sir William Temple, 'Upon Ancient and Modern Learning', in *Miscellanea: The Second Part* (1690), p. 36.
[125] John Seller, *Atlas minimus, or A Book of Geography* (1679), p. 7; below, p. 194.
[126] Below, p. 199. On Scottish philosophical history, see J. G. A. Pocock, *Barbarism and Religion*, ii: *Narratives of Civil Government* (Cambridge, 1999), pp. 171–2; iv: *Barbarians, Savages and Empires* (Cambridge, 2005), pp. 170–2.

downward, from the Fall of Man to the Last Judgement. In the early seventeenth century recent or recently revived apocalyptic interpretations of Scripture were warning that the Second Coming of Christ and the millennium were near at hand,[127] but that was far from implying that previous history had necessarily been one of consistent progress. On the contrary, the powerful myth of the Fall of Man carried with it the implication, not only that it could never be wholly repaired in this world, but that it had been followed by long centuries of decline and decay. It was commonly remarked that there were giants among the ancients, and that people had been taller and lived longer in the time of the Old Testament patriarchs and degenerated since then.[128] In a culture predisposed to find providence at work in all things, such metaphysical speculations were bound to influence assumptions about the direction of historical development, and they had to be confronted, or evaded, by writers who wished to hail a new age of discovery and invention.

Firmly wedded as he was to an apocalyptic view of the present as the last age of a corrupt and declining world, Samuel Purchas could only account for the new discoveries he was celebrating by treating them as exceptional providences. Walter Ralegh adopted a similar view of endless decay since the Fall in his influential *History of the World* (1614), but was nonetheless able to incorporate some hints about forward movement since the Flood. Divine providence alone explained why successive 'kings and kingdoms have flourished and fallen', but ancient states had nonetheless bequeathed a gradual 'restauration of civility' to their successors. The 'many inventions' of the Hebrews and ancient Egyptians had reached Europe via the Greeks, for example, and from them been 'derived to us'.[129] There had been some continuity since the Flood and if not endless progress, at least no consistent downward decline.

Some of the historians of the European Renaissance had been more robust about modern achievements. They initiated the long debate which raged from the fifteenth to the eighteenth centuries about whether the literary and scientific achievements of the ancients were superior to their modern equivalents, their cities more populous and powerful, their peoples healthier and more civilized, their states more stable and longer-lasting. The two sides of the argument were most clearly differentiated in the 'battle of the books' between 'Ancients' and 'Moderns' which raged from the 1690s to the 1750s,[130] but the case for the moderns had been rehearsed much earlier, first in Italy and then, at the end of the sixteenth century, in France, when Bodin, in his manifesto for a new critical history, and Louis Leroy in a

---

[127] See e.g. Capp, *Astrology*, pp. 164–79.

[128] Peter Harrison, *The Fall of Man and the Foundations of Science* (Cambridge, 2007), pp. 11–15, 166–85; Daniel Woolf, *The Social Circulation of the Past: English Historical Culture, 1500–1730* (Oxford, 2003), p. 56.

[129] Victor Harris, *All Coherence Gone: A Study of the Seventeenth Century Controversy over Disorder and Decay in the Universe* (1966), pp. 133–4; Walter Raleigh, *The History of the World* (1614), sigs A2ᵛ, E3ᵛ, pp. 180, 315.

[130] Woolf, *Social Circulation*, pp. 66–9; Joseph M. Levine, *The Autonomy of History: Truth and Method from Erasmus to Gibbon* (Chicago, 1999), chs 4, 5.

treatise on the 'vicissitude' of things, launched a full-scale attack on the notion of an ageing and degenerate world.[131]

Bodin dismissed old myths about past golden ages and the four great empires, and the tendency of old men everywhere to dream that things were once better. Both writers pointed to the new inventions of printing, gunpowder, and the compass as evidence that the present age had advanced beyond that of the Romans. They acknowledged that there were cycles, but some states were now richer and more powerful than any before. All states rose and fell, but each was capable of improving on its predecessors.[132] There was little dispute that ancient philosophers and historians, geographers and physicians, poets and orators, were—in a much used metaphor—the giants on whose shoulders moderns stood, but it was possible that the moderns might not only see further, but make fresh discoveries of their own.

The case made by Bodin and Leroy was quickly borrowed in defence of the moderns in England. There seems to be an echo of it in Camden's expectation, in the Preface to *Britannia* in 1610, that in the future 'another age, and other men' would 'daily find out more' and build on his achievements;[133] and there was another in Gervase Markham's comment that recent English books on husbandry proved that human 'industry' constantly brought forth 'new things, drawing every art and occupation to that height of excellency that the knowledge of our forefathers, compared with the times now present, is but mere ignorance'.[134] They might both have been reading Bacon, who had taken the arguments of the French historians fully on board, and reinforced them with an apocalyptic interpretation of his own which some of Bodin's more scrupulous disciples were quick to deplore.[135]

Bacon referred repeatedly to the millennial prophecy of Daniel that 'many shall run to and fro, and knowledge shall be increased'.[136] He used it in his *Advancement of Learning* (1605), when arguing that 'in these latter times' there had been a 'return of learning' at least as great as that enjoyed by the 'exemplar states' of Greece and Rome, and especially 'in navigation and discoveries' which must lead to 'a further proficiency and augmentation of all sciences'.[137] It occurs again on the title page of *Novum organum* (1620), in which he pointed like Bodin and Leroy to the obstacles standing in the way of advance, including 'the siren song of reverence for antiquity' and the theory of those cycles, 'revolutions of the world's ages', which too often led men to suppose that 'when they get to a certain stage or condition, they cannot go

---

[131] Grafton, *What Was History?*, pp. 174–5; Hans Baron, 'The *Querelle* of the Ancients and Moderns as a Problem for Renaissance Scholarship', *Journal of the History of Ideas*, 20 (1959), pp. 3–22. Cf. Brian P. Copenhaver, 'The Historiography of Discovery in the Renaissance: The Sources and Composition of Polydore Vergil's De inventoribus rerum. I–III', *Journal of the Warburg and Courtauld Institutes*, 41 (1978), pp. 192–214.

[132] Harris, *All Coherence Gone*, pp. 100–4; Grafton, *What Was History?*, pp. 167–9.

[133] Camden, *Britain* (1610), 'To the Reader'. In Gibson's 1695 edition, 'Mr Camden's Preface' was slightly revised, and this statement strengthened to refer to 'another age, a new race of men'.

[134] Gervase Markham, *Markhams farwell to Husbandry* (1620), p. 3.

[135] Grafton, *What Was History?*, pp. 179–80.

[136] Charles Webster, *The Great Instauration: Science, Medicine and Reform, 1626–1660* (1975), pp. 24–5; Daniel 12: 4.

[137] Vickers, *Bacon*, pp. 180, 183–4.

further'. Despite referring himself to cycles in his essay on 'vicissitudes', he had little doubt, when he expanded the *Advancement* in 1623, that the present 'third period' of learning would 'far exceed' those of the Greeks and the Romans.[138]

In 1627, therefore, George Hakewill was able to borrow from Bacon as well as Bodin and Leroy when disputing the 'common error touching Nature's perpetual and universal decay'. Like 'the excessive admiration of antiquity', it only made 'men worse rather than better', encouraging 'country boars' who were easily persuaded that 'nothing can be improved by industry, but all things by a fatal necessity grow worse and worse'. He sometimes sounds like Aubrey and other mid-century advocates of improvement, although he hesitated to replace a vision of perpetual decay with one of perpetual progress. He praised modern inventions, especially in mathematics and natural philosophy, and thought men lived as long nowadays as they had in the distant past, but he saw history as full of cycles. There was 'a vicissitude, an alternation and revolution', 'a wheeling about of all things'. He called it 'a kind of circular progress', but progress might still mean movement backward as well as forward, and Hakewill may not have assumed that one age necessarily advanced further than its predecessors.[139]

When English authors first engaged in the battle of the books in the 1690s, the notion of an upward spiral of progress was articulated with greater confidence by William Wotton, defending the moderns. Responding to William Temple's argument that the ancients were in most respects far superior in learning,[140] Wotton could now cite the great advances in knowledge brought by the 'new philosophy' of the second half of the seventeenth century against him. He admitted that there was no guarantee that knowledge would improve in the next century 'proportionably as it has done in this'. Like Bodin, however, he thought each new revival of learning built on what had gone before, and imagined, like Markham, that in 'some future age, though perhaps not the next, and in a country now possibly little thought of', knowledge might be raised even further 'upon the foundations laid in this our age, to the utmost possible perfection to which it can be brought by mortal men'. He hoped, of course, that the country would be England, and that the men would be successors to Bacon and 'the men of Gresham', Fellows of the Royal Society like himself.[141]

---

[138] Bacon, *Instauratio magna Part II*, pp. 133, 149, 151, 489; Harris, *All Coherence Gone*, p. 131 (quoting 'De augmentis scientiarum', Book VIII, ch. 3). Bacon put forward similar arguments in *Redargutio philosophiarum* (1608): Stephen Gaukroger, *The Collapse of Mechanism and the Rise of Sensibility: Science and the Shaping of Modernity, 1680–1760* (Oxford, 2010), p. 131.

[139] Harris, *All Coherence Gone*, pp. 52, 61, 70–2, 79, 134; Woolf, *Social Circulation*, pp. 57–8; Walsham, *Reformation of the Landscape*, p. 380. Hakewill was attacking Godfrey Goodman's influential *The Fall of Man, or the corruption of nature proved by the light of natural reason* (1616).

[140] Temple admitted that learning and knowledge had made 'mighty progress' in western Europe since 1550, instancing in particular the invention of the compass, 'the greatest improvement' of modern times, but denied that they were at 'a greater height' than they had been in parts of the world much earlier: Temple, 'Upon Ancient and Modern Learning', pp. 38, 48. The early discovery of the compass by the Chinese was still unknown.

[141] William Wotton, *Reflections upon Ancient and Modern Learning* (1694), sig. a3ᵛ, pp. 3, 78, 356, 358. Cf. William Wotton, *A Defense of the Reflections upon Ancient and Modern Learning* (1705). On

The dispute between ancients and moderns was only one of the controversies provoked by seventeenth-century advances in knowledge which prevented any agreement on a single model of historical change. New discoveries in biblical and historical scholarship were simultaneously challenging orthodox views of history hitherto firmly grounded in Scripture. There were arguments reasserting the Aristotelian notion that the world was eternal and therefore without a datable beginning, that there were men before Adam, or at least that the histories of Egyptian and other ancient empires called the usual dating of Creation into doubt.[142] Human history might be far longer and more diverse than previously assumed. Closer observation of the natural world and its geology had a similar effect. In the 1680s Thomas Burnet's *Sacred Theory of the Earth* drew upon Cartesian science to revive older visions of an earth in decay, which would only be restored to its pristine perfection by a future great conflagration at the Last Judgement. Until then it was broken and irreparable.[143]

The Baconians in the Royal Society, eager to demonstrate that their science could be reconciled with religious orthodoxy, were forced to respond. Alongside their new natural history they developed a 'natural theology', sometimes termed 'physico-theology', for the purpose. They could point out that the imperfections of the earth were providentially designed so as to encourage human industry, ingenuity, and improvement, and they found evidence in their researches into the history of population which confirmed that there had been a Creation at more or less the agreed date.[144] But they could never be confident that they had won the argument. As we shall see later, ideas of progressive, perhaps even perpetual, improvement were always contentious and had to be constantly reiterated.[145]

The great increase in knowledge of the past and of the present which occurred in the sixteenth and seventeenth centuries had not produced any simple account of the country's place in the history of the world, therefore, let alone an agreed linear view of secular progress in England such as some of those who contributed to or commented on the revision of Camden's *Britannia* in the years around 1700 were able to envisage. It had, however, increased appreciation of difference, made the past seem more distant, and the rapidity of recent change more visible. Readers of histories were becoming accustomed to novelties, and no longer thought them dangerous by definition, even if they sometimes needed persuasion, in a century of political and intellectual revolutions, that the future would necessarily be better than the present or the past.[146]

---

the Temple–Wotton debate, see Zachary Sayre Schiffman, *The Birth of the Past* (Baltimore, 2011), pp. 224–6.

[142] William Poole, *The World Makers: Scientists of the Restoration and the Search for the Origins of the Earth* (Oxford, 2010), esp. ch. 2; Grafton, *What Was History?*, pp. 245–8.

[143] Walsham, *Reformation of the Landscape*, pp. 381–7.

[144] Walsham, *Reformation of the Landscape*, pp. 376–93; Larry Stewart, *The Rise of Public Science. Rhetoric, Technology, and Natural Philosophy in Newtonian Britain, 1660–1750* (Cambridge, 1992), pp. 33–97; below, pp. 125–6, 197.

[145] Below, pp. 191–5.   [146] Woolf, *Social Circulation*, pp. 70, 393–4.

In the Preface to the first edition of *Angliae notitia* in 1669 Chamberlayne made some effort to defer to historical fashion when he encouraged his readers to compare England's past with its present state, and then look ahead. Using texts like his, they might 'in some measure... foresee, without consulting our astrologers and apocalyptic men, what will be the future state of this nation'. Always instinctively conservative, however, and a future Tory, he was much concerned that they should 'be moved to endeavour the restoration of what was heretofore better, and the abolition of what is now worse'.[147] That health warning, suggesting that the best of possible futures might still lie in restoration of the past, was not removed until the 18th edition in 1694, after there had been another political revolution. By then Chamberlayne faced competition from a rival Whig publication, Guy Miège's *New State of England*, first published in 1691, and he was compelled, like Miège, to record the evidence for the country's recent improvements and its emergence as a world power. Whatever the vicissitudes to be expected in an uncertain future, those at least seemed for the moment to be historical 'matters of fact'.

INFORMATION AND ITS USES

It will be evident from what has been said so far in this chapter that there was a mass of information available for contemporaries to use when formulating their ideas of England. The intellectual explorers responsible for its rediscovery—antiquaries and historians, cartographers and travellers, political writers and *virtuosi*—all had different and sometimes divergent purposes, but they created an intellectual climate in which it was possible for ideas about improvement to acquire momentum. Part of that process was the generation of new knowledge by new kinds of enquiry, and something more needs to be said at the outset about the sort of information which was available for the purpose, and about the ways in which it was already being exploited at the very beginning of the seventeenth century to answer new questions.

Much of the information came from the records of government. Monarchs and their advisers did not need Houghton to tell them that the more they knew about the kingdom, the better it could be managed.[148] As Domesday Book reminds us, English monarchs had been collecting information for purposes of government and taxation for centuries. In the fourteenth century they had a bureaucratic apparatus capable of instituting and sustaining new forms of direct and indirect taxation which foreshadowed the achievements of the fiscal-military states of the later seventeenth and eighteenth centuries; and the impressive figures recorded in the customs accounts of Edward III were still being cited three centuries later as an example of what good commercial policy could achieve.[149] In the first half of the

---

[147] Chamberlayne, *Angliae notitia* (1669 edn), 'To the Reader'. Cf. above, pp. 28–9; below, p. 130.
[148] Above, p. 28.
[149] Olive Coleman, 'What Figures? Some Thoughts on the Use of Information by Medieval Governments', in D. C. Coleman and A. H. John, eds, *Trade, Government and Economy in Pre-industrial England: Essays Presented to F. J. Fisher* (1976), p. 102; Robin E. Glasscock, ed., *The Lay Subsidy of 1334* (British Academy, Records of Social and Economic History NS 2, Oxford, 1975),

sixteenth century, the need to fund wars and secure a religious reformation propelled a search for relevant information of unprecedented scope and vigour. Cardinal Wolsey's surveys of military and taxable resources in the 1520s, Thomas Cromwell's enquiries into the wealth of the church, the institution of parish registers in 1538, and the Chantry Commissions of the 1540s, not to mention the first commission into the consequences of enclosure in 1517, together amounted to a multiple information-gathering exercise which no later sixteenth- or seventeenth-century government matched, even in the Interregnum.

With more settled government after the 1560s, the impetus to innovations of this kind waned, but they had a lasting effect which proved to be cumulative. Part of it rested literally on accumulation: of those government documents which were coming to be better organized through such means as the State Paper Office by the end of the century, and which William Cecil sought to understand through abstracts of muster rolls, customs accounts, and subsidy lists, but scarcely had the time to digest. Their sheer volume left his son Robert in 'blindness and uncertainty' when he tried to fathom 'all these collections' on the royal finances and 'in a wood indeed' when it came to interpreting the records of the royal forests.[150] Nevertheless, there was also an infectious, if slowly developing, drive to list things and add to the store. Ordered information was integral to the government of an established church whose bureaucratic structure allowed it to count the number of households in every parish in 1563 and the number of communicants in 1603;[151] and it was equally important for civil governors of counties and towns listing taxpayers, alehouses, vagrants, and paupers, in an effort to exercise surveillance and local control.[152]

Some of this activity was prompted initially by central government. Wolsey's surveys of corn supplies during the bad harvests of the 1520s, for example, inaugurated a whole series of local surveys and listings of mouths to be fed, and then of the poor and sometimes of whole populations of parishes and towns, beginning with a census in Coventry in 1523 and continuing right through to the 1690s and beyond.[153] Still more impressive as evidence of how a mounting appetite for organized information was stimulated and fed are the London bills of mortality, which were used as a means of monitoring threats to public health in the

pp. xv–xvii; below, p. 134. On the fiscal-military state, see John Brewer, *The Sinews of Power: War, Money and the English State, 1688–1783* (1989); Michael J. Braddick, *State Formation in Early Modern England, c.1550–1700* (Cambridge, 2000), pp. 177–280.

[150] Paul Slack, 'Government and Information in Seventeenth-Century England', *P&P* 184 (Aug. 2004), pp. 38–40; Palliser, *Age of Elizabeth*, pp. 319–20.

[151] Alan Dyer and D. M. Palliser, eds, *The Diocesan Population Returns for 1563 and 1603* (British Academy, Records of Social and Economic History NS 31, Oxford, 2005).

[152] Paul Griffiths, 'Local Arithmetic: Information Cultures in Early Modern England', in Steve Hindle et al., eds, *Remaking English Society: Social Relations and Social Change in Early Modern England* (Woodbridge, 2013), pp. 135–63. On the uses of organized information by trading companies, see Miles Ogborn, *Indian Ink: Script and Print in the Making of the English East India Company* (Chicago, 2007), pp. 69–80.

[153] Paul Slack, *From Reformation to Improvement: Public Welfare in Early Modern England* (Oxford, 1999), pp. 67–8; Mary H. M. Hulton, ed., *Coventry and its People in the 1520s* (Dugdale Society xxxviii, 1999), p. 18.

metropolis. Originating again with Wolsey as simple lists of deaths from plague, by the early seventeenth century they comprised a weekly and then annual series of the number of burials and baptisms in each parish, both of them printed, with totals given for the city as a whole, and the causes of death itemized after 1629.[154] It was a weekly effort in data collection far more elaborate than practical utility might have dictated.

Not all the information in lists and surveys was summed, or aggregated, though there are some exceptions besides the bills of mortality.[155] Neither was it commonly analysed for purposes other than the immediate ones of identifying who could be conscripted, taxed, prosecuted as an encloser, given poor relief, or isolated or avoided in an epidemic of plague. Gradually, however, figures began to be compiled, manipulated, and cited for novel purposes. In the middle of the century current customs and subsidy accounts were being compared with those of Edward III's reign, in an effort to measure changing taxable capacities and the balance between exports and imports, and a Master of the Mint was able to use records of recent debasements of the coinage to calculate the size of the money supply, a subject of as much subsequent interest as the balance of trade.[156] By 1600 there was no lack of information which might furnish economic data once there was a motive to look for it.

Part of the motivation came from the texts on reason of state referred to earlier, which showed that the relative power of different states could only be understood by measuring their wealth and populations, as well as their size. Bodin's *Six Books of the Republic* (1576), translated into English in 1606, had tried to compare some of the resources of England and France, and argued that those like Aristotle who warned against the dangers of large and unruly populations were wholly mistaken. No commonwealths were 'more rich nor more famous in arts and disciplines that those which abound most with citizens'.[157] Botero's *Reason of State*, first published in 1589, and his *Cause of the Greatness of Cities* of 1588, which was also translated into English in 1606, had more to say about the economic benefits of multitudes. The wealth and power of a city or state sprang from two sources: the 'virtue generative of men', which increased population, and 'the virtue nutritive of cities', which produced enterprise and commerce. If the 'industry and art of man' was applied to a fertile soil, as it was in China, 'the propagation of mankind' might increase 'without end... *in infinite*', because it delivered a surplus above subsistence requirements.[158]

---

[154] Paul Slack, *The Impact of Plague in Tudor and Stuart England* (1985), pp. 148–9; J. C. Robertson, 'Reckoning with London: Interpreting the Bills of Mortality before John Graunt', *Urban History*, 23 (1996), pp. 328–31.

[155] Slack, 'Government and Information', p. 40.

[156] R. H. Tawney and Eileen Power, eds, *Tudor Economic Documents* (3 vols, 1924), i. 178–84; J. D. Gould, *The Great Debasement: Currency and the Economy in Mid-Tudor England* (Oxford, 1970), p. 54; C. E. Challis, *The Tudor Coinage* (Manchester, 1978), pp. 244–7; below, pp. 83–5, 183.

[157] Bodin, *Six Bookes*, Book 6, p. 654; Book 5, p. 571.

[158] Giovanni Botero, *A Treatise concerning the causes of the Magnificencie and greatness of Cities* (1606), pp. 78, 91–2; Romain Descendre, *L'État du monde: Giovanni Botero entre raison d'état et*

The influence of these authors, and others writing in the same vein, was quickly apparent in English tracts on commerce which began to stress its political value by referring to 'policies of state', 'matter of state', and occasionally 'reason of state' itself.[159] But their full import was first appreciated, predictably, by Bacon.[160] In 1607, in an unfinished essay prompted by the recent union of the Crowns of Scotland and England, he referred to judgements grounded on 'reason of estate' about the greatness of states and what made one more powerful than another. With British circumstances clearly in mind, he disagreed with Botero on points of emphasis. He thought too much importance had been ascribed to 'largeness of territory', 'treasure or riches', the 'fruitfulness of the soil, or affluence of commodities', and the strength of towns. For Bacon, the true greatness of a state consisted 'essentially in population and breed of men' and their 'military disposition', and he added, with an eye on England, 'commandment of the sea'.[161] In his 1612 essay 'Of the true greatness of kingdoms', however, he took all the usual criteria into account and showed how they might be measured. 'The greatness of a state in bulk or territory doth fall under measure; and the greatness of finances and revenue doth fall under computation. The population may appear by musters, and the number of cities and towns by charts and maps.'[162] He was sketching out a programme for the computation of political and economic resources almost as ambitious as his programme for a complete natural history.

As Bacon indicated, parts of it had been embarked upon already. The number of towns or cities had long been used to compare one state with another, although there was little agreement about what constituted a town.[163] Muster rolls listing all able-bodied males were a better demographic indicator, and they had been cited by John Hales in 1548 when arguing that the English population was declining, and by Thomas Wilson in 1601 when he disputed one of Botero's conclusions.[164] But Botero had been the first to use the term 'population' to indicate a new and quantifiable field of enquiry, and Bacon seems to have been the first to use the word in English to refer to a measurable quantity at all. His contemporaries

---

géopolitique (Geneva, 2009), pp. 153–4, 158; Markku Peltonen, *Classical Humanism and Republicanism in English Political Thought, 1570–1640* (Cambridge, 1995), p. 200.

[159] e.g. John Keymer, *Original Papers regarding Trade in England and Abroad drawn up by John Reymer*, ed. M. F. Lloyd Prichard (New York, 1967), p. 35; Gerard Malynes, *Englands View* (1603), p. 134; Thomas Mun, *England's Treasure by Forraign Trade* (1664), p. 88; Phil Withington, *Society in Early Modern England: The Vernacular Origins of Some Powerful Ideas* (Cambridge, 2010), p. 229; below, pp. 78, 94.

[160] Peltonen, *Classical Humanism*, pp. 190–228.

[161] Peltonen, *Classical Humanism*, p. 197; Bacon, *Works*, ed. James Spedding et al. (7 vols, 1859–64), vii. 48–9. Bacon's stress on military valour was consistent with his view of the stages of civilization: below, p. 193.

[162] Vickers, *Bacon*, p. 302.

[163] *Tudor Economic Documents*, iii. 7; Wilson, 'State of England', pp. 11–12; Botero, *Travellers Breviat*, pp. 13, 19.

[164] BL, Lansdowne MS 238, fo. 309ᵛ; *SCED*, p. 751. Cf. Harrison, *Description*, p. 235. Wilson, who was soon to be an energetic keeper of the State Papers, seems to have been inspired by Botero to undertake other investigations, and at his death left an unfinished work on the revenues of different states: *ODNB sub* Wilson.

preferred to talk about 'populousness' and the 'fewness' or 'multitude' of people, which indicated a relative quality rather than a quantity.[165] What was in prospect now was a wholesale search for data, open to the application of 'number, weight and measure', and capable of giving what Bacon called 'precise definition' not only to the population, but to the power, wealth, and economy of a state.[166]

That was the agenda in the second half of the seventeenth century for William Petty's political arithmetic, and he and his successors, especially Gregory King and Charles Davenant, were to attempt to realize it by producing 'observations', 'computations', 'calculations', or 'exact accounts' in the ways described in later chapters. They were often frustrated. Determined research might discover relevant information in publications or the archives of governments, but it was rarely complete or totally fit for purpose. In the absence of complete national census until 1801, calculating the size of the population proved particularly difficult. Proxies had to be found. The number of parishes would have been a guide, but there was much uncertainty even about that before 1600. Machiavelli had said there were 52,000, probably using a fourteenth-century chronicle which in fact referred to the number of villages and hamlets. Botero knew that there were 9,725, an accurate figure, and commented, perhaps ironically, that the number seemed to have shrunk since 1500 when there were 'reckoned to be 40,000 parishes'. When an archbishop gave yet another number to Queen Elizabeth, she was said to have exclaimed, 'Jesus! Thirteen thousand!'[167]

Whatever the number, however, it could only be converted into total population by making an assumption about the average population of a parish and multiplying up. The methodology was well known. It had been used by Simon Fish in 1529 when he tried to demonstrate the ruinous cost to the whole kingdom of supporting begging friars by guessing how much was paid on average in each of 52,000 parishes; and it was to be used for similar purposes again, with a corrected figure of 9,725, in the 1620s.[168] Uncertainty about either the base figure or the average clearly had the potential to multiply and magnify error. Even in the later seventeenth century when political arithmeticians had better demographic indicators available than the number of parishes, they still needed more or less well-informed multipliers to turn the numbers of burials in London's bills of mortality, or of households in the hearth tax records after 1662, into total populations.[169]

Conclusions from political arithmetic were even more hazardous when its practitioners tried to go further and measure much more than the population in

---

[165] Descendre, *L'État du monde*, pp. 164–8; Paul Slack, *'Plenty of People': Perceptions of Population in Early Modern England* (Stenton Lecture 2010, University of Reading, 2011), pp. 3–4. Bacon had used the word in 1593, before his essays on sedition and the greatness of kingdoms: Francis Bacon, *Early Writings 1584–1596*, ed. Alan Stewart with Harriet Knight (*OFB* i, Oxford, 2012), p. 358.

[166] Webster, *Great Instauration*, p. 351. Cf. Bacon, *Instauratio magna Part III*, p. xxxiii.

[167] Paul Slack, 'Measuring the National Wealth in Seventeenth-Century England', *EcHR* 57 (2004), p. 609; Palliser, *Age of Elizabeth*, p. 320.

[168] Slack, 'Measuring the National Wealth', p. 611. For a similar calculation, in 1552, see *Tudor Economic Documents*, iii. 56–7. In 1576 Bodin referred to an author of estimates of this kind as a 'multiplier', *Six Bookes*, Book 6, p. 671.

[169] Below, pp. 118–19, 140, 195.

order to demonstrate, as Petty intended, 'the uses of knowing the true state of the people, land, stock, trade etc.' As we shall see, that involved educated guesswork, and the deliberate use of hypotheses to fill gaps in the evidence; and where the evidence was plentiful, it rarely led to agreed conclusions.[170] Far from discrediting the whole enterprise, however, inconclusive, partial, and questionable conclusions encouraged the production of yet more calculations to answer if not satisfy critics. Some of them were published, many remained in manuscript, like Gregory King's breakdown of the expenditure of as many as eleven different socio-economic classes on eight different kinds of food and drink.[171] By the end of the century there was a palpable excitement in deriving new knowledge, about things previously unknown and scarcely thought about. Counting, as Dr Johnson remarked to Boswell, had the great attraction that it seemed to bring 'everything to a certainty, which before floated in the mind indefinitely'.[172]

The outcome by 1700 was often more information than could easily be digested or put to any practical purpose. The more that was known about the balance of trade or the size of the money supply, for example, the less easy it was to arrive at any certainty about them. Information might be out of date, as it was when the framers of new constitutions during the Interregnum tried to redistribute parliamentary seats according to the distribution of wealth, and relied for guidance on tax assessments which were a century old.[173] Measurement and calculation were sometimes essential for governments when they designed new taxes or had to pacify Ireland and colonize plantations, but then they were tools guiding the implementation of policy; they only rarely determined it.[174] Number and measure could never become the beacons of political enlightenment which Bacon and Petty envisaged.

New kinds of information had their greatest impact outside the narrow confines of executive government, in that wider public sphere of which both governors and political arithmeticians were a part, and to which they had constantly to relate. Since some of the information concerned matters commonly considered to be state secrets, *arcana imperii*, there were some early doubts about whether it should be widely known.[175] But it was impossible to prevent leakage of information in a political world where the boundaries between the state and the public were much less clearly drawn than they were to be later, and where people like Bacon and Petty were sometimes agents of government and sometimes putting pressure on it from outside. In these circumstances governments might make a virtue of necessity, as

---

[170] Petty, *EW*, i. 313. Cf. below, pp. 121–2.
[171] Adam Fox, 'Food, Drink and Social Distinction in Early Modern England', in Hindle et al., eds, *Remaking English Society*, p. 169.
[172] Andrea A. Rusnock, *Vital Accounts: Quantifying Health and Population in Eighteenth-Century England and France* (Cambridge, 2002), p. 1.
[173] Slack, 'Government and Information', pp. 49–50.
[174] Above, p. 26; below, pp. 97–8, 180–4.
[175] Cf. below, pp. 104, 119–20, 175. For Bacon's distrust of the free exchange of information, see Thomas Leng, 'Epistemology: Expertise and Knowledge in the World of Commerce', in Philip J. Stern and Carl Wennerlind, eds, *Mercantilism Reimagined: Political Economy in Early Modern Britain and its Empire* (Oxford, 2014), pp. 106–7.

Robert Cecil did in 1610 when he opened up discussion of the king's finances in a remarkable 'consultation' with the House of Commons whose 'general object', he said, was 'public utility'.[176] In 1576 Bodin had commented on similar discussions of royal finances in the French Estates General and published many of the figures with critical commentary;[177] and books like his, in England even more than in France, made such information widely available. By 1700, when English parliaments were meeting annually and a popular press was reflecting and contributing to their debates, there was a surfeit of information about the state of the nation open to scrutiny, and one of the themes of this book is the emergence of a large educated public, curious and informed, not least about political economy.

A century earlier, in 1600, that new branch of natural science was still in its infancy, and public appreciation of its novelty only just beginning. Yet there were signs of what was to come, and two of them in particular, in 1603 and 1607, can be used to show that both the ambition to create useful knowledge and a more limited capacity to deliver it were already evident. One of them involved the use of multipliers to arrive at measurements of the comparative strength of different states such as Bodin and Botero had envisaged. The second showed how quantitative evidence could contribute to discussion of a domestic issue of pressing political concern: in this instance, the practices of enclosing and engrossing land, especially when they led to depopulation.

In 1603, in *Englands View*, Gerard Malynes set out to 'give some glance of comparison between England and France, the greatest kingdom of Europe', by publishing figures for their size and population. A prolific writer on commerce of whom we will hear more in the next chapter, he had been reading Bodin and other 'politicians', as he described them, and was trying to compete. France was obviously much bigger, but Malynes computed the difference precisely. England and Wales covered just over 34 million acres and France a little over 91 million; and their populations were just under 17 million for England and 22 million for France. The exercise is remarkable as the first comparative 'computation' of its kind published in England.[178]

It was less remarkable for its accuracy. The calculations of acreages were not wildly wrong, but a little too low. Malynes may have copied the numbers from other publications, or applied some geometry of his own to current maps.[179] His population figure for France, computed from some source which gave him the number of families, was probably a little too large. Other estimates published at the time suggested something closer to 15 million.[180] The population Malynes gave for England, on the other hand, was far too large, since it cannot have exceeded 5 million in 1603. The error arose from his application of large multipliers to the

---

[176] Slack, 'Government and Information', pp. 59–61. For similar problems about the use of information in Venice, see Filippo de Vivo, *Information and Communication in Venice: Rethinking Early Modern Politics* (Oxford, 2007), pp. 15–17, 57–8, 238.
[177] Bodin, *Six Bookes*, Book 6, pp. 671–2.   [178] Malynes, *Englands View*, pp. 135–6.
[179] Slack, 'Measuring the National Wealth', pp. 610–11, 626 n. 91.
[180] Botero, *Magnificencie and greatness of Cities*, p. 82; Nicolas Besongne, *The Present State of France* (1671), p. 14.

old figure of 52,000 'villages and settlements' given in that medieval chronicle. When Malynes repeated the exercise in 1622, however, this time using the much smaller number of parishes, he adjusted his multipliers in order to reach the same conclusion as before.[181] He was clearly determined not to show England too far behind France, and as uncritical in his use of very large numbers as many of his contemporaries. Original as his exercise was, it is hardly surprising that later political arithmeticians, who had better sources at their disposal, never referred back to it.

The second early use of deliberately computed numerical data for purposes of comparison and political argument was far more modest in ambition, but no less designed to make a particular case. It occurred in 1607, in the crisis caused by riots against enclosures and the conversion of arable to pasture by improving landlords in Northamptonshire and neighbouring counties. The disturbances in what became known as the Midland Rising involved crowds numbering hundreds and occasionally thousands, needed a military force to suppress them, and provoked a number of sermons and tracts on the danger of popular insurrection, including Bacon's essay of 1612 on 'seditions and troubles'.[182] The government had to react to both the causes and the consequences, and in July 1607 a short discussion paper, 'A Consideration in Question...touching Depopulation', was prepared to assist the Privy Council in its deliberations.[183] The author was probably Robert Cotton, an antiquary and MP just beginning to have some political influence, and displaying here his facility in using learned argument and quantified evidence to reach conclusions his audience wanted to hear.[184]

He framed his analysis in the terms set by reason of state. Enclosure and depopulation presented multiple and sometimes contradictory threats to the two overriding concerns of government, 'security of state' and 'increase of wealth and people'; and there were no obvious remedies easily available. Security was clearly threatened by the 'domestic commotions' created by enclosure, and perhaps by any depopulation which might follow from it and reduce the number of men capable of bearing arms. But the hedges of enclosed land had the advantage of impeding the movement of any potential invading army, and unenclosed land was well known to encourage a surplus population of the poor and the idle, ripe only for sedition. Too many people would 'surcharge the state' and would need to be 'vented' by wars or colonies if domestic commotion was not to follow.

The perception that England was overpopulated had been a commonplace in the literature on overseas plantations from Hakluyt onwards, and the Midland Rising

---

[181] Malynes, *Lex mercatoria*, p. 235.
[182] Robert B. Manning, *Village Revolts: Social Protest and Popular Disturbances in England 1509–1640* (Oxford, 1988), pp. 229–46; John E. Martin, *Feudalism to Capitalism: Peasant and Landlord in English Agrarian Development* (London, 1983), pp. 161–79; Steve Hindle, 'Imagining Insurrection in Seventeenth-Century England: Representations of the Midland Rising of 1607', *History Workshop Journal*, 66 (2008), pp. 36–40.
[183] *SCED*, pp. 107–9; E. F. Gay, 'The Midland Revolt and the Inquisitions of Depopulation of 1607', *TRHS* ns 18 (1904), p. 217 n. 5.
[184] It is not absolutely certain that Cotton was the author, and the paper has never been attributed to him. In addition to the clean copy of it in BL, Cotton MSS, Titus F. iv, fos 322–3, however, there is what appears to be a draft, initialled by Cotton, in BL, Cotton MSS, Faustina, C. ii, fos 190–6.

accentuated it.[185] Bacon referred to it in 1612 in his essay on seditions, and a Northamptonshire magistrate had spoken about its effects in his own county in parliament in 1604.[186] The 1607 Consideration tried to add precise evidence by claiming that there were forty more baptisms than burials every week even in London, 'a place more contagious than the country'. The author would have needed to choose carefully among London's bills of mortality to find many weeks with a surplus as big as that,[187] but the implication was clear. Depopulation, which had always been the primary argument against enclosure, was no longer the threat it had appeared to be when the population was much lower, in the early sixteenth century; there were now too many people not too few.

When Cotton considered the relationship between 'wealth and people', he was again sceptical about the supposed damage done to both of them by enclosure, and this time he 'proved' his case by comparative numbers arranged in a striking table. The idea for it probably came from Sir Thomas Smith's 'Discourse of the Common Weal', written in 1549 and published in 1581, which had cited Essex, Kent, and Devon, to demonstrate that counties 'where most enclosures be are most wealthy', without giving any figures.[188] The table in the Consideration presented numbers calculated from subsidy returns and muster rolls from two contrasting counties, open-field Northamptonshire and enclosed Somerset. They showed that Somerset paid more than twice as much in taxation and had ten times as many men qualified for military service as Northamptonshire.

The evidence might again have been suspect to any critical eye, since the paper made no allowance for the fact that Somerset was twice as big as Northamptonshire. Even if it had, however, Cotton could have claimed that the message about population remained the same: the enclosed county was the one which furnished more people useful to the state.[189] As for wealth, he repeated the argument of the 'Discourse' that farmers and tenants on enclosed land had much stronger incentives to maximize their private 'profit' than those who held land in common. They should generally be left free to respond to local incentives determined by the quality of their land and their access to markets, and also by relative prices which reflected the growth of demand for grain as against the demand for wool since Henry VIII's reign.[190]

The paper recognized that something must be done to pacify popular complaint, perhaps by new laws or commissions of enquiry which had been tried before; but their focus should be on particular counties and especially those, like

---

[185] Above, p. 31; below, pp. 78–9.
[186] Vickers, *Bacon*, p. 368; Manning, *Village Revolts*, pp. 241–2.
[187] Roger Finlay, *Population and Metropolis: The Demography of London 1580–1650* (Cambridge, 1981), pp. 155–6, gives some annual figures.
[188] [Thomas Smith], *A Discourse of the Common Weal*, ed. Elizabeth Lamond (Cambridge, 1954), p. 49.
[189] In fact, by 1720 one-third of the parishes of Northamptonshire were enclosed but only 10 per cent of the county's population lived there: J. M. Neeson, *Commoners: Common Right, Enclosure and Social Change in England, 1700–1820* (Cambridge, 1993), pp. 58–9.
[190] The argument here (*SCED*, p. 108) is not entirely clear, and hence difficult to confirm or refute from what is now known about relative prices. The general point about relative price incentives had also been made in the 1549 'Discourse' (pp. 52–3), and been recognized by Bacon: below, p. 59.

Northamptonshire, where tenants had been ejected and engrossing visibly led to depopulation. Farmers elsewhere could be left 'to their choice' of styles of husbandry without any 'inconvenience in the state'. In August 1607 a new enclosure commission was indeed appointed, the last on any large scale, and its enquiries were limited to seven Midland counties, most of them the site of recent commotions.[191] We have no means of knowing whether the paper had been read by the Council in July, and it was never published, but it plainly reflected the direction in which educated opinion was moving.[192] The aim now was to balance two apparently incompatible goals: to ensure, as the Consideration pithily put it, that 'the poor man shall be satisfied in his end: habitation; and the gentleman not hindered in his desire: improvement'.

As we shall see later, both the relationship between improvement and habitation and that between wealth and people remained sources of contention into the eighteenth century, whenever improvements in agriculture or the growth of towns and trade seemed to be preventing population growth.[193] I cite the Consideration of 1607 here, however, because it was precocious in other respects, in the relative sophistication of its economic and political analysis as much as in its use of quantitative evidence. The document shows that it was already possible for a scholar as well read and politically engaged as Cotton to use information as a tool for understanding the kingdom and how its economy should be managed. Although the name had not yet been invented, he was thinking like a political economist.

In 1609 Sir Thomas Bodley made a similar point when writing to Bacon about the great strides which were currently being taken towards an advancement of learning. Bacon's claim to be inventing a wholly new kind of natural science seemed to him to ignore the many discoveries and inventions which were 'daily being brought to light by the enforcement of wit or casual events'. Bodley had no doubt that there was already an 'open highway to knowledge'.[194] It was to open much wider in the 1640s, when Bacon's posthumous influence had its full impact, and the advancement of learning became part of a deliberate quest for improvement of every kind.[195] But the highway was being cleared of obstacles during Bacon's lifetime; and we should look in the next chapter at the period before 1640, and ask what had impeded, and what promoted, improvement then.

---

[191] Joan Thirsk, 'Enclosing and Engrossing', in Joan Thirsk, ed., *Agrarian History of England and Wales*, iv: *1500–1640* (Cambridge, 1967), pp. 235–6.

[192] The Privy Council records for this period no longer survive. On the trend of opinion, see below, pp. 58–9; Maurice Beresford, 'Habitation versus Improvement: The Debate on Enclosure by Agreement', in F. J. Fisher, ed., *Essays in the Economic and Social History of Tudor and Stuart England in Honour of R. H. Tawney* (Cambridge, 1961), pp. 40–69.

[193] Below, pp. 139–40, 199, 235.

[194] Deborah Harkness, *The Jewel House: Elizabethan London and the Scientific Revolution* (New Haven, 2007), pp. 250–1; Francis Bacon, *The Advancement of Learning*, ed. Michael Kiernan (*OFB* iv, 2000), p. xxxiii.

[195] Cf. Harkness, *Jewel House*, p. 250; Bacon, *Instauratio magna: Part II*, p. xxiii.

3
Elizabethan Foundations 1570–1640

The principal impediment to the deliberate pursuit of improvement before 1640 was the need to justify it as contributing to the public good, and not simply to the private profit of those involved in it. That was a challenge faced by improvers of all kinds throughout the period covered by this book, but it was particularly prominent at the beginning, in a society firmly attached to the idea of a stable and harmonious body politic or commonwealth, a state never perhaps attainable in perfect form, but more evident in the past than the present and now patently under threat from the naked pursuit of profit by improving landlords and avaricious merchants. In the cultural climate of sixteenth-century England, a political elite which had absorbed the civic humanism of the Renaissance was always conscious that private enterprise might corrupt civic virtue, and its economic and social consequences require reformation in the interests of public welfare and the common weal.[1]

At the same time, however, governments and their agents found themselves having to come to terms with social and economic change, and often actively promoting it, especially when it offered opportunities for public reformation and served their own private interests at the same time. There were conspicuously mixed motives when monarchs and members of parliament tried to increase their landed incomes while seeking to regulate enclosure and provide social welfare for the poor, just as there were when merchants and entrepreneurs licensed by the Crown and looking for profits opened up new avenues for overseas trade and established plantations in Ireland and America. By the end of Elizabeth's reign there were many actors engaged in enterprises ostensibly for the public good which encouraged rather than resisted change. They were laying the foundations for some basic protections of the public welfare, notably in the shape of the Elizabethan poor law, while leaving ample room for innovations which could be represented as improvements.

The result was an unstable and often contested balance between public and private goals of the kind envisaged in Robert Cotton's paper on enclosure in 1607. It challenged old assumptions, not only about how public and private interests

---

[1] For changing concepts of the 'common weal', see Mark Knights and the Early Modern Research Group, 'Commonwealth: The Social, Cultural, and Conceptual Contexts of an Early Modern Keyword', *Historical Journal*, 54 (2011), pp. 659–87; Phil Withington, *Society in Early Modern England* (Cambridge, 2010), pp. 138–44; Paul Slack, *From Reformation to Improvement* (Oxford, 1999), pp. 5–28; and on its fifteenth-century origins, below, pp. 260–1.

interacted, but about how the national economy functioned and how it could and should be managed. Although change of that kind came much more rapidly after the political revolutions of the 1640s, it will be argued in the final section of this chapter that it accelerated in the 1620s, when public debate about how to react to an economic crisis suggested that there were limitations to what political manipulation of the economy could be expected to achieve, that private enterprise must therefore be allowed freer rein, and that it was likely to contribute just as effectively to the public good. In the circumstances of the 1620s, the notion that a stable and harmonious commonwealth could somehow be recreated by political reformation looked decidedly anachronistic.

REFORMING THE COMMONWEALTH

Half a century earlier, in the 1560s, reformation of the commonwealth still retained most of the rhetorical force it had acquired from humanists and social critics since the later fifteenth century. They had examined the condition of England and found its social and moral fabric undermined by avarice and a decline in civic virtue, its towns, agriculture, and population all 'in sore decay', 'in desolation, decay, ruin', and the whole commonwealth in need of 'speedy reformation'.[2] In Edward VI's reign, when wholesale reformation seemed for a moment possible, the rhetoric was deployed on all sides, in government, by religious radicals and Utopian visionaries, and by the leaders of popular rebellions in 1549 who appealed to the common weal just as their predecessors had done in the Pilgrimage of Grace in 1536. The commotions of 1549 naturally evoked a political backlash and showed that reformation was more easily talked about than achieved, but they also produced new analyses of the problems of the commonwealth. The most acute and influential of them, Thomas Smith's 'Discourse of the Common Weal', had issued the salutary warning that there was no prospect in the real world of ensuring that 'all covetousness...be taken from men'. Since all of them were 'naturally covetous of lucre', the profit motive must be harnessed for the public good. To suppose otherwise was to be seduced by what Smith later called 'feigned common wealths', Utopias 'such as never was nor never shall be'.[3]

After 1558 William Cecil, Smith's friend and patron and 'the very Cato of the commonwealth' as Elizabeth's chief minister, had to face political realities when considering proposals and projects for reform of the common weal, many of them written at his instigation, some of them annotated in his own hand.[4] Projects needed to engage private interests if they were to succeed, and once embarked upon they might have novel and unintended consequences, rather than conserve the

---

[2] Slack, *Reformation to Improvement*, pp. 6–8.
[3] [Thomas Smith], *A Discourse of the Common Weal*, ed. Elizabeth Lamond (Cambridge, 1954), pp. 121–2; Sir Thomas Smith, *De republica Anglorum*, ed. Mary Dewar (Cambridge, 1982), p. 144.
[4] Withington, *Society*, p. 149. On proposals for reform sent to Cecil, see Joan Thirsk, *Economic Policy and Projects* (Oxford, 1978), pp. 52–3, 68–70, 86–7.

status quo. Some of the 'devices for reformation' submitted early in Elizabeth's reign were of such sweeping scope that they were scarcely practical propositions at all. A set of 'considerations' intended for parliament in 1559, for example, sought to prevent the advance of 'new men . . . coveting to be hastily in wealth and honour', 'to acquaint men with virtue again', and to achieve a 'reformation of religion' and 'the amendment of manners'.[5] In 1558 Armagil Waad, one of Cecil's clients, had similarly listed 'the wealth of the meaner sort' and their contempt of the nobility among the 'distresses of the common wealth', and advocated a committee of experts in law, divinity, and commerce to consider at length 'all manner of means whereby the common weal may be reformed, benefitted, and kept in good order'.[6] Thomas Smith would have welcomed such a device, but it was a means of kicking insoluble issues into the long grass.

Waad also made two more immediate proposals for political action, both ostensibly practicable, neither of them predictable in its effects. One was designed to boost government revenues by cultivating some of the forests and wastes of the realm, now 'barbarous and barren for want of culture', which only required 'good husbandry' to be 'reduced to fertility'. It was an early recognition that the Crown might copy private landowners and profit from improvement, and a forerunner of similar schemes later which caused more political trouble than they were worth in financial terms.[7] Waad's second proposal was for a recoinage which he expected to bring an end to inflation. When implemented in 1560 it restored confidence in the currency but was only a temporary antidote. As prices continued to rise, a paper sent to Cecil in 1588 proposed a repeat of the medicine, but without any confidence that it would work. Inflation was produced by the 'natural disposition of the time and revolution of things within this variable world', and could not be remedied 'until by the like motion, vicissitude, and altercourse, things in this variable world' turned again to 'where they were'.[8] Until that great turn of the historical cycle, if it ever came, governments had to accept that they were at the mercy of events, and deal with the consequences as best they could.

Cecil had reflected on that lesson in the 1560s, in one of the greatest economic crises of the century, when English cloth exports collapsed because of war in the Low Countries. In a memorandum of his own, written in 1564, he contemplated the consequences for every aspect of the economy.[9] Some of them he thought wholly beneficial. No attempt should be made to restore the country's dependence on a single outlet for its principal manufactures. The monopoly control on exports to Antwerp enjoyed by the Merchant Adventurers Company had created multiple

---

[5] R. H. Tawney and Eileen Power, eds, *Tudor Economic Documents* (3 vols, 1924), i. 325–30. On this document, whose author is unknown, see Robert Tittler, *Nicholas Bacon: The Making of a Tudor Statesman* (1976), pp. 105–6. For another 'device for reformation' see BL, Lansdowne MS 19, no. 41, fo. 89.
[6] TNA, SP Domestic, 12/1, fos 145–54; *ODNB sub* Waad.   [7] Below, pp. 60–1.
[8] BL, Lansdowne MS, 55 no. 12, fos 31–2.
[9] *Tudor Economic Documents*, ii. 45–7. On the document and the circumstances, see G. D. Ramsay, *The City of London in International Politics at the Accession of Elizabeth Tudor* (Manchester, 1975), pp. 274–5 and *passim*.

imbalances in the commonwealth. There were already too many merchants speculating on the foreign exchanges, and damaging the balance of trade by importing 'unnecessary trifles' and useless luxuries. There were far too many clothworkers at home who were well known to be disorderly and 'of worse condition to be quietly governed than the husbandmen'. Those now unemployed as a result of the slump could be sent 'into Ireland to help the peopling of the countries there'. It would be 'no evil policy' either if a fall in textile production and the demand for wool restrained the conversion of arable to pasture and the notorious decay of tillage.

Cecil's comments were sometimes wholly reactionary, as when he argued that it would be no bad thing if English exports of wool rather than cloth were restored to their medieval volume. But he also welcomed the production of the new draperies, the light cloths exported to southern Europe, because they minimized the country's dependence on the old draperies, 'fine white cloths', which went to Antwerp. He was explicit that the time was ripe to look forward rather than backward, to 'attempt . . . an alteration', rather 'than to make a reverse, without any fruit to be had or gathered of these troubles now passed'; and he was farsighted in some of the alterations he envisaged. Over the next half-century England's foreign trade was restructured and diversified by merchants seeking new markets; production of the new draperies was expanded by Protestant refugees from the Low Countries whose migration to towns like Canterbury, Colchester, and Norwich Cecil encouraged;[10] and overseas plantations did in the end take off some of the excess population at home.

That mixture of instinctive conservatism with an often grudging willingness to adjust to the times was characteristic of policy-makers engaging in social and economic engineering for the benefit of the commonwealth in Elizabeth's reign. The Statute of Artificers of 1563 could never have put a stop to social mobility and imposed a 'uniform order' on occupational distribution in the way some legislators hoped, and its provisions regulating wages had gradually to be adjusted to fit industry as well as agriculture, and to accommodate the reality of inflation while retaining customary notions of justice and hierarchy.[11] The Usury Act of 1571 which legislated for a maximum legal rate of interest for the first time was a similar case of the reluctant sacrifice of commonwealth ideals. The growing practice of money-lending showed earlier laws prohibiting it to be wholly ineffective, but respectable opinion remained hostile. It was against the law of God and a deterrent to charity, and Cecil thought it would add further to the infinite numbers of merchants and financiers pestering the commonwealth. The statute was passed only after long debate and in amended form, by a parliament forced to tolerate interest within narrow limits, but not yet ready firmly to approve it.[12]

---

[10] Thirsk, *Economic Policy and Projects*, pp. 43–4; Nigel Goose, 'Introduction', in Nigel Goose and Lien Luu, eds, *Immigrants in Tudor and Early Stuart England* (Brighton, 2005), pp. 14–29.

[11] S. T. Bindoff, 'The Making of the Statute of Artificers', in Bindoff et al., eds, *Elizabethan Government and Society: Essays Presented to Sir John Neale* (1961), p. 57; Donald Woodward, 'The Background to the Statute of Artificers: The Genesis of Labour Policy, 1558–63', *EcHR* 2nd ser. 33 (1980), pp. 32–44; Slack, *Reformation to Improvement*, p. 58.

[12] Norman Jones, *God and the Moneylenders: Usury and Law in Early Modern England* (Oxford, 1989), pp. 34–65 (Cecil's views are cited at p. 55). Cf. below, pp. 86–7.

Cecil himself was much happier with novelties when they seemed to suit commonwealth purposes, especially if he could find English or foreign precedents for them. A statute of 1563 making the consumption of fish rather than meat compulsory on Wednesdays was no 'innovation', he claimed, because there was earlier legislation to similar effect, and its purpose was to 'restore' private piety and the prosperity of the country's fisheries and ports. He approved the first English state lottery, established in 1588, because it was based on a foreign model and designed to finance the repair of Dover and other harbours, although it failed to raise anything like the sums expected.[13] Much more effective in attracting private investment for public purposes were patents of monopoly, giving exclusive rights to the inventors or first users of new technologies and products, a device widely used on the Continent. First employed in England in 1552, it was enthusiastically promoted by Cecil once he was persuaded that those who discovered 'things useful to the public' needed some incentive to undertake the risks involved in their development.[14]

Intended to cover 'all good sciences and wise and learned inventions tending to the benefit of the commonwealth', patents were an endlessly flexible instrument applied to projects which served multiple purposes.[15] They increased the country's military capabilities by introducing skills like gun-making, iron-founding, and the manufacture of saltpetre. They promoted the domestic production of expensive imports such as pins, oil, white soap, and clear glass. They encouraged the immigration of foreign artisans, skilled in dyeing, weaving fustians, or knitting stockings, and promised to provide employments for the poor. When they were first introduced, patents were an integral part of an Elizabethan policy for importing new industries, which might flourish as the new draperies did, and attracting foreign engineers, like those who worked on repairing harbours and fortifications and were indispensable when draining fens and marshes. The government was borrowing and copying European expertise and inaugurating a process of international technology-transfer which was to transform the character of the English economy by the end of the seventeenth century.[16] It was also giving public approval to private enterprises which turned out to have consequences scarcely compatible with the ideal of a common weal.

---

[13] Slack, *Reformation to Improvement*, p. 24; David Dean, 'Elizabeth's Lottery: Political Culture and State Formation in Early Modern England', *Journal of British Studies*, 50 (2011), pp. 587–611. The repair of Dover harbour, 'the largest civil engineering project of the sixteenth century', had to be funded from royal licences and a parliamentary levy on shipping: Ian W. Archer, 'Social Order and Disorder', in Paulina Kewes et al., *The Oxford Handbook of Holinshed's Chronicles* (Oxford, 2013), pp. 389–90; Eric H. Ash, *Power, Knowledge, and Expertise in Elizabethan England* (Baltimore, 2004), pp. 55–86.

[14] Thirsk, *Economic Policy and Projects*, pp. 52–3.

[15] Deborah E. Harkness, *The Jewel House* (New Haven, 2007), pp. 143–5, 154–5. Thirsk, *Economic Policy and Projects*, provides the fullest account of their economic importance. On early patents, see also Christine MacLeod, *Inventing the Industrial Revolution* (Cambridge, 1988), pp. 10–13; William Hyde Pryce, *The English Patents of Monopoly* (Cambridge, Mass., 1906), pp. 7–9.

[16] Ash, *Power, Knowledge, and Expertise*, pp. 32–4, 215–16; Christine MacLeod, 'The European Origins of British Technological Predominance', in Leandro Prados de la Escosura, ed., *Exceptionalism and Industrialisation: Britain and its European Rivals, 1688–1815* (Cambridge, 2004), pp. 111–26. Cf. below, pp. 230–1.

The consequences might have been predicted from the example of agrarian improvement, a parallel process of innovation which had begun much earlier under its own steam and was only slowly being accepted by governors as something they could not prevent and which it might be in their interest to promote. Armagil Waad's 1558 proposal for the cultivation of forests and wastes pointed in that direction, but the enclosure and improvement of occupied land was still tainted by the assumption that it inevitably led to the conversion of arable to pasture and hence depopulation. It was a charge often repeated but difficult to sustain in the century after 1550 when the population of England was visibly rising again. There had been more substance to it in the later fifteenth century when aspiring peasants were already busy engrossing farms, and there were acquisitive 'yeoman-merchants' like John Heritage in Gloucestershire, neither a lord nor a peasant but in effect a capitalist entrepreneur, buying up land and allegedly evicting tenants.[17] John Hales may well have been right to claim in 1549 that the 'chief destruction of towns and decay of houses' in the countryside had come before 1485.[18]

By 1600 a more persuasive complaint about agrarian improvement was the one articulated by Thomas Wilson in his 'State of England'. The pursuit of profit had destroyed the whole balance and harmony of the commonwealth. Gentlemen had lost all interest in the martial arts, learnt as well as any farmer 'how to improve their lands to the uttermost', and become good husbandmen instead. Yeomen had been able to exploit long leases and rising prices, and begun 'to manure and enclose and improve their grounds'. All of them had ceased to practise hospitality, and left a growing number of poor cottagers and labourers wholly dependent on 'small wages'.[19] Wilson was describing a process of social polarization in the countryside which benefited larger landowners, who were more efficiently managing their estates, and yeomen with some security of tenure who took an increasing share of the profits.[20] Improvements in their material wealth were conspicuous when they had to testify in court to how much they were worth in moveable property in the early seventeenth century. By 1650 yeomen were reporting sums ten times larger than their grandfathers or great-grandfathers had done a century earlier, and the wealth of gentlemen (excluding their land) seems from the same evidence to have risen threefold over the same period. The material wealth of husbandmen had only kept pace with inflation.[21]

---

[17] Christopher Dyer, *A Country Merchant, 1495–1520: Trading and Farming at the End of the Middle Ages* (Oxford, 2012), pp. 11, 34, 99; Jane Whittle, *The Development of Agrarian Capitalism: Land and Labour in Norfolk, 1440–1580* (Oxford, 2000), pp. 305–15.

[18] Joan Thirsk, 'Enclosing and Engrossing', in Joan Thirsk (ed.), *Agrarian History of England and Wales*, iv: *1500–1640* (Cambridge, 1967), p. 213.

[19] Thomas Wilson, 'The State of England Anno Dom. 1600', ed. F. J. Fisher, in *Camden Miscellany XVI* (Camden Soc. 3rd ser. 52, 1936), pp. 18–20, 38–9.

[20] Joan Thirsk, 'Making a Fresh Start: Sixteenth-Century Agriculture and the Classical Inspiration', in Michael Leslie and Timothy Raylor, eds, *Culture and Cultivation in Early Modern England: Writing and the Land* (Leicester, 1992), pp. 15–18; Richard W. Hoyle, 'Introduction', in Richard W. Hoyle, ed., *Custom, Improvement and the Landscape in Early Modern Britain* (Farnham, 2011), pp. 9–14.

[21] Alexandra Shepard and Judith Spicksley, 'Worth, Age, and Social Status in Early Modern England', *EcHR* 64 (2011), pp. 516–17, 528. Much of the new wealth of yeomen seems likely to

Elizabethan Foundations 1570–1640

In these circumstances the rhythm of conversion from arable to pasture and hence depopulation depended on relative price incentives, and the statutes in favour of tillage which had begun in the 1480s lost much of their purpose. In 1593 parliament at last repealed some of them, probably because, as Bacon pointed out, the recent 'surcharge of people' had increased demand for grain and done more to 'entice' farmers to convert pasture back to tillage than 'all the penal laws' enacted for that purpose.[22] Whenever there were agrarian disturbances or bad harvests, there was pressure to revive them, and they were reinstated in 1597 and 1601 for that reason, with Bacon himself now speaking in their favour because they were for 'the benefit of the commonwealth'. But opinion was changing. Henry Jackman echoed the Discourse of the Common Weal in arguing that 'men are not to be compelled by penalties but allured by profit to any good exercise' in 1597, and Ralegh insisted in 1601 that the best course was to 'leave every man free' to use his land as he wished, 'which is the desire of a true Englishman'.[23]

After the Midland Rising in 1607 government enthusiasm for regulation rapidly waned. The tillage statute of 1563 was repealed in 1624 and a final commission of enquiry in the 1630s was more interested in levying fines on offenders than forcing them to restore pasture to arable. Charles I's Attorney General was able to assure him that enclosed counties were three times more valuable than the rest, and that a few local trouble-makers should not be permitted to oppose an 'improvement' which would 'effect so public a good';[24] and one projector saw the possibility of 'a speedy and general enclosure in all parts of the kingdom' because even the 'vulgar and common sort of people' could see the benefits.[25] Although the common sort were still resisting enclosure and conversion in the Midlands in the 1650s, a preacher there in 1635 had to admit that some local farmers were ready to 'go to the Devil for enclosing' because they went 'for as good ground as is in England'.[26]

What remained, however, was the general and persistent suspicion that the devil of corruption was at work in any pursuit of private profit which paid no regard to the public good. In 1597 one member of parliament had directed his rhetoric, not

---

have been in the farm equipment which went with intensive agrarian improvement. Cf. below, p. 160.

[22] Francis Bacon, *Early Writings 1584–1596*, ed. Alan Stewart with Harriet Knight (*OFB* i, Oxford, 2012), pp. 357–9. For the history of legislation against enclosure see Thirsk, 'Enclosing and Engrossing', pp. 213–38.

[23] Andrew McRae, *God Speed the Plough: The Representation of Agrarian England, 1500–1660* (Cambridge, 1996), pp. 7–12; Slack, *Reformation to Improvement*, p. 59; David M. Dean, *Law-Making and Society in Late Elizabethan England: The Parliament of England, 1584–1601* (Cambridge, 1996), p. 272. On Bacon's view of enclosure, see below, p. 235.

[24] Thirsk, 'Enclosing and Engrossing', pp. 236–8; Maurice Beresford, 'Habitation versus Improvement', in F. J. Fisher, ed., *Essays in the Economic and Social History of Tudor and Stuart England* (Cambridge, 1961), pp. 49–51; Hoyle, 'Introduction', pp. 22–3.

[25] BL, Royal MS 18A. xxv, fo. 1ᵛ; Beresford, 'Habitation versus Improvement', p. 54.

[26] Joan Thirsk, 'Agricultural Policy: Public Debate and Legislation', in Joan Thirsk, ed., *Agrarian History of England and Wales*, v: *1640–1750* (2 vols, Cambridge, 1984–5), ii. 318–21; Joseph Bentham, *The Christian Conflict: A Treatise* (1635), p. 317. For continuing opposition to local depopulation, see also Robert Powell, *Depopulation Arraigned* (1636), pp. 6–7; Christopher W. Brooks, *Law, Politics and Society in Early Modern England* (Cambridge, 2008), p. 200.

against all agrarian improvement but against 'corrupt improvement', an 'unnatural and cruel improvement' at the hands of landowners motivated by 'self-love', pursuing their own 'profit and pleasure', and hence creating a 'more brutish land' unable to 'fructify to the common good'.[27] Agrarian projectors tried to defend themselves. John Norden argued that enclosure of the royal forests was for 'the good of the commonwealth' because their 'plantation' would bring 'the former unprofitable inhabitants to a civil and religious course of life'.[28] Schemes for the 'improvement and enclosure' of waste grounds, marshes, and fens were similarly intended to populate them with civilized and self-sufficient farmers and make 'the commonwealth greatly enriched and bettered';[29] and one of the foreign engineers involved in draining the fens thought it 'one of the wonders of the world' that they had 'lain so long neglected ... among a nation of so politic a government'.[30] Yet claims to superior civility cut little ice when the Crown expected to receive part of the profits, as it did in these cases, and when it could blatantly set up a royal commission on the topic of 'projects and improvement of the king's revenue' as it did in 1612.[31]

By then the patents and monopolies which William Cecil encouraged as a stimulus to new start-up companies had been similarly corrupted. They were used as indirect taxes on established trades and manufactures which the patentees could claim to have 'improved ... by private industry'.[32] They might serve to reward energetic gentry like Sir Arthur Heveningham who enraged more established families in Elizabethan East Anglia by raising local rates under authority of patents to repair highways and harbours.[33] After 1580 most projectors and patentees were no longer inventors and skilled craftsmen but courtiers and speculators, the 'blood-suckers of the commonwealth' making 'themselves rich and all the realm poor' who were attacked in parliament in 1601 and again in the 1620s.[34] Commenting on a project for licensing alehouses in 1634, the Attorney General thought it 'much better that ... there be little noise made of any further intention than reformation', but if 'so considerable a yearly sum may be added to the king's

---

[27] T. E. Hartley, ed., *Proceedings in the Parliaments of Elizabeth I* (3 vols, Leicester, 1995), iii. 215–21. The author of these notes for a speech in the Commons has not been identified, and it is not certain that it was ever delivered: *Proceedings in the Parliaments of Elizabeth I*, iii. 181.

[28] Richard Hoyle, 'Disafforestation and Drainage: The Crown as Entrepreneur', in Richard Hoyle, ed., *The Estates of the English Crown 1558–1640* (Cambridge, 1992), pp. 360–1.

[29] Hoyle, 'Disafforestation and Drainage', p. 378; John Cramsie, *Kingship and Crown Finance under James VI and I, 1603–1625* (Woodbridge, 2002), pp. 37–8.

[30] BL, Lansdowne MS 74, no. 65, fo. 180ʳ.

[31] Joan Thirsk, 'The Crown as Projector on its Own Estates, from Elizabeth I to Charles I', in Hoyle ed., *Estates of the English Crown*, p. 300.

[32] Thirsk, 'The Crown as Projector on its Own Estates', pp. 300–1; Cramsie, *Kingship and Crown Finance*, p. 36. On the continuing importance of such private–public partnerships in government finance, see Joel Mokyr, *The Enlightened Economy: An Economic History of Britain, 1700–1850* (New Haven, 2009), pp. 392–4.

[33] A. Hassell Smith, *County and Court: Government and Politics in Norfolk, 1558–1603* (Oxford, 1974), pp. 123–4, 229–34, 252–65; Diarmaid MacCulloch, *Suffolk and the Tudors* (Oxford, 1986), pp. 259–72.

[34] Thirsk, *Economic Policy and Projects*, pp. 57–60, 98–100; Cramsie, *Kingship and Crown Finance*, pp. 29–35, 164; Brooks, *Law, Politics and Society*, pp. 197–200.

revenue, I see no inconvenience in it'.[35] In this case the government was taxing an acknowledged threat to the commonwealth which it had no other means of controlling, just as it was when earlier efforts to stop the growth of London failed, and there were fines for new buildings instead. A proclamation drafted for that purpose by Bacon in 1615 vainly alleged that the aim was 'a public reformation' of the city not 'any private benefit' to the king: 'the reformation rather than the profit was sought.'[36]

The scores of new projects, multiplying 'like the frogs of Egypt', many of them 'ridiculous, many scandalous, all very grievous', as Clarendon later observed, did lasting damage to the government's ability to regulate some of the largest and most controversial economic enterprises of the time, where it acted as a player when it might have been an arbiter.[37] It could not hope to settle arguments about whether projects for draining the fens were based on 'wholesome counsel' or simply a 'goodly pretence and show of a common good' when Charles I himself became one of the undertakers in the Bedford Level and (as it turned out) had little reward except the reputation of engaging in 'arbitrary government and tyranny itself'.[38] There was no prospect of such a government implementing projects for 'speedy reformation' of something as complicated as the coal trade, despite popular pressure for controls on marketing and quality to protect consumers, at a time when coal was overtaking timber as the primary source of the country's thermal energy.[39] Neither could it hope to persuade local governors to cooperate in imposing burdensome regulations on the new draperies, the one rapidly expanding sector of the textile industry, when its projects for 'remedy and reformation' were designed to bring some 'improvement' to the customs revenue as well as to 'improve our native commodities'.[40] Crown manipulation of projects for profit had given improvement a bad name.

No one was better placed to act as a participant-observer and pass judgement on these events than Francis Bacon. As one of the Crown's 'commissioners for suits', charged with vetting projects and patents of monopoly, as a warden in the unsuccessful company to regulate the new draperies, and as a private investor in Elizabethan mining enterprises, he knew at first hand that conflicts between private and public interests imposed constraints on what was politically possible for even

---

[35] Slack, *Reformation to Improvement*, pp. 73–4; Brooks, *Law, Politics and Society*, pp. 412–13.

[36] James F. Larkin and Paul L. Hughes, eds, *Stuart Royal Proclamations* (2 vols, Oxford, 1973–83), i. no. 152, pp. 345, 347; Slack, *Reformation to Improvement*, p. 55. Cf. below, pp. 144, 147.

[37] Thirsk, 'Crown as Projector', p. 352; Brooks, *Law, Politics and Society*, p. 200. According to Clarendon, projects had raised as much as £200,000 a year but only £1,500 of that had come to the king.

[38] H. C. Darby, *The Draining of the Fens* (2nd edn, Cambridge, 1956), p. 49 n. 2; Frances Willmoth, *Sir Jonas Moore: Practical Mathematics and Restoration Science* (Woodbridge, 1993), p. 94.

[39] John U. Nef, *The Rise of the British Coal Industry* (2 vols, 1932), ii. 214–16, 240–51; *Tudor Economic Documents*, i. 273; John Hatcher, *The History of the British Coal Industry*, i: *Before 1700* (Oxford, 1993), pp. 41–50.

[40] Michael Zell, 'Walter Morrell and the New Draperies Project, c.1603–1631', *Historical Journal*, 44 (2001), pp. 656–72. For a similar case, see David Cressy, *Saltpeter: The Mother of Gunpowder* (Oxford, 2013), pp. 106–18.

the best intentioned governors.[41] As a minister of the Crown under James I he had learnt that he served a monarch who in any case had no time for proposed 'commonwealth commissions' and no interest in cooperating with parliament in pursuing commonwealth ideals. As early as 1620 he had concluded that appeals to the common weal had lost all utility since everyone now thought 'care of the commonwealth' was 'but a pretext in matters of state'. When members of the government in the 1630s began to talk instead about 'the public good', because it carried fewer connotations of consensus and left more room for royal command, they were signalling that efforts to find agreement on reform of the commonwealth were a thing of the past.[42]

In one respect only had Crown and parliament between them managed to implement commonwealth ideals by erecting institutions to protect the body politic which were of lasting importance, and that was achieved before the end of Elizabeth's reign because critical circumstances made it necessary. The threats to public order and public welfare presented by epidemics of plague and by bad harvests in the sixteenth century had produced instructions to magistrates on how to respond. Finally codified in printed 'Books of Orders' issued under authority of the royal prerogative, they were enforced whenever these crises occurred, in the case of plague from 1578 to 1666, in the case of dearth between 1586 and 1630.[43] The steady growth in the number of vagrants, beggars, and paupers during half a century of population growth and inflation had similarly led parliament in 1597 and 1601 to build upon earlier local collections for the relief of the poor and legislate for local rates for that purpose. It is likely that at least half the parishes in England were raising poor rates by 1640, and almost all of them were doing so by 1700.[44] As they developed over time these rudimentary defences against the hazards of an insecure environment produced a formal apparatus for social welfare unlike anything else in Europe, and a permanent reminder that private interests must sometimes be subservient to the public good.

None of them were popular with those who had to pay for them or suffer their consequences. The dearth orders were most radical in their potential impact on the normal workings of a market economy, which helps to explain why they were abandoned as soon as circumstances permitted. Explicitly directed against the 'covetousness and uncharitable greediness' of farmers and corn-dealers who preferred 'their own private gain above the public good', they tried to ensure that all available grain reached local and poorer consumers and not simply those who were

---

[41] Cesare Pastorino, 'The Mine and the Furnace: Francis Bacon, Thomas Russell, and Early Stuart Mining Culture', *Early Science and Medicine*, 14 (2009), pp. 641–50; Zell, 'Walter Morrell', p. 663; Cramsie, *Kingship and Crown Finance*, pp. 164–9.

[42] Slack, *Reformation to Improvement*, pp. 59–61, 75.

[43] Paul Slack, 'Books of Orders: The Making of English Social Policy, 1577–1631', *TRHS* 5th ser. 30 (1980), pp. 1–22. There was a separate Book of Orders on poor relief in 1631, on which see also B. W. Quintrell, 'The Making of Charles I's Book of Orders', *EHR* 95 (1980), pp. 553–72.

[44] Marjorie Keniston McIntosh, *Poor Relief in England 1350–1600* (Cambridge, 2012), pp. 129–30, 254, 273–88; Paul Slack, *Poverty and Policy in Tudor and Stuart England* (1988), pp. 113–37; Steve Hindle, *On the Parish? The Micro-Politics of Poor Relief in Rural England c.1550–1750* (Oxford, 2004), pp. 227–99 (esp. pp. 251–6).

able to pay inflated prices. They may well have mitigated the effects of bad harvests in some parts of the country, but were unable to prevent famine in pastoral areas like the north-west in 1586–7 and 1597–8 and widespread distress and popular disorder elsewhere. They were fully enforced for the last time in 1630, when they were already being adjusted to local conditions by magistrates enforcing the poor laws; and by the middle of the century they were unnecessary as agrarian productivity increased.[45]

The plague orders, imposing quarantine on infected households in order to prevent the spread of infection, lasted much longer, until the disease disappeared from England, not because they were popular or had much immediately demonstrable effect, but because of the magnitude of the threat against which they were directed. Although often criticized as counterproductive, they were part of a European campaign against contagion, developed by a process of trial and error which might itself be termed improvement, and which paid dividends. It was sometimes locally effective in protecting one town against infection from another, and it prevented the return of plague to most of western Europe after 1720, thanks to the erection of a *cordon sanitaire* between the Habsburg and Ottoman empires and the coordinated imposition of quarantine on shipping from infected ports.[46]

The impact of statutory poor relief has proved more difficult to define because it grew in less coordinated fashion and in response to local conditions and expectations in every parish. Despite their unpopularity and repeated attempts to keep costs down, poor rates transferred ever larger sums from the relatively well off to the disadvantaged. By 1750 they amounted to nearly £700,000 a year across the kingdom, redistributing around 1 per cent of the national income to something like 8 per cent of the population.[47] Contemporaries argued that relief on such a scale limited economic growth by encouraging idleness and inhibiting labour mobility, but it seems more probable that it provided some modest incentive to enterprise and innovation by reducing risk and contributing to social stability. Since the largest slice of relief was directed towards the elderly poor, it may also have indirectly supported a demographic regime of late marriage and low fertility which was one of the conditions for economic growth.[48] Its impact on attitudes was more direct and inescapable. The Elizabethan poor law demonstrated week by week and

---

[45] Paul L. Hughes and James F. Larkin, eds, *Tudor Royal Proclamations* (3 vols, New Haven, 1969), ii. 532, iii. 194; Slack, *Reformation to Improvement*, pp. 62–6. Cf. Steve Hindle, 'Dearth and the English Revolution: The Harvest Crisis of 1647–50', *EcHR* 61 (2008), Special Issue 1, pp. 64–98; below, pp. 81–2.
[46] Paul Slack, *The Impact of Plague in Tudor and Stuart England* (1985), pp. 207–26, 313–26; Geoffrey Parker, *Global Crisis* (New Haven, 2013), pp. 629–30. Cf. below, pp. 223–4.
[47] Paul Slack, *The English Poor Law 1531–1782* (Basingstoke, 1990), p. 30; Joanna Innes, 'The "Mixed Economy of Welfare" in Early Modern England: Assessment of the Options from Hale to Malthus (*c.*1683–1803)', in Martin Daunton, ed., *Charity, Self-Interest and Welfare in the English Past* (1996), pp. 144–8; below, pp. 176, 239.
[48] Mokyr, *Enlightened Economy*, pp. 440–5; Richard M. Smith, 'Charity, Self-Interest and Welfare: Reflections from Demographic and Family History', in Daunton, ed., *Charity, Self-Interest and Welfare*, pp. 38–9; Slack, *Poor Law*, p. 55.

year after year that the poor had a statutory entitlement as well as a moral right to an adequate standard of living.[49]

Statutory poor relief was not the only source of social welfare, however. It was supplemented by private charity which, although difficult to measure, probably contributed as much in cash terms as the poor rates. Voluntary and casual almsgiving to neighbours and beggars in the streets is unquantifiable, but must always have been very large. Better documented are the formal philanthropic bequests which grew in number and aggregate value across the period. Despite the Protestant insistence that good works were no infallible route to salvation, men and women with accumulated capital to spare sought to perpetuate their name and merit by founding charitable trusts for purposes defined and limited by statutes in 1597 and 1601.[50] They included funds for what would soon be termed improvements, for the repair of roads, bridges, and harbours as well as churches, and for the supply of fresh water to towns.[51] Many more of them were endowments for the relief of the poor in one form or another, for hospitals and almshouses, doles for widows and orphans, and the more novel purposes of workhouses and the training of poor apprentices, as well as for free schools and scholarships to the universities. From the 1690s there were also subscription charities for similar purposes, harnessing the charitable instincts and interests of a wider public. The result was a 'mixed economy of welfare', large and flexible enough to serve multiple purposes and allow the poor and disadvantaged some freedom of choice between public and private sources of support.[52]

Like public funding from the poor rates, private philanthropy was always open to the charge that it consumed wealth better used for other purposes, and in the case of endowments for education and training that they were either a disincentive to self-help and industriousness, or, on the contrary, encouraged the young to rise above their station and put them in the wrong social or occupational niches. When Bacon argued in his essay 'Of Riches' that 'glorious gifts and foundations' were prone to 'putrify and corrupt' the commonwealth, he had in mind Thomas Sutton's bequests for the London Charterhouse in 1611, which he opposed because there were already too many grammar schools depriving the country of husbandmen and apprentices, and creating more scholars 'than the state can prefer or employ'.[53] He

---

[49] Slack, *Reformation to Improvement*, pp. 163–4. Cf. Lorie Charlesworth, *Welfare's Forgotten Past: A Socio-legal History of the Poor Law* (Abingdon, 2010), pp. 35–8.

[50] W. K. Jordan, *Philanthropy in England 1480–1660: A Study of the Changing Pattern of English Social Aspirations* (1959), p. 243 and *passim*; Ralph Houlbrooke, *Death, Religion and the Family in England 1480–1750* (Oxford, 1998), pp. 118–33; Ilana Krausman Ben-Amos, *The Culture of Giving: Informal Support and Gift-Exchange in Early Modern England* (Cambridge, 2008), pp. 115–22. On the statutes, see Gareth Jones, *History of the Law of Charity, 1532–1827* (Cambridge, 1969), pp. 22–6; McIntosh, *Poor Relief*, pp. 288–93, 301–2.

[51] Jordan, *Philanthropy*, pp. 278–9; W. K. Jordan, *The Charities of London, 1480–1660* (1960), pp. 202–6; W. K. Jordan, *The Forming of the Charitable Institutions of the West of England* (Transactions of the American Philosophical Society ns. 50, Part 8, Philadelphia, 1960), pp. 34–6.

[52] Innes, 'Mixed Economy of Welfare', *passim*.

[53] Vickers, *Bacon*, p. 412; Jordan, *Philanthropy*, p. 285 n. 3; Francis Bacon, *The Letters and Life*, ed. James Spedding (7 vols, 1861–74), iv. 252–3; *ODNB sub* Sutton. The same point, that educational provision had created too many 'penmen of all sorts', was made by *Britannia languens* (1680), in

was indulging in hyperbole. The foundation of three or four hundred new grammar schools between 1540 and 1640 can scarcely have had so large an effect on social and occupational mobility, any more than it can on its own explain the great growth in literacy which occurred over the same period.[54]

By the eighteenth century the level of literacy was sufficiently high in Britain and other parts of western Europe for a modern economic historian to argue, as Bacon might have done, that there had been an 'overinvestment' in education if it is judged solely in terms of what was needed for economic development.[55] But literacy was a private as much as a public good, giving access to religious instruction and entertainment as well as useful skills, and it was a consumer good acquired by all who could afford to invest in it. Public educational provision was therefore the product rather than the cause of a society in which, as one local historian complained in the 1630s, everyone was 'of an aspiring mind';[56] and if the institutions of social welfare made a contribution to that, it was again indirect, by reducing risk and sustaining the nuclear family and the relatively high living standards that came with it. Aspiring and educated minds, moreover, were themselves public goods in a society whose economic improvement and political stability depended on the circulation of information and on the sharing of knowledge, moral assumptions, and new aspirations of all kinds.

Private philanthropy and the public provision of social welfare were both products of that society, and in one important respect they contributed directly to what an economist would call the human and social capital of the nation. By increasing the number of small associations of private citizens active for the public good and by adding to their responsibilities, they encouraged the revival of the kind of 'civil society' which had promoted the culture of the commonwealth in the fifteenth century and been weakened by the dissolution of guilds and fraternities during the Reformation. In the seventeenth century every aspect of social welfare was being managed by corporate bodies, by parish vestries, charitable trusts, civic corporations, and companies of merchants, whose collective cultures communicated and sustained shared values. These were all 'little commonwealths', no more capable of providing speedy reformation for the ills of the larger commonwealth

---

J. R. McCulloch, ed., *Early English Tracts on Commerce* (Cambridge, 1952), pp. 356–7. For some evidence of an overproduction of graduates by 1640, see Mark H. Curtis, 'The Alienated Intellectuals of Early Stuart England', *P&P* 23 (Nov. 1962), pp. 25–43; Lawrence Stone, 'The Educational Revolution in England, 1560–1640', *P&P* 28 (July 1964), pp. 41–80.

[54] David Cressy, *Literacy and the Social Order: Reading and Writing in Tudor and Stuart England* (Cambridge, 1980), p. 165 and *passim*. On the difficulties of defining and measuring literacy, see also Keith Thomas, 'The Meaning of Literacy in Early Modern England', in G. Baumann, ed., *The Written Word: Literacy in Transition* (Wolfson College Lectures 1985, Oxford, 1986), pp. 97–131; Adam Fox, *Oral and Literate Culture in England, 1500–1700* (Oxford, 2000), pp. 406–9.

[55] Robert C. Allen, Tommy Bengtsson, and Martin Dribe, eds, *Living Standards in the Past: New Perspectives on Well-Being in Asia and Europe* (Oxford, 2005), p. 12. Mokyr, *Enlightened Economy*, pp. 239–40, points out that literacy rates were lower in Britain than some other parts of western Europe in the eighteenth century.

[56] Tristram Risdon, *A Chorographical Description or Survey of the County of Devon* (1811), p. 10 (written in the 1630s). Cf. Keith Thomas, *The Ends of Life: Roads to Fulfilment in Early Modern England* (Oxford, 2009), pp. 30–3.

than the Crown and its agents, but more consistent in deploying their own ideals of civic virtue and the public good, and better equipped to adjust them over time so as to fit changing circumstances.[57]

PRIVATE ENTERPRISE FOR THE PUBLIC GOOD

Private initiatives for the public good were equally conspicuous in English enterprise overseas, in colonial and commercial ventures in Ireland and America, and in the new markets opened up for English trade in Europe and the East. There too the Crown depended on groups of agents and investors, often appropriately termed 'adventurers' and 'undertakers', whose primary interest was profit and whose ostensible purpose was the promotion of the national interest. In their own little commonwealths, in English enclaves planted abroad and new companies and associations at home, they naturally used the familiar language of reformation, civic virtue, and the common weal, to justify their undertakings; but they were consciously embarking on innovations which were initially precarious and required prolonged and repeated effort if they were to deliver public as well as private benefit. They were engaged in improvement.

Persistent rebellion made sixteenth-century Ireland the most urgent target for improvement in the English style by English hands. The rhetoric employed by projects for cultivating the wastes, royal forests, and fens of England transferred naturally to proposals for Irish plantations, in Ulster in the 1570s, Munster in the 1580s, and Ulster again—and most successfully—after 1607. A prospectus for the first Ulster plantation, in which Sir Thomas Smith was engaged, explained that the 'enterprise of peopling and replenishing' that part of Ireland 'with the English nation' would take off England's surplus population, bring honour to the kingdom, and be 'very profitable to them that are doers therein'. As for the native Irish, 'so barbarous a nation' would be brought to the civility which came 'more by keeping men occupied in tillage, than by idle following of herds'.[58] Although cattle-farming made better economic sense in much of Ireland, the same prospect was held out in projects for Munster. If 'the soil of the land' was restored to its 'natural fruitfulness' by English farmers, the countryside would be 'regenerate and born of new', and 'reformation breed competent wealth'.[59] Ulster was pictured in similar terms in 1610 as a land of 'pleasant fields and rich grounds' left desolate and 'unmanured', and crying out for 'profitable improvement'.[60]

---

[57] Slack, *Reformation to Improvement*, pp. 148–56; Withington, *Society*, pp. 155–6, 184–6; below, p. 260.

[58] [I.B.], *A Letter sent by I.B. gentleman* (1571), sigs Aii^r, Ci^v, Di, Ei^r, Fi^v; Nicholas Canny, *Making Ireland British, 1580–1650* (Oxford, 2001), pp. 121–2; Christopher Maginn, *William Cecil, Ireland, and the Tudor State* (Oxford, 2012), pp. 86–8.

[59] Canny, *Making Ireland British*, pp. 117, 132, 134.

[60] Thomas Blenerhasset, *A Direction for the Plantation in Ulster* (1610), sigs A2^v, B1^r. Munster had also been described as not 'manured' by the Irish: Canny, *Making Ireland British*, p. 129.

A land half empty and devastated by rebellion and war, moreover, was one which could be regarded as 'vacant' and so under Roman law open to conquest and colonization.[61] It was precisely the same as the ancient Britain which the Romans had found, a 'savage wilderness', according to Edmund Spenser, 'unpeopled, unmanured, unproved'.[62] Writers of Smith's generation were well aware that Roman precedents also offered ample warning about the dangers of corruption inherent in imperial adventures, but they were nonetheless proposing to build an empire.[63] It would have new towns, like Smith's proposed new capital of 'Elizabetha' in Ulster, and a new social order of English lords, freeholders, and copyholders, each occupying a measured number of acres in Munster.[64] Only in the second Ulster plantation after 1607 did such careful planning bear permanent fruit, thanks to heavy migration from Scotland as well as England and financial backing from the London livery companies. By 1640 Ulster was not quite the urban and commercialized province which Francis Bacon had hoped to see, but the amount of land under tillage had notably increased, and there were new towns like Londonderry and a much smaller Charlemont.[65] It had been subjected to the 'general improvement' which one projector advertised in 1610 in order to 'let all the inhabitants of spacious Britain know' what was best for the public and also 'their own good'.[66]

The pursuit of empire in America followed a similar pattern, deploying the same arguments to sustain momentum when initial expectations were disappointed, and persuade the English public that the effort involved was all for their own and the public good. The two avenues for plantation were also in competition with one another for adventurers, investors, and emigrant labour, each offering an alternative when the other failed to deliver. In 1580 Munster was presented as a much closer and more attractive proposition than any 'Cathay or Terra Florida' across the Atlantic, and when that ran into difficulties, Richard Hakluyt claimed that there was no need to 'meddle with the state of Ireland' since there was 'the great and ample country of Virginia' waiting for English colonization.[67] In America as in Ireland there were also the same anxieties about the corruption which might come with any colonizing activity generated by avarice and which might explain its early

---

[61] Canny, *Making Ireland British*, pp. 133–4, 204.
[62] David Armitage, *The Ideological Origins of the British Empire* (Cambridge, 2000), pp. 53–4; Canny, *Making Ireland British*, pp. 121–2.
[63] Markku Peltonen, *Classical Humanism and Republicanism in English Political Thought, 1570–1640* (Cambridge, 1995), pp. 75–9; Maginn, *Cecil*, pp. 161–2; Withington, *Society*, pp. 202–5.
[64] Canny, *Making Ireland British*, pp. 121–2, 130–1, 200; S. J. Connolly, *Contested Island: Ireland 1460–1630* (Oxford, 2007), p. 179.
[65] Connolly, *Contested Island*, pp. 301–6; William J. Smyth, *Map-Making, Landscapes and Memory* (Cork, 2006), p. 98; Canny, *Making Ireland British*, pp. 199–242; Aidan Clark with R. Dudley Edwards, 'Pacification, Plantation, and the Catholic Question, 1603–23', in T. W. Moody et al., eds, *A New History of Ireland*, iii: *Early Modern Ireland 1534–1691* (Oxford, 1976), p. 204; Bacon, *Letters and Life*, iv. 116–25.
[66] Blenerhasset, *Direction*, sigs B2ᵛ, C4ᵛ, D2–3ʳ.
[67] Canny, *Making Ireland British*, p. 117; Smyth, *Map-Making*, p. 421. For overlapping interests and personnel in Irish and American plantation schemes, see Smyth, *Map-Making*, pp. 423–31.

failure in Virginia and the Chesapeake, just as it did in Munster.[68] In the American case, however, there were differences, quite apart from those created by relative distance from home, which made public justification for plantation all the more necessary, and the notion of improvement even more essential to it.

Since the English had no prior sovereignty to defend across the Atlantic as they had in Ireland, their claims to dominion could only rest on arguments derived from Scripture and from natural and Roman law about rights to trade between peoples, to bring barbarous peoples to civility and Christianity, and especially to populate *terra nullius*, vacant land. An Oxford professor of civil law anticipated Samuel Purchas when he explained in 1589 that 'God did not create the world to be empty', and hence 'the seizure of vacant places is regarded as a law of nature'.[69] Needing to distinguish their own ventures from the Spanish conquests, and as conscious as some Spanish jurists and theologians had been that native Americans had rights of their own and should be led rather than coerced into Christian civility, some Elizabethan writers on plantation had doubts about whether land already occupied could properly be described as vacant.[70] But North America was much more thinly populated than the territories seized by Spain and its land by comparison wholly untouched by intensive farming of any kind. Once native Americans began to behave as hostile savages, English planters who regarded themselves as industrious farmers and not avaricious conquistadors turned readily to a discourse of improvement which allowed them to treat as vacant any land left uncultivated and not profitably used.[71]

That was the case for dominion presented by propaganda for American plantations in the 1620s, when John Donne's sermon to the Virginia Company deduced from 'the law of nations' that 'all places' must be 'improved, as far as maybe, to the best advantage of mankind in general'.[72] In the 1630s the point was reiterated with respect to New England, where William Wood thought 'improvements' would make the land at least as fertile as that of old England, and John Cotton imagined 'a vacant soil' giving rights of possession to whoever imposed 'culture and husbandry upon it'.[73] As a Massachusetts court later admitted, it followed that if 'any of the Indians' had 'subdued' the land, 'they had a just right to it', but native populations were being decimated by the diseases Europeans brought with them. 'The Lord hath cleared our title to what we possess', John Winthrop claimed after a

---

[68] Andrew Fitzmaurice, *Humanism and America* (Cambridge, 2003), pp. 37–9; Canny, *Making Ireland British*, p. 120.

[69] Christopher L. Tomlins, *Freedom Bound: Law, Labor, and Civic Identity in Colonizing English America, 1580–1865* (New York, 2010), p. 23; Fitzmaurice, *Humanism and America*, pp. 140–6; Anthony Pagden, *Lords of all the World: Ideologies of Empire in Spain, Britain and France c.1500–c.1800* (New Haven, 1995), pp. 76–86; above, pp. 31–2. The relevant biblical texts, often cited, were Genesis 1: 28, 9: 1.

[70] Armitage, *Ideological Origins*, pp. 92–6; Fitzmaurice, *Humanism and America*, pp. 67–87. Cf. Nicholas Canny, 'A Protestant or Catholic Atlantic World? Confessional Divisions and the Writing of Natural History', *Proceedings of the British Academy*, 181 (2012), pp. 100–3.

[71] Tomlins, *Freedom Bound*, p. 97; J. H. Elliott, *Empires of the Atlantic World: Britain and Spain in America, 1492–1830* (New Haven, 2006), pp. 85–9.

[72] Tomlins, *Freedom Bound*, pp. 143–7. Cf. Fitzmaurice, *Humanism and America*, pp. 141–3.

[73] Elliott, *Empires of the Atlantic*, p. 91; John Cotton, *God's Promise to his Plantation* (1630), p. 5.

Elizabethan Foundations 1570–1640 69

smallpox epidemic in 1634, and the natives who remained had no 'settled habitation', and no enclosures or 'tame cattle to improve the land'. They had left it 'open to any that could and would improve it'.[74]

As in Ireland, success in American plantations depended as much on capital investment and manpower as on rhetoric, however, and they were slow in coming. Whatever doubts there might still be about the damage done to colonial adventures by avarice, John Smith knew from his campaign to get support for Virginia in 1616 that no 'other motive than wealth' would 'ever erect . . . a commonwealth' there. It nevertheless failed to attract emigrants in any great number to the Chesapeake or New England before 1630, or the substantial sums required by the new companies set up to recruit them.[75] Even the best advertised of them, the Virginia Company, had to rely on lotteries for ready cash, and it raised only £37,000 in total stock, a trifling sum by comparison with other commercial ventures of the time. It had 560 gentry among its investors, including influential politicians like Sir Edwin Sandys, but few of the great city merchants joined them, and all of them were motivated by a sense of public duty rather than any prospect of the financial return which landowners and commercial tycoons could more easily obtain from investments elsewhere. The merchants who ultimately took control of commerce with America were smaller traders seizing opportunities across the Atlantic because they were unable to break into the commercial monopolies of the city elite.[76]

The thousands of emigrants who went to the colonies and ensured their permanence after the collapse of the Virginia Company in 1624 were entrepreneurial in a similar way, searching for roads to self-improvement denied to them at home. Many of the 20,000 who migrated to New England in the 'great migration' of the 1630s were hoping to build godly communities of a kind no longer tolerated in Britain, and at least a quarter of them went out as indentured servants expecting ultimately to rise in a new world. They had better chances of success than the much larger number, perhaps 50,000, who moved to the Chesapeake and the West Indies in the same decade, aspiring to become tobacco planters and finding a much less healthy environment when they got there.[77] When emigration from Britain

---

[74] Paul Warde, 'The Idea of Improvement *c*.1520–1700', in Hoyle, ed., *Custom, Improvement and the Landscape*, p. 133; Elliott, *Empires of the Atlantic*, p. 66; Richard Drayton, *Nature's Government* (New Haven, 2000), p. 56. The governor of Connecticut in 1647 claimed title to New Haven from a patent of James I, but also through 'purchase from the natives and . . . quiet possession and improvement many years': *The Journal of John Winthrop, 1630–1649*, ed. Richard S. Dunn and Laetitia Yeandle (abridged edn, Cambridge, Mass., 1996), p. 331.

[75] Fitzmaurice, *Humanism and America*, p. 194; David Cressy, *Coming Over: Migration and Communication between England and New England in the Seventeenth Century* (Cambridge, 1987), pp. 4–7.

[76] Robert Brenner, *Merchants and Revolution: Commercial Change, Political Conflict, and London's Overseas Traders 1550–1653* (Princeton, 1993), pp. 92–112. On the character of investment in companies, see also Theodore K. Rabb, *Enterprise and Empire: Merchant and Gentry Investment in the Expansion of England, 1575–1630* (Cambridge, Mass., 1967).

[77] Cressy, *Coming Over*, pp. 68–70. On the demography of Virginia, see Carville V. Earle, 'Environment, Disease and Mortality in Early Virginia', in Thad W. Tate and David L. Ammerman, eds, *The Chesapeake in the Seventeenth Century: Essays on Anglo-American Society and Politics* (New York, 1979), pp. 96–125. Sugar was not introduced into the West Indies until the 1640s: Brenner, *Merchants and Revolution*, p. 161.

declined at the end of the century, the farming communities and small towns of the north were able to grow, largely through natural increase, while the plantation societies of the south became dependent on slave labour from Africa.

Improvement nonetheless remained the public touchstone for success on both sides of that great social and economic divide. In Boston in 1702 Cotton Mather attributed the failure to 'people and improve' other parts of New England to the pursuit of 'worldly interests' and ignorance of the 'nobler designs of Christianity', but he had no doubt that one of those designs was to make the whole earth 'profitable for the use of men', 'tilled and improved', and not left to 'lie waste without any improvement'. Three years later a historian of recent improvements in Virginia complained that they were being hindered by too easy a reliance on tobacco-growing. New crops like sugar and new manufactures were needed to 'set all hands industriously to work' and promote 'the improvement of trade' across the whole of the empire.[78] Public and private improvement had not yet gone far enough.

The language of improvement was less prominent, and the investment of merchant capital much greater, in the enterprises which opened up England's trade to the east, and did more than plantations in the west to transform the country's economic prospects. The founders of the Levant Company for trade with Turkey in 1592 and the East India Company in 1600 were a closely knit group of merchants, rich enough to bear the risks involved in long-distance commerce, influential enough to acquire monopoly powers to protect their investments, and spectacularly successful in acquiring access to new markets. Few of them were Merchant Adventurers, who continued to deal largely with northern Europe; but neither were they newcomers to overseas commerce, like the colonial traders with America, since they had already been engaged in new trading ventures with Russia and Spain. 'Born rich', as one of their critics observed, and 'adding wealth to wealth', by the 1630s they had the commanding position in the City of London which the Merchant Adventurers had earlier enjoyed.[79]

Their hefty profits came from imports rather than exports, from silks, currants, spices, and the other 'unnecessary trifles' once imported from Antwerp about which William Cecil had complained. They could now be obtained in larger quantity closer to their sources, and match rapidly rising consumer demand generated by the rising incomes of the landed classes at home.[80] Agrarian improvement indirectly stimulated an import-led restructuring of overseas commerce. While English cloth exports were only saved from complete collapse by new draperies going to the Mediterranean, English imports multiplied at least threefold between the 1570s and 1630s, and the re-export of part of them to other parts of Europe was already

---

[78] Cotton Mather, *Magnalia Christi Americana* (1702), Book I, pp. 147, 152; Robert Beverley, *The History and Present State of Virginia* (1705), title page, Book I, pp. 21, 37, 59, 94; Book II, p. 77. For similar comments on Virginia to the Board of Trade in 1697, see Carole Shammas, 'English-Born and Creole Elites in Turn-of-the-Century Virginia', in Tate and Ammerman, *Chesapeake in the Seventeenth Century*, p. 287. Cf. Elliott, *Empires of the Atlantic*, pp. 166–9; below, pp. 98, 254–5.

[79] Brenner, *Merchants and Revolution*, pp. 11–13, 61–74 (quotation at p. 73).

[80] Brenner, *Merchants and Revolution*, pp. 11, 16, 25–33, 41–5; Ralph Davis, *English Overseas Trade 1500–1700* (1973), pp. 30–1.

making up for some of the outflow of bullion needed to pay for them. Over the same period the tonnage of English shipping had also increased threefold. By 1640, when England's commercial competitors, especially the Dutch, were at war with one another, there was, according to Sir Thomas Roe, 'a general opinion that the trade of England was never greater'.[81]

None of this guaranteed public approval for the companies and merchants who were responsible for it. Their privileges and profits simply increased the hostility aroused by other corporations and monopolies, and caused a Commons committee on free trade, led by Edwin Sandys, to conclude in 1604 that it was 'against the natural right and liberty of the subjects of England' to restrict commerce 'into the hands of some few, as now it is'. Trade could only ensure the stability and strength of the realm if it encouraged 'the more equal distribution of the wealth of the land' by circulating freely, like 'the equal distributing of nourishment in a man's body'.[82] The East India Company, exporting treasure in return for useless luxuries, was especially vulnerable to the charge that it was 'gainful to the adventurers, ... with public detriment to the state', but it was only the most exposed of the several privileged companies and societies which were alleged to be responsible for 'the common wealth being made private'.[83]

The fact that the privileges came from a government looking to its own profit made matters worse by exacerbating conflicts between rival interest groups to no useful purpose. The issue of a patent for a tobacco monopoly in 1620 only strengthened Sandys's opposition to such devices since it threatened his own interests in the Virginia Company;[84] and the Cockayne project of 1614, designed for 'the better improvement' of the old draperies by insisting they were dyed and finished before export, was wholly counterproductive. It meant the suspension of the privileges of the Merchants Adventurers and a victory for their rivals, the London Clothworkers Company, and it discredited the Council as a reliable manager of the country's economy by causing widespread unemployment in the textile industry.[85] Although the courts generally upheld corporate privileges, especially those of the major overseas trading companies, they had to listen to arguments that 'private societies and companies' were doing more harm than good to the commonwealth.[86]

---

[81] C. G. A. Clay, *Economic Expansion and Social Change: England 1500–1700* (2 vols, Cambridge, 1984), ii. 164–5, 187; Brenner, *Merchants and Revolution*, pp. 24–8, 42.

[82] *SCED*, p. 437. For earlier criticism of corporations, see *Tudor Economic Documents*, iii. 267–8.

[83] Timothy Fitzgerald, *Discourse on Civility and Barbarity: A Critical History of Religion and Related Categories* (Oxford, 2007), p. 208 (citing Purchas); Robert Kayll, *The Trades Increase* (1615), pp. 53, 55.

[84] Cramsie, *Kingship and Crown Finance*, pp. 167–8; Theodore K. Rabb, *Jacobean Gentleman: Sir Edwin Sandys, 1561–1629* (Princeton, 1998), pp. 229–30. On similar conflicts between 'concessionary interests' affecting debate on free trade, see Robert Ashton, *The City and the Court, 1603–1643* (Cambridge, 1979), pp. 83–120; Robert Ashton, 'Conflicts of Concessionary Interest in Early Stuart England', in Coleman and John, eds, *Trade Government and Economy in Pre-industrial England*, pp. 113–31.

[85] *Acts of the Privy Council*, xxxi: *1615–1616*, p. 386; G. D. Ramsay, 'Industrial Discontent in Early Elizabethan London: Clothworkers and Merchants Adventurers in Conflict', *London Journal*, 1 (1975), pp. 227–39.

[86] Brooks, *Law, Politics and Society*, p. 387. Stephen D. White, *Sir Edward Coke and the Grievances of the Commonwealth* (Manchester, 1979), pp. 284–90, provides a vivid illustration of the number of

Those private societies, separately at odds with one another, had to defend themselves by showing that they were all promoting the public good. Like the London craftsmen of popular literature in the 1590s with claims to the gentility of their trades, the thousand or so merchants active in overseas trade at the beginning of the seventeenth century aspired to the honourable state and civic values of the landed elite from which many of them came;[87] and some of them were advertising their virtues in print. In 1635 William Scott described the diligence, industriousness, honesty, and thrift necessary for any successful tradesman, and pointed out that 'it is likely he that hath done well for himself, will know how to do well for the public good'.[88] In a similar guide for overseas merchants in 1638 Lewes Roberts insisted that theirs was 'an art or science invented by ingenious mankind for the public good, commodity, and welfare of all commonwealths'. Addressing the Long Parliament a few years later, and hoping that it would preserve existing company privileges, he noted that in other countries like France overseas commerce was thought 'base and sordid'. In England only domestic trade was best 'left to the poor and common people to enrich themselves by'; overseas trade had earned 'respect and honour' because success in it required all the attributes of 'true nobility'.[89]

Although there was some validity to the comparison with France,[90] the consequence of such special pleading was less to make mercantile enterprise and acquisitiveness wholly respectable than to persuade the political elite that they were indispensable to the country's success in the international competition for sea-borne trade. In 1601 the secretary to the Merchant Adventurers demonstrated with a wealth of detail how his company's 'almost incredible trade and traffic' with Europe benefited English navigation and shipping, 'the incomes and customs of the Prince', and 'the state and commonwealth' in general;[91] and the detailed defences of the East India Company by Dudley Digges and Thomas Mun showed that the same applied to the new long-distance trades which were equally of 'service to the state', and 'the very touchstone of a kingdom's prosperity'.[92] In 1625 the merchant seaman John Hagthorpe described the 'politic secret war' then under way between England, Holland, and Spain, all striving to 'beat each other out of their trades',

economic disputes and projects which Coke had to deal with as an attorney and as law officer of the Crown.
[87] Laura Caroline Stevenson, *Praise and Paradox: Merchants and Craftsmen in Elizabethan Popular Literature* (Cambridge, 1984), pp. 180–210; Richard Grassby, *The Business Community of Seventeenth-Century England* (Cambridge, 1995), pp. 56–7, 115–18, 169–70.
[88] William Scott, *An Essay of Drapery, 1635*, ed. Sylvia L. Thrupp (Boston, 1953), pp. 8, 30–2, 36, 39–40. On these 'bourgeois virtues' see also Grassby, *Business Community*, pp. 283–97; Brooks, *Law, Politics and Society*, pp. 390–1; below, pp. 240–1.
[89] Lewes Roberts, *The Merchants Mappe of Commerce* (1638), p. 12; Lewes Roberts, *The Treasure of Traffike* (1641), in McCulloch, ed., *Early English Tracts on Commerce*, pp. 82–4.
[90] Cf. above, p. 34; below, p. 247.
[91] John Wheeler, *A Treatise of Commerce*, ed. George Burton Hotchkiss (New York, 1931), pp. 338, 341, 440. Much of the treatise was based on an original tract by George Nedham, a Merchant Adventurer writing much earlier: G. D. Ramsay, *The Politics of a Tudor Merchant Adventurer: A Letter to the Earls of Friesland* (Manchester, 1979), pp. 39–40.
[92] Withington, *Society*, pp. 227–30; Thomas Mun, *A Discourse of Trade from England unto the East-Indies* (1621), in McCulloch, ed., *Early English Tracts on Commerce*, p. 5.

and holding 'the whole state of things as it were in balance'; it would only be resolved by 'some inevitable destiny'.⁹³ Despite all the opprobrium directed against them, the literature on commerce and companies had established that England's political and economic destiny depended on the skill and enterprise of those involved in them.

A manuscript by John Keymer, circulating in 1622, made the case so persuasively that it was subsequently attributed to Walter Ralegh, and published and cited under his name during the Anglo-Dutch wars of the 1650s and 1660s.⁹⁴ Keymer acknowledged the objections made against private companies and their profits, but if the English were to beat the Dutch it was necessary 'to allure and encourage the people for their private gain to be all workers and erectors of a commonwealth'. It would be necessary also to harness the expertise of merchants, who alone had reliable knowledge about overseas markets, to guide diplomacy and policy, preferably through a standing council of trade, a 'State Merchant'. Then government would be better informed, trade, navigation, and domestic industry all be improved, and the whole kingdom become 'powerful and rich'. As an Elizabethan spokesman for the Merchant Adventurers had argued, merchandising brought with it 'wealth and knowledge'.⁹⁵

Most of that wealth and knowledge was being generated in London, as its population rose from less than 100,000 in 1570 to nearly 400,000 by 1640.⁹⁶ The rapid growth of the metropolis was the most remarkable of the phenomena which shook the old balance of the commonwealth and which had to be defended against its critics in the name of the public good. When provincial ports objected to the capital's increasing monopoly of England's overseas trade, the city's Jacobean counsel borrowed the metaphor of circulation from the Commons committee on free trade in 1604, and turned it to a different purpose. London had become the 'university' which trained all the nation's merchants, and was therefore responsible for the 'advancement... of merchandising and trading', which was the life-blood nourishing 'the whole body of the kingdom'.⁹⁷ The argument, like the metaphor, needed constant reiteration for the rest of the century, but it already had some substance. London was emerging as a single national market determining economic behaviour across a large part of the provinces, dictating fashion, distributing

---

⁹³ John Hagthorpe, *Englands-Exchequer: Or A Discourse of the Sea and Navigation* (1625), p. 7; *ODNB sub* Hagthorpe.
⁹⁴ Sir Walter Raleigh, *Observations touching Trade and Commerce with the Hollander* (1653), which had been published under another title in Raleigh, *Judicious and Select Essays and Observations* (1650). Cf. the appendix to William Carter's *England's Interest Asserted, in the Improvement of its Native Commodities* (1669), 'Some Collections of Sir Walter Rawley's presented to King James'. On attributions to Ralegh, see also below, p. 96.
⁹⁵ John Keymer, *Original Papers regarding Trade in England and Abroad*, ed. M. F. Lloyd Prichard (New York, 1967), p. 36; *ODNB sub* Keymer; Ramsay, *Politics of a Tudor Merchant Adventurer*, p. 55.
⁹⁶ Jeremy Boulton, 'London 1540–1700', in Peter Clark, ed., *The Cambridge Urban History of Britain*, ii: *1540–1840* (Cambridge, 2000), pp. 316–17.
⁹⁷ Brooks, *Law, Politics and Society*, p. 387; *SCED*, pp. 437–8.

information, and creating and communicating knowledge.[98] Its occupational diversity is evident from a list of new city companies proposed for incorporation in the 1630s, which ranged from comb-makers, glass-sellers, and distillers, to musicians, gun-makers, pin-makers, brick- and tile-makers, and vintners;[99] and London was the place where specialist skills and new technologies, in medicine, mathematics, navigation, and accountancy as well as commerce and industry, fed off one another, and were improved. Like Antwerp a century earlier and Amsterdam at much the same time, it was becoming an 'information economy' which spurred ingenuity, innovation, and enterprise.[100]

In order to 'stir up the sharp wits' of Londoners to further endeavours 'for the public and common good' in 1594, Hugh Plat set out many of the 'ingenious devices' and 'profitable inventions' already available to them in his *Jewell House of Art and Nature*. There were useful techniques and new discoveries to be exploited in husbandry, medicine, metallurgy, and chemistry, and Plat claimed for all of them the legitimacy of 'experiments' based on 'the infallible grounds of practice'.[101] He was echoing Fitzherbert who had written much earlier about 'the practice or knowledge of an husbandman well proved' being better than 'the science... of a philosopher not proved'; and he was anticipating Lewes Roberts who was to make similar claims in 1638 about the importance for knowledge of the 'experience' and 'experiments' of merchants.[102] In the 1570s, however, Plat's mathematical friends John Dee and Thomas Digges were already talking about 'experimental science' and the public utility of 'experimental actions' in language not very different from Bacon's.[103] Bacon's great achievement was to build a comprehensive programme of experimental philosophy on these slender foundations, but it seems entirely reasonable to argue, as several historians have done, that they influenced his thinking and that Bodley was right to complain that Bacon had overstated the contempt of his contemporaries for 'experiments familiar and vulgar'.[104] Some of those

---

[98] Cf. below, pp. 145–9. The classic account of London's importance is E. A. Wrigley, 'A Simple Model of London's Importance in Changing English Society and Economy, 1650–1750', in E. A. Wrigley, *People, Cities and Wealth* (Oxford, 1987), pp. 133–56.

[99] Brooks, *Law, Politics and Society*, p. 198.

[100] Harold J. Cook, *Matters of Exchange: Commerce, Medicine, and Science in the Dutch Golden Age* (New Haven, 2007), pp. 42–81.

[101] Harkness, *Jewel House*, pp. 212–14; Hugh Plat, *The Jewell House of Art and Nature* (1594), sigs A2r, B1r; *ODNB sub* Plat.

[102] John Fitzherbert, *The Boke of Husbandry* (?1533), fos 63v, 89v–90r; Roberts, *Merchants Mappe*, sigs A2, A5–6.

[103] Harkness, *Jewel House*, p. 218; John Dee, *The Elements of Geometrie of... Euclide* (1570), sigs Aiii, Civ; Thomas Digges, *An Arithmeticall Militare Treatise, named Stratioticos* (1579), sig. Aii.

[104] Harkness, *Jewel House*, pp. 241–53; Ash, *Power, Knowledge and Expertise*, pp. 187–98; above, p. 52. On Bacon's interest in the mechanical arts, see Julie Robin Solomon, *Objectivity in the Making: Francis Bacon and the Politics of Inquiry* (Baltimore, 1998), pp. 72, 114; Paolo Rossi, *Francis Bacon: From Magic to Science* (1968), pp. 8–11. This is not to suggest that Bacon had anything approaching the face-to-face interaction with practical experimenters of Samuel Hartlib, still less of Robert Hooke: Mark Jenner, '"Another Epocha"? Hartlib, John Lanyon and the Improvement of London in the 1650s', in Mark Greengrass et al., eds, *Samuel Hartlib and Universal Reformation: Studies in Intellectual Communication* (Cambridge, 1994), p. 355; Rob Iliffe, 'Material Doubts: Hooke, Artisan Culture and the Exchange of Information in 1670s London', *British Journal for the History of Science*, 28 (1995), pp. 285–318.

contemporaries were enthusiastic if unsystematic experimenters, benefiting from the knowledge economy inseparable from a large commercial metropolis.[105]

This was the London whose growth Elizabeth's government had been unable to prevent, and which James I and Charles I wanted to see improved, rebuilt in brick rather than wood, and transformed into an 'imperial city' fit for a British monarch. One of the proclamations imposing new building regulations which Bacon drafted in 1615 listed the 'ornaments' of the city which could already compete with those of Paris. They included 'Britain's Burse' (Robert Cecil's New Exchange) with its bustling shops, the New River bringing fresh water into a crowded metropolis, and even Sutton's Charterhouse, which Bacon had recently condemned as a waste of riches.[106] All these were examples of private enterprise serving public needs, and so were the brick buildings which filled much of the West End by 1650, funded by private property-owners and speculative builders catering for the latest fashion among the landed and commercial elite. London was being improved, but not by royal command. When Charles I tried to impose a unified political structure on an ever larger metropolis in the 1630s, he predictably failed in face of local vested interests.[107] London was left to be managed, as much of the rest of the economy was managed, by a host of privileged corporate bodies in the City and its suburbs, sometimes collaborating but more often contending with one another in an untidy structure which left more room for piecemeal innovation than centrally planned reformation.

Whether they were welcomed or simply tolerated, changes of the magnitude of London's growth and the foundation of an overseas empire of land and trade necessarily had a cultural effect. The language repeatedly used to justify them began to encourage as well as endorse the kinds of private and corporate behaviour which created them. When merchants described their status and activities in terms of honour and public utility they were appropriating the values of a landed and military aristocracy and extending their application. When they deplored avarice and elevated the virtues of diligence and thrift above the vices of idleness and profligacy they were not simply reacting to their critics, but following the literature of agrarian improvement and claiming moral approval for modes of behaviour essential to success in trade. It has been pointed out that English merchants in the years around 1600 were borrowing terms like 'provident' from the vocabulary of religious piety and applying them to business, and hence by 'rhetorical manipulation' shaping its moral identity. The application of 'improvement' in contexts far removed from its agrarian origins was another example of the same tactic.[108] We have seen that the term was already being extended to confer approval on adventures in Ireland and America, and on the manufacture of old and new draperies and the advance of commerce. Its reach was to expand much further later in the century,

---

[105] Cf. below, pp. 229–31.
[106] *Stuart Royal Proclamations*, i. 345–7. On Britain's Burse, see Ian Archer, 'Material Londoners?', in Lena Cowen Orlin, ed., *Material London, ca. 1600* (Philadelphia, 2000), pp. 174–5.
[107] Slack, *Reformation to Improvement*, pp. 70–4; Ashton, *City and Court*, pp. 163–7.
[108] Skinner, 'Moral Principles and Social Change' and 'The Idea of a Cultural Lexicon', in *Visions of Politics* (3 vols, Cambridge, 2002), i. 147–54, 173–4. Cf. below, p. 166.

but its increasing employment was an indication of alterations in the intellectual climate. As the shape of the economy changed, what one of William Cecil's projectors called 'the natural disposition of the time' was changing with it.[109]

ECONOMIC THINKING AND THE CRISIS OF THE 1620s

Since most sectors of the English economy were not only changing rapidly between 1570 and 1640 but becoming the subject of debate and enquiry, it is natural to ask whether there was as yet any appreciation of how they interacted with one another sufficiently comprehensive and coherent to constitute economic understanding. In the new information economy of London, had the economy itself already become a separate compartment of knowledge with its own modes of argument? A short answer might be that it could not have done, since 'economy' (or 'oeconomy' as it was usually spelt) had not yet acquired its modern meaning, and there was no other term which could serve the same purpose.[110] A longer and more positive answer involves some account of how the use of the word, and the use of terms associated with it, evolved, sometimes again through metaphorical extension, and how economic crisis and debate in the 1620s accelerated the process.

All early modern discussion of economy rested on two classical texts, Xenophon's *Oeconomicus* and a work usually (though wrongly) attributed to Aristotle, the *Oeconomica*.[111] Both were about the management of households and how to make their property, including land, more profitable; and oeconomy still retained that primary application in the eighteenth century.[112] But it was easily extended by analogy, to 'the oeconomy of a common-wealth', as it was by Thomas Hobbes,[113] and the notion of 'political economy' might seem to follow naturally. The term was first used in print in France in two treatises about national reconstruction after the French Religious Wars, the best known of them Antoine de Montchrétien's *Traicté de l'oeconomie politique* of 1615.[114] William Petty seems to have been in search of it

---

[109] Above, p. 55.

[110] For a powerful argument to this effect, see Margaret Schabas, *The Natural Origins of Economics* (Chicago, 2005), pp. 1–11. Historians have nevertheless tried to describe economic thinking before 1600, including Diana Wood, *Medieval Economic Thought* (Cambridge, 2002) and Neal Wood, *Foundations of Political Economy: Some Early Tudor Views on State and Society* (Berkeley, 1994); and literary scholars have looked for its reflection in the imaginative literature of the period, e.g. David Landreth, *The Face of Mammon: The Matter of Money in English Renaissance Literature* (Oxford, 2012); Jonathan Gil Harris, *Sick Economies: Drama, Mercantilism and Disease in Shakespeare's England* (Philadelphia, 2004).

[111] Andrea Finkelstein, *The Grammar of Profit: The Price Revolution in Intellectual Context* (Leiden, 2006), pp. 71–4. On *Xenophons Treatise of Householde* (1532), see above, p. 18. The pseudo-Aristotelian work was not translated but received detailed comment in John Case, *Thesaurus oeconomiae* (Oxford, 1597).

[112] Karen Harvey, *The Little Republic: Masculinity and Domestic Authority in Eighteenth-Century Britain* (Oxford, 2012), pp. 24–59. Cf. Brooks, *Law, Politics and Society*, p. 352.

[113] Thomas Hobbes, *Leviathan*, ed. Noel Malcolm (3 vols, Oxford, 2012), ii. 378 (ch. 23).

[114] C. Théré, 'Economic Publishing and Authors, 1566–1789', in Gilbert Faccarello, ed., *Studies in the History of French Political Economy: From Bodin to Walras* (1998), p. 2. The term had been used earlier in Louis Turquet de Mayerne's *La Monarchie aristodémocratique* (Paris, 1611), probably written

when he referred to 'political oeconomies' and 'public oeconomy' in the 1670s, but it did not catch on in either England or France until the later eighteenth century.[115] In the absence of a special terminology, therefore, seventeenth-century writers discussing the wealth of a nation or nations in sometimes novel ways had to use language and concepts familiar from other contexts, from arguments about wealth and riches, or trade and commerce.

They took the classical authors as their authorities, and, like scholastic commentators earlier, reinterpreted their works to fit contemporary conditions. 'Riches (as Aristotle hath defined) are either natural or artificial,' Gerard Malynes declared in 1603.[116] Natural riches lay in the earth, the land and its produce, and the minerals that rested beneath it; artificial riches were things derived from land by industry and art, harvested or turned into commodities by labour. Land was therefore primary, 'the fountain and mother of all the riches and abundance of the world', as Lewes Roberts explained in 1641,[117] and wealth was treated less as a reified abstraction than as a concrete part of the natural and physical world, to be interpreted in the terms appropriate to natural philosophy. Since even precious metals were 'vegetable things', as Malynes pointed out, all the fruits of the earth might be described in the language of contemporary alchemists as products of nature, capable of being transformed and perfected by human intervention.[118] It followed that the ways to wealth and prosperity were often pictured in organic terms, which proved to be as easily adapted to fit arguments from reason of state as they had earlier been to notions of the body politic.

In a discourse on 'natural and political' bodies in 1606, for example, Edward Forset visualized a country's prosperity as springing from 'a vegetable power' inherent in the monarch, who functioned like the soul in the body, ensuring 'the flourishing and felicity of a commonweal', and 'the happiness of the subject and the subject's welfare', as well as 'the security of the Prince'. Using an image he may have borrowed from Botero, he visualized the Prince engaging in a universal 'generating and propagating', which embraced every aspect of the economy, people, trade, production, and consumption alike. He must increase his dominions and populate colonies, but also 'cherish in the subjects an appetite of acquiring commodities', and found market towns for 'digesting' and spreading them to 'all parts of the realm'.[119] In 1641 Roberts similarly extended his definition of 'the abundance,

---

around 1590, and discussed in Hugh Trevor-Roper, *Europe's Physician: The Various Life of Sir Theodore de Mayerne* (2006), pp. 124–5, 155–7.

[115] Below, p. 123; Julian Hoppit, 'The Contexts and Contours of British Economic Literature, 1660–1760', *Historical Journal*, 49 (2006), p. 80. The subtitle of the first substantial British text on the topic, Sir James Steuart's *An Inquiry into the Principles of Political Oeconomy, being an Essay on the Science of Domestic Policy in Free Nations* (2 vols, 1767), indicates the domestic associations which 'oeconomy' still retained.

[116] Gerard Malynes, *Englands View* (1603), p. 118. Cf. Mun, *Discourse of Trade*, p. 40.

[117] Roberts, *Treasure of Traffike*, p. 61.

[118] Carl Wennerlind, *Casualties of Credit: The English Financial Revolution, 1620–1720* (Cambridge, Mass., 2011), pp. 49–51; Schabas, *Natural Origins*, pp. 2–3; below, pp. 108–9.

[119] Edward Forset, *A Comparative Discourse of the Bodies Natural and Politique* (1606), pp. 4, 13–14, 96; above, p. 45. For sentiments similar to Forset's, see Keymer, *Original Papers*, pp. 62–3; below, p. 109.

plenty, and riches of an estate or nation' beyond natural and artificial commodities to their 'profitable use and distribution... by commerce and traffic', which was essential to the wealth, strength, and safety of a kingdom.[120] Much of this was commonwealth rhetoric familiar since the fifteenth century,[121] but commerce within and between nations was becoming central to it. As Edward Misselden observed in 1622, 'matters of state and of trade' were 'involved and wrapped up together'.[122]

Economic change since the fifteenth century had also altered the character of economic discussion by drawing attention to the several features of an economy whose interaction made its management for agreed purposes ever more difficult. English writers followed scholastic commentators in Europe who were struggling with the intricacies of market prices and the function of money and interest rates in a changing commercial world.[123] Well before William Harvey's discovery of the circulation of the blood made the metaphor popular, money was being described in physiological terms, nourishing every part of the body politic, but needing to be held in balance since too much of it could be as dangerous as too little.[124] Trying to account for price inflation in the sixteenth century, and to explain how debasement of the coinage could have started a chain reaction, Thomas Smith not only referred to circular motion but employed the mechanical metaphor of wheels and clocks, one of the first uses of what became a popular analogy.[125]

Population growth, that other conspicuously novel feature of the later sixteenth century, similarly prompted questions about the economic consequences of demographic fluctuations, and showed that the labour supply, like the money supply, could be too large as well as too small. As Bacon and others appreciated in 1607, and with Aristotle again in mind, popular disorder and unemployment followed if 'the population of a kingdom' exceeded 'the stock of the kingdom which should maintain them'.[126] Emigration to the plantations was an obvious safety-valve, and one of its advocates, Richard Eburne, spotted the potential link between rising population and inflation in 1624, when he argued that a 'diminution of the people' would 'easily cause... the prices of all things to fall of themselves'.[127] Other commentators thought surplus labour could be better employed at home, improving 'the natural commodities of this realm by industry and increase of arts' and thus add to 'the stock of the kingdom'. If full employment could be delivered, 'increase

---

[120] Roberts, *Treasure of Traffike*, pp. 60, 111.
[121] Wood, *Foundations*, pp. 40–1, 66–8; Wood, *Medieval Economic Thought*, p. 123. For an example from the 1540s, see *Tudor Economic Documents*, iii. 314.
[122] Edward Misselden, *Free Trade, or The Meanes to Make Trade Florish* (1622), sig. A3ᵛ.
[123] Barry J. Gordon, *Economic Analysis Before Adam Smith: Hesiod to Lessius* (1975), pp. 240–71.
[124] Wennerlind, *Casualties of Credit*, pp. 38–9. Cf. above, p. 73.
[125] Wood, *Foundations*, pp. 212–14. For later uses of the clock image, see Roger Fenton, *A Treatise of Usurie* (1611), p. 2; David Pennington, 'Beyond the Moral Economy: Economic Change, Ideology and the 1621 House of Commons', *Parliamentary History*, 25 (2006), p. 224; Gerard Malynes, *A Treatise of the Canker of Englands Common wealth* (1601), p. 95.
[126] Paul Slack, *'Plenty of People': Perceptions of Population in Early Modern England* (Stenton Lecture 2010, Reading, 2011), pp. 3–4.
[127] Richard Eburne, *Plaine Path-Way to Plantations* (1624), p. 10. I owe this reference to Keith Thomas.

of wealth and people', the economic desiderata of the Council paper on enclosure in 1607, might go hand in hand.[128]

Of the many interrelated strands determining England's economic fortunes, however, the most intricate technically was foreign exchange; and it was the first to be discussed with any sophistication in England because exchange rates governed the level of English exports, particularly of textiles, and hence the level of domestic employment in industry. An early Elizabethan 'memorandum' on the subject, once attributed to Thomas Gresham because it reflected first-hand knowledge, explained in some detail how exchange rates might be manipulated by merchants, either 'against a common wealth', if they simply pursued their private profit, or 'for a common wealth' if they acted in the public interest with an eye to 'impoverishing... foreign countries' and 'enriching our own'.[129] At the turn of the century Gerard Malynes, who had experience in the mint, laid claim to the same territory, advertising in print his expert understanding of how 'the wealth of a realm may increase or decrease', depending on 'the course of commodities, money, and exchange'.[130] It might seem that modes of thinking about the economy were responding to changing circumstances and fashions of political argument.

Yet there was one further ingredient, inherited from the past, which complicated contemporary discourse and inhibited any rapid recognition of the economic primacy of considerations of profit and power, whether for individuals or nations. For Malynes and his contemporaries, in Europe as well as England, economic relationships like all others had to be judged primarily in terms of justice and equity, and with reference to the Aristotelian distinction between two kinds of justice—distributive and commutative—which had shaped ethical assumptions for centuries.[131] Distributive justice respected inequalities in status, power, property, and wealth, and was sometimes called the best or superior equity because it justified their perpetuation along with the obligations that went with them. Commutative justice, on the other hand, was sometimes termed corrective or rectificatory because it treated unequals equally; and that was supposed to govern the market exchange of goods, including their international exchange, where the value of things exchanged must be equal, whatever the status of the parties involved. Never easy to define, the distinction nevertheless meant that there were two concepts of equity to be employed in assessing the merits of the status quo and of public and private actions which might disturb them.

---

[128] Mun, *Discourse of Trade*, pp. 46–7; above, p. 50. Cf. Keymer, *Original Papers*, pp. 36–7.
[129] Raymond de Roover, *Gresham on Foreign Exchange* (Cambridge, Mass., 1949), pp. 300–1, 305. For similar comments, see Ramsay, *Politics of a Tudor Merchant Adventurer*, p. 64.
[130] Malynes, *View*, p. 118.
[131] Malynes, *View*, pp. 4–5; Aristotle, *Nicomachean Ethics*, 5.2–5. The influence of the two kinds of justice on contemporary economic thought is discussed in Finkelstein, *Grammar of Profit*, pp. 235–81, and illustrated in Gordon, *Economic Analysis before Adam Smith*, pp. 241–4, 271; Wennerlind, *Casualties of Credit*, pp. 9–10, 34–6; Lianna Farber, *An Anatomy of Trade in Medieval Writing: Value, Consent, and Community* (Ithaca, NY, 2006), pp. 2–4, 23–4; Craig Muldrew, *The Economy of Obligation* (Basingstoke, 1998), pp. 43–4. For later discussion of the distinction, see Izhak Englard, *Corrective and Distributive Justice: From Aristotle to Modern Times* (Oxford, 2009).

The two concepts of justice owed much of their continuing appeal to their utility in other areas of contemporary debate. In political discourse, for example, commutative justice was thought to be characteristic of democracies or 'popular' governments, like that of the highly commercialized Dutch Republic, while distributive justice was the distinguishing mark of aristocracies, although supposedly 'mixed' monarchies like England and France might be thought to combine the two and so be superior to both of them.[132] The distinction was also relevant in mathematics when it was applied to the measurement of proportion in different contexts. Bacon noted that there was 'a true coincidence between commutative and distributive justice, and arithmetical and geometrical proportion';[133] and different forms of taxation could be judged equitable because they were either arithmetic and commutative, as in the case of poll taxes and excises, or geometric and distributive, as with graduated rates on property.[134] Such fine distinctions had understandably proved difficult to apply, however, when judging other areas of economic activity, like usury, or the determination of 'just' prices and wages; and they seemed increasingly irrelevant to any understanding of how commercial markets operated in the real world.[135]

Malynes nevertheless treated the two concepts of justice as a yardstick by which to measure proportion and balance in a stable commonwealth. A just and geometrical distribution of property 'by number, weight and measure' ensured 'harmony' and 'good concord', 'equality and concord', within the nation, with everyone living 'contentedly and proportionably in his vocation'.[136] Commutative justice, on the other hand, guaranteed another 'kind of equality' in commerce, especially between countries and continents where there should be 'common intercourse and . . . a friendly correspondence'.[137] The latter ideal was far removed from the realities of international commerce governed by competition between the trading nations of Europe, and it was equally unreal in domestic commerce where exchange, as Hobbes observed, was not determined by commutative justice but by contract, between parties with unequal power in the marketplace.[138] There were occasional references to the two concepts of justice in other economic contexts after 1600, as in Robert

---

[132] See e.g. Jean Bodin, *Six Bookes of a Common-weale* (1606), pp. 775–6; Gerard Malynes, *Consuetudo vel lex mercatoria* (1622), p. 235; Roger Coke, *A Discourse of Trade* (1670), p. 71.

[133] Francis Bacon, *The Advancement of Learning*, ed. Michael Kiernan (*OFB* iv, Oxford, 2000), p. 77. In 1671 Roger Coke cited the two kinds of justice in a similar context, while claiming that Aristotle himself, unlike the moderns, 'never understood one proposition either in geometry or number': Coke, *Discourse of Trade*, pp. 70–1.

[134] See for example Petty, *EW*, i. 94; Davenant, *PCW*, i. 141–3; below, pp. 94, 180.

[135] Cf. Jones, *God and the Moneylenders*, pp. 8–10; John T. Noonan, *The Scholastic Analysis of Usury* (Cambridge, Mass., 1957), pp. 199–201; Muldrew, *Economy of Obligation*, pp. 44–5; Finkelstein, *Grammar of Profit*, pp. 240–3; Raymond de Roover, 'The Concept of the Just Price: Theory and Economic Policy', *Journal of Economic History*, 18 (1958), pp. 418–34.

[136] Gerard Malynes, *Saint George for England, Allegorically described* (1601), sigs A4ᵛ, A6ᵛ, p. 14; Malynes, *View*, p. 4; Malynes, *Lex mercatoria*, sig. A5ᵛ, p. 119.

[137] Malynes, *View*, p. 5; Malynes, *Lex mercatoria*, pp. 229–30. Cf. Andrea Finkelstein, *Harmony and the Balance* (Ann Arbor, 2000), pp. 94–6; Thomas Leng, 'Commercial Conflict and Regulation in the Discourse of Trade in Seventeenth-Century England', *Historical Journal*, 48 (2005), pp. 937–9; below, p. 190.

[138] Hobbes, *Leviathan*, ed. Malcolm, ii. 228–30; Muldrew, *Economy of Obligation*, p. 324.

Powell's attack on enclosures in 1636, but they were becoming a mode of rhetoric rather an analytical tool. They added little to economic understanding but gave some intellectual polish to a more imprecise, and much more widely shared, conviction that economic activity had to be judged by its contribution to the general good.[139] They endorsed popular assumptions that the economy had a moral dimension.

Deep-seated notions of a 'moral economy' can be identified in many different forms and at different social levels. When Malynes and other authors attacked monopolies by citing Aristotelian concepts, and Edward Coke opposed some guild restrictions because at common law no one could be prohibited from working in a lawful trade,[140] they were making the same points about justice and equity. Preachers against enclosure and low wages in the textile industry appealed to 'religion and humanity' and the right of every man to an adequate livelihood, and Londoners objected to the New River bringing piped water to citizens who paid for it, because what was 'intended for a public good' was being 'converted to a private gain'.[141] The people who protested against fraudulent practices in the marketing of essentials like corn and coal, or against the removal of common rights in royal forests and the fens, occupied the same moral ground, and so did participants in the disturbances which occurred whenever the harvest failed and which led E. P. Thompson to his classic definition of the 'moral economy of the crowd'.[142]

In the case of food riots, those involved were able to invoke the support of the Book of Orders regulating the marketing of corn, first issued under the royal prerogative in 1586, but that was only a temporary support, dispensed with after the 1630s by central and local governments who thought famine more effectively prevented by the normal workings of the market and the targeting of poor relief to those most at risk. The moral economy of the crowd then had to rely, as it had done before 1586, on arguments from the 'law of necessity' which was generally acknowledged to take precedence over property rights when people were starving.[143] Yet similar arguments from natural justice could have been directed against

---

[139] Powell, *Depopulation Arraigned*, pp. 48–9; Brooks, *Law, Politics and Society*, pp. 343–4.

[140] Malynes, *Lex mercatoria*, pp. 229–30; Brooks, *Law, Politics and Society*, p. 389. Cf. Gordon, *Economic Analysis before Adam Smith*, pp. 266–7; David Harris Sacks, 'The Greed of Judas: Avarice, Monopoly, and the Moral Economy in England, ca.1350–ca.1600', *Journal of Medieval and Early Modern Studies*, 28 (1998), pp. 263–8, 271–7.

[141] Sacks, 'The Greed of Judas', pp. 286–7; Patrick Collinson, 'Christian Socialism in Elizabethan Suffolk: Thomas Carew and his *Caveat for Clothiers*', in Carole Rawcliffe et al., eds, *Counties and Communities: Essays on East Anglian History Presented to Hassell Smith* (Norwich, 1996), pp. 165–7; Mark S. R. Jenner, 'From Conduit Community to Commercial Network? Water in London 1500–1725', in Paul Griffiths and Mark S. R. Jenner, eds, *Londinopolis: Essays in the Cultural and Social History of Early Modern London* (Manchester, 2000), pp. 257–8.

[142] E. P. Thompson, 'The Moral Economy of the English Crowd in the Eighteenth Century', *P&P* 50 (February 1971), pp. 76–136. For later work on various species of crowd behaviour, see, e.g., John Walter, 'Grain Riots and Popular Attitudes to the Law: Maldon and the Crisis of 1629', in John Brewer and John Styles, eds, *An Ungovernable People: The English and their Law in the Seventeenth and Eighteenth Centuries* (1980), pp. 47–84; Keith Lindley, *Fenland Riots and the English Revolution* (1982). For complaints about market abuses, see Muldrew, *Economy of Obligation*, pp. 48–9; Nef, *British Coal Industry*, ii. 242–4, 259–62.

[143] John Bohstedt, *The Politics of Provisions: Food Riots, Moral Economy, and Market Transition in England, c.1550–1850* (Farnham, 2010), pp. 9, 14–15, 48–52; above, pp. 62–3.

the Book of Orders itself. All kinds of state regulation of the economy by means of statutes, proclamations, guilds, and monopoly companies were open to attack on moral grounds because they deprived men of the right to 'the free exercise of their industry' in order to gain a livelihood.[144] The charge was particularly common in the early seventeenth century because the Crown was misusing the royal prerogative for its own profit, but there was in any case mounting evidence that there were strict limits to what could be achieved by any government engaging in economic management. Aristotelian concepts of justice had much to say about the damaging consequences of economic change, but they contributed nothing at all to contemporary understanding of the processes which caused it and how they could be manipulated to public advantage.

Advances in economic thinking of that kind tended to come only at moments of crisis, when economic and political circumstances conspired to push questions about economic regulation up the political agenda, and prompted public debate and controversy about them. One such episode, as William Cecil's papers show, had occurred in the mid-sixteenth century with the collapse of the Antwerp market for English cloth.[145] A second, of much greater intellectual consequence, came in the early 1620s, in the wake of the disruption caused by the Cockayne scheme,[146] when bad harvests hit the domestic economy, and war and currency manipulations in Europe depressed foreign markets for English textiles still further. The outcome was the rejection of a prescription for government intervention in the interests of the common weal proposed by Malynes, and the triumph of an argument about the balance of trade formulated by Thomas Mun and Edward Misselden, which emphasized the primacy of market forces. It was a dispute about the causes of a commercial and monetary crisis which had affected every part of the economy, and its result marked a turning point in the character of English economic discourse.[147]

It was not arrived at painlessly, but it came relatively quickly under pressure from a government looking desperately for advice on how to react, and quickly discovering that there was no single simple solution. Sub-committees of the Privy Council took evidence on the collapse of the exchanges and the 'decay of clothing', and the House of Commons had its own 'Great Committee' on the scarcity of money and its impact on trade.[148] The Commons identified as many as fourteen causes of the crisis; the Council could only reduce the number to nine. The immediate result, in October 1622, was the creation of a standing Commission on Trade, which had terms of reference as wide ranging as the causes and consequences of the immediate crisis, and fifty-one members drawn from every conceivable political and

---

[144] e.g. *SCED*, p. 437.
[145] Above, pp. 55–6; B. E. Supple, *Commercial Crisis and Change in England 1600–1642* (Cambridge, 1964), pp. 197–8. For similar episodes later, see below, pp. 95, 132–3, 180–4.
[146] Above, p. 71.
[147] For a full account of the circumstances and debates, see Supple, *Commercial Crisis and Change*, and on the intellectual consequences, Mary Poovey, *A History of the Modern Fact* (Chicago, 1998), pp. 66–91; Finkelstein, *Harmony and the Balance*, pp. 26–97; Wennerlind, *Casualties of Credit*, pp. 30–43.
[148] Supple, *Commercial Crisis and Change*, pp. 54–5, 66–7, 232–3, 268–9.

commercial interest group, with a quorum headed by the President of the Council, Henry Montagu, who had been Recorder of London and was himself a director of the Virginia Company.[149] It survived until the death of James I in 1625 and was the first of a series of Councils and Boards of Trade.

No one could say that the issues were not being fully ventilated and all kinds of experts consulted. In addition to Malynes, who was busy writing *Lex mercatoria*, the most famous merchants' handbook of the century, they included members and agents of the great trading companies like Mun and his Hackney neighbour Misselden, Robert Cotton, the antiquary who had written the 1607 discourse on enclosure, and Ralph Maddison, an improving landlord and investor in coal mines with ambitions to be an author as successful as Malynes.[150] In one form or another, the advice they gave was soon available in print, and their divided opinions and arguments were open to public scrutiny. Since there was no unanimity, their deliberations had no immediate effect on government policy. The economy recovered under its own steam, as English exports revived and the rate of exchange improved, and not because of any 'act of state' as Maddison later claimed. But Maddison was right when he pointed to the novelty of an unprecedented 'public agitation' which had informed public and political opinion about the economic state of the nation.[151]

The range of information being collected is reflected in the instructions given to the Commission on Trade in 1622. As might have been expected, they referred to shipping and the fisheries, to the old and new draperies, to declining rents and customs and a deficient money supply, and they included controversial items like the export of treasure to the East Indies, the need for sumptuary laws, and 'whether it be necessary to give way to a more open and free trade or not'. The most remarkable feature of the document, however, was its recognition that economic policies could not be expected to cure all economic ills at a stroke, but must change with the times. Since it was 'impossible to foresee' the outcome of any political intervention, there was no 'expectation of a total and absolute reformation of every part' of the Commission's agenda 'all at once'. The factors involved in commerce, 'the occurrents of trade', were many and variable, 'and must be directed and governed as times and occasions shall serve and do vary'. The aim was piecemeal and gradual improvement, 'the improvement of our native commodities', and in effect the improvement of England's economic performance generally.[152]

Equally novel was an instruction which was to be repeated in the terms of reference of all later Boards of Trade into the eighteenth century. In order to assess

---

[149] SCED, pp. 2–3, 16–23; Supple, *Commercial Crisis and Change*, pp. 66–7; Astrid Friis, *Alderman Cockayne's Project and the Cloth Trade* (Copenhagen and London, 1927), pp. 424–5; *ODNB sub* Henry Montagu.
[150] See *ODNB* on all of these; and in addition, on Cotton, above, p. 50; Wm A. Shaw, ed., *Select Tracts and Documents Illustrative of English Monetary History 1626–1730* (1896), pp. 1–38; Supple, *Commercial Crisis and Change*, pp. 170, 186, 268; and on Malynes and his *Lex mercatoria* (1622), below, p. 109; *The Correspondence of John Wallis*, ed. Philip Beeley and Christoph J. Scriba, vol. iii (Oxford, 2012), p. 17.
[151] Supple, *Commercial Crisis and Change*, pp. 188–9, 268.   [152] SCED, pp. 16–23.

the country's economic performance over time, the commissioners should 'diligently observe the true balance of the trade of this kingdom', and to that end use 'all records and writings, as you shall find needful for your better information'.[153] The balance of trade was a recent invention. Although the general need to boost exports and reduce imports was well appreciated by writers on commerce from at least the fourteenth century,[154] and an effort had been made to calculate the difference between them in 1563,[155] the first careful measurement, for the years 1612–14, was undertaken by Lionel Cranfield and others with access to the customs records in 1615.[156] There was talk in government circles about 'balancing' trade flows in the context of the Cockayne scheme, and Bacon was one of the first to refer explicitly to a 'balance of trade' in 1616.[157] But it needed the debates in Council committees in 1622 and 1623 to make the term familiar, and the publications of the chief participants to make the concept and the methodology a permanent topic of public discourse. They effectively created for the first time something that looks recognizable as English political economy, with its own founding texts, organizing concepts, and particular mode of discourse.

In order for that to happen, Malynes's simplistic view of how to manage exchange dealings by government fiat had first to be rejected and cleared out of the way. In a Council sub-committee in 1622, he argued, as he had done in his publications, that there was a single prime mover in the great clock of international commerce, the exchange rate, which was being manipulated by European financiers and speculators to England's disadvantage. The obvious remedy was for the English government to declare that exchange should only take place at par—at the 'true and intrinsic value of your monies', 'value for value', or *par pro pari*, as true commutative justice dictated.[158] It was already abundantly evident, however, that financial markets were more powerful than monarchs, and Malynes's proposed antidote had been tried in 1576 and found unworkable.[159] The Council had to turn to another group of experts led by Mun, who had been pondering the role of supply and demand in international trade while writing his defence of the East India Company in 1621. They pointed out that the exchange rate was wholly secondary, a function

---

[153] *SCED*, pp. 20, 23.

[154] Arthur B. Ferguson, *The Articulate Citizen and the English Renaissance* (Durham, NC, 1965), p. 94; G. A. Holmes, 'The "Libel of English Policy"', *EHR* 76 (1961), pp. 193–216.

[155] *Tudor Economic Documents*, i. 178–84; Lawrence Stone, 'Elizabethan Overseas Trade', *EcHR* 2nd ser. 2 (1949–50), pp. 34–6; W. M. Ormrod, 'The English Crown and the Customs, 1349–63', *EcHR* 2nd ser. 40 (1987), p. 39 and n. 65.

[156] Menna Prestwich, *Cranfield: Politics and Profits under the Early Stuarts* (Oxford, 1966), pp. 120–1, 182–3; A. P. Newton, ed., *Calendar of the Manuscripts of Lord Sackville*, i, *Cranfield Papers 1551–1612* (HMC lxxx, 1942), p. 289; *SCED*, pp. 454–7.

[157] *SCED*, pp. 457, 459–61; *Acts of the Privy Council 1615–16*, p. 188; Bacon, *Letters and Life*, ed. Spedding, vi. 22 (and cf. v. 198). For the methodology, borrowed from double-entry book-keeping, see Finkelstein, *Harmony and the Balance*, pp. 89–97.

[158] Supple, *Commercial Crisis and Change*, pp. 186–7; Finkelstein, *Harmony and the Balance*, p. 44; Malynes, *Canker of Englands Common wealth*, p. 99. Cf. De Roover, *Gresham on Foreign Exchange*, pp. 157–9.

[159] Thomas Wilson, *A Discourse upon Usury*, with an introduction by R. H. Tawney (1925), pp. 149–51; Supple, *Commercial Crisis and Change*, p. 189.

of supply and demand for different currencies which was in turn primarily determined by the course of trade in commodities. Any effort to fix it artificially was likely to be damaging and prejudicial to commerce. If one wanted to understand net movements of coin and bullion, and the influx or outflow of treasure, attention should therefore focus, not on the intricacies of exchange dealings, but on the balance of payments, and what lay behind that was the balance of trade.[160]

A large part of Mun's achievement is attributable to his deliberate avoidance of the scholastic 'sophistry' of Malynes and his use of plain language to explain the 'plainness' of commercial practice in immediately intelligible terms. The 'conjuring' involved in Malynes's *par pro pari* panacea had been banished for good, Mun announced in *England's Treasure by Forraign Trade*, the famous tract which became one of the all-time best-sellers of English economic literature, largely written in the 1620s.[161] It was not published until 1664 after Mun's death, but his conclusions had almost instantly become public, thanks to Misselden's *Circle of Commerce, or The Ballance of Trade* (1623), which trumpeted the utility of the balance in less plain but equally potent language. It was an 'excellent and political invention', which showed, like a pair of scales, 'the difference of weight in the commerce of one kingdom with another'. He used it to compare figures for 1621–2 with those for 1613–14 and found 'a great underbalance of trade with other nations'. 'We felt it before in sense', he observed, 'but now we know it by science. We found it before in operation, but now we see it in speculation.' The depression of the 1620s had given birth to a new intellectual construct, a term 'of art or science' which Maddison could assume was 'well known to many' by 1640.[162]

Part of the appeal of this new piece of speculative science was the model it created of an economic machine which was autonomous and largely self-regulating.[163] According to Mun, the balance was governed 'by a necessity beyond all resistance'. For Misselden trade had a 'natural liberty' which would 'not endure the command of any, but of God alone'.[164] Neither of them wanted to argue that government had no economic role at all to play, in protecting the country's overseas commerce, for example, or encouraging and improving domestic manufactures for export; but radical and rapid interference was likely to do more harm than good. Economic

---

[160] Supple, *Commercial Crisis and Change*, pp. 94–6, 187–8, 269–70; Finkelstein, *Harmony and the Balance*, pp. 77–8. In 1597 a Spanish Jesuit, Luis Molina, had expressed similar doubts about attempts to interfere with exchange rates because they disrupted normal market mechanisms, but Mun and Misselden may not have been aware of his work: Gordon, *Economic Analysis before Adam Smith*, pp. 240–1.

[161] Thomas Mun, *England's Treasure by Forraign Trade* (1664), pp. 45, 48, 87–8; Supple, *Commercial Crisis and Change*, pp. 213–16; Hoppit, 'Contexts and Contours', pp. 97 n. 55, 109. On Mun's plain style see Poovey, *History of the Modern Fact*, p. 68; Miles Ogborn, *Indian Ink* (Chicago, 2007), p. 137.

[162] Edward Misselden, *The Circle of Commerce: Or the Ballance of Trade* (1623), sig. A4ʳ, pp. 116, 119–21, 130; Ralph Maddison, *Englands Looking In and Out* (1640), sig. a3ʳ (addressing MPs).

[163] Cf. Joyce Oldham Appleby, *Economic Thought and Ideology in Seventeenth-Century England* (Princeton, 1978), pp. 47–51.

[164] Mun, *England's Treasure*, p. 87 (based on Mun's memorandum of 1623: Supple, *Commercial Crisis*, pp. 96n., 270); Misselden, *Circle of Commerce*, p. 112. For the context see Poovey, *History of the Modern Fact*, pp. 75–6; Finkelstein, *Harmony and the Balance*, pp. 70–1.

policy should be tempered by circumstances, by 'the state of times and trades';[165] and the balance was a diagnostic tool ideal for that purpose, a quantitative indicator of the economic health of the nation which required constant examination. Like a thermometer, it was a pointer to dangerous symptoms which might need attention, but it was not in itself a cure.

The balance of trade was never as precise a scientific instrument as its inventors imagined. A 'true' balance of trade, or the 'perfect' balance of trade demanded from later Boards of Trade, proved to be a chimera. There were theoretical pitfalls in trying to calculate a trade balance based on visibles in a complex international trading system; and the calculation depended upon reconciling values and quantities in ratebooks and customs accounts, which, as Sir Thomas Roe explained to parliament in 1641, made any conclusion 'very uncertain'.[166] If it could have been achieved, it would in any case have provided a snapshot of only part of the economy at a particular moment. It was constantly used and often debated and referred to, but could never have been the narrow intellectual straitjacket in a 'mercantilist system' its later critics supposed.

When it was used for economic diagnosis, moreover, the lessons drawn from it had to be trimmed to the reality of the times from the start. Commentators on the balance of trade were unanimous in emphasizing the damage done by domestic consumer demand for luxury imports. Misselden attributed much of the 'underbalance of trade' to the 'prodigality' of those who 'excel in excesses', and Malynes thought 'superfluous commodities' responsible for 'the commonwealth's destruction', not least because they only made people 'discontented... with their state, always beholding how much inferior they are unto others'.[167] Even Mun, an East India merchant importing foreign luxuries, thought that consumers should somehow be persuaded to 'bridle' and 'moderate' their appetite for them.[168] Throughout the century there were repeated calls for sumptuary laws of the kind which had been repealed in 1604 because they were unenforceable, but they never returned. After a trade boom in the 1630s, when imports of wine and other commodities rose rapidly to unprecedented levels, Lewes Roberts articulated political reality. Some imports, he admitted, served 'more for pomp and show, than for need and use'; but it was difficult to prohibit them in 'an age or kingdom of peace and plenty', and far better to tax them by imposing heavy import duties, which would equally ensure that the commonwealth would be 'both improved and benefitted by this chief and good husbandry'.[169]

The legal enforcement of other ethical restraints on economic behaviour was similarly being eroded. The practice of usury was still as morally suspect as it had

---

[165] Misselden, *Circle of Commerce*, pp. 118–19.
[166] *SCED*, p. 41; Slack, 'Government and Information', pp. 52–3.
[167] Edward Misselden, *Free Trade* (1622), pp. 96–7; Malynes, *Saint George*, pp. 42, 48. The first passage from Malynes is underlined in a seventeenth-century hand in the Bodleian copy.
[168] Mun, *Discourse of Trade*, pp. 45–6. Cf. Mun, *Treasure*, p. 60.
[169] Roberts, *Treasure of Traffike*, pp. 76–7. Cf. below, pp. 131, 142. For the large imports of wine and tobacco in the 1630s, see Phil Withington, 'Intoxicants and the Early Modern City', in Steve Hindle et al., eds, *Remaking English Society* (Woodbridge, 2013), pp. 149–51.

been when it was legalized in 1571, but the balance of opinion had altered in the interim. In 1603 Malynes had predictably condemned it out of hand because it caused 'discord', 'some few waxing thereby too rich, and many extreme poor';[170] but in 1622 he conceded that it was a sin 'rather in the conscience than in the act'. Misselden similarly hedged his bets, acknowledging that interest rates were a viper consuming the bowels of the commonwealth while explaining that their level was determined by the money supply. Mun, however, positively welcomed them because they enabled 'younger and poorer merchants to rise in the world and to enlarge their dealings'.[171] Bacon took Mun's argument much further. Although William Cecil, his uncle, had thought there were too many merchants and artisans for the good of the commonwealth, Bacon was convinced that high interest rates meant there were now far too few. He wanted the maximum legal rate reduced from 10 to 5 per cent in order to stimulate 'all industries, improvements, and new inventions', 'industrious and profitable improvements' of every kind.[172] In the event a more cautious parliament in 1624 reduced the rate to 8 per cent, and, agreeing with Malynes, deliberately left the sinfulness of usury 'to be determined by divines'. After that even divines had to accept, as one of them said, that usury was 'a necessary evil', given 'the wretchedness of the world'.[173]

If we put the Usury Act alongside the repeal of one of the tillage statutes in the same year, we can begin to accumulate evidence of a radical shift in English assumptions about the economy in the 1620s. We might add to the list parliament's consideration of a bill to repeal old legislation against middlemen in 1621, and its failure ever to agree on methods of controlling the new draperies, the one expanding sector of the textile industry. If we include the final attempt rigorously to enforce the book of dearth orders in 1630s, it would seem that there was a general loosening of a regulatory framework designed to inhibit rapid economic change.[174]

Despite the simultaneous attack on the privileges of monopoly companies like the Merchant Adventurers, this did not amount to a new policy of free trade. It was dictated less by economic principle than by the proven impracticality or political unpopularity of closer regulation. It applied only to domestic commerce and was accompanied by increasing recognition of the need for protection against foreign competitors.[175] In 1622 the Commission on Trade was asked to consider whether a

---

[170] Malynes, *View*, p. 5. Cf. Malynes, *Saint George*, passim.

[171] Jones, *God and the Moneylenders*, pp. 160–3; Finkelstein, *Harmony and the Balance*, p. 61; Mun, *England's Treasure*, pp. 58–9.

[172] Vickers, *Bacon*, pp. 421–3. Cf. Jones, *God and the Moneylenders*, pp. 179–86. In fact, any effort to reduce interest rates artificially was more likely to benefit the rich, and especially landowners, who had better security than young aspiring tradesmen. See the discussion of the 1714 reduction to 5 per cent in Peter Temin and Hans-Joachim Voth, *Prometheus Shackled: Goldsmith Banks and England's Financial Revolution after 1700* (Oxford, 2013), p. 94.

[173] Jones, *God and the Moneylenders*, p. 175; Robert Bolton, *A Short and Private Discourse... concerning Usury* (1637), p. 29.

[174] Slack, *Reformation to Improvement*, pp. 57–8, 64–5; Thirsk, 'Enclosing and Engrossing', p. 236; Pennington, 'Beyond the Moral Economy', p. 218; J. P. Cooper, 'Economic Regulation and the Cloth Industry in Seventeenth-Century England', *TRHS* 5th ser. 20 (1970), pp. 73–85. Cf. Supple, *Commercial Crisis and Change*, pp. 247–9.

[175] Clay, *Economic Expansion and Social Change*, ii. 199–201.

law was needed to prohibit imports by foreign merchants, and there was a proclamation defending the Eastland Company against Dutch competition in the Baltic trade which pointed in the same direction and foreshadowed the later Navigation Acts.[176] These were early moves towards a pattern of policy-making evident in the 1650s and even more pronounced in the 1690s, which combined protections for overseas commerce with free trade at home because both were in the national interest.[177]

In this as in other respects the 1620s were formative for English economic thinking. They laid the foundations for what can accurately be called a 'mercantilist' consensus in English public discourse.[178] It was not, as some of its critics and historians later argued, an economic frame of mind shaped solely by considerations of bullion and treasure. But the focus on the balance of trade underlined lessons from reason of state about the wealth of a nation determining its power relative to its competitors, and identified commerce, and particularly sea-borne and international commerce, as the prime mover and creator of England's greatness. Publicity for the balance had shown further how trade affected every part of the domestic economy. Misselden made the point neatly: 'When trade flourisheth, the King's revenue is augmented, lands and rents improved, navigation is increased, the poor employed. But if trade decay, all these decline with it.' That being so, merchants were right to claim that their private gain had public utility: 'Is not the public involved in the private, and the private in the public? What else makes a common-wealth but the private wealth . . . of the members thereof in the exercise of commerce amongst themselves and with foreign nations?'[179]

Once that was accepted, arguments for a well-balanced harmonious commonwealth, with wealth justly distributed and merchants kept firmly in their place, might still claim the moral high ground, but they held onto it ever more precariously. The victory of Mun and Misselden over Malynes had not banished morality from the economy, and they would have been the first to declare any such suggestion absurd. Justice and equity remained aspects of the public and private face of commerce into the eighteenth century, not only in the moral economy of the crowd, though that continued, but more prominently and with greater effect in the everyday morality of relationships in a market society necessarily based at every level on trust and credit.[180] Ethical conceptions of economic relationships were upheld in the law courts, where people's rights to a livelihood and adequate subsistence were defended against monopolistic guild restrictions, in petitions for and against economic proposals in parliament, and in the private behaviour of

---

[176] *SCED*, p. 21; Supple, *Commercial Crisis and Change*, pp. 88–9.
[177] David Ormrod, *The Rise of Commercial Empires: England and the Netherlands in the Age of Mercantilism, 1650–1770* (Cambridge, 2003), pp. 343–5; below, pp. 185–6.
[178] Cf. Michael J. Braddick, *State Formation in Early Modern England c.1550–1700* (Cambridge, 2000), p. 431.
[179] Finkelstein, *Harmony and the Balance*, pp. 60–1, 286 n. 49.
[180] Craig Muldrew, 'Interpreting the Market: The Ethics of Credit and Community Relations in Early Modern England', *Social History*, 18 (1993), pp. 169, 177–8. Cf. Margaret Schabas and Neil De Marchi, 'Oeconomies in the Age of Newton', *History of Political Economy*, 35 (2003), Supplement 1, p. 5.

property-owners whose attachment to the values of charity and civic virtue equally impeded any wholesale triumph of economic laissez-faire.[181] When Mun dismissed Malynes's argument about justice as mere sophistry, however, he removed one of the intellectual supports on which the notion of a moral economy depended, and substituted an argument about England's treasure and trade which encouraged an economic conception of the purpose and power of the state. The discourse of commonwealth was being replaced by a discourse of commerce and commodities.[182]

It must be admitted that to call this kind of discourse economic thinking is in some respects premature. Closer attention to trade and its measurement left other parts of the economy awaiting the same kind of analysis and debate. In 1641 Lewes Roberts deplored the fact that land, the agrarian economy, had not yet been subjected to the same 'inquisition' as artificial riches and the commercial exchange of the commodities they provided, despite its importance for national improvement.[183] That was soon to be remedied by Hartlib and his colleagues in the 1650s; and in the 1660s Petty was to turn his attention to the measurement, not of trade, but of more fundamental and more abstract economic entities like the value of labour and population, and the national wealth to which they contributed.[184]

The role of government as an economic agent, that vital aspect of political economy, also remained unresolved. The issue had been ventilated in the exchange between Malynes and Misselden about the 'Circle of Commerce' in 1623. Malynes pictured kings and princes sitting in the circle's centre, 'at the stern of trade', because only their 'absolute government' could determine where the public interest lay when merchants were arguing solely for their own private benefit. Misselden preferred to emphasize the necessity for monarchs to collaborate with mercantile interests and even mobilize public debate in what he called 'the parliamentation and consultation' of all parts of the commonwealth about the 'causes and remedies' operating in commerce.[185] He thought that process had already begun in 1622, but there was no prospect of its continuing under Charles I, when most of the attempts by absolute government to regulate the economy were blatantly self-interested.[186] In 1640 the parliamentary and consultative option had still to be properly tested.

---

[181] Brooks, *Law, Politics and Society*, pp. 387–91; Harvey, *Little Republic*, pp. 61–3; below, pp. 205–6. Cf. Leslie Hannah, 'The Moral Economy of Business: A Historical Perspective on Business and Efficiency', in Peter Burke et al., eds, *Civil Histories* (Oxford, 2000), pp. 301–19; Jean-Pierre Hirsch, *Les deux rêves du commerce: Entreprise et institution dans la région lilloise (1780–1860)* (Paris, 1991), pp. 432–4; Deirdre N. McCloskey, *The Bourgeois Virtues: Ethics for an Age of Commerce* (Chicago, 2006).
[182] Cf. Craig Muldrew, 'From Commonwealth to Public Opulence: The Redefinition of Wealth and Government in Early Modern Britain', in Hindle et al., *Remaking English Society*, pp. 321–3. I have benefited from discussing this point with John Walter. It will be clear that some of the shifts towards secular concerns which Steve Pincus attributes to the 1650s seem to me to have occurred a generation earlier: 'From Holy Cause to Economic Interest: The Study of Population and the Invention of the State', in Alan Houston and Steve Pincus, eds, *A Nation Transformed: England after the Restoration* (Cambridge, 2001), pp. 272–98.
[183] Roberts, *Treasure of Traffike*, p. 61.   [184] Below, pp. 106–8, 120–3.
[185] Ogborn, *Indian Ink*, pp. 132–4.
[186] For criticisms of Caroline economic policies in the Long Parliament, see the Grand Remonstrance (1641), clauses 27–32, in Samuel Rawson Gardiner, ed., *The Constitutional Documents of the Puritan*

'Public agitation' in the 1620s had publicized new ways of thinking about the wealth of the nation and how it might be increased which were scarcely consistent with the notion of a body politic kept in equilibrium by a paternalistic sovereign; and it had placed them firmly on the political agenda. But it had not determined which among the several potential drivers of economic improvement should have the upper hand, Crown or parliament or private investors and the little mercantile commonwealths engaged directly in commercial enterprise. In 1649, when military force and political revolution erected a republican 'Commonwealth of England', a new generation of Baconian reformers may well have thought that the political issue had now been solved, and that the road to national improvement was at last wide open.

Revolution 1625–1660 (3rd edn, Oxford, 1906), p. 212; Thirsk, 'Enclosing and Engrossing', p. 23; Thirsk, 'Crown as Projector', pp. 351–2; Cressy, *Saltpeter*, pp. 122–3.

4
Revolutions 1640–1670

As the restoration of the monarchy in 1660 demonstrated, none of the revolutions after 1640 brought the 'healing and settling' of political divisions that contemporaries hoped for. The expectations they aroused nonetheless had permanent effects, and not the least of them was the creation of a culture of improvement with elements attractive to all parties and the capacity to evolve and thrive after the Restoration. Constant argument and public debate about the purposes and institutions of government left behind an enduring consensus that, however it was interpreted, change for the better, intellectual and material progress, was not only achievable but would bring greater human happiness.

The people chiefly responsible, Samuel Hartlib and the circle of writers he attracted around him, drew on a diverse inheritance of projects and proposals from the past, in Europe as well as England, in an effort to solve the urgent problems of a revolutionary moment. Their publications and their correspondence allow their interests and activities to be fully documented, and testify to their ingenuity, and to that of Hartlib in particular, in devising a rhetoric of improvement which gave them a common purpose.[1] Their ambition, which they passed on to the Royal Society of the Restoration, was the simultaneous pursuit of economic betterment and a Baconian advancement of learning, and William Petty, the most versatile member of the circle, brought something new to both of them by devising a new kind of political economy, one which embraced the whole wealth of the nation and not just its balance of trade.[2] In the 1660s much of the rhetorical equipment of Hartlib's improvement, including its aspiration to national or even universal happiness, remained ill defined and visionary: that was part of its general appeal. Thanks to Petty's political arithmetic, however, material progress—if and when it occurred—could be measured.

---

[1] Hartlib's papers, now in Sheffield University Library, were used in G. H. Turnbull, *Hartlib, Dury and Comenius: Gleanings from Hartlib's Papers* (Liverpool, 1947), but first fully exploited by Charles Webster, *The Great Instauration: Science, Medicine and Reform, 1626–1660* (1975). References which follow (prefaced *HP*) are to *The Hartlib Papers* (2nd edn on CDROM, HROnline, University of Sheffield, 2002).

[2] Petty also left a large archive, now in the British Library, and listed in *Petty Papers: Additional Manuscripts 72850–72908, Additional Charters 76966–76990* (British Library, 2000).

GREAT EXPECTATIONS

For more than a decade after 1640 successive political revolutions raised expectations that change for the public good must surely at last be delivered by political reformation. The dawn of a new age seemed to be heralded by the successes of the Long Parliament in the 'wonderful year' of 1641, by its victory in war in 1646, by the erection of a republic by a rump of it in 1649, and finally, for some, by Cromwell's dissolution of the Rump Parliament in 1653.[3] Fuelled by news-sheets and pamphlets in unprecedented profusion—as many as 4,000 titles and 4 million copies were printed in 1642[4]—the agenda for reformation expanded and grew ever more ambitious. While John Milton expected liberty of conscience and freedom of expression to bring the 'reforming of Reformation itself' in 1641, Hartlib thought England had been 'new begotten again unto the world of nations' and was ready to embark on a 'perfect reformation' which had been inconceivable in 1639, one which would mean an 'improvement of all human affairs in all persons everywhere'.[5] In *The Parliament's Reformation* (1646) and *England's Reformation* (1647), he saw the prospect of 'a reformation of this state' which would advance learning, increase the national wealth, and make everyone, even the poor, 'much bettered and improved above what now they are'. As late as 1653, though probably not for very long afterwards, he was still hoping for 'a more speedy reformation in time to come' than any which had so far occurred.[6]

In the creative chaos that came with civil wars and religious and political revolutions, when even soldiers claimed to be actors in 'this kind of reformation' and instruments 'for the public good',[7] and a freer press debated the fundamental liberties of the people and questioned the existing distribution of property and power, familiar issues about economic and social betterment were reinterpreted. Many advocates of improvement were careful to avoid extremes like the 'levelling spirit now abroad' and the 'Utopian fiction' of social equality,[8] but their proposals were presented in a similar language of heightened expectation. In 1645 Thomas Johnson renewed the attack on the privileges of the Merchant Adventurers on the grounds that trade was a 'native freedom' and it was the 'birthright' of every Englishman to exercise his 'invention and industry' for his own 'enjoyment and

---

[3] Hugh Trevor-Roper, *Religion, the Reformation and Social Change* (1967), pp. 264–5, 275.
[4] Joad Raymond, *Pamphlets and Pamphleteering in Early Modern Britain* (Cambridge, 2003), pp. 90, 163; Michael Braddick, *God's Fury, England's Fire: A New History of the English Civil Wars* (2008), p. 173.
[5] Blair Worden, *The English Civil Wars 1640–1660* (2009), p. 79; *Samuel Hartlib and the Advancement of Learning*, ed. Charles Webster (Cambridge, 1970), pp. 93, 97; *HP*, 30/4/6A; Webster, *Great Instauration*, p. 50 (citing Comenius, writing c.1642).
[6] *Hartlib and the Advancement of Learning*, pp. 93, 111–39; Samuel Hartlib, *A Discoverie for Division or Setting out of Land* (1653), sig. A2^r.
[7] Barbara Donagan, *War in England 1642–1649* (Oxford, 2008), pp. 260, 291. Creative chaos is Braddick's phrase, *God's Fury*, p. 591.
[8] Joseph Lee, *A Vindication of a Regulated Inclosure* (1656), p. 29; John Cooke, *Unum necessarium: or The Poore Mans Case* (1648), p. 36; John Gurney, *Brave Community: The Digger Movement in the English Revolution* (Manchester, 2007), pp. 107–8, 137–8.

improvement'.⁹ All monopoly companies encroached on 'the liberty of the subjects', Cheney Culpeper told Hartlib in 1646. Now that the monopoly of power by the king was being pulled down, the monopolies of the church, the law, and trade must inevitably follow, 'and thus will Babylon tumble, tumble, tumble, tumble'.¹⁰

A speedy reformation and improvement of social welfare was similarly expected by several writers in 1649 and shortly afterwards. After the disappointment of earlier efforts to 'put life' into the poor laws by books of orders and local experiments with workhouses, Rice Bush saw the prospect of 'progress' by means of free schools and employment schemes and all magistrates striving to 'improve their interest for the public good'. Henry Robinson was chiefly concerned with the advancement of trade and navigation, but he added reformed hospitals and schools to his list of 'new inventions and improvements' which would ensure that everyone could at last 'enjoy that liberty which we have so dearly purchased'. He had more to say about the need to remedy a deficient money supply by means of a bank on the Dutch model, but that too could be presented in the same light.¹¹ Addressing a new Council of Trade on the subject of money and banks in 1650, William Potter argued that the time was ripe to copy the Dutch. Surely now, when the 'idol' Charles I had at last been pulled down, providence must 'bless this nation with greater riches and prosperity than ever before'.¹²

Even *Waste Land's Improvement* now offered a solution to all the problems of the commonwealth, according to the author of a tract presented to a parliamentary committee on trade in 1653. If all wastes, fens, and forests were enclosed, tilled, manured, and improved, they could be tenanted by the poor and all who had suffered in the wars, soldiers and civilians alike. There would be no need for the burdensome new taxes of excises and assessments on land, all the state's debts could quickly be paid from rents expected to amount to £100,000 a month, and the 'glory of the nation' would be increased by a standing army and navy funded in perpetuity. In so painless and practicable a Utopia the English, with their reputation as 'an ingenious and industrious people', would convert 'desolate wastes into fruitful fields and our wide howling wildernesses into comfortable habitations', and 'enjoy at last some benefit by all our revolutions, transplantings, and overturnings in authorities'.¹³

There was nothing at all new about arguments for improvements in agriculture, commerce, and navigation, or for freer trade at home and an increase in the money

---

⁹ Thomas Johnson, *A Discourse Consisting of Motives for the Enlargement and Freedome of Trade* (1645), pp. 2–3; Thomas Johnson, *A Plea for Free-Mens Liberties* (1646), sigs A2, A3ᵛ; Thomas Leng, *Benjamin Worsley (1618–1677): Trade, Interest and the Spirit in Revolutionary England* (Woodbridge, 2008), pp. 61–2. Some of Johnson's language (e.g. *Discourse*, p. 50) had much in common with that of the Leveller William Walwyn (e.g. his *Manifestation* (1649), *passim.*).
¹⁰ 'The Letters of Sir Cheney Culpeper, 1641–1657', ed. M. J. Braddick and Mark Greengrass, in *Camden Miscellany XXXIII* (Camden Society 5th ser. 7, 1996), pp. 269–70.
¹¹ Slack, *From Reformation to Improvement* (Oxford, 1999), pp. 77–8; Henry Robinson, *Briefe Considerations, Concerning the advancement of Trade and Navigation* (1649), sig. A2, p. 2; Henry Robinson, *Certain Proposalls in order to the Peoples Freedome* (1652), pp. 18, 23–6.
¹² William Potter, *The Key of Wealth* (1650), p. 83.
¹³ *SECD*, pp. 135–40. The tract is signed 'E.G', but his identity is unknown.

supply, which would increase the national wealth. In the last great enclosure controversy, in south Leicestershire in 1652–4, there was some reference to the radical concept of a 'Norman Yoke', inflicted on commoners ever since 1066, but the winning arguments were what they had been in 1607, that land in private hands was better managed than land held in common, and that enclosed counties were plainly richer than others.[14] So ingrained already was the language of agrarian improvement that even someone as hostile to profit motives as Gerrard Winstanley, leader of the 'true Levellers' or 'Diggers', found it necessary to insist that their purpose was to 'quietly improve the waste and common land' when digging it up at St George's Hill, Surrey, and so ensure that the whole community benefited from 'all sorts of commodities'.[15] The case for greater freedom of trade was justified, as it had been in 1604, by the prosperity of countries where 'the wealth of the land is more equally distributed', and opposed once again, by Henry Parker in 1641, because 'public reason of state' showed that merchants contributed more value to the common wealth than mere husbandmen or even clothiers.[16]

While not changing the content of familiar arguments, however, revolutions and overturnings since 1640 had given some of them fresh and immediate relevance. The old standards of justice and proportion were being applied to judge whether new forms of taxation were equal and equitable when the assessment was shown to fall more heavily on some parts of the country than others, and excises were condemned because they spared the landed interest and hit 'the middle sort of people', 'the principal fountain from whose industrious streams flowed the riches of the commonwealth'.[17] Their novelty sharpened such perceptions, in the same way that political revolutions and a Cromwellian standing army reinvigorated classical republican notions of civic virtue and the dangers of corruption.[18] 'When governments do change, reasons do change', Ralph Maddison remarked in 1655. Circumstances altered cases and gave old arguments new momentum.[19]

Having had experience on committees and councils of trade in 1650 as well as in 1622, Maddison had particularly in mind the changes which came with the shift from monarchy to republic in 1649, and which transformed relations with the

---

[14] Joseph Lee, *Considerations Concerning Common Fields* (1654), pp. 13, 21, 39; Joseph Lee, *Vindication*, pp. 28–9, 35; Silvanus Taylor, *Common-good, or The Improvement of Commons, Forrests, and Chases, by Inclosure* (1652), pp. 5, 12, 23, 33. Cf. above, p. 51. On the radical myth of the Norman Yoke, see Christopher Hill, *Puritanism and Revolution* (1958), pp. 50–122.

[15] Andrew McRae, *God Speed the Plough* (Cambridge, 1996), p. 138. Like English colonizers in Ireland (above, p. 66), they represented themselves as improving, cultivating, and manuring. For other references to improving and manuring, see Gerrard Winstanley, *The Complete Works*, ed. Thomas N. Corns, Ann Hughes, and David Loewenstein (2 vols, Oxford, 2009), i. 31, 41, 46; ii. 243, 261. Walter Blith, the great apostle of agrarian improvement, commented acidly that there were 'thousands of places more capable of improvement than theirs': Gurney, *Brave Community*, p. 138.

[16] Johnson, *Discourse*, p. 23; Henry Parker, *Of a Free Trade* (1648), p. 31.

[17] Thomas Burton, *Diary*, ed. John Towill Rutt (4 vols, 1828), ii. 230–1; Philo-Dicaeus, *The Standard of Equality in Subsidiary Taxes & Payments, or A Just and strong Preserver of Publique Liberty* (1647), sig. B4; Zachary Crofton, *Excise Anatomiz'd and Trade Epitomiz'd* (1659), pp. 2, 5, 9.

[18] Cf. Jonathan Scott, *Commonwealth Principles: Republican Writing of the English Revolution* (Cambridge, 2004), passim.

[19] Ralph Maddison, *Great Britain's Remembrancer Looking In and Out* (1655), sig. A2ᵛ.

Dutch, making that neighbour republic more than ever a model to be emulated and a rival to be feared. He reminded his readers that in the 1620s, under a 'kingly government', they had learnt about the workings of foreign exchange and the balance of trade. Now, in a Commonwealth, new avenues for economic improvement had opened up to them, like the banks and free ports of the United Provinces.[20] They should look there for examples of best practice, as other writers were doing. 'Though Holland seem to get the start of us, yet we may so follow as to stand upon their shoulders, and so see further,' Hugh Peter insisted in 1651. England must encourage the further 'improvement of nature' in agriculture and internal navigation, and 'London in particular' be improved by broader streets, cleaned and paved, houses of brick and stone, plentiful almshouses and hospitals, a fire brigade, and a large Thames-side quay like that in Rotterdam.[21] All these were the fruits of broadly distributed wealth, and could be achieved in England, John Cook explained, now that it was a 'true commonwealth' like Holland, where no one was 'exceeding rich, nor any beggars permitted', although there were 'different degrees among them, lords and others'.[22]

Yet the Dutch were also competitors, and likely to monopolize all the trade of Europe unless, as Maddison commented, they were 'incorporated one nation with us', a prospect they had refused to contemplate.[23] The failure of proposals for a union of the two republics had led instead to the Navigation Act of 1651, directed against the Dutch carrying trade, and to the first Anglo-Dutch War which followed.[24] That conflict preoccupied the parliament and councils of the Rump to the exclusion of almost everything else. Like the economic crisis of the early 1620s it prompted a fresh turn in economic policy, not this time its radical reconstitution, but its redirection towards aggressive commercial imperialism and the expectation that command of the world's commerce would solve all economic problems at home.

Though never successful in healing and settling the nation, and therefore incapable of creating a bank without the credit political settlement might have given it, the Rump could claim to have built upon some of the initiatives of kingly government in the 1620s. Besides the Navigation Act, there was a further reduction in the maximum rate of interest from 8 per cent to 6 per cent in 1651, and there was a new Council of Trade with only fifteen members which seemed more likely to be effective than its predecessor in 1622.[25] Since it had no executive authority of its own, however, the Council's reports and deliberations had little immediate effect on policy, but they reflected trends in political opinion which shaped it. On this

---

[20] Maddison, *Remembrancer*, sig. A2ᵛ; above, p. 83.
[21] Slack, *Reformation to Improvement*, pp. 78, 83; Robinson, *Certain Proposalls*, pp. 8–9, 19.
[22] Cooke, *Unum necessarium*, p. 36.
[23] Maddison, *Remembrancer*, pp. 37, 42.
[24] On the origins of the Dutch War and the policies of the Rump, see Steven C. A. Pincus, *Protestantism and Patriotism: Ideologies and the Making of English Foreign Policy, 1650–1668* (Cambridge, 1996), pp. 40–75.
[25] Leng, *Worsley*, pp. 60–1; J. P. Cooper, 'Social and Economic Policies under the Commonwealth', in G. E. Aylmer, ed., *The Interregnum: The Quest for Settlement, 1646–1660* (1972), pp. 121–42.

occasion, there was little disagreement about what was needed to ensure, as the Council's instructions put it, that the commerce and commodities of England and its plantations were 'multiplied and improved' in order to 'supply the commonwealth ... with whatsoever it necessarily wants'.[26]

The Secretary to the Council, Benjamin Worsley, who had Hartlib's son as his clerk, could have written that instruction himself. In the summer of 1649, when he was drafting papers on how to control and improve Virginia, he showed Hartlib an earlier treatise arguing that imports from the plantations, along with improvements in agriculture and the fisheries, might make England 'in a few years' the richest and happiest country in the world. He had learnt from the Dutch that imports were as important for the economy of any nation as exports, 'whether they be commodities for pleasure or necessity', provided a good proportion of them was re-exported. By 1652 he was warning that too rigid an attachment to the balance of trade prevented England rivalling the other republic as 'a rich and general magazine or store ... for other nations'. England should beat the Dutch at their own game, and prevent them 'engrossing the universal trade, not only of Christendom but indeed of the greater part of the known world'.[27]

When writing in 1649 about the liberties of the people, including their freedom from the 'yoke of foreign dominion', Henry Robinson had drawn the same conclusion. In the contest between Holland and England, whichever gained the greatest trade would 'get and keep the sovereignty of the seas, and consequently the greatest dominion of the world'.[28] The observation was often repeated,[29] and it was given its classic formulation in 1650, in words attributed to Walter Ralegh:

> whosoever commands the sea, commands the trade; whosoever commands the trade of the world commands the riches of the world, and consequently the world itself.[30]

The maxim may not have been penned by Ralegh himself, but his name gave it the resonance of past Elizabethan glories and made it commonplace in later statements about the nation's manifest destiny.

Other old texts were also being polished up as propaganda against the Dutch, like John Boroughs's *Sovereignty of the British Seas*, written in the 1630s and printed for the first time in 1651, with its references to King Edgar, the founder of the

---

[26] *SCED*, p. 502; Leng, *Worsley*, pp. 60–70; Charles Webster, 'Benjamin Worsley: Engineering for Universal Reform from the Invisible College to the Navigation Act', in Mark Greengrass et al., eds, *Samuel Hartlib and Universal Reformation* (Cambridge, 1994), 231–3.

[27] Webster, 'Benjamin Worsley', pp. 227–8; Leng, *Worsley*, pp. 35–6, 63, 76–9.

[28] Robinson, *Briefe Considerations*, pp. 1–2.

[29] For an example, see John Evelyn, *Navigation and Commerce, their Original and Progress* (1674), p. 15.

[30] Sir Walter Raleigh, *Judicious and Select Essayes and Observations* (1650), p. 20, in an essay on 'the invention of ships'. The essay was often republished, but this seems to be its first appearance in print. The Athenian empire was the usual model for such observations: Jonathan Scott, *When the Waves Ruled Britannia: Geography and Political Identities, 1500–1800* (Cambridge, 2011), pp. 45–6; Vickers, *Bacon*, pp. 402–3. In his *History of the World*, Part I, Book III, ch. VI, Ralegh refers to Athenian navigation without coining this famous aphorism. On the differences between versions of texts attributed to Ralegh in 1650 and 1653, see Anna R. Beer, *Sir Walter Ralegh and his Readers in the Seventeenth Century* (1997), pp. 158–9, 164–6; and on another attribution, above, p. 73.

Anglo-Saxon navy, as 'Emperor' of the British Seas. In 1652 the Council of State paid Marchamont Needham £200 to translate John Selden's *Mare clausum* (1635) so as to transfer to the Rump his arguments for the king of 'Great Britain' having lordship of the seas around as 'an inseparable and perpetual' component of his sovereignty.[31] Here was a vision of a world-wide empire of the seas, and one made practicable by the navy which the Rump had created, which the author of *Waste-Land's Improvement* wanted to see perpetuated, and which proved able to defeat the Dutch, seize Jamaica from Spain, and, for a time, take command of the whole Mediterranean.[32]

The army, a much less popular candidate for perpetuation, had meanwhile laid foundations for a British empire which fully incorporated Scotland and Ireland as earlier writers had hoped.[33] In 1652 Silvanus Taylor was able to celebrate Cromwell's conquest of both of them as the work of divine providence, forging a united commonwealth out of the three former kingdoms and opening up the prospect of 'such variety of improvements' that no other nation of the world could rival it.[34] The improvements took time to manifest themselves, especially in Scotland. Once it was occupied and a new regime imposed, English improvements could be attempted as they had been before, often by Scottish lowlanders trying to bring economic development and civility to the highlands.[35] They achieved little before 1660. Although Cromwell's soldiers were thought by Defoe to have first brought agrarian improvement to Inverness, the English agent sent north to manage the excise found little sign of it in 1656 even in the lowlands. He was amazed at 'the barrenness of the country' and the poverty of a people 'generally affected with sloth... rather than any dextrous improvement of their time'. The Scots pursued colonial opportunities abroad, including further migration to Ulster, but the intellectual elite of Edinburgh did not import London's improvement culture until the 1680s.[36]

What made a difference in Ireland was the military imposition of a new landed and cultural aristocracy by plantations far more aggressive than those seen earlier in Munster and Ulster. As Clarendon remarked, Cromwell's Irish settlement was one no 'virtuous prince and more quiet times' could ever have accomplished, confiscating half the island and transferring about a third of it, some seven million acres, from Catholic to Protestant ownership.[37] The need to repay loans advanced by

---

[31] David Armitage, *The Ideological Origins of the British Empire* (Cambridge, 2000), pp. 118–19.
[32] N. A. M. Rodger, *The Command of the Ocean: A Naval History of Britain, 1649–1815* (2004), pp. 1–49. Cf. N. A. M. Rodger, *The Safeguard of the Sea: A Naval History of Britain 660–1649* (1997), pp. 432–4.
[33] Armitage, *Ideological Origins*, pp. 57–9.   [34] Taylor, *Common-good*, pp. 23–4.
[35] Julian Goodare, *The Government of Scotland, 1560–1625* (Oxford, 2004), pp. 64, 106, 170–1, 225–7; Christopher A. Whatley, *The Scots and the Union* (Edinburgh, 2006), p. 115.
[36] Daniel Defoe, *A Tour Through the Whole Island of Great Britain* (2 vols, 1962), ii. 407–8 (and p. 330 for a similar comment about Ayrshire); Frances Dow, *Cromwellian Scotland, 1651–1660* (Edinburgh, 1979), p. 171; Allan I. Macinnes, *The British Revolution 1629–1660* (2005), pp. 233–4; below, pp. 161, 163, 189–90, 233–4.
[37] Derek Hirst, *Dominion: England and its Island Neighbours 1500–1707* (Oxford, 2012), p. 225; Karl S. Bottigheimer, *English Money and Irish Land: The 'Adventurers' in the Cromwellian Settlement of Ireland* (Oxford, 1971), pp. 3, 139–40. Cf. below, pp. 161, 163, 188–9, 232.

London 'adventurers' to suppress the Irish rebellion had already made it necessary to survey and identify 2.5 million acres in 1642.[38] The far larger task of mapping the whole country ten years later was undertaken first by Worsley, who had seen the rebellion as a military surgeon in Dublin and now served as secretary to the parliamentary commission, and then more comprehensively and quickly by Petty, who went out as chief physician. Petty's 'Down Survey', so called because it was laid down in a map, measured the shape and size of Ireland, or indeed any country, for the first time, and distinguished between 'all lands as is profitable and all lands which is unprofitable'.[39] Worsley and Petty both profited from the spoils, Petty on a grander scale, and they were joined by other members of the Hartlib circle who already had land there, like the Boyles, or who rushed out to exploit evident opportunities, like Gerard Boate, author with his brother of *Irelands Natural History* (1652), dedicated to Cromwell and a Baconian exercise not yet matched for England itself.[40]

Military conquest had made Ireland the 'gallant country for improvement' advertised in a London newspaper in 1656, a territory where exponents of Baconian utility could show their mettle. As Petty discovered, imposing English improvement on an Irish estate proved to be easier said than done, needing the landlord's 'daily presence and inspection' to get it in and off the ground. But improvement was a cult, even a creed, of conquerors and colonists, much as it was for planters in the British West Indies who even at this early stage were taking surveys and counting heads and boasting about new techniques of sugar production.[41] For Petty himself Ireland was a 'white paper',[42] an open invitation to further intellectual enquiry at least as much as an opportunity for material profit. The most obvious effect of colonial experiments in improvement may well have been a cultural one, and felt first back home, in the metropolis and the Royal Society of London, where it reinforced the sense that improvement was at last carrying all before it.

From that point of view, the importance of the overturnings of the 1640s and early 1650s lay in the stimulus they gave to personal mobility and the exchange of ideas, across Europe as well as the British Isles. Without that, the Hartlib circle could never have come into being and contributed to English cultural capital in the way that it did. The intellectual inspiration came from three exiles, Hartlib himself who had moved as a student from Polish Prussia to Cambridge in 1625–6 and

---

[38] J. H. Andrews, 'How Many Acres? A Cartometric Exercise of 1642', *Irish Geography*, 34 (2001), pp. 1–10; William J. Smyth, *Map-Making, Landscapes and Memory* (Cork, 2006), pp. 166–7.

[39] Leng, *Worsley*, pp. 18–19, 80–2; William Petty, *The History of the Survey of Ireland, commonly called the Down Survey*, ed. T. A. Larcom (Dublin, 1851), p. 13. A full account of the survey is in Smyth, *Map-Making*, pp. 172–96.

[40] Toby Barnard, *Improving Ireland? Projectors, Prophets and Profiteers, 1641–1786* (Dublin, 2008), pp. 19–33; Toby Barnard, 'The Hartlib Circle and the Cult and Culture of Improvement in Ireland', in Greengrass, ed., *Samuel Hartlib and Universal Reformation*, pp. 283–6.

[41] *The Public Intelligencer*, 24–31 Mar. 1656; Barnard, *Improving Ireland?*, pp. 13, 30–5, 68. On the West Indies, see e.g. Richard Ligon, *A True and Exact History of the Island of Barbados* (1657), pp. 86–91; Robert V. Wells, *The Population of the British Colonies in America before 1776: A Survey of Census Data* (Princeton, 1975), pp. 7–9.

[42] Petty, *EW*, i. 9.

returned permanently to England in 1628, John Dury who had been minister to the Presbyterian congregation in Hartlib's home town of Elbing until 1630, and Jan Comenius, the greatest scholar among them, who left Bohemia for Poland in 1620 and was only temporarily in England in 1641–2.[43] All three had seen the destructive effects of war in Germany and shared hopes that states and principalities would unite in a programme of political reconstruction and public education which would restore piety and social harmony. They had read the encyclopedic scholarly works published in 1627 by Joseph Mede, Hartlib's Cambridge friend, and Johann Heinrich Alsted, Comenius' teacher, and they had absorbed their apocalyptic expectations of an imminent fulfilment of the millennium, when the Christian perfection of the world and universal peace would at last be achieved. Inspired equally by Bacon's prospectus for a 'Great Instauration' of learning, they were able to incorporate all this into a 'pansophic' vision of universal enlightenment and reformation. In 1642 Hartlib summarized its essentials in his English translation of Comenius' *Pansophiae Prodromus* (1639), and in doing so referred for the first time in print to improvement. Properly 'used and improved', the advancement of learning would bring 'universal knowledge of all things' and 'serve to the improvement of our age'.[44]

That all-embracing ambition, taken partly from English authors and offering immediate possibilities for reformation, had particular appeal after 1642 when civil war brought the prospect of 'England turned Germany'.[45] It was far from being a coherent programme for action, but Hartlib had the capacity through correspondence and personal persuasion to assemble a loose coalition behind the banner of improvement for the public and universal good. Through his Cambridge contacts he had acquired friends like John Pym and patrons like the Earl of Warwick who were to be powerful parliamentary figures in the 1640s; he added to their number men influential in the civic politics of London, like Thomas Andrewes, leader of the colonial merchants who seized power in the City in 1649 and provided useful contacts there;[46] and he attracted into his circle people closer to him in their aspirations who made their own contribution to what improvement, in practice or in prospect, came to mean.

An early member, and the first to have an influence outside the circle through his publications, was Gabriel Plattes. A man of obscure origins, who died in penury supported by Hartlib in 1644, he might have been the son of a small farmer, as he once claimed, or of an immigrant foreign artisan. He had certainly accumulated a vast knowledge of experiments in husbandry, mining, and metallurgy by 1639

---

[43] *Hartlib and the Advancement of Learning*, ed. Webster, 'Introduction', pp. 1–72; Greengrass et al., eds, *Samuel Hartlib and Universal Reformation*, 'Introduction', pp. 13–18.

[44] *Hartlib and the Advancement of Learning*, ed. Webster, p. 33; Paul Warde, 'The Idea of Improvement, c.1520–1700', in Richard W. Hoyle, ed., *Custom, Improvement and the Landscape in Early Modern Britain* (Farnham, 2011), p. 138.

[45] Ian Roy, 'England Turned Germany? The Aftermath of the Civil War in its European Context', *TRHS* 5th ser. 28 (1978), pp. 127–44; Donagan, *War in England*, pp. 28–32, 215–16.

[46] *Hartlib and the Advancement of Learning*, ed. Webster, pp. 25–6; Webster, 'Benjamin Worsley', pp. 229–30; Slack, *Reformation to Improvement*, pp. 85–6; Robert Brenner, *Merchants and Revolution* (Princeton, 1993), pp. 545–6, 587–9.

when he published *A Discovery of Infinite Treasure, hidden since the worlds beginning*. Its purpose was to reveal the potential of 'new inventions and improvements' to make 'this country the paradise of the world'. If encouraged, they would deliver 'the future happiness and flourishing estate of the kingdom'; if neglected, people should consider 'in what case their posterity will be' in two or three centuries without them.[47] Two years later, in *Macaria*, he offered members of the Long Parliament a model Utopia he hoped they would follow in order to 'lay the corner stone of the world's happiness'. Macaria had a 'Great Council' collecting knowledge of the 'inventions of all the world', like the Fellows of Solomon's House in Bacon's *New Atlantis* (1627), and the parliament of England was urged to copy them. There should be a 'Society of Experimenters', a college of experts in medicine, and 'under-councils' on husbandry, fishing, trade, and new plantations, whose pursuit of improvements would bring greater 'plenty and prosperity' and—contrary to the opinion of some divines—deliver it 'before the day of Judgement'.[48]

Other members of the group, from different backgrounds, played variations on the same themes. Cheney Culpeper, a Kentish squire of declining fortunes (despite his faith in agrarian innovation), read and admired Plattes, but thought his proposals might have been further 'improved to the public good' if they had given the coming millennium greater prominence. Culpeper had concluded from his wider reading, part of it in The Hague where he had been a diplomatic agent, that the prophet Elijah would soon return and usher in Bacon's Great Instauration and universal enlightenment.[49] Henry Robinson, a successful merchant with experience in the Low Countries and Italy who had been drawn into the circle by his publications on liberty of conscience, would not have gone so far. He had written about trade in the conventional terms of reason of state as secretary to the East India Company in 1641, and he had anticipated Worsley when he envisaged England becoming an 'emporium' where other nations might find 'foreign commodities of all sorts'. By 1650, however, even he was presenting proposals to the Rump for inventions and improvements which would herald nothing less than 'the perfection of human society'.[50]

Robinson was satirized by a Leveller, William Walwyn, as 'one of our inventing innovating travellers',[51] but that description applied more accurately to younger members of the circle still with careers to make, especially the two great rivals Worsley and Petty. Having worked in Dublin and Amsterdam before 1650 and acquired a library of alchemical and heterodox literature to rival Culpeper's, Worsley had developed his own unorthodox piety. It included the expectation of

---

[47] *ODNB*, sub Plattes; Charles Webster, *Utopian Planning and the Puritan Revolution: Gabriel Plattes, Samuel Hartlib and MACARIA* (Oxford, 1979), pp. 14–16; Gabriel Plattes, *A Discovery of Infinite Treasure* (1639), sigs A3–A4, pp. 20, 86–7, 89.

[48] [Gabriel Plattes], *A Description of the Famous Kingdome of Macaria* (1641), sig. A2, pp. 3–6, 11, 13; Vickers, *Bacon*, pp. 471, 484, 487.

[49] 'Letters of Culpeper', pp. 123, 129, 131–48, 203–4, 208.

[50] Henry Robinson, *Englands Safetie in Trades Increase* (1641), p. 20; Robinson, *Certain Proposalls*, p. 20; Henry Robinson, *The Office of Adresses and Encounters* (1650), p. 2; *ODNB*, sub Robinson. On his connections with Culpeper, see 'Letters of Culpeper', pp. 267, 317, 359.

[51] *Hartlib and the Advancement of Learning*, ed. Webster, p. 202.

Christ's Second Coming 'very shortly', and he managed, with some difficulty, to combine that with his life-long ambition to magnify 'the empire of England' in the 'interest of state', and at the expense of the Dutch. Culpeper spotted the tension between the two in 1646 and advised him without success to play down his economic nationalism and concentrate on the benefits of a millennium for all mankind.[52]

Precisely the same inconsistency is evident in Petty's thinking before 1660, but Petty was more circumspect and deliberate in his publications, reserving his more apocalyptic speculations for private papers and correspondence. He also had a more innovative intellect which distanced him further from Hartlib than 'very dear Worsley'. It had been shaped by an education in mathematics and medicine in a Jesuit college in Caen and then in the Netherlands, and by conversations with Hobbes and Gassendi in Paris, before Culpeper introduced him to the circle in 1647. Nonetheless, Petty's first significant publication was an *Advice* to Hartlib in 1648 about an advancement of learning whose twin and perhaps incompatible goals were to make the English Commonwealth as rich as that of the Dutch and at the same time to bring it 'nearer to perfection' for 'the general good and comfort of all mankind'.[53]

Trying similarly to hold together the disparate aspirations of his colleagues and gain support from hard-headed politicians who alone could deliver the reforming state crucial to their purposes, Hartlib was always conscious of the need to trim his rhetorical sails to the winds of established opinion. Even in the heady days of great expectation in 1649 he was cautious about what he said in print. He took careful note of how Peter Chamberlen avoided any reference to 'the City of Peace' in a tract on poor relief 'lest people should be too much frightened', while noting with some satisfaction that Chamberlen's proposals might indeed lead to 'the whole world' at last being able to 'live in plenty and peace &c, and all wars cease'.[54] Millenarian expectations cooled after 1653, and lost any political traction they might once have had. In 1659, perhaps as a desperate last throw, Chamberlen ventured to publish his vision of 'collegiate habitations' across the nation, combining schools, workhouses, and hospitals, and bringing universal happiness; but by then an enthusiast like John Beale was retreating to the ideal of a Christian Utopia in small select communities, because so little had come from talking 'big of reforming laws and making whole nations churches, and of erecting the kingdom of Christ all over the world'.[55]

A decade earlier, however, the prospect of a state-directed Baconian reformation in England leading to a more enlightened world had seemed real. Then Petty had

---

[52] Leng, *Worsley*, pp. 25, 37, 47, 75–7, 97, 152; 'Letters of Culpeper', p. 244.

[53] Ted McCormick, *William Petty and the Ambitions of Political Arithmetic* (Oxford, 2009), pp. 42, 52; William Petty, *The Advice of W.P. to Mr Samuel Hartlib* (1648), in *Harleian Miscellany* (10 vols, 1808–13), vi. 142, 156. On Petty's early career, see McCormick, *Petty*, pp. 14–39, and on his *Advice*, McCormick, *Petty*, pp. 66–74.

[54] *HP*, 28/1/17A, 19B; Slack, *Reformation to Improvement*, p. 83.

[55] Slack, *Reformation to Improvement*, p. 78; J. C. Davis, *Utopia and the Ideal Society: A Study of English Utopian Writing, 1516–1700* (Cambridge, 1981), pp. 316, 330.

translated for Hartlib a paragraph from Bacon's *Novum organum*, published in Latin in 1620, about 'noble inventions' and their universal and permanent utility. It may have first caught Petty's attention because it came in a section in which Bacon had cited a comment from Aristotle to the effect that nature could only be manipulated for useful purposes by understanding it and complying with its constraints; and that became Petty's scientific motto: *res nolunt male administrari*, 'things will have their course' and 'nature would not be cozened'.[56] The passage he translated for Hartlib would also have given both of them cause for reflection, because it contrasted political improvements, which were transient and benefited only a few, with discoveries like printing, gunpowder, and the compass, which were 'new creations', 'imitacious of God's own works', whose benefits 'extend to all mankind universally' and last forever. Bacon had ended with a warning of particular resonance in a decade of political revolutions:

Moreover, the reformation of states in civil affairs for the most part is not compassed without violence and disturbances. But inventions make all men happy without either injury or damage to any one single person.[57]

For all the sense of urgency which was invested in the possibility of political reformation in England in the 1640s and early 1650s, Hartlib and his friends knew it could only be the beginning. The ultimate goal was wisdom and its pursuit through learning and invention, an ongoing process only ended when the world ended. Until then the prospect, universal in its scope and appeal, was the one which the French encyclopedists in the eighteenth century drew from Bacon's *Organum*, a prospect of mankind's betterment and its increasing happiness.[58] Hartlib and Petty learnt from the same source that improvement was infinite and found themselves moving from a politics of speedy reformation towards one of gradual enlightenment.[59]

INFINITE IMPROVEMENT

The practical political project on which all members of Hartlib's circle were able to agree, since it encapsulated most of their preoccupations, was the 'Office of Address'. First advertised in print 1647, it was presented to parliament as a proposal for 'an office of public address in spiritual and temporal matters' which would enhance the 'glory of God and the happiness of this nation'. An early draft had included a reference to the promise of a glorious and happy future held out by the 'miracle' of England's deliverance from four years of civil war, but that had been

---

[56] Francis Bacon, *The Instauratio magna Part II*, ed. Graham Rees (*OFB* xi, Oxford, 2004), p. 195 ('one cannot govern nature save by complying with her'); Petty, *EW*, i. 9n. Aristotle's original is *Metaphysics*, Book XII.10. On other possible sources and their significance, see Julie Robin Solomon, *Objectivity in the Making* (Baltimore, 1998), pp. 136–7.
[57] McCormick, *Petty*, p. 45; Bacon, *Instauratio magna Part II*, p. 193.
[58] Bacon, *Instauratio magna Part II*, p. xxvii and n. 20; below, pp. 252–4.
[59] The distinction is drawn by Braddick, *God's Fury*, pp. 590–1.

removed, perhaps out of deference to an audience tired of hearing claims about special providences.[60] As finally published, the proposal was ambitious enough.

It retained something from the plans of Dury and Comenius for colleges to promote a union of churches and universal learning, and from the 'little academy' which Hartlib had run for a few months in Chichester twenty years earlier, designed to 'advance piety, learning, morality and other exercises of industry, not usual then in common schools'. The immediate and apparently successful model, however, was the *Bureau d'adresse* founded by Théophraste Renaudot in Paris in 1629–30, which combined an advice bureau, medical centre, employment exchange, and pawnshop for all who cared to attend, with a public discussion group for learned and publicly minded citizens.[61] Since London was already 'a brave centre for all kinds of correspondency', as Culpeper observed to Hartlib in 1646, it must surely be able to follow suit and use all channels of communication to promote every aspect of the public good.[62]

The English Office of Address was to have two parts, one for spiritual, the other for temporal affairs, and in effect distributed the activities of Renaudot's *Bureau* between them. The first was an office for 'Communications', furthering religious understanding and the advancement of learning, but also encouraging 'ingenuities' and 'the most profitable inventions'.[63] As elaborated by Petty in his *Advice* to Hartlib in 1648, it owed much more to Bacon and Plattes than to Dury and became an immense enterprise to 'increase and improve what is good'. There should be academies, 'literary workhouses' like Hartlib's in Chichester, but dedicated particularly to mathematics and geometry since these were 'the best grounded part of speculative knowledge'. There must be a college for tradesmen, complete with botanical and anatomical theatres, and a teaching hospital, in order to ensure that 'the wits and endeavours of the world' were directed to developing 'such arts as are yet undiscovered'. Petty wanted new institutions giving anyone who might benefit the education he had enjoyed, and opening up opportunities for talent. No longer should there be people 'now holding the plough which might have been made fit to steer the state'.[64]

The second office was one for 'Accommodations', for the exchange of information about business and employment opportunities. Closer to Renaudot's model, it was more easily realized. Hartlib and Dury handed over its management to Henry Robinson, and he set it up as an 'Office of Addresses and Encounters' in Threadneedle Street, deliberately and appropriately 'close to the old Exchange' in the City. Although little is known about its operations, it seems to have been both a shop for second-hand goods and an employment exchange, charging a small entry fee to those who could afford it. But it was also intended as a place where investors could

---

[60] *Hartlib and the Advancement of Learning*, ed. Webster, pp. 119, 205; Webster, *Great Instauration*, pp. 67–77.
[61] *Hartlib and the Advancement of Learning*, ed. Webster, pp. 7, 44–6. On Renaudot, see Howard M. Solomon, *Public Welfare, Science, and Propaganda in Seventeenth Century France: The Innovations of Théophraste Renaudot* (Princeton, 1972).
[62] 'Letters of Culpeper', p. 266.   [63] Webster, *Great Instauration*, p. 69.
[64] Petty, *Advice*, pp. 142–6, 153–4.

learn about business ventures, and even as a marriage market where people could be informed about the 'persons and portions' of potential partners. Robinson was able to advertise it in language as visionary as Petty's. 'In these later times' divine providence had raised mankind from its first 'irrational and brutish' state, and now there was an opportunity for everyone to enjoy 'peace, plenty and contentation [contentment] to perpetuity'. It was necessary to begin in a small way by facilitating contacts and contracts, since part of his design depended ultimately on political support and investment, and until that was assured its details must remain for the moment a state secret, an *arcanum imperii*. In the end, however, his office would meet all conceivable requirements and 'extend as far as human necessity, which is little less than infinite'.[65]

Although Petty claimed not to have 'leisure to frame Utopias', and to have confined himself to what was eminently practicable in the real world,[66] neither office could achieve what it promised without ample funding. The small fees Robinson charged were trivial, but anything larger would have left him open to the accusation of profiteering; and the whole notion of private profit gained from enterprises for the public good had been permanently tainted by the activities of projectors over the past half-century. As early as 1634 Dury warned Hartlib that each of them might be classed as a 'subtle projector' out only for his own good, and it was a charge members of their circle could never evade. There was an obvious contradiction too between their desire to publicize new inventions and improvements for the general benefit, and the determination of those who promoted them to protect their secrets and potential profits from would-be competitors. Plattes concealed the details of his new seed-setting engine and Worsley those of his new method for making saltpetre. Potter refused to be specific about his proposed bank, insisting that he did not seek a patent but was only worried that his plan would be stolen by other countries. Although Culpeper regretted that some of his colleagues were not 'of a communicative disposition', both he and Hartlib recognized that there must be some 'sensible inducements towards all enterprises'. In effect the enterprises of all of them were vulnerable to the objection which Henry More made against Petty's 'great projects', that they were lacking in 'sincerity and untainted morality'.[67]

The only answer to their dilemma, as Hartlib appreciated, was substantial public funding. The support they got was never enough. Hartlib looked for the kind of backing he supposed Cardinal Richelieu had given to Renaudot in Paris, so that new knowledge and inventions could be 'publicly made use of as the State should think most expedient'. Thanks to the influence of his friends in the Rump and its committees, he was given an annual pension of £100 in 1649 in recognition of his

---

[65] Robinson, *Office of Adresses and Encounters*, pp. 1–6; *Hartlib and the Advancement of Learning*, ed. Webster, p. 205.
[66] Petty, *Advice*, p. 148.
[67] Koji Yamamoto, 'Reformation and the Distrust of the Projector in the Hartlib Circle', *Historical Journal*, 55 (2012), pp. 381, 385–6, 389–91; Greengrass et al., eds, *Samuel Hartlib and Universal Reformation*, 'Introduction', p. 20. For Potter, see *Key of Wealth*, sigs A2ᵛ, B1ʳ. Renaudot was vulnerable to similar criticism: Solomon, *Public Welfare*, pp. 57–8.

role as 'Agent for the Advancement of Universal Learning'. On the strength of it, he was sometimes called 'State Agent for Universal Learning', but it was an empty title and the pension was always in arrears.[68] When the Rump put Crown lands up for sale, it was persuaded to hold back the premises at Vauxhall which had been Charles I's ordinance factory and to keep them for 'practical experiments for the use of the Commonwealth' like those which the Marquis of Worcester had sponsored there before the civil war. They may have started again in a small way by 1660, but Vauxhall did not develop into anything like the 'College for Inventions and Advancement of all Mechanical Arts and Inventions' Hartlib hoped for.[69]

The grand designs for the Office of Address in which he placed such faith had achieved next to nothing, apart from Robinson's temporary operation in Threadneedle Street, and the private meetings of experimental philosophers in Oxford and London which had to wait until 1662 for the public encouragement the Royal Society gave them. It had seemed possible that some of Hartlib's educational aspirations might be realized by proposals for a new university college in Durham, where, with Cromwell's backing, the premises formerly occupied by the dean and chapter were allocated to what might have been Petty's 'literary workhouse'. It took six years of planning and petitioning to get it off the ground in 1657 and it had been dissolved by 1660.[70] A scheme for free medical services for County Durham got nowhere at all in the face of resistance from local magistrates.[71] Hartlib's only substantial political success came in London, on the more familiar territory of social welfare, where he and his allies were able to bring earlier projects to fruition and found a Corporation for the Poor in 1647 which centralized welfare provision in the parishes of city.

Hartlib hoped that it might do much more. He wanted reformed workhouses, no longer called houses of correction but 'nurseries' or 'magazins of charity'. They would 'civilize' the children starving and begging in the streets who were illustrated in one of his publications, 'pictures of pity to every pious heart'. There were twice-weekly meetings for the purpose, advertised by the mayor as open to anyone 'well-affected to so pious and charitable a work'. If London became a 'leading example', parts of the vision of the Office of Address might be achieved, and then the whole country might in the end become 'a city of God'. In 1657 Hartlib was informed that another mayor hoped to see a 'charity school' in every London ward so that 'Christ's kingdom in little ones may more and more be advanced', but there seem to have been none of them before 1660. There was a new surveyor-general responsible for cleaning and paving the streets and supported by local rates, but that service was not taken over, as had once been planned, by the Corporation for the Poor.[72]

---

[68] Slack, *Reformation to Improvement*, p. 84; Webster, *Great Instauration*, pp. 69–72.
[69] Webster, *Great Instauration*, pp. 347–8, 363–7; below, p. 237.
[70] Webster, *Great Instauration*, pp. 233–42. Cf. the better-supported plans for Trinity College Dublin: Webster, *Great Instauration*, pp. 230–2.
[71] David Harley, 'Pious Physic for the Poor: The Lost Durham County Medical Scheme of 1655', *Medical History*, 37 (1993), pp. 148–66.
[72] Slack, *Reformation to Improvement*, pp. 79, 82, 85–7, 89; Mark Jenner, '"Another *Epocha*"? Hartlib, John Lanyon and the Improvement of London in the 1650s', in Greengrass et al., eds, *Samuel Hartlib and Universal Reformation*, pp. 343–56.

The new Corporation never became the model for a welfare state with free schools and medical services which the Hartlib circle envisaged. For a time its new workhouse was able to expand on traditional mechanisms of social welfare, setting hundreds of poor on work, some of them in their own homes, and educating as many as eighty of their children who were sent there to be taught.[73] But the benefits were confined to the inner city. The Rump Parliament failed to pass any of the bills which came before it for the reform of poor relief across the whole nation, and the Corporation collapsed with the return to the king of the properties it occupied in 1660.[74] A republican regime with more urgent priorities like the contest with the Dutch, and chronically short of the cash to pay for them, proved no more likely to fund expensive projects for the public good than the monarchy which had preceded it.

The lasting impact of the Hartlib group therefore lay less in concrete political achievements than in their influence on public attitudes, and sometimes on private practices. It was greatest in agriculture, the arena where the private profits made from inventions and improvements had for a century been defended because they promoted the general good. Agrarian improvement now became wholly respectable. One of Hartlib's colleagues noted that 'poor people' still opposed enclosure of the commons, but few others did so, and none of the writers from Plattes onwards who advocated new methods of cultivation, new crops, improved ploughs, and the drainage of the fens, in which Hartlib invested himself. The most influential of them, Walter Blith, visited the Lincolnshire fens to see the consequences of drainage for himself, and changed his tune from scepticism to approval between the two editions of his *English Improver* in 1649 and 1652.[75] A self-styled 'lover of ingenuity', Blith challenged his readers to fulfil the Creator's design and bring barren land to 'so vast an improvement as all the world admires and subsists from'. Himself a former Cromwellian captain who had bought Crown lands, he encouraged his fellow soldiers who had done the same to engage in 'great improvement by wise management', and some of them claimed in 1660 to have made substantial investments doing exactly that (Figure 3).[76] Like similar veterans in Inverness, they were no doubt exceptional, and whether they or any other of Blith's readers succeeded in introducing new practices depended as much on the quality of their land as on the books they read. There were nevertheless enough improving farmers in the 1650s for a

---

[73] Valerie Pearl, 'Puritans and Poor Relief: The London Workhouse 1649–1660', in Donald Pennington and Keith Thomas, eds, *Puritans and Revolutionaries: Essays in Seventeenth-Century History Presented to Christopher Hill* (Oxford, 1978), pp. 206–32.

[74] Cooper, 'Social and Economic Policies under the Commonwealth', p. 129. The Corporation was revived in the 1690s: below, p. 177.

[75] Samuel Hartlib, *Samuel Hartlib his Legacie* (2nd edn, 1652), p. 41; Frances Willmoth, *Sir Jonas Moore: Practical Mathematics and Restoration Science* (Woodbridge, 1993), pp. 96–7.

[76] Walter Blith, *The English Improver Improved* (1652), title page; Michael Leslie, 'The Spiritual Husbandry of John Beale', in Michael Leslie and Timothy Raylor, eds, *Culture and Cultivation in Early Modern England: Writing and the Land* (Leicester, 1992), p. 156; Ian Gentles, 'The Management of the Crown Lands, 1649–60', *Agricultural History Review*, 19 (1971), pp. 37–41.

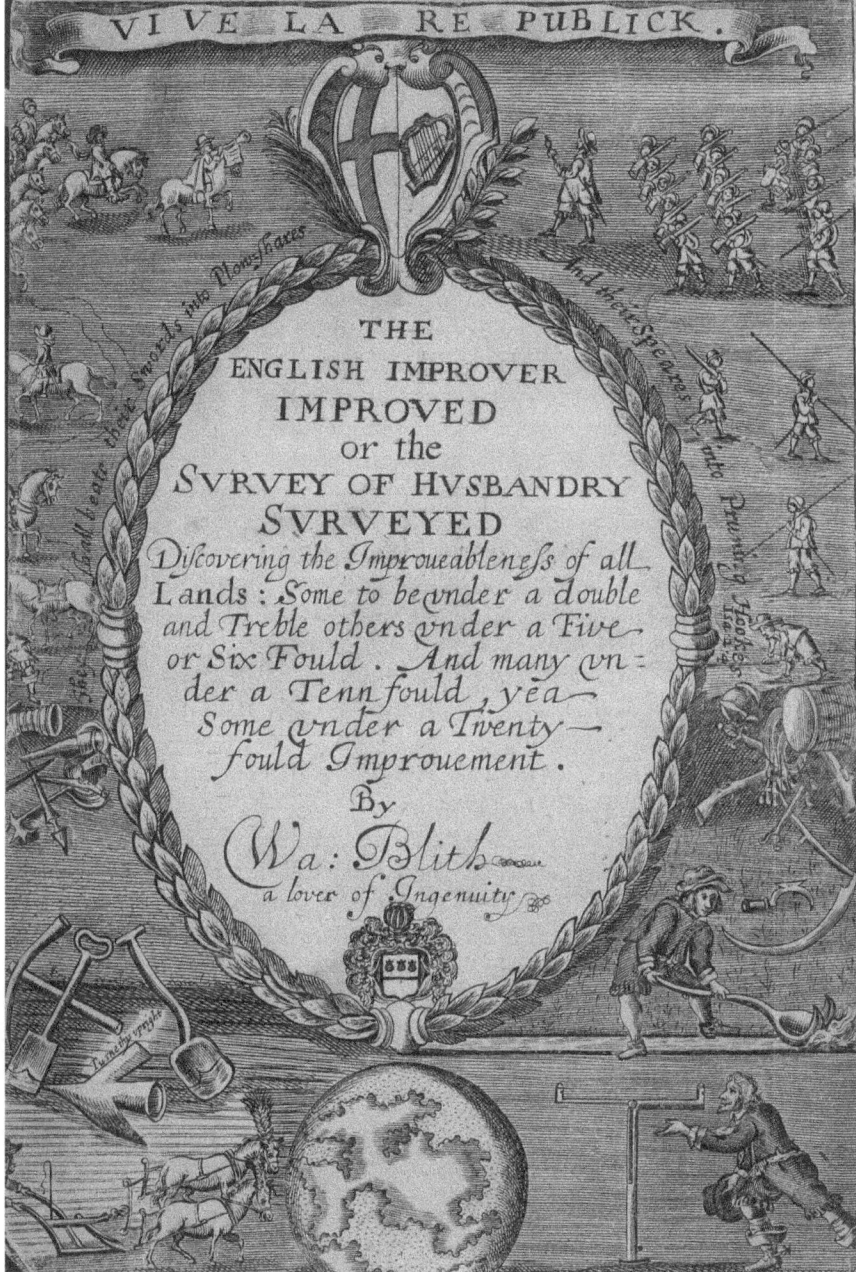

Figure 3. Swords turned into ploughshares, and improvements in husbandry multiplied: Walter Blith, *The English Improver Improved* (1652).

nephew of John Beale on the Welsh borders to style them the 'order of Sir Francis' in recognition of their Baconian inspiration.[77]

Changing cultural attitudes as well as practices was bound to be a slow process, as Hartlib acknowledged. 'The major part of the people', he observed, were 'wonderfully wedded to old customs' and not yet 'easily won to any new course, though never so much to their own profit'.[78] But later commentators had no doubt that Hartlib and his circle had changed all that. According to John Houghton in 1682, people in the 1640s had been 'so averse' to 'all care of improvements' that they treated anyone proposing them as if he were a projector coming 'from the fens to borrow five shillings to purchase five thousand pounds yearly'. Since the civil wars, however, there had been 'great improvement made of lands', especially by gentlemen farmers, 'such an improvement as England never knew before'.[79] John Aubrey made the same point in the 1670s and extended it to improvements of all kinds. Before 1649 it had been 'held a strange presumption' for anyone 'to attempt improvement of any knowledge whatsoever, even of husbandry itself; they thought it not fit to be wiser than their fathers, and not good manners to be wiser than their neighbours'. Such prejudices, he implied, were now things of the past.[80]

As determined improvers themselves, neither Aubrey nor Houghton was an unprejudiced observer, but there was substance to their historical impressions. Thanks to Hartlib and the writers around him, by 1660 there were books about the improvement of everything, not just agriculture but manufactures and trade, social welfare and public and private health, the economy and the international standing of the nation.[81] However empty his title, after 1649 the republic's new Agent for the Advancement of Universal Learning had accelerated a process of cultural change which ensured that 'what is wanting in the present age' would not, as he had once feared, be wanting in the future.[82]

It was implicit in the thinking of Hartlib's circle, moreover, that improvement was not something to be achieved once and for all; it came, as he said of Renaudot's enterprises, 'piece-meal'.[83] It had a coherent purpose, and its impact was cumulative, but it was endless. It was, as its advocates often declared, a force of infinite potential. Central to this mentality was the notion that the earth itself was a treasure-house of unlimited wealth, providentially designed to be unlocked by

---

[77] Leslie and Raylor, eds, *Culture and Cultivation*, 'Introduction', pp. 4–5; Joan Thirsk, 'Agricultural Innovations and their Diffusion', in Joan Thirsk, ed., *Agrarian History of England and Wales*, v: *1640–1750* (2 vols, Cambridge, 1984–5), ii. 546–8. For a full account of agrarian proposals and their effects, see Thirsk, 'Agricultural Innovations', pp. 552–81; Webster, *Great Instauration*, pp. 469–83.

[78] Hartlib, *Legacie*, sig. B2^v.

[79] John Houghton, *A Collection for the Improvement of Husbandry and Trade*, ed. Richard Bradley (4 vols, 1727–8), iv. 80; Thirsk, 'Agricultural Innovations', p. 561.

[80] Michael Hunter, *John Aubrey and the Realm of Learning* (1975), p. 42. They were not: see below, p. 235.

[81] See, for example, Thomas Moffett, *Healths Improvement* (1655), a new title, to suit the times, added to a tract largely written c.1595.

[82] Hartlib, *Legacie*, sig. B2^v.

[83] *Hartlib and the Advancement of Learning*, ed. Webster, pp. 121, 205 n. 33.

mankind. In 1655 one of the contributors to Hartlib's *Legacie* explained to his readers that 'the earth, by reason of the divine benefaction, hath an infinite multiplicative virtue, as fire and the seeds of all things have'.[84] The idea was popular in contemporary works on the land and its wealth in minerals, including an essay by Malynes on metals which Hartlib reprinted.[85] It could be found in the alchemical works which had appealed to Plattes and to John Winthrop Junior, who carried improvement along with alchemy back to New England in 1643; and it inspired Culpeper's notion of a 'vegetative principle' at work in an animist chemical universe. It was an elaboration of the organic model of growth which Edward Forset, and Botero before him, had hinted at when talking about the wealth of a state.[86]

The economic implications of this concept were brought out most forcefully by writers in Hartlib's circle who looked to banks as solutions to the country's continuing liquidity problems. Robinson thought a bank on the Dutch model would be 'capable of multiplying the stock of the nation, for as much as concerns trading, *in Infinitum*: in brief, it is the Elixir or Philosopher's Stone'. William Potter's proposed bank, which would issue paper money on the security of land rather than specie (like the bank in Holland), had an even greater 'capacity of enriching this nation' (Figure 4, overleaf). It was a key of wealth unlocking untold riches. It would increase the money supply tenfold or even more, and stimulate the 'multiplication, improvement, [and] distribution' of commodities of every kind. The nation's stock would increase year by year, 'nothing interrupting . . like fire that wants no fuel . . . *ad infinitum*'. Within forty years the people of England could be worth 'five hundred thousand times more than now', provided there continued to be enough commodities in the world for sale and exchange.[87]

Potter's proviso reflected the conventional contemporary view that the world's natural resources and therefore its treasures and trade were necessarily finite. They might increase for a time but in the end the land would be unable to sustain further growth in either population or productivity. As we shall see later, that was a prospect which continued to preoccupy political economists into the eighteenth century, and it might seem a formidable obstacle to the case for infinite material progress.[88] For the moment, however, two other contemporary assumptions moved that prospective end-point into the far distance. The first was the conviction

---

[84] Simon Schaffer, 'The Earth's Fertility as a Social Fact in Early Modern Britain', in Mikuláš Teich et al., eds, *Nature and Society in Historical Context* (Cambridge, 1997), p. 127.

[85] Alix Cooper, '"The Possibilities of the Land": The Inventory of "Natural Riches" in Early Modern German Territories', *History of Political Economy*, 35 (2003), Supplement 1, pp. 133–40; Carl Wennerlind, 'Credit-Money as the Philosopher's Stone', *History of Political Economy*, 35 (2003), Supplement 1, pp. 235–47. Cf. Wennerlind, *Casualties of Credit*, pp. 44–79.

[86] Walter W. Woodward, *Prospero's America: John Winthrop Jr, Alchemy and the Creation of New England Culture, 1606–1676* (Chapel Hill, NC, 2010), pp. 73–5; 'Letters of Culpeper', pp. 135–6, 144; above, pp. 45, 77. Cf. Plattes, *Discovery of Infinite Treasure*, sig. C3.

[87] Wennerlind, *Casualties of Credit*, pp. 33–4, 68–70, 77; William Potter, *Key of Wealth*, pp. 2–3, 5–6; William Potter, *The Trades-Man's Jewel* (1650), p. 16. Potter may have been influenced by Culpeper: 'Letters of Culpeper', pp. 132–3. Cf. Webster, *Great Instauration*, pp. 450–1.

[88] Below, pp. 197–9.

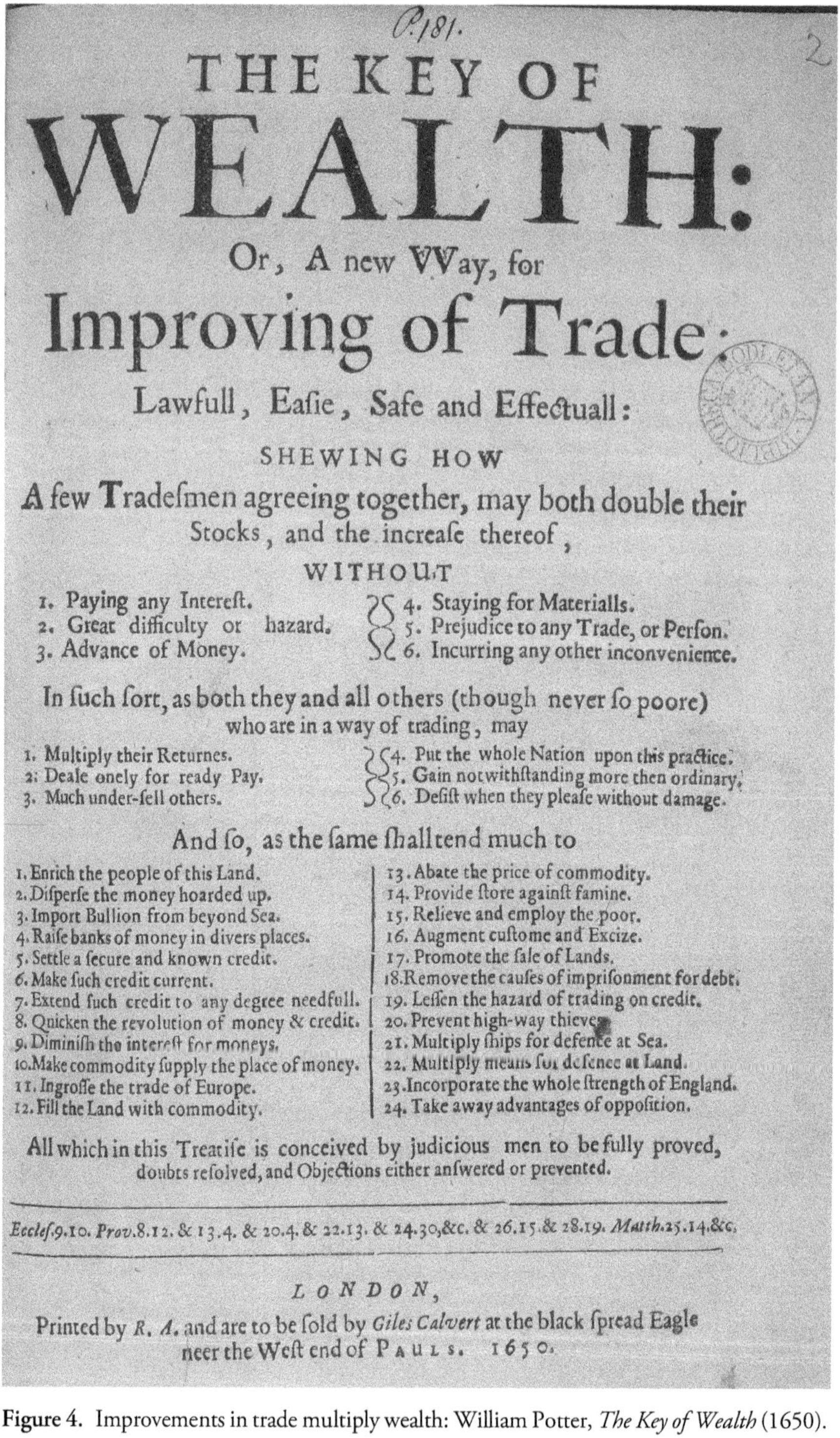

THE KEY OF WEALTH:

Or, A new Way, for

Improving of Trade:

Lawfull, Easie, Safe and Effectuall:

SHEWING HOW

A few Tradesmen agreeing together, may both double their Stocks, and the increase thereof,

WITHOUT

1. Paying any Interest.
2. Great difficulty or hazard.
3. Advance of Money.
4. Staying for Materialls.
5. Prejudice to any Trade, or Person.
6. Incurring any other inconvenience.

In such sort, as both they and all others (though never so poore) who are in a way of trading, may

1. Multiply their Returnes.
2. Deale onely for ready Pay.
3. Much under-sell others.
4. Put the whole Nation upon this practice.
5. Gain notwithstanding more then ordinary,
6. Desist when they please without damage.

And so, as the same shall tend much to

1. Enrich the people of this Land.
2. Disperse the money hoarded up.
3. Import Bullion from beyond Sea.
4. Raise banks of money in divers places.
5. Settle a secure and known credit.
6. Make such credit current.
7. Extend such credit to any degree needfull.
8. Quicken the revolution of money & credit.
9. Diminish the interest for moneys.
10. Make commodity supply the place of money.
11. Ingrosse the trade of Europe.
12. Fill the Land with commodity.
13. Abate the price of commodity.
14. Provide store against famine.
15. Relieve and employ the poor.
16. Augment custome and Excize.
17. Promote the sale of Lands.
18. Remove the causes of imprisonment for debt.
19. Lessen the hazard of trading on credit.
20. Prevent high-way thieves.
21. Multiply ships for defence at Sea.
22. Multiply means for defence at Land.
23. Incorporate the whole strength of England.
24. Take away advantages of opposition.

All which in this Treatise is conceived by judicious men to be fully proved, doubts resolved, and Objections either answered or prevented.

Ecclef.9.10. Prov.8.12. & 13.4. & 20.4. & 22.13. & 24.30,&c. & 26.15.& 28.19. Matth.25.14.&c.

LONDON,
Printed by R. A. and are to be sold by Giles Calvert at the black spread Eagle neer the West end of PAULS. 1650.

Figure 4. Improvements in trade multiply wealth: William Potter, *The Key of Wealth* (1650).

that, even if the amount of trade in the world was finite and competition for it therefore a 'zero-sum game', England could seize an ever larger slice of it. The second was the observation that the earth's resources and the productive capacity of its population were far from being fully utilized. Until they were, there were limitless possibilities for increase.

Writers inside and outside Hartlib's circle were therefore able to argue that fuller exploitation of the plantations would bring the nation 'continual increase of wealth and trade' and 'continual increase of power by sea and land'. Freer trade at home was presented as a means to increase employment and commerce, and it was left to its most eloquent opponent, Henry Parker, to point out that while trade was a public good, it was 'not amongst things infinitely good; ... there are bounds and degrees beyond which it may not be diffused, nor can be enlarged, without disprofit'.[89] The growth of large cities was similarly a way of 'advancing the inland trade of the nation', not only for Robinson, who wanted villages and small towns gathered into larger conglomerations, but for James Harrington, who observed that great cities brought a general 'increase in fruitfulness' to their hinterlands and designed his ideal commonwealth, 'Oceana', as 'a commonwealth for increase'.[90] Harrington's publisher, John Streater, welcomed a growing population, 'the increase of mankind', as an opportunity for increasing trade and manufactures. 'Infinite' numbers of people could be profitably employed and make 'the best improvement of all the commodities of the country'.[91] Hartlib was far from alone in thinking that well-ordered commonwealths were those which offered 'infinite welfare, benefit and wonderful advantage' for everyone, though only he perhaps would have hailed the cultivation of silkworms as an invention providentially designed to bring 'infinite wealth and happiness'.[92]

For Hartlib, as for Bacon, the prospect of happiness which necessarily came with new inventions and the material progress they created extended to all mankind. The process might begin in England, but it would ultimately make 'all the world ... enhappined'.[93] Robinson thought the new liberties enjoyed by the people of England could never satisfy them, 'much less ... really make them happy', unless they were prosperous, and Plattes looked ahead to a time when 'this land and so consequently other kingdoms' might 'live in worldly happiness and prosperity for ever hereafter ... whereby *Utopia* may be had really, without any fiction at all'. Plattes had taken the title for his own utopia, 'Macaria', from the Greek for happiness.[94]

---

[89] Thomas Leng, '"A Potent Plantation Well Armed and Policeed": Huguenots, the Hartlib Circle, and British Colonization in the 1640s', *William and Mary Quarterly*, 3rd ser. 66 (2009), pp. 179, 189; Parker, *Free Trade*, p. 20.
[90] Robinson, *Briefe Considerations*, p. 4; Robinson, *Certain Proposalls*, pp. 8–9; *The Political Works of James Harrington*, ed. J. G. A. Pocock (Cambridge, 1977), pp. 160, 469–70.
[91] [John Streater], *Observations Historical, Political and Philosophical, Upon Aristotle's first Book of Political Government* (1654), no. 4, pp. 19, 20; no. 1, pp. 1, 6.
[92] Samuel Hartlib, *A Rare and New Discovery* (1652), title page, 'To the Reader', pp. 4–5, [13].
[93] *HP*, 28/1/19B. On Bacon, see above, p. 102.
[94] Robinson, *Certain Proposalls*, p. 2; Robinson, *Office of Adresses*, pp. 2–3; Webster, *Great Instauration*, p. 32; *Hartlib and the Advancement of Learning*, p. 79.

Happiness had a strategic as well as rhetorical function in the language of the Hartlib group. The Achilles heel in any argument for material prosperity and progress was always that it left its proponents open to the charge of endorsing avarice. Like the profit motives of projectors, appetites for wealth and consumer satisfactions were morally suspect. Hartlib and his associates had to be careful when explaining that their inventions and experiments might possibly be 'lucriferous' as well as 'luciferous', profitable as well as enlightening, and their acknowledged goal, like that of Thomas Mun, was moderation in consumption and a 'sufficiency' of 'moderate riches', not an excess of them.[95] An appeal to happiness, however, was even more useful in settling, or rather evading, that familiar issue. A more grandiose aspiration than moderation, it nevertheless had precisely the right connotations as a modest expectation legitimately open to all. Bound in harness together, 'happiness and welfare' or their synonyms became a cliché of political and economic commentators in the second half of the century.[96]

Happiness itself, of course, was susceptible to divergent interpretations, and they were quickly articulated. Samuel Gott, an acquaintance of Hartlib's disenchanted by the 'pamphlets and stories' of an age driven to 'extremes', found the 'true happiness of man' only in Christian faith and the pursuit of temperance and moderation. 'Happy without holiness, that is happy without happiness,' agreed Richard Baxter in 1659.[97] Felicity also retained much of its Aristotelian association with the contemplative life, as the debates about the meaning of happiness in More's *Utopia* and seventeenth-century poetry on the theme of the happy man show.[98] As Bodin had recognized, that was not easily reconciled with the acquisitive drive of a market society in which people and nations took pleasure in what another Utopian writer, Peter Plockhoy, called 'troublesome riches', 'the multiplying of super-necessary things', in 1660.[99] The tension is evident in Walwyn's works, with their repeated allusion to 'the advancement of a communitive happiness', a shared contentment. For him 'vain and ridiculous

---

[95] Petty, *Advice*, p. 154; Greengrass et al., eds, *Samuel Hartlib and Universal Reformation*, 'Introduction', p. 19; Kevin Dunn, 'Milton among the Monopolists', in Greengrass et al., eds, *Samuel Hartlib and Universal Reformation*, pp. 180–1; James R. Jacob, 'The Political Economy of Science in Seventeenth-Century England', *Social Research*, 59 (1992), pp. 516–18; above, p. 86. On lucriferous/luciferous, see below, p. 167.

[96] See, for example, Cooke, *Unum necessarium*, pp. 29, 57, 64; Samuel Fortrey, *Englands Interest and Improvement* (Cambridge, 1663), sig. A3ᵛ; Carew Reynell, *The True English Interest or An Account of the Chief National Improvements* (1674), p. 1; Joseph Trevers, *An Essay to the Restoring of our Decayed Trade* (1675), p. 19; Andrew Yarranton, *England's Improvement by Sea and Land* (1677), p. 7; George Everett, *The Path-way to Peace and Profit* (1694), p. 5; 'Philalethes', *The Plain Man's Essay for England's Prosperity* (1698), p. 6.

[97] Davis, *Utopia*, pp. 142–6, 152; Richard Baxter, *A Holy Commonwealth* (1659), pp. 27, 208.

[98] Eric Nelson, *The Greek Tradition in Republican Thought* (Cambridge, 2004), pp. 12–13, 46–8; Thomas More, *Utopia*, ed. Edward Surtz and Jack H. Hexter (New Haven, 1965), pp. 160–5; Maren-Sofie Røstvig, *The Happy Man: Studies in the Metamorphoses of a Classical Ideal* (2 vols, Oslo and Oxford, 1954–8), vol. i.

[99] Bodin, *Six Bookes*, pp. 4–7; Pieter Corneliszoon Plockhoy, *An Invitation to the aforementioned Society or Little Common-Weath* (1660), pp. 3, 11. On Bodin, cf. Nelson, *Greek Tradition*, p. 98, and on Plockhoy, Davis, *Utopia*, pp. 331–8.

follies' like 'new inventions' in buildings, furniture, and dress were obstacles to true felicity.[100]

Nevertheless, happiness was acquiring new prominence and fresh connotations in the middle of the century which conspired to accelerate its long evolution from a rare experience, the fruit of supreme virtue scarcely attainable in this world, to a commonplace mixture of physical well-being and psychological content. It had obvious political utility in the troubles of the 1640s and after, not just to radicals and improvers but to Charles I, promising 'the great comfort and happiness' of his subjects by means of the parliaments of 1640 and setting a model for all his successors;[101] and it offered a tempting way through the thickets of religious dispute too. In 1639 Robert Crofts's *Terrestriall Paradise or Happinesse on Earth* claimed that no one had yet written a book on 'the enjoying of earthly happiness' including 'riches, honour, and pleasures', and set out to fill the gap.[102] Crofts soon provoked critical reactions, including one from Gott.[103] But his theme was consistent with the subsequent rise of 'content' as a religious goal, gradually replacing a godly piety based on suffering and affliction, and producing by the 1670s a large learned and popular literature on the routes to real happiness, which included Richard Allestree's best-seller, *The Art of Contentment* (1675).[104] It had often been said in the past that material comfort and economic well-being were also ways to happiness, or at least preconditions for it, as when Forset visualized princes promoting growth and hence the happiness and welfare of their subjects.[105] After 1660, however, Robinson's prospect of 'the happy estating' of the nation in perpetuity could be advanced with favourable currents of pious opinion alongside it.[106]

Hartlib and his colleagues had given a powerful push along that road by constructing a rhetoric of endless improvement which covered all their different, occasionally discordant, aspirations. Plattes summarized their common purpose with the popular

---

[100] *The Writings of William Walwyn*, ed. Jack R. McMichael and Barbara Taft (Athens, Ga, 1989), pp. 82–3, 335–6, 345. By 1652, however, Walwyn seems to have come to terms with economic appetites and their contribution to 'the increase of wealth and plenty': *The Writings of William Walwyn*, p. 449.

[101] James F. Larkin and Paul L. Hughes, eds, *Stuart Royal Proclamations* (2 vols, Oxford, 1973–83), ii. 753, 755. There had been similar claims by monarchs or their spokesmen in earlier parliaments, but political use of the word seems to me much more common from the 1640s onwards. Cf. Darrin M. McMahon, *The Pursuit of Happiness: A History from the Greeks to the Present* (2006), pp. 176–7.

[102] Robert Crofts, *The Terrestriall Paradise, Or, Happinesse on Earth* (1639), 'To the Reader', p. 1. Cf. Robert Crofts, *Paradise Within Us: Or, The happie Mind* (1640); Robert Crofts, *The Way to Happinesse on Earth: Concerning Riches, Honour, Conjugall Love, Eating, Drinking* (1641).

[103] Samuel Gott, *An Essay of the True Happines of Man* (1650), pp. 20, 28; William Fenner, *A Treatise of the Affections* (1650), p. 63 (first published in 1641). On the latter see Christopher Tilmouth, *Passion's Triumph over Reason: A History of the Moral Imagination from Spenser to Rochester* (Oxford, 2007), pp. 164–6.

[104] Ann Thompson, *The Art of Suffering and the Impact of Seventeenth-Century Anti-Providential Thought* (Aldershot, 2003), pp. 111–28; Blair Worden, 'The Question of Secularization', in Alan Houston and Steve Pincus, eds, *A Nation Transformed: England after the Restoration* (Cambridge, 2001), pp. 38–40. On Allestree, see McMahon, *Pursuit of Happiness*, pp. 118–90. For examples of other works on the theme, see Joseph Glanvill, *The Way of Happiness* (1670); Joseph Alleine, *The True Way to Happiness* (1675).

[105] Above, p. 77. For later discussion of the theme, see below, pp. 151–3, 166, 206–7, 220–1.

[106] Robinson, *Office of Adresses*, p. 3.

image of the beehive, that 'well ordered flourishing commonwealth', whose happily industrious workers for the general good were able to 'metamorphise' private greed into 'good husbandry and godly providence' for the general good.[107] The beehive was a political metaphor much used in the seventeenth century, and other writers represented the workers as toiling for a monarch (still usually identified as a king) and not for the 'reformed commonwealth' to which Hartlib pointedly referred in the title of his detailed account of the hive in 1655.[108] Petty incorporated a hive into his coat of arms in 1677, and gave it a distinctive twist with his motto 'ut apes geometriam', 'as the bees practise geometry'. Like him, bees were experts in mathematics, and also, he added in a later letter, in 'political arithmetic' and in 'seeking out riches'.[109] Whether monarchists or republicans, they were always builders of communities and creators of wealth, and perfect improvers.

Like the image of the beehive, Hartlib's vision of improvement survived the Restoration with most of its elements still intact. At the beginning of 1660 his correspondents were lamenting the disappointment of their 'professions to a reformation in these last eighteen years', and some of them looking to their realization, as Beale had already done, in 'cloisters' and small communities, not across the whole kingdom. By the end of the year, however, they were more sanguine as the London meetings of *virtuosi* gained favour at court and acquired the influence which produced the Royal Society, chartered in 1662 'for the improving of natural knowledge'.[110] Its founders, Beale and Boyle prominent among them, pursued most of the Baconian programme for the advancement of learning and useful knowledge as Hartlib's colleagues had interpreted it, and some of them were able to retain apocalyptic expectations of what it might achieve.[111]

Their immediate aims, however, were more utilitarian and ostensibly more practicable. They hoped to complete the earlier projects for a history of trades by collecting detailed accounts of new inventions and manufacturing techniques; and their 'Georgical Committee' distributed questionnaires in order to compile 'a good history of agriculture and gardening in order to improve the practice thereof'. They were disappointed. The Committee received only eleven responses to its enquiries, and their own publications were not written, as Walter Blith's had been, in a plain language intelligible to working farmers but in a style which might impress a sophisticated audience that was reading the *Philosophical Transactions* and Thomas Sprat's *History of the Royal Society*, published in 1667.[112] Improvement was preached as never before, but to an educated and already half-converted public.

---

[107] Jacob, 'Political Economy of Science', pp. 511–12.
[108] Timothy Raylor, 'Samuel Hartlib and the Commonwealth of Bees', in Leslie and Raylor, *Culture and Cultivation*, pp. 91–129. Cf. below, pp. 207–8.
[109] Petty's Latin version was 'ut apes arithmeticem politicam facere' and 'ut apes opes quaerere': *The Petty–Southwell Correspondence 1676–1687*, ed. Marquis of Lansdowne (1928), pp. 226, 229.
[110] Hartlib and the Advancement of Learning, pp. 196–7, 200, 207.
[111] Michael Hunter, *Science and Society in Restoration England* (Cambridge, 1981), pp. 115–18. On the various versions of apocalyptic thinking after 1660, see Warren Johnston, *Revelation Restored: The Apocalypse in Later Seventeenth-Century England* (Woodbridge, 2011).
[112] Hunter, *Science and Society*, pp. 91–102; Thirsk, 'Agricultural Innovations and their Diffusion', pp. 560–6.

In the twenty years after 1660 there were more publications with improvement in their title than in the twenty years before 1640, including in 1663 *The Improvement improved* by the parliamentary engineer Andrew Yarranton, and *Englands Interest and Improvement* by the courtier Samuel Fortrey. Yarranton was another former parliamentary soldier who had bought Crown lands, and he wrote initially about clover, but he was equally interested in the improvement of navigation and the fisheries; and Fortrey, while concentrating on trade, referred to other improvements which advanced 'the welfare and happiness' of the kingdom.[113] In 1661 there had been three tracts on equally varied improving themes. Francis Cradocke returned to Potter's proposal for a land bank in *Wealth Discovered*, which praised Hartlib and the 'new inventions' he sponsored, while being careful to repudiate allegations that the author was the son of a Cromwellian radical, 'Cradock the Preacher'.[114] Thomas Powell's *Humane Industrie*, an account of 'manual arts' and their 'progress and improvement', discussed techniques like glass- and watch-making. John Evelyn's *Fumifugium* urged parliament to embark on the 'improvement of public works' in order to 'perfectly improve and meliorate the air' around London and hence its 'health and felicity'.[115]

Unlikely as it might have seemed later, Charles II himself could be presented as a patron of social and environmental reform and national improvement in the 1660s. He wrote to the City in 1660 to support public subscriptions to employ the poor of London on hemp and benefit the fisheries and the navy, and Hartlib took careful note of the fact as a propitious omen for the future.[116] When Charles laid out St James's Park according to the latest fashion in 1661, that too was taken by Edmund Waller to be an exemplar of his dedication to an improving mission. Waller imagined the king there, contemplating his capital, 'the seat of empire', ruminating upon 'rising kingdoms and... falling states', and fortifying his determination to

> Reform these nations, and improve them more
> Than this fair park, from what it was before.[117]

In 1663 Fortrey understandably thought it was 'the genius and disposition of the times to study more the interest and improvement of the nation' than it had ever

---

[113] Slack, *Reformation to Improvement*, pp. 88, 96 n. 89; Warde, 'Idea of Improvement', pp. 143–4; Andrew Yarranton, *The Improvement improved* (1663), apparently a revision of a lost earlier tract; Fortrey, *Englands Interest and Improvement*, sig. A3ᵛ, p. 3.

[114] Francis Cradocke, *Wealth Discovered* (1661), 'Preface', pp. 1–2, 41, 43. On Cradocke, see Wennerlind, *Casualties of Credit*, pp. 115–17, and on the preacher, *ODNB sub* Samuel Cradocke.

[115] Thomas Powell, *Humane Industrie: or, A History of most Manual Arts Deducing the Original, Progress and Improvement of them* (1661); John Evelyn, *Fumifugium* (1661), sigs A4, a1. On the latter see Mark Jenner, 'The Politics of London Air: John Evelyn's *Fumifugium* and the Restoration', *Historical Journal*, 38 (1995), pp. 535–51.

[116] Hunter, *Science and Society*, p. 130; W. H. and H. C. Overall, eds, *Analytical Index... to the Remembrancia... of the City of London* (1878), pp. 143–4; HP 27/21, fo. 1ʳ.

[117] Alastair Fowler, ed., *The New Oxford Book of Seventeenth-Century Verse* (Oxford, 1991), pp. 399–400. On public images of Charles II, see Kevin Sharpe, *Rebranding Rule: The Restoration and Revolution Monarchy, 1660–1714* (2013), pp. 106–7, 121.

been.[118] Designed now more for the benefit of the nation than that of the individual or the world, improvement had become part of the collective mentality of the cultural and political elite.

Hartlib's correspondents necessarily adjusted to the new climate, some of them as active Fellows of the Royal Society. Worsley was never elected a Fellow, but he was still pursuing his own economic projects and personal religious enlightenment, and still convinced that governments had a duty to promote 'the good and welfare of every particular person'. He acquired a minor political role, as a member of the new Council of Trade of 1668, much larger and more unwieldy than its 1650 predecessor, and then as assistant secretary to a Council for Foreign Plantations in 1670. He owed what influence he had chiefly to writings on the plantations, whose improvement would support 'the empire of England' in a world 'vastly different' from what it had been in his youth.[119]

In the new world of the Restoration his rival Petty was a far more influential figure. Now the social and intellectual peer of the *virtuosi* and their colleague in the Royal Society, endlessly inventive and adept at publicizing new projects, he never acquired the political power he craved because it was difficult to take him entirely seriously. In 1663 and 1664 his invention of a 'double-bottomed' boat, a kind of catamaran he thought would revolutionize the navy, won more ridicule than public applause. None of its three prototypes, named *Invention, Invention II,* and *Experiment,* ever went into production, and Charles II spent 'a happy hour or two' in Petty's presence laughing at that and other activities of a Royal Society which spent its time 'only in weighing of air . . . and doing nothing else'. The king, who enjoyed Petty's company more than many of his contemporaries, summed him up with some accuracy as a man not 'contented to be excellent', but always 'aiming at impossible things'.[120]

One of those things Petty did achieve, and it was of genuinely revolutionary importance. In the decade after 1660, through published tracts and papers circulated to friends, he transformed the character of English political economy. He sometimes still referred to his hopes for a 'universal reformation',[121] but his focus was now on the wealth and welfare of England, and the use of political arithmetic to measure them.

POLITICAL ECONOMY

Political arithmetic was defined by Charles Davenant in 1698 as 'the art of reasoning by figures, upon things relating to government'. He attributed the

---

[118] Fortrey, *Englands Interest and Improvement*, 'To the Reader'.
[119] Leng, *Worsley*, pp. 151–7, 163–5; *SCED*, pp. 533–7.
[120] McCormick, *Petty*, pp. 150–6; *Petty–Southwell Correspondence*, p. 281. For a description of Petty as the most 'grating man' in the three kingdoms see Barnard, *Improving Ireland?*, p. 61; and for later ridicule of the double-bottom, in Dublin, K. Theodore Hoppen, ed., *Papers of the Dublin Philosophical Society, 1683–1709* (2 vols, Irish Manuscripts Commission, Dublin, 2008), ii. 909.
[121] Petty, *EW*, i. 26 (1662).

name and its 'rules and method' to the 'excellent wit' of Petty, who had applied it to the topics in which Davenant was most interested, revenue and trade.[122] When Petty's *Political Arithmetick* was first published in full, three years after his death in 1687, his son pointed out that Petty intended his new 'piece of science' to do more than that. He had wanted to explore 'the perplexed and intricate ways of the world' and especially those that concerned 'the glory of the Prince' and 'the happiness and greatness of the people'.[123] Political arithmetic was a tool applicable to almost everything.

While writing the work in the 1670s Petty explained his purpose in correspondence. His aim was to go beyond the vacuous rhetoric of 'sermocinations' and 'euphonical nonsense' and investigate the realities of a world where nature must have its course—*res nolunt male adminstratrari*. It was a world in which there was 'a political arithmetic and a geometrical justice' still to be discovered and interpreted. In the text itself, he explained his methodology. He had used 'only arguments of sense' to arrive at 'observations or positions expressed by number, weight, and measure'. Some of them might not be 'true, certain and evident'; they might rather be 'suppositions'. But they were intended 'at worst...to show the way to that knowledge I aim at'. They were 'a specimen of the political arithmetic I have long aimed at'.[124] He may have invented the name in or around 1670, but the methodology was already being used in works he wrote or influenced in the 1660s, and it had been in gestation before then.

It evidently owed something to Bacon, to earlier notions of justice and proportion and the utility of numbers, and to Petty's mathematical education. His arguments with Cartesian philosophers about axioms and hypotheses had persuaded him of the utility of applied mathematics, the 'mixed mathematics' which he wanted taught in one of the colleges of Hartlib's Office of Address.[125] In the 1650s, as Aubrey recalled, he had pestered James Harrington's Rota Club 'with his arithmetical proportions, reducing polity to numbers', and he saw in numbers and proportions an objective means of reconciling extremes of political opinion. In 1649 he had suggested to Hartlib that parliament should hold popular referenda on the 'fundamental' matters it was debating, so as to leave 'no room for any factions', and there was never any prospect of that.[126] He thought nonetheless that 'two extravagant contrary suppositions' might easily be reconciled, and 'some solid and consistent conclusion' arrived at, if numbers were applied to the issues at stake, and

---

[122] Davenant, *PCW*, i. 128. [123] Petty, *EW*, i. 239–40.
[124] Petty, *EW*, i. 239–40n., 244–5. Cf. McCormick, *Petty*, pp. 175–7. Davenant illustrated the use of 'hypotheses' in *PCW*, i. 251; ii. 175–6, 184. The methodology is considered in A. M. Endres, 'The Functions of Numerical Data in the Writings of Graunt, Petty and Davenant', *History of Political Economy*, 17 (1985), pp. 245–64.
[125] Petty, *Advice*, pp. 144–5, 157; above, pp. 79–80; McCormick, *Petty*, pp. 53–6, 61–2. On contemporary interest in mixed mathematics, see R. W. Serjeantson, 'Proof and Persuasion', in Katharine Park and Lorraine Daston, eds, *The Cambridge History of Science*, iii: *Early Modern Science* (Cambridge, 2006), pp. 163–4.
[126] John Aubrey, *Brief Lives*, ed. Oliver Lawson Dick (3rd edn, 1958), p. 240; *HP*, 28/1/10.

not only in politics.¹²⁷ Contentious hypotheses about the state of the economy and the general welfare of the nation might be tested against the available evidence, the arguments for and against weighed and balanced, and new more certain knowledge arrived at. Political arithmetic could become an instrument for economic prediction and even for social engineering. 'Little small threads of mathematics' could be put to 'vast uses'.¹²⁸

The first published exercise in political arithmetic, though not given that name, was John Graunt's *Natural and Political Observations . . . upon the Bills of Mortality* (1662). The work was often attributed to Petty, and he may have influenced some of the remarks in its conclusion, about the potential of such studies of the 'land and hands' of a territory to make both trade and government 'more certain and regular' and even 'balance parties and factions both in church and state'.¹²⁹ Petty also contributed data from the parish registers of his birthplace, the small town of Romsey in Hampshire. But Petty would never have had the patience to compile most of the data, the tables and numbers, or to construct the averages, ratios, and extrapolations from which Graunt derived carefully qualified conclusions. The major part of the text, if not all, was Graunt's.¹³⁰

Presenting the book to the Royal Society, which rewarded him with a Fellowship on the strength of it, Graunt claimed with appropriate modesty to have used only 'the mathematics of my shop-arithmetic' in his calculations.¹³¹ Like others in the Hartlib circle, however, he was a more substantial figure in the City and more adventurous intellectually than that suggests. A London haberdasher by trade, he had risen to prominence as an officer in the militia, and later became warden of his guild and an investor in the New River Company, although he died in debt. His intellectual curiosity led him from orthodox Calvinism to Socinianism, and finally (much to Petty's dismay) to Catholicism. But by then it had prompted him to look at the London bills of mortality which others had sometimes used as indicators of demographic change, and to exploit them much more creatively. He began by tabulating and comparing annual totals of baptisms and burials, much as he might draw up a shopkeeper's accounts of income and expenditure, but he was able to derive from them the first calculations of mortality and fertility rates and the first life-tables, and to found a new science, statistical demography.¹³²

---

¹²⁷ Petty, *EW*, ii. 470. For similar aspirations in the use of number, see Roger Coke, *A Discourse of Trade* (1670), Epistle Dedicatory; William Letwin, *The Origins of Scientific Economics* (New York, 1964), pp. 120–2.

¹²⁸ *William Petty on the Order of Nature: An Unpublished Manuscript Treatise*, ed. Rhodri Lewis (Tempe, Ariz., 2012), 'Introduction', p. 1, citing Petty, *The Discourse . . . Concerning the Use of Duplicate proportion* (1674), sigs A5ᵛ–A6ʳ.

¹²⁹ Petty, *EW*, ii. 395–7. See also Petty, *EW*, ii. 377 and note. On the concept of balance, see Philip Kreager, 'New Light on Graunt', *Population Studies*, 42 (1988), pp. 138–40.

¹³⁰ On Graunt's career and relations with Petty, see McCormick, *Petty*, pp. 131–5; and in addition to works cited there, Philip Kreager, 'John Graunt', in K. Kempf-Leonard, ed., *Encyclopedia of Social Measurement* (2005), ii. 161–6.

¹³¹ Petty, *EW*, ii. 323.

¹³² *ODNB* sub Graunt; Webster, *Great Instauration*, pp. 444–6. On earlier use of the bills, see above, pp. 45, 51.

Perhaps most immediately striking was his demonstration that the populations of London and of England and Wales were both much lower than many had previously supposed. He used three different methods to arrive at a figure of 384,000 for the parts of the city covered in the bills, and 460,000 for the whole built-up area including Westminster and some of the out-parishes. Although perhaps a little too large, the numbers were much more accurate than the 'millions of people' Graunt had heard Londoners talk about.[133] He also compared his London data with those of a 'country parish', Romsey, and deduced that natural increase outside London allowed the populations of both the provinces and the metropolis to grow, though the second had been growing much more rapidly than the first. More ingenious still was his use of acreages, vital statistics, and taxation records in order to derive a population for the whole of England and Wales, and conclude that it was about six and half million (6,400,000).[134] That was probably a million too many, to judge from the estimates of modern historians, but the careful, modest paragraphs in which Graunt explains his methodology still communicate a sense of the excitement involved in discovering 'some truths and not commonly believed opinions', 'a new thing', 'clear knowledge'.[135]

Not surprisingly, the book had an immediate impact, running to five editions by 1676, and earning a review in the French *Journal des sçavans* in 1666, perhaps contributed by Petty, which prompted the first bills of mortality in Paris.[136] In England it created an appetite for the collection of demographic data by *virtuosi* and scholars gathering materials for the natural history of their own localities, and sometimes encouraged a narrow antiquarianism.[137] Graunt himself, however, collected statistics from collaborators, about Cranbrook and Tiverton, for his 1665 edition, and hoped for more from parishes in Northumberland, Cheshire, Norfolk, and Nottinghamshire in order to achieve a national coverage.[138] He had shown that partial information could lead to knowledge of the whole, and his findings were cited for more than half a century in most discussions of population, local, national, and global.

In the conclusion to his *Observations*, Graunt wondered whether knowledge of such matters was 'necessary to many' or only fit for 'the sovereign and his chief ministers', since they concerned matters of state. His own publication might seem

---

[133] Petty, *EW*, ii. 331, 371, 383–6. A French text, translated in 1615, had given a reasonably accurate figure of 350,000 for London's population: Pierre d'Avity, *The Estates, Empires & Principalities of the World* (1615), p. 6. If Graunt knew of it, he may have used it as a benchmark, but there is no evidence that he did. Graunt's methodology is succinctly summarized in Jacques Dupâquier, 'Londres ou Paris? Un grand débat dans le petit monde des arithméticiens politiques (1662–1759)', *Population*, 53 (1998), pp. 311–12.

[134] Petty, *EW*, ii. 329, 369–72, 389–93.   [135] Petty, *EW*, ii. 334, 397.

[136] Petty, *EW*, ii. 422; *Le Journal des sçavans pour l'année MDCLXVI* (new edn, Paris, 1729), pp. 214–16 (2 Aug. 1666); Ian Hacking, *The Emergence of Probability* (Cambridge, 1975), pp. 102–9. For French concerns about population at this time, see J. J. Spengler, *French Predecessors of Malthus: A Study of Eighteenth-Century Wage and Population Theory* (Durham, NC, 1942), pp. 16, 24–5, 34–7; Jacques and Michel Dupâquier, *Histoire de la démographie* (Paris, 1985), pp. 69–70; below, p. 136.

[137] Above, pp. 20–1; Slack, 'Government and Information', p. 45.

[138] Petty, *EW*, ii. 398–9. Petty had already provided data from the Dublin bills of mortality.

to have answered his question, but he left it open for further 'consideration'.[139] Petty, however, was simultaneously writing his *Treatise of Taxes* (1662), in which he offered plentiful observations for government and public alike, not only about the excise, of which Graunt was to be one of the collectors, and the hearth tax, whose introduction Petty may have suggested, but also on other current issues like free ports, monopolies, the growth of London, and 'new inventions'. His mode of analysis here was not that of political arithmetic, but of geometric proportion, which showed how the distribution of parishes, professions, or taxes might be adjusted to fit the distribution of population or social classes, and be a tool both for analysis of the social structure and for social and economic reform, for policy.[140] It opened up prospects for the future, on which Petty's opinions differed radically from those of Graunt.

Since there was only 'a certain proportion of work to be done' in England, and only 'a certain proportion of trade in the world', Graunt had been inclined to suppose that the country was already overpopulated, while leaving it to others to enquire into 'the fundamental trade, which is husbandry and plantation' and determine whether there was a need for further emigration.[141] Petty, however, thought it 'a false opinion that our country is fully peopled', and looked instead, as Streater had done, to increase the skills and industry of the people at home.[142] In one of his manuscripts, he cited Graunt's figure for population to support his contention that if the land was 'improved to the utmost of known husbandry', many more hands could be employed in producing 'exquisite manufactures and the extraordinary productions of art' for export. England might even support 'a city like London planted in every twenty miles square thereof'.[143] There was no hint of a finite amount of work to be done or an immediate ceiling to growth there, but rather a new emphasis on the productive capacity of labour.[144]

What remained to be determined was how far both population and productivity might grow. That depended partly on knowing the relationship between land, 'the mother' of wealth, and labour, its 'father and active principle', and led him in the *Treatise* to try to calculate the land–labour ratio, though with incomplete success.[145] But it also depended on knowing the total product of the nation, the

---

[139] Petty, *EW*, ii, 397.
[140] Petty, *EW*, i. 25–40, 60–1, 74–5, 91–5; McCormick, *Petty*, pp. 135–47; Richard Bonney, 'Early Modern Theories of State Finance', in Richard Bonney, ed., *Economic Systems and State Finance* (Oxford, 1995), p. 178.
[141] Petty, *EW*, ii. 353–4, 371–2.
[142] Petty, *EW*, i. 21–2. For an instance of the 'false opinion', see John Bland, *Trade Revived* (1659), p. 11.
[143] *The Petty Papers*, ed. Marquis of Lansdowne (2 vols, 1927), i. 208, which seems from its content to be of 1662 or a little later.
[144] Cf. Craig Muldrew, 'From Commonwealth to Public Opulence: The Redefinition of Wealth and Government in Early Modern Britain', in Steve Hindle et al., eds, *Remaking English Society* (Woodbridge, 2013), pp. 322–3.
[145] Adam Fox, 'Sir William Petty, Ireland, and the Making of a Political Economist, 1653–87', *EcHR* 62 (2009), pp. 393–4; Petty, *EW*, i. 68, 181; BL, Petty Papers, Add. MS 72865, fo. 6ʳ; McCormick, *Petty*, pp. 140–2. Cf. Finkelstein, *Harmony and the Balance*, pp. 116–18; Marian Bowley, *Studies in the History of Economic Theory before 1870* (1973), pp. 84–6.

national wealth, and the relative contribution made to it by its component parts. That was what Petty attempted to calculate in *Verbum sapienti*—a word to the wise—written in 1665, a work which presented the first ever set of national accounts, for England or any other country.[146] It contained numbers, but the arithmetic involved was much simpler than Graunt's and the figures were put together with fewer evidential bricks and much more conjectural straw. They showed the utility of hypotheses and proportions. The result, presented in a few brief paragraphs, was a major innovation in political economy and the second wholly original product of political arithmetic after Graunt's.[147]

The crucial foundation was Petty's perception, influenced perhaps by Graunt's shopkeeper's arithmetic, that the income and expenditure of a nation, like that of a household or a business, could be compared as in a set of accounts; and that they must over time balance one another. He seems to have had the first glimmer of the idea in 1662, when he observed in the *Treatise* that 'the first thing to be done is to compute what the total of the expense of this nation is by particular men upon themselves'.[148] Now he set up a model on the supposition that total national expenditure and total income were precisely the same, and that allowed him to fill gaps in the information available to him on either side of the national account. There was some circular reasoning involved, and much guesswork, since he knew only the population (from Graunt), the general level of labourers' wages, and the value of land (from tax returns), but very little about income derived from investments in commerce. Nonetheless, his model gave him a plausible figure for the total income and expenditure of the nation of £40 million a year, the first ever calculation of Gross Domestic Product (GDP). That sum could be capitalized, and show that the total wealth of the nation must be £667 million. More than that, since land produced £8 million a year and other kinds of capital, including stock in trade, a further £7 million, then labour must contribute £25 million to the total product, and be worth more than trade and land put together. The rival contending interests of merchants and landowners, each claiming to be custodians of the national wealth, were both cut down to size.

The contribution made by labour, on the other hand, was elevated, and Petty's model gave him the first calculation of what he called 'the value of the people'. National income per capita, or GDP per head, was £6 13s. 4d.; and in capital terms people were worth on average £69, and the workers among them, three million out of six million, £138 each. Various calculations followed. Petty could show the total cost in lost labour attributable to the plague then raging in London (£7 million, he thought), and conclude that the epidemic was 'a great loss to the kingdom' and not,

---

[146] Petty, *EW*, i. 103–10. On the method, see Paul Slack, 'Measuring the National Wealth in Seventeenth-Century England', *EcHR* 57 (2004), pp. 612–16, and on its importance, Richard Stone, *Some British Empiricists in the Social Sciences, 1650–1900* (Cambridge, 1997), pp. 26–31; Angus Maddison, *Contours of the World Economy, 1–2030 AD: Essays in Macro-Economic History* (Oxford, 2007), pp. 253–4.
[147] Petty seems to have given Graunt a manuscript copy: Antoin E. Murphy, *The Genesis of Macroeconomics: New Ideas from Sir William Petty to Henry Thornton* (Oxford, 2009), pp. 32–3.
[148] Petty, *EW*, i. 91.

as some thought, 'a seasonable discharge of its pestilent humours'.[149] The real 'supernumeraries' in the population were not the poor, who could be set to work and more than pay their way, but members of the professions like the clergy, whose number ought to be proportionate to the population they served.[150] Above all, Petty concluded that the value of labour per head could be increased, since current wages were low and earned by 'a very gentle labour' by that half of the population which was employed. If food and other essentials could be produced by fewer hands, all of them 'labouring harder', and if the remainder were employed in manufactures—'the compendium and facilitations of art'—then the national wealth could be increased without any increase in the population at all.[151]

Petty attempted no calculation at this stage of what the addition to the national wealth might be. He confined himself to saying it might and could increase until it was greater than that of 'any of our neighbour states'. Only then could the English rest from 'this great industry', contemplate their achievement and the finer things of life, and enjoy the pleasures of the intellect which extended '*ad infinitum*'.[152] Five years later, however, in *The Political Anatomy of Ireland*, finished in 1672 and summarized in a paper on 'the improvement of Ireland' for the Irish Council of Trade in 1676, he applied the methodology of *Verbum sapienti* to Irish circumstances, and measured what might be achieved there. The country that had once been his empty piece of paper and a mere 'embryo' was now a 'political animal'. It was 'scarce twenty years old' since the Cromwellian settlement, but it was ripe for dissection.[153] Petty referred again to the need to calculate the ratio between land and labour, which was 'the most important consideration in political oeconomies'; and he was now able to compare the population densities of England and Ireland with those of France and Holland. Ireland was 'much under-peopled', since there were ten acres per head, compared with four in England and France, and only one in Holland. The present population of just over a million was also unproductive, generating a national income per head of no more than £4 a year compared with nearly £7 in England. If all of them, including the impoverished native Irish, were 'well employed', Petty calculated that GDP in Ireland could rise by £1 million a year, 25 per cent, without any increase in population.[154]

Petty returned to his figures for England in *Political Arithmetick*, written at the same time as his *Political Anatomy*, and tried to quantify the potential for growth in the national income there. Looking at levels of income per head in Ireland and Holland, he was fully persuaded that more arts and manufactures would push England further towards the second and far away from the first. In *Verbum sapienti* his calculations assumed that the people enjoyed no surplus income over necessary expenditure, no 'superlucration', as he called it, above subsistence. He now thought it might be possible for the English to double their earnings per head, and hence

---

[149] Petty, *EW*, i. 108–10.   [150] Petty, *EW*, i. 25, 79.
[151] Petty, *EW*, i. 118. An earlier manuscript by Petty, on 'trade and its increase', had come to a similar conclusion: *Petty Papers*, i. 214.
[152] Petty, *EW*, i. 119–20.   [153] Petty, *EW*, i. 129. Cf. McCormick, *Petty*, pp. 187–92.
[154] Petty, *EW*, i. 181, 214–23.

superlucrate £25 million or so a year in total. That would double the capital wealth of the nation—its 'whole stock and personal estate'—in a matter of five or six years. It sounds like some of William Potter's extravagant predictions, and Petty had to confess that his was a pipe-dream, a supposition 'which I wish were true, but find no manner of reason to believe'; but he thought a more modest surplus of £2 million a year easily attainable.[155]

Broad-brush and constantly shifting numbers of this kind were no doubt one of the things which persuaded Adam Smith that he had 'no great faith in political arithmetic'. For a classical economist there were obvious holes in Petty's analysis, including, in *Verbum sapienti*, the absence of any allowance for savings and investment and hence capital accumulation. His calculations of superlucration and surplus made no allowance either for diminishing marginal returns.[156] Yet Petty had moved decisively away from some of the concerns of earlier writers on the economy. His model had indicated, if not quite as eloquently as Adam Smith's prose, that the vital balance for the wealth of a nation was 'the balance of the annual produce and consumption'.[157] Although he did not dispense wholly with balance-of-trade theory, his interest in consumption as a spur to production led him to suggest in the *Political Anatomy* that there was a need to 'beget a luxury' in Irish 'plebeians, so as to make them spend, and consequently earn double what they at present do ... to the great enrichment of the commonwealth'.[158] His interest in income and expenditure per head, the added value contributed by every citizen, had also directed attention to the wealth of the people as much as to that of the nation, and led him to appreciate, as early as 1662, that 'men are really and actually rich according to what they spend and enjoy in their own persons'.[159] Their material comforts must increase if there was to be an increase in national wealth, and material progress should be a central concern of 'political oeconomies' and 'public oeconomy' if there was to be a 'policy which tends to peace and plenty'.[160]

The full impact of Petty's concept of political economy was delayed because he published none of the works I have been considering, other than the *Treatise of Taxes*, in his lifetime. *Political Arithmetick* was first published in an authorized version in 1690, *Verbum sapienti* along with the *Anatomy of Ireland* in 1691. It may be that, having burnt his fingers with publicity for the 'double-bottom' which created such ridicule, he became sensitive about public pronouncements. But all these works circulated in manuscript, were copied to his friends like Graunt, or delivered to appropriate politicians in Dublin or Westminster, and information from them leaked out. An unauthorized version of *Political Arithmetick* appeared in print in 1683, and in 1677 Thomas Sheridan knew enough of *Verbum sapienti* to

---

[155] Petty, *EW*, i. 108, 307–8. Cf. Petty, *EW*, ii. 564. On Petty's concept of a surplus, see Tony Aspromourgos, 'The Invention of the Concept of Social Surplus: Petty in the Hartlib Circle', *European Journal of the History of Economic Thought*, 12 (2005), pp. 1–24.
[156] Adam Smith, *The Wealth of Nations Books IV–V*, ed. Andrew Skinner (1999), p. 114 (Book IV, ch. V); Stone, *British Empiricists*, p. 31. Cf. *Nehemiah Grew and England's Economic Development*, ed. Julian Hoppit (Records of Social and Economic History ns 47, British Academy, 2012), p. xxxviii.
[157] Smith, *Wealth of Nations*, p. 76 (Book IV, ch. III). [158] McCormick, *Petty*, p. 192.
[159] Petty, *EW*, i. 16, 91. [160] Petty, *EW*, i. 130, 181; ii. 481; Fox, 'Petty', p. 401.

summarize some of Petty's figures, which were carefully noted down by John Locke.[161] Both texts were of obvious interest to Petty's friends and admirers, like Pepys and Peter Pett, and to later political arithmeticians, notably Gregory King and Davenant, who were critics acknowledging that they stood on Petty's shoulders and so were able to see further.[162]

King introduced some necessary additions, correcting Petty's omission of savings and investment in the calculations in *Verbum sapienti*, and usefully subdividing Petty's hold-all category of the income from 'the labour of people' so as to distinguish profits from commerce and services. But King's basic approach was the same, and his final calculation of the national income (£43.5 million a year in 1688) was very close to Petty's. Both King and Davenant thought Petty had massaged his figures and magnified the strength and wealth of England in order to please a royal audience, but recognized that they were based on incomplete evidence. Although their own conclusions were better grounded, they too were bound, as Davenant admitted, to be 'hypothesis', 'a guess founded upon probable grounds'. Davenant's justification was the same as Petty's and an indirect tribute to him:

In the art of reasoning upon things by figures, it is some praise at first to give only an imperfect and rough draught and model, which, upon more experience and better information, may be corrected.[163]

The ways in which Petty's model was corrected and the circumstances which made correction necessary will be considered in later chapters. *Political Arithmetick* and the *Political Anatomy* were themselves influenced by circumstances after 1670, and they will also be referred to again along with Petty's later writings. In all of them, however, there was the same intellectual creativity as that shown in the model of the national wealth first unveiled in *Verbum sapienti* in 1665, and a consistency of argument which deserves emphasis here. One of his aims was always to show that an increase in population and productivity was not only desirable, but could be achieved by political intervention. He embarked, for example, on a series of proposals for transplanting thousands of families, primarily between England and Ireland, a process which would enable whole countries to be 'transmuted', as if by alchemy, into productive hives of industry and plenty. He acknowledged that such schemes might be thought 'impracticable and intolerable', and one of them he presented in *Political Arithmetick* as a 'jocular and perhaps ridiculous digression', a 'dream or reverie'.[164] It was in fact a thought experiment designed to indicate the whole thrust of his argument.

---

[161] Petty, *EW*, ii. 640; Thomas Sheridan, *A Discourse of the Rise and Power of Parliaments* (1677), pp. 185–7; BL, Add. MS 15642, pp. 100–1, 106 (Locke's 1679 Journal). There is an MS copy of *Verbum sapienti* in Beinecke Library, Yale University, Osborn b. 395, which appears to be additional to those cited in Petty, *EW*, i. 100. For Petty's connections with Sheridan, see McCormick, *Petty*, pp. 264, 275–6.
[162] Cf. below, pp. 163–8.
[163] Davenant, *PCW*, i. 128–30, 251; Slack, 'Measuring the National Wealth', pp. 622–6; below, pp. 163–4, 180.
[164] Fox, 'Petty', p. 397; Petty, *EW*, i. 285; McCormick, *Petty*, pp. 168–205; Sabine Reungoat, *William Petty: Observateur des Îles Britanniques* (Paris, 2004), pp. 177–203. Petty elsewhere defined

If the three kingdoms of England, Scotland, and Ireland were treated as one, and the people of the Scottish highlands and most of those in Ireland were transplanted to the Scottish and English lowlands, England would be 'thicker peopled' and the wealth of the kingdom increase, perhaps by as much as £3 million a year. There might be even greater benefits if the English in North America were repatriated. The profit would come partly through a 'very little addition' of labour into agriculture on the most fertile land, and much more from moving hundreds of thousands of Irish and Scots 'from the poor and miserable trade of husbandry to more beneficial handicrafts', especially in towns and cities.[165] The other two of the three kingdoms would become more like England, and England more like Holland, fully peopled, urban, and industrious. The processes of economic and material improvement were to be accelerated by demographic and social engineering.

Proposals like these demonstrate that for Petty political arithmetic and political economy were not solely, or even primarily, modes of enquiry and discovery indulged in for their own sake, but tools to be used for a purpose. His numbers and models could be based on guesswork as often as on empirical investigation because they were guides to political action as much as to the knowledge he aimed at.[166] But there is more to be said about them than that. In one of his transplantation proposals, in 1686, perhaps intended for the eyes of James II, the political appeal was particularly evident. He envisaged that there might be an exchange of half a million English for half a million Irish. In Ireland 'all improvements' would be encouraged; in England the proportion of Catholics would rise from 1 in 200 to 1 in 11 or 12. It would please the king, but it would deliver Petty's own hopes for the future, not only bring 'real liberty of conscience in both kingdoms' but further 'the design of multiplying mankind' by fully peopling them 'in a short time'.[167]

In the 1680s, as we shall see in a later chapter, Petty was much preoccupied with measuring the multiplication of mankind over the centuries from the Creation and forward to the end of the world. It was a theme susceptible to analysis by political arithmetic, by showing how long it had taken for population to double, and extrapolating that process into the future.[168] The methodology was used by Graunt as well as Petty and their successors, and one of its functions was to give them another reason to look to the future, to the very long run. By comparing one country with another and one part of the national wealth with another, political economy enabled them to predict what the future wealth of the kingdom might be; political arithmetic gave them an instrument to predict demographic change. In both cases they were predisposed to expect growth.

They were led to that expectation by what they took to be God's providential design. Petty was at one with his contemporaries in being interested in current

---

'ridiculous' as something valuable which others treated with contempt: *William Petty on the Order of Nature*, ed. Lewis, p. 120 n. 84.

[165] Petty, *EW*, i. 285–90, 301–2. [166] Cf. McCormick, *Petty*, pp. 298–300.

[167] Reungoat, *Petty*, pp. 190–2; *Petty Papers*, i. 57–67. For the context of these proposals, see Scott Sowerby, *Making Toleration: The Repealers and the Glorious Revolution* (Cambridge, Mass., 2013), pp. 258–9.

[168] Below, pp. 197–8.

theological disputes and the efforts to resolve them by means of what became natural theology. Graunt, for example, was eager to demonstrate that his findings about the growth rate of global population, when projected backwards, showed that mankind could not be more than 5,610 years old. They therefore confirmed the conventional date of Creation and refuted those who thought the world much older and that there had been men before Adam.[169] His *Observations* showed also that there was a small surplus of men over women, and hence that polygamy was 'useless to the multiplication of mankind', contrary to much contemporary speculation.[170] Although better disposed towards polygamy as a solution to low population than Graunt, and greatly interested in the supposed marriage practices of Californian Indians, Petty took equal delight in employing his arithmetic to silence 'Scripture-scoffers and Pre-Adamites', and even to answer sceptics who supposed that there would soon be no room on the earth 'for all the bodies that must rise at the last day'.[171]

After he had been accused of atheism in 1675, as a result of one of his essays on atomic theory, he kept his religious speculations to private papers and correspondence, but he was in any case hostile to the disputes of religious sects, their 'gibberish denominations and uncertain phrases', which had caused so much strife in England and across Europe.[172] Public religion should stick to the essentials of Christianity, and freedom of conscience was not only necessary intellectually, but useful in the 'advancement of trade'. Like many others, he noted how much the Dutch had gained from it, and in *Political Arithmetick* commented that entrepreneurial instincts were everywhere the characteristic of 'the heterodox part' of any society. Unlike some of his contemporaries, and later theorists like Weber, he did not attribute them solely to Protestantism. They were not 'fixed to any species of religion as such'.[173] They sprang from local circumstances.

That eirenical attitude towards religion had its political advantages, especially when there was a Catholic monarch, but it owed as much to Petty's intellectual background in the Hartlib circle, with its early interest in a union of churches, as to his political economy. It was also part of the interest he shared with his friends and contemporaries in eschatology, the whole shape and direction of sacred history from the Creation to the last things, and how the world would be at the final Day of Judgement. That was something more fundamental and far more important than the issues which divided religious denominations. He had little time either for the millenarian speculations which were raised by the Great Plague in London in 1665, when there was again 'a time of great expectations' about 'miracles' and 'strange alterations in the world', and when Beale and others in the Royal Society pondered

---

[169] Petty, *EW*, ii. 331, 388. Cf. above, p. 42.

[170] Petty, *EW*, ii 329, 378. For later interest in polygamy, see Faramerz Dabhoiwala, *The Origins of Sex: A History of the First Sexual Revolution* (2012), pp. 215–31; below, pp. 137–8.

[171] McCormick, *Petty*, pp. 239–40; *Petty Papers*, ii. 52–4, 57–8 (discreetly written in Latin); Petty, *EW*, ii. 466–7, 477–8; *Petty–Southwell Correspondence*, pp. 91–3. Cf. Reungoat, *Petty*, pp. 261–7.

[172] McCormick, *Petty*, pp. 224–6, 241.

[173] Petty, *EW*, i. 263–4; Fox, 'Petty', pp. 400–1. For arguments that Catholics were unsuited to trade, see below, p. 135; Peter Pett, *The Happy Future State of England* (1688), pp. 107–11.

once more the role of divine intervention in a scientific universe.[174] Petty would have thought such things temporary and idle fancies. But he never lost his interest in eschatology, God's ultimate purpose in the world, and it became more obsessive in the 1680s when he was pondering the implications of the multiplication of mankind.

A foretaste of it can be found in the conclusion of *Verbum sapienti* when he asked what mankind should 'busy itself about' once peace and plenty were achieved. They should think about the 'works and will of God' since it was the 'natural end of man in this world' to prepare for 'spiritual happiness in that other world' to come. Those were the 'motions of the mind' which delivered infinite pleasure.[175] His standpoint was more explicit in 1685, in his reply to Robert Southwell's question about what would happen when the earth was fully peopled. Petty responded that the world would end with the full discovery of God's wisdom, the perfection of all knowledge; and that could only be achieved through dense populations 'in great cities and cohabitations of men' where 'art and sciences are better cultivated than in deserts'.[176] It was a prospect inherited from Bacon and others who took the prophecies of Daniel as guides to the future, and it accorded with some of the arguments in Matthew Hale's *Primitive Origination of Mankind* (1677), which influenced Petty and other Baconians like Southwell.[177] It was the vision of a road to the future whose course Petty used his political arithmetic and projections of demographic and economic growth to chart, and it inspired the whole of his political economy. Material progress was the necessary precondition for wisdom and endless pleasure.

Petty's writings, those he published and those he did not, display a striking uniformity of tone and purpose over the whole of his career. In his *Advice* to Hartlib in 1647, he was already using Holland as the model of what states and peoples could achieve, when everyone was 'set on work, barren grounds made fruitful, wet dry, and dry wet, when even hogs and more indocile beasts shall be taught to labour, when all vile materials shall be turned to noble uses, when one man or horse shall do as much as three, and everything be improved to strange advantages'. That owed something to Gabriel Plattes;[178] but Petty was still thinking in similar terms thirty years later when writing for private circulation a treatise 'Of the Scale of Creatures', from man down to 'the smallest and simplest animal that man can discern', a theme which Hale and other natural theologians were also exploring in their analyses of the works of Creation.[179]

---

[174] Peter Elmer, *The Miraculous Conformist: Valentine Greatrakes, the Body Politic, and the Politics of Healing in Restoration Britain* (Oxford, 2013), pp. 75, 143.
[175] Petty, *EW*, i. 119–20.
[176] McCormick, *Petty*, p. 289; *Petty–Southwell Correspondence*, pp. 213–15.
[177] Webster, *Great Instauration*, pp. 22–5; Harrison, *Fall of Man*, pp. 186–91. On Hale, and Southwell's interest in him, see Beinecke Library, Yale University, OSB MS 41, Box 3; below, p. 140.
[178] Petty, *Advice*, p. 156; Plattes, *Discovery of Infinite Treasure*, p. 92.
[179] McCormick, *Petty*, pp. 226–30. The introduction to *William Petty on the Order of Nature*, ed. Lewis, explores the origins and later influence (through its manuscript circulation) of this treatise. Cf. below, p. 200.

There Petty was concerned to show that the distinction between men and animals was one of degree. Some animals had near-human attributes without any relation to their size, including those bees whose inventiveness and 'admirable operations' made them second only to man 'in respect of... spirituality and power'. Man, however, was nearest in likeness to God and capable of moving ever nearer:

> I do not only compare man with the inferior creatures, ... but I do also compare the highest improvements of mankind in his mass with the rudest condition that man was ever in: thereby inferring that if man hath improved so much in the several past centuries and ages of the world, how far he might proceed in six thousand years more....

Mankind would advance further, 'towards the top of the great scale' of Creation.[180]

When Thomas Sheridan learned from what he knew of *Verbum sapienti* that labour was far more valuable than either land or trade, and concluded that man was 'as improvable' as land 'if not more', he had grasped something, but only part, of Petty's message. Always aiming at impossible things, 'examining all possible contingencies' and deducing 'nothing which is not demonstration', Petty was far more than the expert in 'manufactures and improvement of trade' whom John Evelyn admired.[181] The improvements he envisaged, and whose trajectory his political economy was designed to illuminate, were infinite. They were perhaps the greatest of the great expectations engendered by the English revolutions of the mid-seventeenth century. A century later, in 1759, Dr Johnson took a decidedly sceptical view of them:

> When the philosophers of the last age were first congregated into the Royal Society, great expectations were raised of the sudden progress of useful arts; the time was supposed to be near when engines should turn by a perpetual motion, and health be secured by the universal medicine.... But improvement is naturally slow. The society met and parted without any visible diminution of the miseries of life.[182]

Improvement was naturally slow, but Petty's vision extended over centuries rather than decades, and his rewriting of political economy turned the material progress expected by Hartlib's generation into something whose advance or retreat could be demonstrated and tracked over time.

---

[180] *William Petty on the Order of Nature*, ed. Lewis, pp. 96–7, 119–21.
[181] Sheridan, *Rise and Power of Parliaments*, p. 188; *The Diary of John Evelyn*, ed. E. S. de Beer (6 vols, Oxford, reprint 2000), iv. 58.
[182] Ernest B. Gilman, *Plague Writing in Early Modern England* (Chicago, 2009), p. 27, citing *The Idler*, 88 (1759).

5

Wealth and Happiness 1670–1690

For more than a decade after 1660 infinite improvement was a promise for the future which bore little relationship to current realities. When there were wars with the Dutch, new commercial threats from the France of Louis XIV, and every sign of economic depression at home, many writers about improvement and trade were anxious to reverse economic decline not accelerate economic growth; and the evidence which people like Petty presented in an effort to contradict them was either dismissed as partial and insubstantial or interpreted as proof of national degeneration and decay. The prospect of wealth and happiness, material progress, seemed to be receding into the far distance.

That climate of opinion was substantially altered, though not entirely turned around, by two things. One was mounting evidence, especially in the long interval of peace after the end of the third Anglo-Dutch War in 1674, that trade, wealth, and plenty were in reality increasing. The second was a style of argument, derived largely from shifts in European moral and political philosophy, to the effect that self-interest and the private appetites for profit which delivered material comforts were to be welcomed because they promoted the public good, not condemned as evidence of a corrupt and declining world. This was the case propounded in England at the end of the 1670s by John Houghton and Nicholas Barbon, who stressed the economic utility of consumer demand for luxuries in growing cities like London and its contribution to general wealth and happiness.

The argument that the growth of London drained away more and more of the riches of the provinces, and the consumption of luxuries ever more of the country's treasure, was never silenced. Houghton and Barbon faced immediate and hostile criticism, and their case remained contentious. But it also won fresh support, and by 1690 the fact that England's wealth had grown and grown rapidly since the middle of the century was incontrovertible. It remained to be seen whether that could continue when there was war against France, and if it did, whether it would make the nation happy as well as prosperous.

DECAY OR PROSPERITY?

When it was written in the early 1670s, *Political Arithmetick* was designed as an intervention in a debate about the condition of England. In his Preface, Petty listed the 'persuasions which I find too current in the world', and which he wanted to dispel: that rents had fallen, money was in short supply, the country 'underpeopled', and trade in lamentable decay, and that 'the whole kingdom grows every day poorer and poorer'. He was referring directly to the 'dismal suggestions' made by Roger Coke in a treatise of 1670 on 'the decay of the strength, wealth and trade of England'; and several others were making the same points, all of them summarized in *The Grand Concern of England Explained* in 1673, which concluded that England had been 'far richer and more populous' forty or fifty years before.[1]

Some of these writers were welcoming the Restoration as an opportunity for political and religious reform and rejecting most of what had happened since 1640, but all of them were looking for means to improve the national economy and convinced that they were crucial to the country's well-being. Edward Chamberlayne, who may have been the author of the *Grand Concern*, published an agenda of proposals 'beneficial for England' for the attention of parliament in *England's Wants* in 1667. Many of them were about improving the moral condition of the nation, partly through a fully reformed established church with a largely celibate clergy, and partly by means of sumptuary laws and heavy taxes on 'all such commodities as occasion either excess or luxury, wantonness, idleness, pride or corruption of manners'. Yet this future Tory also included on his list items picked up from Hartlib and his colleagues: hospitals and pawnshops, salaried physicians, a uniformity of measures, calendar reform, and even encouragements to the conversion of the Jews.[2] He referred again to what was wanting in 1669, in the first edition of *Angliae notitia*, and only gradually adopted a more optimistic stance as edition succeeded edition over the next thirty years.[3]

Roger Coke was a man of similar stamp. As attached as Chamberlayne to the landed interest and the established church, and nearly as prolific in his publications, he also aspired, though unsuccessfully, to be a second Petty. He deployed some arithmetic and a large array of axioms and theorems to arrive at predetermined propositions, often passionately held and sometimes contrary to what was now received opinion. He was fiercely opposed to the Navigation Acts, for example, arguing in *England's Improvements* (1675) that the English should follow the Dutch and allow free access by foreign merchants to English ports. Yet his aim was wholly conventional, to 'demonstrate that England is capable of greater wealth and strength than the United Netherlands (or perhaps any country else)'. What prevented that was 'the vast and wild prodigality of vain men and women'. 'Our affluence, luxury, and irregular management of trade renders us poorer, and in a worse condition, than if we had no trade at all.' Although clearly not something

---

[1] Petty, *EW*, i. 241–2; Roger Coke, *A Discourse of Trade* (1670), title page, 'Preface to the Reader', largely reprinted in Roger Coke, *A Treatise Wherein is demonstrated, That the church and state of England, are in equal danger with the trade of it* (1671); *The Grand Concern of England Explained* (1673), p. 5.

[2] Edward Chamberlayne, *England's Wants: or several proposals probably beneficial for England* (1667), title page, pp. 4–8, and *passim*. On the authorship of the *Grand Concern*, see Paul Slack, 'The Politics of Consumption and England's Happiness in the Later Seventeenth Century', *EHR* 122 (2007), p. 612.

[3] Edward Chamberlayne, *Angliae notitia* (1669 edn), pp. 69, 84–5; above, p. 43.

Coke welcomed, that was one of the first uses of the word 'affluence' to refer to England's material wealth.[4]

The contentious items in the grand concerns of England were therefore still what they had been in the 1620s and 1650s. Attention focused again on the balance of trade, and it is no accident that Mun's *England's Treasure* was published at last in 1664.[5] Coke referred to Mun, and also to Fortrey's calculation in *Englands Interest and Improvement* in 1663 that the adverse balance of trade with France amounted to as much as £1,600,000 a year. The figure was reduced slightly in a 'Scheme of Trade' presented to parliament in 1674, and both figures, though wholly misleading, were much cited for half a century afterwards.[6] They sustained continuing hostility to all things French; and they gave the support of spurious numbers to persistent complaints about the 'vices tending to prodigality' which Locke thought ruinous to trade in 1674, and about the 'extravagant habits and expenses of all sorts of persons' which the *Grand Concern* called 'high living'.[7]

According to Lord North in 1669, the taste for 'high living' had infected even the nobility and gentry, who were deluded into thinking the indulgences of 'self-love' a 'great happiness'. They led only to idleness, waste, and a universal 'excess in bravery of apparel'. When everyone was living 'above or beyond' their means, the answer once again was frugality like that of the Dutch and the 'moderation' in expenditure which North saw as the key to the 'oeconomy' of a nation just as it was to that of a household.[8] In 1666 Charles II himself was persuaded to set an example and try to turn the tide against French fashions when he sponsored and (for a time) sported the new gentleman's waistcoat of good English cloth, and in doing so effectively invented the three-piece suit.[9] Some members of parliament wanted the restoration of sumptuary laws in order to encourage English textile production, reinforce social distinctions in dress, and restrain the appetites of a people who had become 'wanton in their plenty' since 'the restoration of the king'.[10] Majority opinion continued to favour instead heavy customs duties on luxuries, because they were likely to be more effective, and tariff barriers against French imports, which tended in practice to fluctuate with changes in English foreign policy.[11]

---

[4] Roger Coke, *England's Improvements* (1675), pp. 113–15; Coke, *Discourse*, pp. 19, 46–7, 77; below, p. 170.

[5] On the circumstances, see Steven C. A. Pincus, *Protestantism and Patriotism* (Cambridge, 1996), p. 257.

[6] Coke, *Discourse*, pp. 15, 38; Samuel Fortrey, *Englands Interest and Improvement* (Cambridge, 1663), pp. 22–5; D. C. Coleman, 'Politics and Economics in the Age of Anne: The Case of the Anglo-French Trade Treaty of 1713', in D. C. Coleman and A. H. John, eds, *Trade, Government and Economy in Pre-industrial England* (1976), pp. 189, 196–7.

[7] John Locke, *Political Essays*, ed. Mark Goldie (Cambridge, 1997), pp. 221–2; *Grand Concern*, pp. 49–50.

[8] Dudley, Lord North, *Observations and Advices Oeconomical* (1669), pp. 47, 65–8, 107; *The Use and Abuses of Money, And the Improvements of it* (1671), p. 7.

[9] David Kuchta, *The Three-Piece Suit and Modern Masculinity: England, 1550–1850* (Berkeley, 2002), pp. 80–1, 88.

[10] *Journals of the House of Lords*, 12 (1826), p. 228 (1668); Anchitell Grey, ed., *Debates of the House of Commons from the year 1667 to the year 1694* (10 vols, 1763), v. 155 (1678). Cf. *Journals of the House of Commons*, 8 (1803), pp. 467–8 (1663).

[11] C. D. Chandaman, *The English Public Revenue, 1660–1688* (Oxford, 1975), pp. 14–21.

Paradoxical though it may seem, there was a similarly familiar complaint, sometimes voiced by the authors who condemned extravagance, that there was not enough money around.[12] The argument was made more persuasive by the agrarian depression of the 1660s, when falling grain prices and declining rents hit the landed interest hard and reinforced perceptions that all aspects of the economy, commerce, money, and land, were closely connected. Between 1668 and 1670 committees of the Lords and Commons and a new Council of Trade considered the decay of rents along with that of trade, and consulted experts like Worsley and Josiah Child, a merchant and government contractor who still had his fortune to make in the East India Company but now made his mark as a political economist. It was a 'parliamentation and consultation' about economic causes and remedies like that of the 1620s, and one in which Child saw himself in the role of a second Malynes.[13]

Child's remedy for current ills was a deceptively simple one, another reduction in the maximum rate of interest. Low interest rates were the '*causa causans* of all the other causes' of the wealth of the Dutch Republic. When they had been lowered in England, in 1624 and 1651, they had multiplied the number of merchants and increased the 'riches and splendour of this kingdom . . . to above four (I might say above six) times so much as it was'. The trade of England since 1624, Child assured the Lords committee, had not been 'decayed in gross, but increased'. A further reduction would boost it again.[14] It was politically impossible in wartime when the government's need for loans kept interest rates high;[15] and there was in any case an argument of principle against Child's panacea, just as there had been against that of Malynes in 1622, and on the similar ground that low interest rates were 'not a cause but an effect of great riches'.[16]

The strength of the arguments was less important, however, than the fact that they immediately became public and had as great an impact on opinion outside parliament as the debates of the 1620s. In 1668 Child published his *Brief Observations concerning Trade, and Interest of Money*, which drew heavily on Petty and Worsley and had been written in 1665; and he reprinted alongside it Thomas Culpeper's old *Tract against Usurie* of 1621, perhaps to show how things had changed since then. Culpeper's son, Sir Thomas the younger, did the same, and added his own essay arguing that lower interest rates would solve all problems at once, increase the price of land, promote agrarian improvement, leave 'no common fields unenclosed', and even, he told the Lords committee in 1669, encourage

---

[12] e.g. *Use and Abuses of Money*, pp. 22–5.
[13] *ODNB sub* Child; William Letwin, *The Origins of Scientific Economics* (New York, 1964), pp. 4–10; above, p. 89.
[14] Josiah Child, *Brief Observations concerning Trade, and Interest of Money* (1668), pp. 7–8; *SCED*, p. 69.
[15] C. G. A. Clay, *Economic Expansion and Social Change: England 1500–1700* (2 vols, Cambridge, 1984), ii. 274–5.
[16] *SCED*, p. 73, which may have been a quotation from the title page of *Interest of Money Mistaken, . . . proving that the abatement of interest is the effect and not the cause of the riches of a nation* (1668). Petty assumed similarly that there was a 'natural' interest rate independent of government fiat: Petty, *EW*, i. 48. Cf. above, pp. 84–5.

frugality.[17] There were naturally hostile responses from those who thought the economy in precipitous decline, one of them insisting that ever since the interest rate was reduced in 1651 'trade hath been lost and the gentry ruined'.[18]

Public argument could never have settled disputes about the state of England; it simply exacerbated them. Child himself had second thoughts by 1673 when he told Petty he now thought the country 'in a declining condition' compared with what it had been, although he still hoped his proposal would be accepted so that 'our children may see this island more wealthy, strong, and populous than ever their fathers did'.[19] Publicity about interest rates and their consequences nonetheless added weight to the view that money must circulate if all sectors of the economy were to flourish, and that economic growth or decline affected all of them simultaneously. Fortrey had shown in 1663 that England's capacity to compete with France and the Dutch Republic depended as much on increasing the productivity of land and the output of manufactures as on low interest and tariff barriers.[20] Worsley told the House of Lords committee discussing the decay of rents from land in 1669 that they should be equally concerned with encouraging manufactures and exports which would increase the stock of the kingdom and bring down interest rates. In *England's Improvement Reviv'd* (1670) John Smith argued that economic revival depended, not only on improvements in trade and navigation, but on land, every acre of which must be 'improved that is capable of improvement' in order to boost productivity and population.[21]

In one of his reports back to the Doge and Senate in 1674 the Venetian ambassador remarked that the House of Commons was filled with 'persons skilled in economics',[22] but the point had wider application. Representatives of every vested interest had learnt which rhetorical buttons to press when arguing their case. Monopoly trading companies were anxious to show that they benefited employment in domestic manufactures as well as strengthening the country against its rivals, and clothiers faced with French and Dutch competition cited Child and Ralegh to demonstrate the importance of their industry for foreign trade and navigation.[23] By the 1670s even the stocking-makers of Leicester were able to resist the imposition of a local monopoly with the argument that it was not 'the curious making of a few stockings, but the general making of many that is most for

---

[17] Letwin, *Origins of Scientific Economics*, pp. 5–6, 15–19; Joyce Oldham Appleby, *Economic Thought and Ideology in Seventeenth-Century England* (Princeton, 1978), pp. 88–9; SCED, p. 73.
[18] *Use and Abuses of Money*, p. 15.
[19] BL, Petty Papers, Add. MS 72850, fo. 132.
[20] Fortrey, *Englands Interest*, pp. 6–8, 14–20, 25–9, 32–3, 40–2.
[21] SCED, pp. 69, 71; John Smith, *England's Improvement Reviv'd* (1670), pp. 1, 11, 13. Cf. Worsley's proposals to the Council of Trade in 1668, in SCED, pp. 533–6.
[22] Steve Pincus, *1688: The First Modern Revolution* (New Haven, 2009), p. 89.
[23] [Robert Ferguson], *The East-India-Trade: A Most Profitable Trade* (1677), p. 6; Philopatris, *A Treatise wherein is demonstrated . . . That the East-India trade is the most national of all foreign trades* (1681), p. 4, 7, 12–13; *Certain Considerations Relating to the Royal African Company of England* (1680), pp. 1, 5; William Carter, *England's Interest Asserted, in the Improvement of its Native Commodities* (1669), pp. 31–3 and *passim*.

the public good, for that sets more people on work'.[24] The nation was full of would-be political economists.

London was already that in 1664, according to the French historiographer Samuel Sorbière, in a report on his visit to England which added further fuel to the debate about England's decay. He found the English 'naturally lazy', spending 'half their time in taking tobacco' in coffee-houses, but they were 'all the while exercising their talents about the government, talking of new customs, of the chimney money [the hearth tax], the management of the public treasure, and the lessening of trade; and so looking back, calling to mind the strength of their fleets in Oliver's time'.[25] Although Sorbière praised the Royal Society and had 'great expectations' of Boyle and his colleagues, Thomas Sprat, the Society's historian, was driven to respond in an open letter to Christopher Wren. If Sorbière had visited the coal pits of Newcastle or the lead mines of Derbyshire, or seen the clothworkers of the north or of the west, the plough-lands of Devon, or the orchards of Hereford, not to mention the new rivers of the Fens and the tin mines of Cornwall, he would have known that the English were industrious not idle. As for the complaint of declining trade, 'the public and cheerful voice of all Englishmen' was about its improvement since 1660 and the country's capacity now to outdo the Dutch in every respect. Reminding Wren of a conversation in the 1650s, when they had agreed that the Rome of Augustus was the place they would most wish to have lived in, Sprat now thought the England of Charles II, with its science and improvements, and its capital being wondrously rebuilt after the Fire, was far superior, and the best of all conceivable worlds.[26]

Sprat sounded more like Bacon exaggerating the achievements of James I than an unprejudiced observer of a country whose domestic and naval disasters in the 1660s contrasted starkly with the days of Cromwell.[27] He can scarcely have convinced many of his readers that England's present prospects were rosy. But they had been learning about England's economic history in the longer term, and there they could find some reassurance. It was still possible to look back to the benchmark of Edward III's reign, as William Temple did in 1673, and regret that there was not now a similarly healthy balance of trade; but Sheridan used the same evidence only to argue that England's trade in 1677 was thirty times greater than it had been then.[28] In the second half of the seventeenth century, thanks as much to the works

---

[24] Mary Bateson et al., eds, *Records of the Borough of Leicester* (7 vols, 1899–1974), iv. 563–8.
[25] Samuel Sorbière, *Relation d'un voyage en Angleterre* (Cologne, 1666), p. 107 (first published in Paris, 1664); English translation from Samuel Sorbière, *A Voyage to England . . . As also Observations on the same Voyage, by Dr Thomas Sprat* (1709), p. 54.
[26] Thomas Sprat, *Observations on Monsieur de Sorbier's Voyage into England* (1668 edn), pp. 76–8, 139–42, 242–9.
[27] Above, p. 75.
[28] William Temple, *Observations upon the United Provinces of the Netherlands*, ed. Sir George Clark (Oxford, 1972), p. 121; Thomas Sheridan, *A Discourse of the Rise and Power of Parliaments* (1677), pp. 212–13. For similarly divided views on commercial trends since the fourteenth century, see W.S., *The Golden Fleece* (1656), 'To the Reader'; *Britannia languens*, in J. R. McCulloch, ed., *Early English Tracts on Commerce* (Cambridge, 1952), pp. 397–9; Chamberlayne, *Angliae notitia* (1700 edn), pp. 65–6.

of Harrington and Bacon as to those of Ralegh, the origins of England's commercial success were more often traced to the sixteenth century than the fourteenth.

A writer in 1668, for example, was convinced that employment had increased since 1485, when the idle retainers of nobles were dismissed, that trade had flourished ever since 1545 when popery was banished, and that England was now much richer than Scotland because it was no longer dominated by its nobility and gentry.[29] A few years later Ambrose Barnes, the ex-Republican merchant of Newcastle, had learnt from reading Slingsby Bethel that 'popish religion creates an unaptness for trade', and from Harrington that in Henry VIII's reign the wings of the nobility and clergy had been clipped, and landed property redistributed to 'the people'.[30] By 1680 a preacher addressing the merchant seamen of London was able to cite Bacon's favourite prophecy from the Book of Daniel about the advancement of knowledge, and apply it to 'the great improvement' of navigation in recent centuries. He could assume that there was no need in front of this particular audience to elaborate on the 'modern improvements' in English commerce which had brought so many benefits.[31]

There was a related interest in the comparative economic development of other countries over time and what that had to teach about England's prospects. Child had made his point about low interest rates by looking at their general level elsewhere. They were lower in the United Provinces and Italy than in England, of course, a little higher (7 per cent) in France, but 12 per cent in Scotland and Ireland and 10 or 12 per cent in Spain where the people were 'poor, despicable, and void of commerce'.[32] In 1675 Lord Treasurer Danby commissioned and carefully annotated a report on excise taxes and their virtues, which opened with a short history of trade, beginning with the ancient cities of Greece and the Levant, now wholly in decline under the Ottoman yoke, and ending with England and Holland. Since Elizabeth's reign England had been greatly enriched by commerce, its customs duties had vastly increased, and every county now had gentry families whose wealth had originated in trade. The best example of the benefits of modern commerce, however, was Holland, a 'little spot of unhealthful land' now with flourishing towns, ample public revenues thanks to modest excise taxes, and 'infinite numbers of people' thriving 'like bees in a hive'. The lesson for Danby, and for English improvers, was that trade, industry, and people 'beget each other'.[33]

The author had clearly been reading William Temple's *Observations upon the United Provinces of the Netherlands*, published in 1673, an account whose political insights and literary merits gave it instant appeal and lasting fame. A diplomat as well travelled as Petty, and another Irish landowner, Temple naturally drew the

---

[29] *Interest of Money*, pp. 10, 18–20.
[30] *Memoirs of the Life of Mr. Ambrose Barnes*, ed. W. H. D. Longstaffe (Surtees Society 50, 1867), pp. 47, 213; Slingsby Bethel, *The Present Interest of England Stated* (1671), p. 21.
[31] Richard Holden, *The Improvement of Navigation a great Cause of the Increase of Knowledge* (1680), pp. 3–4, 8.
[32] Child, *Brief Observations*, p. 9.
[33] Beinecke Library, Yale University, OSB MSS 6, Danby Papers, series II, Box 2, file 31, fos 1–4. The author of the paper is unknown.

same contrast as Petty did between Ireland and Holland. He had no interest in 'computations' and no need for them. Any educated observer could see that the key to the 'rise and progress of the United Provinces' was the 'great multitude of people crowded into small compass of land' who were necessarily induced and habituated to parsimony, industry, and labour. In Ireland, by contrast, 'the largeness and plenty of the soil and scarcity of people' produced only idleness. The whole of history, from Tyre, Athens, and Carthage, to the rise of Venice, showed that environment was the 'original of trade'.[34]

In the following year John Evelyn brought something from Temple to the title and content of his *Navigation and Commerce, their Original and Progress*. He added a longer history of the growth of commerce, particularly in England and Holland, and an emphasis on control of the oceans and therefore of the world borrowed from the aphorism attributed to Ralegh; and he concluded much as Temple had done, but with one significant reservation. He agreed that it was not 'the vastness of territory but the convenience of situation' which 'improved a nation'. But he denied that a 'multitude of men' necessarily enriched a country: it was not their number that mattered, but 'their address and industry'.[35] Evelyn's qualification about multitudes pointed to the one immediate, conspicuous, and persistent obstacle which seemed to lie in the way of England's improvement and progress. The population was no longer rising, and it was not to increase significantly again for half a century.

That watershed in the country's demographic history was a phenomenon common across northern Europe. It was evident in the Dutch Republic after 1670, and although it was not a matter for much anxious comment in France until the eighteenth century, Colbert was already pursuing policies there in the 1660s which were intended to stimulate population growth.[36] What drew attention to the problem in England in the same decade was the fall in agricultural rents and rise in the cost of the labour which made 'want of people' a recurrent complaint in all economic debate, inside and outside parliament.[37] Its causes seemed equally obvious: the great plague of 1665 and civil and foreign wars; emigration overseas, which we now know to have been at its greatest between the 1630s and 1660s; but also a decline in fertility which was attributed both to family limitation and to 'a spirit of madness' preventing people marrying at all,[38] another trend which

---

[34] Temple, *Observations upon the United Provinces*, pp. 7, 108–10. On Temple's Irish interests, see William Temple, *An Essay upon the Advancement of Trade in Ireland* (?Dublin, 1673), 2–3; Toby Barnard, *Improving Ireland?* (Dublin, 2008), pp. 75, 87.

[35] John Evelyn, *Navigation and Commerce, their Original and Progress* (1674), pp. 15–16.

[36] Jan de Vries and Ad van der Woude, *The First Modern Economy: Success, Failure, and Perseverance of the Dutch Economy, 1500–1815* (Cambridge, 1997), p. 689; Leslie Tuttle, *Conceiving the Old Regime: Pronatalism and the Politics of Reproduction in Early Modern France* (Oxford, 2010), pp. 50–4, 151; Andrea A. Rusnock, *Vital Accounts* (Cambridge, 2002), pp. 179–80. There was some suspicion in England that other countries were also complaining of decay in the 1660s: Smith, *England's Improvement*, p. 1.

[37] e.g. *SCED*, pp. 68, 77, 80, 85.

[38] *SCED*, pp. 742, 758–60. Carew Reynell (*SCED*, p. 758) put some numbers to the losses, and while exaggerating casualties in the civil war and from plague, seems to have estimated the number of emigrants more accurately. For modern calculations of the latter, see E. A. Wrigley and R. S. Schofield,

seems to have been at an unusual peak in mid-century.[39] Thomas Sheridan's observation in 1677 that England was 'already very much underpeopled' had become a commonplace.[40]

Underpopulation was consequently used to justify a host of proposals often advocated for other reasons. None of them got very far in the real political world. Some commentators wanted controls on emigration, a striking reversal of earlier opinion which saw it as a necessary outlet for a population surplus, but they were never imposed given the continuing need to populate old and new plantations. There was much more pressure for encouragements to immigration and for legislation permitting naturalization, and want of people was an often cited argument for greater religious toleration, but such devices were controversial. James II was not being entirely disingenuous when he instructed his judges to say the Declaration of Indulgence of 1687 was 'a certain means to make these kingdoms more populous, and by consequence to be the chiefest place for trade in the Christian world', but that did not make the Declaration popular.[41] Letters of denization helped to bring an estimated 50,000 French Protestants into England between 1660 and 1700, but xenophobia kept a Naturalization Act off the statute book until 1709.[42]

Any effort to stop parish officers preventing 'poor people from marrying' in order to avoid extra burdens on the poor rates met with equally strong resistance. 'Those that are like[ly] to multiply may do it for the king, but not for Middle Claydon,' the parson of that parish effectively retorted in 1671.[43] There were calls for tax and other incentives to encourage marriage and childbearing of the kind which were being implemented in France and which had been advocated in England in the 1530s, at the end of an earlier period of demographic stagnation. They were given no serious parliamentary hearing before the 1690s.[44] Suggestions that the marriage laws should be reformed to make divorce easier or even legitimize bigamy (for men), which Locke thought might produce 'a multitude of strong and healthy people', naturally got nowhere at all. When Michael Mallet proposed a bill to

---

The Population History of England 1541–1871: A Reconstruction (Cambridge, 1989), pp. 220–1, 224; Nicholas Canny, ed., *The Oxford History of the British Empire*, i: *The Origins of Empire* (Oxford, 1998), pp. 138–9, 176–7, 222.

[39] On proportions not marrying, see E. A. Wrigley et al., *English Population History from Family Reconstitution, 1580–1837* (Cambridge, 1997), p. 195; Richard Grassby, *Kinship and Capitalism: Marriage, Family and Business in the English-Speaking World, 1580–1740* (Cambridge, 2001), p. 78; Lawrence Stone, *The Family, Sex and Marriage in England 1500–1800* (1977), pp. 44–7. For similar complaints in the 1690s, see below, pp. 195–6.

[40] Sheridan, *Rise and Power of Parliaments*, p. 179.

[41] BL, Add. MS 32523, fo. 54ᵛ. Cf. Andrew Browning, ed., *English Historical Documents, 1660–1714* (1953), pp. 387, 396; Scott Sowerby, *Making Toleration* (Cambridge, Mass., 2013), pp. 62–3.

[42] Daniel Statt, *Foreigners and Englishmen: The Controversy over Immigration and Population, 1660–1760* (Newark, Del., 1995), pp. 34–7 and *passim*.

[43] SECD, p. 759; Steve Hindle, *On the Parish?* (Oxford, 2004), pp. 349–51. Cf. *Britannia languens*, p. 350; Davenant, *PCW*, ii. 192; BL, Lansdowne MS 691, fo. 104ᵛ.

[44] SCED, p. 759; Paul Slack, *'Plenty of People': Perceptions of Population in Early Modern England* (Reading, 2011), p. 5; below, p. 181. Harrington may have revived interest in this topic: James C. Riley, *Population Thought in the Age of the Demographic Revolution* (Durham, NC, 1985), pp. 65–7.

legalize bigamy in 1675, 'pretending' that 'it was for peopling the nation', but probably to enable the king to beget a legitimate heir, MPs understandably doubted his sanity.[45]

Given the constant insistence on the need for hands if not lands for England's improvement, however, it was entirely rational to question whether trade and wealth were increasing, or ever could increase, as fast the advocates of present or future prosperity claimed. Never confident that population growth in England was either likely or necessary, John Graunt persuaded Pepys in 1663 that 'the trade of England is as great as ever it was, only in more hands'. Landowners facing prolonged agrarian depression would not have agreed, and neither would cloth merchants suffering from a long decline in the traditional textile industry, one of them complaining as late as 1675 that trade had decayed since 1640 and men were living in an 'iron age'.[46] We now know that all sectors of English trade were growing after 1660, but Sprat was almost alone in citing 'the perpetual advancement of the customs' against the 'universal murmur that trade decays' in 1667.[47]

In the 1670s, however, and particularly during the great commercial boom which came with peace in 1674, there was mounting optimism that trade was increasing and the nation's prosperity with it.[48] As circumstances changed, all factions and parties, Court and Country and then Tory and Whig alike, began to accuse one another of exaggerating decay and to make the contrary case for mounting prosperity. Slingsby Bethel may have been a little premature in 1673 when he argued that Dutch trade had already passed its zenith and England was able to surpass them 'without fighting them'. Petty was certainly premature when claiming in *Political Arithmetick* that England had grown so wealthy that even France might be outdone without the necessity of war.[49] But Carew Reynell, despite his fears about depopulation and its consequences for the landed interest, warmly welcomed the growth of 'great and glorious London' and was confident about England's 'infallible advance' towards 'infinite wealth and greatness'; and Bethel, the Interregnum radical, fully agreed. London was becoming 'the emporium and great mart of Europe', and the English were set to become 'the richest people in the universe'.[50]

---

[45] John Locke, *Political Essays*, ed. Mark Goldie (Cambridge, 1997), pp. xxvi, 255–7 (and cf. p. 253); Basil Duke Henning, *The House of Commons 1660–1690* (3 vols, 1983), iii. 9. For later interest in bigamy and its justification, see Mark Knights, *The Devil in Disguise: Deception, Delusion, and Fanaticism in the Early English Enlightenment* (Oxford, 2011), pp. 125–40.

[46] *The Diary of Samuel Pepys*, ed. R. C. Latham and W. Matthews (11 vols, 1970–83), iv. 22; Joseph Trevers, *An Essay to the Restoring of our Decayed Trade* (1675), p. 20.

[47] Thomas Sprat, *History of the Royal Society*, ed. Jackson I. Cope and Harold Whitmore Jones (1959), pp. 400–1. He may have taken the point from Clarendon: S. C. A. Pincus, 'Popery, Trade and Universal Monarchy: The Ideological Context of the Outbreak of the Anglo-Dutch War', *EHR* 107 (1992), pp. 15–16.

[48] On the commercial boom, see Clay, *Economic Expansion and Social Change*, ii. 181; Chandaman, *English Public Revenue*, pp. 32–5.

[49] Slingsby Bethel, *Observations on the Letter Written to Sir Thomas Osborn* (1673), pp. 8, 11; Petty, *EW*, i. 278, 284, 296–7. Cf. Steve Pincus, 'From Butterboxes to Wooden Shoes: The Shift in English Sentiment from Anti-Dutch to Anti-French in the 1670s', *Historical Journal*, 38 (1995), pp. 333–61.

[50] Carew Reynell, *The True English Interest or An Account of the Chief National Improvements* (1674), title page, sigs A5–6ʳ; Slingsby Bethel, *An Account of the French Usurpation upon the Trade of England* (1679), pp. 21, 24.

There were real conflicts of interest between defenders of land and trade, just as there were between Tories who mocked 'the improvers of England' as a 'fanatic crew' and their Whig targets who retorted that Tories were papists.[51] But much of the noise made about trade, as John Collins pointed out, was propaganda, trying to drive a wedge between 'the gentry or Country party' and 'the merchants and trading part of the nation'.[52] It was not an expression of different economic perceptions. By the end of the 1670s it was usual for defenders of trading companies to emphasize that the interests of land and commerce were the same, and during the Exclusion Crisis it was a Tory supporter of the landed interest who argued that the country 'never had a greater trade... never more shipping, never more merchants'.[53] So unanimous was the confidence in England's thriving commerce by 1677 that Sheridan thought it necessary to warn his contemporaries that they had 'wrong notions of providence' because so many of them were beginning to assume that supremacy in world trade would fall into the country's lap as a matter of course.[54]

If there was no longer much dispute about the country's economic success, however, there remained considerable disquiet about its consequences, and about two of them in particular. Everyone agreed that the most dangerous of them was the debauchery and extravagance which might ultimately threaten commerce and industry. Even as vociferous a supporter of trade as Bethel thought he should set an example of frugality rather than 'riotous wasting' by cutting down on the traditional hospitality expected of him as Whig Sheriff of London in 1680–1.[55] The second, more contentious, consequence, which Bethel welcomed, was the growth of London itself. The case against it was still what it had been in 1673, when the new coaches on city streets, which Child cited as evidence of rising wealth, were taken as symbols of all that was worst about metropolitan excess, and attacked at length by the author of the *Grand Concern*.[56] By 1680, however, party divisions had increased Tory hostility to a metropolis dominated by Whigs, and it had a weight of provincial opinion behind it. One writer argued that although there had never been more money in the nation, it was not 'equally divided', and the fault lay with 'the gentry and nobility living so much at London' and spending their rental incomes there. Another favoured a tax on new building near a city whose continued expansion made 'the nation weak and rickety', and he returned to the powerful

---

[51] [Andrew Yarranton], *A Coffee-House Dialogue* (1679); Andrew Yarranton, *England's Improvements Justified* (1680).
[52] John Collins, *A Plea For the bringing in of Irish Cattel* (1680), p. 19. Cf. Richard Allen, *Insulae fortunatae* (1675), p. 10.
[53] Ferguson, *East-India-Trade*, p. 4; John Nalson, *The Present Interest of England; or, A Confutation of the Whiggish Conspiratours Anty-Monyan Principle* (1683), p. 4, 23–4. For the rage of party in parliament and the press during the Exclusion Crisis, see Mark Knights, *Politics and Opinion in Crisis, 1678–81* (Cambridge, 1994), pp. 168–74.
[54] Sheridan, *Rise and Power of Parliaments*, p. 226.
[55] Bethel, *Present Interest*, pp. 12–13.
[56] Child, *Brief Observations*, pp. 7–9; *SCED*, pp. 379–86. For later reference to coaches, see below, p. 192.

point made by the *Grand Concern* that metropolitan growth threatened to 'depopulate' the whole country.[57]

No one took the demographic argument more seriously than Petty, the man who measured and publicized the growth of London, who included its 'coaches and splendour of equipage exceeding former times' among 'the improvements of England', and who welcomed its consumer demand because, like the 'luxury' he wanted to see among the Irish, it stimulated economic growth.[58] Yet Graunt had shown that London was wholly dependent on immigration from the provinces for its rising population, thanks to high urban mortality and unusually low fertility, which he attributed to metropolitan 'intemperance' and 'fornications' and, less conventionally, to 'the anxieties of the mind of men of business'.[59] If London was to grow further, it was vital that the population of the rest of England should continue to grow, and Petty used records from the hearth tax and the Compton Census of communicants in 1676 in an effort to persuade himself that it was doing so. His conclusion, that the English population had risen from Graunt's 6.4 million in 1662 to 7.4 million in 1682, gave him some reassurance.[60]

He would have found some comfort also in Matthew Hale's demonstration, published in 1677, that the multiplication of mankind was continuing, and no longer threatened by the great plagues and famines of the past. Recent epidemics were 'sharp and speedy' and 'rarely' the cause of major mortality, and famine had 'not been of late times much observed', not least 'because of the great industry of mankind, improving and increasing the fruits of the earth'.[61] Petty would have agreed with most of that, and after the plague of 1665 he had been ready to take bets that the next one would be less severe, especially if he was put in charge of quarantine precautions against it.[62] He was probably fortunate that the wager was never tested. But he knew by 1680 that even without plague urban mortality continued to rise, and he deduced that it was bound to increase as London grew further towards his ideal of being 'the best peopled and wealthiest city of the known world'. He agreed with his contemporaries also that there was no sign of fertility increasing to compensate, either in London or the provinces, and came to think, as we shall see in the next chapter, that a point would come when the population of England could grow no further.[63]

A realistic answer to the problem of a more or less stationary population ought to have been that it was not—or at least not yet—a problem at all. If commerce and wealth were increasing, it showed that productivity per head must be increasing. Evelyn was perhaps hinting at that when he denied that multitudes were essential to

---

[57] *Ananias and Saphira discover'd* (1679), p. 5; Nalson, *Present Interest*, p. 31.
[58] Petty, *EW*, i. 243; above, p. 123.
[59] Petty, *EW*, ii. 371, 373–4. For similar reflections on the causes of low fertility in London, see Gregory King, 'Natural and Political Observations and Conclusions upon the State and Condition of England, 1696', in Peter Laslett, ed., *The Earliest Classics* (Farnborough, 1973), p. 44.
[60] Slack, 'Measuring the National Wealth', p. 618; Petty, *EW*, ii. 460.
[61] Matthew Hale, *The Primitive Origination of Mankind, considered and examined according to the light of nature* (1677), pp. 203, 212–14, 227–9.
[62] *Petty Papers*, ed. Marquis of Lansdowne (2 vols, 1927), i. 26, 39–40; Petty, *EW*, i. 109.
[63] BL, Petty Papers, Add. MS 72866, fos 53–4; below, p. 198.

economic growth, and Sheridan when he remarked that men were as improvable as land if not more so; but no one, least of all Petty, was willing wholly to abandon the view that the multiplication of mankind was both providentially designed and economically beneficial.[64] What was acceptable instead was the proposition that since England was not as densely populated or urbanized as Holland, there was room for both population and productivity per head to increase at the same time.

That was Petty's conviction, and it was expressed with striking force in a paper for the Royal Society on 'the best way of England's improvement' by Reason Mellish, written around 1670. Mellish disputed Graunt's argument that England, given its acreage, could support no more than the present population. The number of people could grow by at least 25 per cent if they were all 'rightly employed'; and if the land was fully improved, as Fortrey had suggested, only 5 per cent of the total need be employed in agriculture. The majority could be engaged in industry and commerce, and the nation 'flow with plenty of everything that can be imagined'. England was 'capable of very great improvement'.[65]

Mellish deduced that immigration must be encouraged if his scenario was to be realized, and others supposed with greater optimism that economic migrants must necessarily be attracted as the economy grew. In 1678 Mark Lewis argued that population would increase wherever there was prosperity, and one would feed off the other. 'Where trade is, there is employment, where employment is, thither the people will resort, where people are there will be consumption of commodity... and trading will flourish.' Having perhaps read one of Petty's manuscripts, he had learnt that 'the consumption of the people is not so much as the product of their labours, but they are really the riches and strength of a nation: the more the merrier, like bees in a hive, and better cheer too.' Although he claimed that the banks he proposed would 'much conduce to... civilizing the nation, now degenerate into debauchery', there was no suggestion here that consumer appetites must be restrained.[66] On the contrary, consumption of commodity was the driver of commercial and economic growth. Consumption was presented as an economic virtue.

In 1673 Temple thought such opinions already so deep-seated that they must be rebutted. The example of the United Provinces showed that some of the 'maxims... current in our common politics' in England were wholly false. Chief among them was the proposition that the 'example and encouragement of excess and luxury... is of advantage to trade'. On the contrary, consumer appetites would ruin the nation. They could never be restricted only to domestic manufactures as some of their advocates supposed, and they must necessarily spread by imitation to all sections of society, even those 'by whose industry the nation subsists'.[67] Temple must

---

[64] Above, p. 128. On productivity and population, see also below, pp. 196–9, 238–40.

[65] Royal Society of London, Classified Papers, vol. 10. iii, item 16, fos 362–71. On Mellish, see Henning, *House of Commons 1660–1690*, iii. 48.

[66] Mark Lewis, *Proposals to the King and Parliament: Or A Large Model of a Bank* (1678), pp. 20–1, 25; Petty, *EW*, i. 306. Cf. Mark Lewis, *Proposals to increase Trade* (1677); Carl Wennerlind, *Casualties of Credit* (Cambridge, Mass., 2011), pp. 106–7.

[67] Temple, *Observations upon the United Provinces*, p. 120.

have known that a country as rich in natural resources as England could never be an environment like that of Holland, naturally inclined to frugality, even if it had not had the nobility and court which the Dutch thought responsible for English extravagance and corruption.[68] He might also have read Worsley's paper of 1668 which argued that sumptuary laws, which might have imposed artificial restraints on 'consumption and expense', were impracticable because other nations would retaliate in kind and English exports would decline. Temple's worst fears about current trends in opinion would have been fortified if he had seen an early manuscript of *Political Arithmetick*, where Petty had said that people would 'spend more' as urban populations increased and welcomed the 'greater consumptions' of townspeople as compared with countrymen, who lived 'out of the sight, observation, and emulation of each other'.[69]

The same tune had been sounded much earlier, however, in proposals for banks in the 1650s. Potter, for example, had anticipated Mark Lewis's expectation that an increase in the quantity of money in circulation would make people 'spend upon the comforts of this life', and so multiply employment and benefit everyone.[70] The case was reiterated even more strongly by Robert Verney in 1682, in a proposal for a bank founded on the deposited inventories of clothiers and offering credit in return. It would 'make a greater and quicker trade by increase of consumption', and by that means 'the number and estates of the rich will be augmented, and the number and necessities of the poor diminished'. Even at a time of financial and political crisis in the City of London, Verney was able to conclude with an assurance that any member of Hartlib's circle would have applauded:

To sum up all, whatever increaseth the money, or a valuable credit equivalent, increaseth expense and consumption; and whatever augments consumption, increaseth trade and employs more hands; so that England ... blessed with plenty of staple commodities and provisions, may ... become the emporium of Europe.[71]

Temple's current maxims of political economy remained controversial, but they had become more persuasive as the evidence for an increase in England's trade and prosperity mounted. By 1682, moreover, the case for consumer appetites and the cities where they were bred and flourished had been given even greater prominence by two writers joining in the debate about the condition of England for the first time.

CONSUMERS AND CITIES

In 1677 and 1678 two anonymous tracts defended consumers and cities with unqualified enthusiasm. The first of them was *England's Great Happiness*, which took the form of a dialogue between 'Content' and 'Complaint' where Content had

---

[68] Below, pp. 209, 244.   [69] *SCED*, pp. 535–6; Petty, *EW*, i. 290.
[70] William Potter, *The Key of Wealth* (1650), pp. 5–6; above, p. 109.
[71] Robert Verney, *Englands Interest or the Great benefit to Trade by Banks or Offices of Credit* (1682), pp. 4–7. On the crisis in the City, see Jennifer Levin, *The Charter Controversy in the City of London, 1660–88, and its Consequences* (1969), pp. 17–21, 56.

more space and the better of the argument. The complaints it dismissed were those of Roger Coke and the *Grand Concern* of 1673, that 'high living', consumer extravagance, was the ruin of England and the cause of its decline. On the contrary, Content asserted, 'a general high living' was both evidence of wealth and a stimulus to its acquisition. It made people industrious and 'everyone strive to excel his fellow', so that the English would never be satisfied 'till they engross the trade of the universe'.[72] The author was immediately identified as John Houghton, an apothecary-shopkeeper who supplied customers like Robert Hooke with imported chocolate and tobacco, and who was elected to the Royal Society in 1680.[73]

The second tract, following a few months later, remained anonymous, but it is clear from its content and style that it was the first publication of Nicholas Barbon, son of the Interregnum Puritan Praise-God Barbon, who had studied medicine in Leiden and was now making himself the greatest speculative builder in London. His business concerns were evident in the tract's title: *A Discourse Shewing the Great Advantages that New-Buildings and the Enlarging of Towns and Cities Do Bring to a Nation*. Houghton had praised the recent growth of London, and hoped to see it double in size in the next forty years and 'all the towns in England strive to imitate it'. Barbon had an equivalent message. Great cities were creative concentrations of arts and industry, and their consumer demand increased productivity in the countryside by means of 'improvements... greater than the increase of people' there. The multiplication of mankind combined with ever greater productivity per head brought a 'peculiar happiness', 'happiness and advantage', to town and country alike.[74]

It is likely that the two authors knew one another, perhaps through Hooke, and they were probably acquainted with Petty since both of them were skilled in political arithmetic and showed some knowledge of his findings in their publications.[75] Houghton in particular had absorbed the whole of the vision of improvement which, he said later, had been 'stirred up' by 'the indefatigable pains of Mr Hartlib and some others' and by 'profitable hints' from the Royal Society.[76] He could remember the 1665 plague, when he was apprentice to the master of the London pesthouse, and how the flight of the citizens had taken metropolitan habits to the provinces. That, together with the 'projects and industry' of gentry farmers,

---

[72] *England's Great Happiness; or, A Dialogue between Content and Complaint* (1677), pp. 6–8, 261. Houghton's authorship was acknowledged in a second edition in 1677. He may have taken his title from Richard Allen's *Insulae fortunatae* (1675), a tract 'showing the happiness of these nations' and dismissing talk of the decay of trade.

[73] Miles Ogborn, *Indian Ink* (Chicago, 2007), pp. 196–7.

[74] Nicholas Barbon, *Discourse Shewing the Great Advantages* (1678), pp. 2–4, 10, 13–15; Houghton, *England's Great Happiness*, p. 17. On Barbon's authorship of the *Discourse*, see Paul Slack, 'Perceptions of the Metropolis in Seventeenth-Century England', in Peter Burke et al., eds, *Civil Histories* (Oxford, 2000), pp. 175–6 n. 69.

[75] *The Diary of Robert Hooke 1672–80*, ed. Henry W. Robinson and Walter Adams (1935), pp. 14, 324, 395. Hooke had borrowed a manuscript copy of *Political Arithmetick* in 1678 (*The Diary of Robert Hooke*, p. 380). For the apparent influence of Petty on Barbon, see e.g. *Discourse Shewing the Great Advantages*, p. 14.

[76] John Houghton, *A Collection for the Improvement of Husbandry and Trade*, ed. Richard Bradley (4 vols, 1727–8), iv. 85.

had 'caused such an improvement' there 'as England never knew before'. Since then England had 'increased more in trade . . . than it is probable any nation hath done in like space'. 'Would we but consider the great things we have done', Houghton boasted in 1677, 'it would perhaps make us believe nothing to be impossible.'[77]

Both of them were first provoked into publication, however, by political circumstances, by a swing of opinion against France after 1674, especially in parliament, which directly affected their own interests. It led in 1678 to a prohibition on French imports, including wine, linen, and brandy, which Houghton had defended as necessities.[78] It also led to proposals for new taxes to meet potential military needs for defence (or even war) against Louis XIV, including the revival of a tax on new buildings in London which had been imposed in 1657 and abandoned at the Restoration. Although Petty had already shown that such impositions had no effect in restraining the growth of a metropolis 'already too big for the body', as the *Grand Concern* put it, the proposal was being hotly debated again in 1677.[79] It came to nothing in the end, but not before Barbon had picked up his pen and combined with Houghton in an effort to stem the tide of political opinion.

Their rhetoric was powerful, and in Houghton's case protracted:

If we have great magazines for war, and multitudes of brave ships; if we have a mint employed with more gold and silver than . . . they can well coin; if it be an affront to cause one to drink in any worse metal than silver . . . ; if we have six times the traders and most of their shops and warehouses better furnished than in the last age; . . . if many of our poor cottagers' children be turned merchants and substantial traders; if our good lands be made much better, and our bad have a six-fold improvement; if our houses be built like palaces, over what they were in the last age, and abound with plenty of costly furniture; and rich jewels be very common; and our servants excel in finery the great ones of some neighbour-nations; if we have most part of the trade of the world, and our cities are perhaps the greatest magazines thereof; if after a destructive plague and consuming fire, we appear much more glorious; . . .

then no one could deny that England had 'more wealth now than ever we had at any time before the Restoration' of Charles II. If high living was suppressed, however, and luxury trades destroyed, national decline would inevitably follow. The English would be 'no more than tankard-bearers and ploughmen, and our city of London will in short time be like an Irish hut'.[80]

Houghton did not stop there, however. He elaborated his case in his monthly periodical, *A Collection of Letters For the Improvement of Husbandry and Trade*, which was distributed free to his correspondents and fellow *virtuosi* between 1681 and 1685. Quoting liberally from *England's Great Happiness* in 1682, he again defended high living, and the higher and more widely practised the better; and he

---

[77] Houghton, *Collection*, iv. 51–2, 56; Houghton, *England's Great Happiness*, p. 18.
[78] Slack, 'Politics of Consumption', p. 612; Houghton, *England's Great Happiness*, p. 5.
[79] Slack, 'Perceptions of the Metropolis', pp. 174–5; Carolyn Andervont Edie, 'New Buildings, New Taxes, and Old Interests: An Urban Problem of the 1670s', *Journal of British Studies*, 6 (1967), p. 57; Petty, *EW*, i. 40–1; *Grand Concern*, pp. 4–5.
[80] Houghton, *England's Great Happiness*, title page, pp. 8, 19–20.

confronted the moral charge against it head-on: 'those who are guilty of prodigality, pride, vanity, and luxury, do cause more wealth to the kingdom than loss to their own estates.'[81] National interest thrust conventional morality aside. In 1685 he tried to clarify his argument about consumption being an economic stimulus, describing it now as a motor driving the 'wheel' of the whole economy. He weakened the force of the metaphor by suggesting that the wheel might turn from industry to idleness once plenty was achieved, but he thought its motion could somehow be halted by good political management.[82] In the Tory reaction which followed the Exclusion Crisis he placed his faith firmly in an enlightened monarch supported by a cooperative and generous parliament. It was in the interest of subjects 'plentifully to supply their king'; and their taxes were also 'a great wheel', encouraging the circulation of money, consumption, and economic improvement, and ensuring 'peace and plenty'.[83]

When Barbon presented the growth of cities as similarly a road to 'peace and prosperity' he copied something of Houghton's style. 'If a great city be the glory of a nation', Barbon contended, 'if it renders a people more easily governed and increases the prince's revenue, and if it be impossible for a nation to be either rich, strong or great, without increase of buildings', then it was absurd to seek to prevent urban growth.[84] Barbon did more than indulge in rhetoric, however. A cleverer man than Houghton, he brought out what had been implicit in *England's Great Happiness*, and in so doing answered Houghton's anxiety in 1685 that incentives to industry would lose their edge in conditions of luxury and plenty. The key was 'emulation', which occurred naturally in growing cities and never lost its force:

Emulation provokes a continued industry, and will not allow no intervals or be ever satisfied: the cobbler is always endeavouring to live as well as a shoemaker, and the shoemaker as well as any in the parish. So every neighbour and every artist [artisan] is endeavouring to outvy each other, and all men by a perpetual industry are struggling to mend their former condition; and thus the people grow rich, which is the great advantage of a nation.[85]

Barbon's emulation was a motor of economic growth with perpetual motion and infinite potential, and he too amplified his arguments in later publications. In *An Apology for the Builder* of 1685 and *A Discourse of Trade* of 1690 he showed that the whole nation benefited from the growth of towns through the circulation of their wealth; and the prospects for England extended further still, to the possibility of a 'universal monarchy over the seas, an empire no less glorious, and of much more profit, than of the land'. By 1690 the potential for economic growth was literally infinite: national resources were 'infinite and can never be consumed', and so were 'the wants of the mind'. 'Man naturally aspires' and 'his wants increase with his

---

[81] Houghton, *Collection*, iv. 55.   [82] Houghton, *Collection*, iv. 382–3, 389.
[83] Houghton, *Collection*, iv. 303–20; Houghton, *England's Great Happiness*, pp. 17, 18.
[84] Barbon, *Discourse Shewing the Great Advantages*, pp. 2, 20.
[85] Barbon, *Discourse Shewing the Great Advantages*, p. 5. Houghton almost made the same point when referring in 1677 to everyone striving to exceed his fellow and never being satisfied (*England's Great Happiness*, p. 7, quoted above, p. 143); he was clearer about it in 1696: *Collection*, ii. 7–8.

wishes, which is for everything that is rare, can gratify his senses, adorn his body, and promote the ease, pleasure and pomp of life'.⁸⁶ Emulation therefore remained the key, and its outward and visible sign was conspicuous consumption, of which Barbon was himself a famous practitioner—often 'as fine and as richly dressed as a lord of the bedchamber on a birthday', as Roger North remarked.⁸⁷ The moral point was as clear to Barbon as it had been to Houghton: 'prodigality' was 'a vice that is prejudicial to the man, but not to trade'.⁸⁸

In some respects Houghton and Barbon were doing no more than add new voices to a tide of opinion already under way. In 1683 someone rushed to catch the prevailing wind by publishing *Political Arithmetick* without Petty's authorization, and appended to it an encomium on how the English had improved 'themselves, lands, and all things else', 'a miracle' inconceivable in any other country.⁸⁹ Petty himself published another essay, also in 1683, showing that the population of London was doubling every forty years. It had now reached 670,000, an increase of 50 per cent since 1662, and it was producing all those 'arts of delight and ornament' which were 'best promoted by the greatest number of emulators', in cities.⁹⁰ In *Political Arithmetick* he had taken 'every man desiring to put on better apparel when he appears in company' as an instance of how crowded towns produced incentives to spend; and despite his being much less careful about his personal appearance than Barbon, he lived in some style since his wife, according to Evelyn, 'could endure nothing mean, or that was not magnificent'.⁹¹

Competitive consumption was always an element in Petty's economic thought. Yet he had never welcomed unconstrained consumer appetites with Barbon's relish, criticizing 'extravagant expenses' as much as he condemned sloth, and defining covetousness as 'a desire of more than we want, with wrong to others'.⁹² Barbon, by contrast, threw such caution to the winds and in later publications gave Petty's thinking a much sharper edge by putting the infinite desires of consumers at the core of his conception of economic growth. He helped to make consumption a topic no future political economist could wholly ignore;⁹³ and he had initiated in England a gradual demoralization of luxury whose moral and social consequences proved controversial for more than a century.⁹⁴

---

⁸⁶ Nicholas Barbon, *An Apology for the Builder* (1685), pp. 30, 37; Nicholas Barbon, *A Discourse of Trade* (1690), pp. 5–6, 15. Cf. Nicholas Barbon, *A Discourse Concerning Coining the New Money lighter* (1696), pp. 1, 3, 48. On Barbon's conception of infinity, see Andrea Finkelstein, *Harmony and the Balance* (Ann Arbor, 2000), pp. 211–18.
⁸⁷ Roger North, *The Lives of the Norths* (3 vols, 1890), iii. 56.
⁸⁸ Barbon, *Discourse of Trade*, 62.
⁸⁹ [J.S.] *The Fourth Part of the Present State of England* (1683), pp. 9, 14–15. The compiler, J.S., who signed the initial 'letter to the Reader', might perhaps have been John Seller, who published maps and works on navigation and geography.
⁹⁰ Petty, *EW*, ii. 455–6, 474. Cf. below, p. 198.
⁹¹ Petty, *EW*, i. 290; McCormick, *Petty*, p. 162.   ⁹² Petty, *EW*, i. 254; *Petty Papers*, i. 162.
⁹³ See e.g. Terence W. Hutchison, *Before Adam Smith* (Oxford, 1988), pp. 73–86; Appleby, *Economic Thought and Ideology*, pp. 172–83.
⁹⁴ Christopher J. Berry, *The Idea of Luxury* (Cambridge, 1994), pp. 108–25; Keith Thomas, *The Ends of Life* (Oxford, 2009), p. 138; Joyce Oldham Appleby, 'Consumption in Early Modern Social Thought', in John Brewer and Roy Porter, eds, *Consumption and the World of Goods* (1993),

Barbon's first publication also marked a milestone in English thinking about the metropolis, but that turning point was easier for his contemporaries to arrive at. Government attempts since the 1580s to restrain the growth of London had been wholly futile, and its compensating benefits had become ever more attractive. As early as 1620, Robert Cotton, one of the king's commissioners responsible for imposing some restraint on new buildings, received a paper defending them in terms Barbon would have endorsed. They were 'honourable and useful' because they demonstrated the king's power and 'the wealth of his subjects'. London must be allowed to grow if it was to rival Paris and Madrid, and its consumer demand would benefit the whole country, 'the gentleman, husbandman, the country farmer, and the grazier'.[95] By 1660 there were scores of panoramas and bird's-eye views of the city displaying its splendours, and the number of maps and street-plans showing its increasing size proliferated soon afterwards. The largest and most accurate of them, William Morgan's map published in 1682, advertised a metropolis 'so rich and populous that no prince in Europe commands the like'.[96]

Morgan scarcely exaggerated, since London was about to become the largest capital city in western Europe. Thomas Gainsford had thought it already more populous than Paris as early as 1618, and Howell's great panegyric on *Londinopolis* of 1657 hailed it equally prematurely as the greatest city in Europe. Once Paris had inaugurated its own bills of mortality, political arithmeticians could get to work. In the 1670s Graunt and Locke both calculated that Paris was still ahead, but it had probably fallen behind by 1687 when Petty was able to show that London was not only bigger than Paris, but even—less plausibly—more populous than Paris and Rouen or Paris and Rome put together. His figures were immediately disputed by arithmeticians in France, and their debates continued for another half-century before Paris finally conceded defeat. Petty naturally never had any doubt, and with typical hyperbole, since Constantinople and Edo in Japan were both far bigger, asserted that London was to all appearances 'the greatest and most considerable city of the world' and 'manifestly the greatest emporium'.[97]

More persuasive than anything else in influencing English public opinion, however, was the conspicuous recovery of London after the Great Fire of 1666 which had consumed 13,000 houses, most of them in the inner city, destroyed capital in buildings and stock worth around £8 million, and left 65,000 Londoners

---

pp. 162–73; Maxine Berg and Elizabeth Eger, eds, *Luxury in the Eighteenth Century: Debates, Desires, and Delectable Goods* (Basingstoke, 2003).

[95] BL, Cotton MS Titus B.v, fo. 213. Cf. above, p. 75.

[96] Slack, 'Perceptions of the Metropolis', pp. 170–3. Cf. Peter Borsay, 'London, 1660–1800: A Distinctive Culture?', in Peter Clark and Raymond Gillespie, eds, *Two Capitals: London and Dublin 1500–1840* (Proceedings of the British Academy 107, Oxford, 2001), pp. 191–2; Peter Clark, 'The Multi-centred Metropolis: The Social and Cultural Landscapes of London, 1600–1840', in Clark and Gillespie, eds, *Two Capitals*, pp. 245–7.

[97] Slack, 'Perceptions of the Metropolis', pp. 168–9; Vanessa Harding, *The Dead and the Living in Paris and London, 1500–1670* (Cambridge, 2002), pp. 15–16; Petty, *EW*, ii. 505, 517, 522–40. On the argument about London and Paris, see Jacques Dupâquier, 'Londres ou Paris? Un grand débat dans le petit monde des arithméticiens politiques (1662–1759)', *Population*, 53 (1998), pp. 311–25; Sabine Reungoat, *William Petty: Observateur des Îles Britanniques* (Paris, 2004), pp. 276–9.

homeless, more than double the population of any provincial town. Although the grand plans of Wren, Evelyn, and Hooke for a totally redesigned layout were never implemented, by 1670 most of the houses and shops had been rebuilt and the 'wretched town' which Evelyn had earlier condemned as 'a wooden, northern, and inartificial congestion' was a city of brick, sash-windows, and artificially uniform terraces in the new style which Barbon was adopting along the Strand.[98] There were commissioners for paving and cleaning the streets, ambitious new buildings planned for the city hospitals, and even for a time—because the fire made it necessary—an office for new addresses, in Bloomsbury Square, which Hartlib might have thought heralded greater things. Some of Petty's paper projects for 'the improvement of London', for surveyors, pavements, sewers, and new hospitals, were being realized.[99] The city that emerged, burnt only 'to be burnished', 'not so much ruined as refined', was a powerful testimony to the nation's wealth and economic resilience, and to London's capacity to engage in urban improvement on a scale similar to that being undertaken in Amsterdam at the same time.[100]

As the metropolis grew further to become Defoe's 'great and monstrous thing', filled up with 'sin and seacoal (as it was coarsely expressed)', it continued to attract its critics. Since infant mortality continued to rise, in rich as well as poor neighbourhoods, it remained wholly dependent on immigrants in their thousands every year, many of them driven, it was suggested, by 'discontent, curiosity, hopes of greater wages, or of living lazier lives'.[101] Although there were further proposals for restraints on new buildings in 1689 and 1709, however, after their failure in 1677 that particular battle was effectively over.[102] In *Macaria* Gabriel Plattes had wanted to show 'how great cities which formerly devoured the fatness of the kingdom' might, through economic improvement, 'yearly make a considerable retribution'. Graunt had still been equivocal in 1662, thinking London 'the metropolis of England' was 'perhaps a head too big for the body, and possibly too strong'.[103] But Davenant acknowledged that Barbon had won. Summarizing the two views of the metropolis in 1695, he gave significantly greater space to those who thought 'the growth of London not hurtful to the nation', because there was 'not an acre of

---

[98] Stephen Porter, *The Great Fire of London* (Stroud, 1996), pp. 69–74, 116–51; Frances Harris, *Transformations of Love: The Friendship of John Evelyn and Margaret Godolphin* (Oxford, 2002), p. 30; Elizabeth McKellar, *The Birth of Modern London: The Development and Design of the City, 1660–1720* (Manchester, 1999), pp. 23–5, 50–3, 177, 183–4.

[99] Slack, *Reformation to Improvement*, pp. 90–1, 97–8; Porter, *Great Fire*, p. 84.

[100] Robert Arnold Aubin (ed.), *London in Flames: London in Glory* (New Brunswick, NJ, 1943), p. 55; Jonathan I. Israel, *The Dutch Republic: Its Rise, Greatness and Fall, 1477–1806* (Oxford, 1995), pp. 863–73.

[101] Daniel Defoe, *A Tour Through the Whole Island of Great Britain*, ed. G. D. H. Cole and D. C. Browning (2 vols, 1962), i. 168, 323; Appleby, *Economic Thought and Ideology*, p. 149. On infant mortality, see Vanessa Harding and Philip Baker, *People in Place: Families, Households and Housing in Early Modern London* (Centre for Metropolitan History, 2008), pp. 15, 29. Cf. Gill Newton, 'Infant Mortality Variations, Feeding Practices and Social Status in London between 1550 and 1750', *Social History of Medicine*, 24 (2011), pp. 260–80. I am grateful to Richard Smith for confirmation of the trend in mortality from his current research.

[102] Slack 'Perceptions of the Metropolis', p. 175.

[103] *Samuel Hartlib and the Advancement of Learning*, ed. Charles Webster (Cambridge, 1970), p. 88; Petty, *EW*, ii. 320–1.

land in the country, be it never so distant, that is not in some degree bettered by the growth, trade and riches' of the capital.[104]

The debate about luxury, by contrast, was far from over. What caused most immediate and, as it proved, lasting offence in the arguments of Houghton and Barbon was their unabashed defence of prodigality. Having missed the point in *England's Great Happiness* and welcomed its author to the Royal Society, John Beale was horrified when he read Houghton's periodical in April 1682 and found a 'damnable sentence', the one stating plainly that the vices of prodigality, pride, vanity, and luxury enriched the kingdom. He wrote to Boyle objecting to it with a vehemence which indicated the threat it seemed to present, not only to conventional morality but to the efforts of the Royal Society to show that science and religion were entirely compatible with one another. He also protested directly to Houghton, and demanded a retraction.[105] In the next number of his *Collection*, in May, Houghton held his ground. He quoted liberally from *England's Great Happiness* once again, and while adding that he had no wish to 'encourage sin', commented that there was nothing in Christianity which condemned 'costly apparel' in the courts of princes, or wine at weddings like those in Scripture, or the 'Tory and Whig feasts' of the present. Within those broad limits, he insisted, 'we may go far enough in these things to treble the wealth of this kingdom'.[106]

This was more than a private quarrel between fellow-admirers of Hartlib and friends of science and improvement. In his letter to Boyle, Beale cited with approval a sermon preached before the House of Commons by Edward Stillingfleet, Dean of St Paul's and a natural theologian as keen as Boyle to reconcile Scripture and morality with science. The target of the sermon was plainly *England's Great Happiness*. The Dean acknowledged that peace and prosperity had tended to 'make us a happy nation', but warned that nations were 'more or less happy according to their virtues and vices'; and they were easily led to 'riot and luxury' by 'the heaping up of riches without . . . respect to a general good'.[107] That appeal to the moral consensus was proclaimed and published at the height of panic about the Popish Plot in November 1678.

Its message was repeated in 1683 by another cleric in *Englands Vanity*, a diatribe against 'the monstrous sin of pride in dress and apparel'. There were those who believed 'that superfluity is a necessary evil in a state' and 'that men maintain more by their pride than by their charity'—a clear reference to Houghton and Barbon; and the author acknowledged that the vanity of lords and ladies might indeed encourage industry, since the very pins in a lady's hair passed through two or three working hands before use—a foretaste of Adam Smith. But he deplored the

---

[104] *SCED*, pp. 809–10.
[105] *The Correspondence of Robert Boyle*, ed. Michael Hunter, Antonio Clericuzio, and Lawrence Principe (6 vols, 2001), v. 307–13; Ogborn, *Indian Ink*, pp. 157–60.
[106] Houghton, *Collection*, iv. 62–6.
[107] Edward Stillingfleet, *A Sermon Preached on the Fast-Day, November 13. 1678* (5th edn, 1679), pp. 28, 31–2, 42, 47–8. Stillingfleet was a friend of Matthew Hale and author of *Origines sacrae* (1662), a work on natural theology which went through many editions: William Poole, *The World Makers* (Oxford, 2010), pp. x, 27–8.

disappearance of the three-piece suit of the 1660s, 'perhaps the most grave and manlike dress that ever England saw... brought in too late and... sent out again too soon'; and his primary target was the 'usurpation' of upper-class fashions by plebeians. The whole kingdom was now 'in masquerade', thanks to 'the riots and luxury of a meritorious bewitching world', a 'world of need-nots', as Baxter had recently called it.[108]

Whigs who associated luxury with the corruption of royal courts like that in France, and frugality with the virtues of the ideal commonwealth as exemplified by the Dutch, naturally had no difficulty in joining Beale and the church establishment, and deploying the balance-of-trade argument against prodigality. In 1680 the author of *Britannia languens*, probably a Whig lawyer, engaged in a long discussion of England's decay, quoting Roger Coke and Fortrey as well as Mun, and attributing all of it to a 'consumptive trade', consuming like a contagious disease the whole body politic. It had encouraged vanity, extravagance, and emulation, deterred people from marrying, and led to a vast increase in new buildings in London. He took as his text Temple's attack on the ruinous maxim of current English politics about the benefits of excess and luxury. 'National luxury' had left England 'vicious, soft, effeminate, debauched, dispeopled and undisciplined'.[109]

The apparent unanimity of such otherwise discordant voices tells us something about the novelty of Houghton's rhetoric, but it did not mean that his underlying message was wholly unacceptable. Faced with issues which inextricably mingled commerce with religion and politics, it was possible, even probable, that commentators as well read as these should often be ambivalent, and hold different strands of intellectual enquiry in separate mental compartments. That applies to Beale, and to the author of *Britannia languens*, who drew on some of the insights of improvers like Petty while criticizing others.[110] Temple himself could write about how riches were grounded in commerce—'once in motion, trade begets trade, as fire does fire'—and how population increased in circumstances of ease and plenty; and yet explain with equal facility that 'ease, plenty and luxury' inevitably undermined commonwealths.[111]

When people were in two minds, the balance between them could easily be tilted one way or the other over time as circumstances altered, and the moral principles which all of them inherited were being diluted as the religious foundations on which they rested slowly changed. Moral handbooks like Allestree's *Art of Contentment*, which were more widely read than any tract on commerce, were becoming less rigorous in their demands, and increasingly inclined to regard worldly happiness

---

[108] *Englands Vanity* (1683), title page, pp. 15–16, 19, 23, 29–31, 124–5; Richard Baxter, *A Christian Directory* (1673), part IV, p. 140 (a reference I owe to Keith Thomas). The author of *Englands Vanity*, a 'compassionate conformist', has not been identified.

[109] *Britannia languens*, pp. 287, 301, 330, 346, 374–6, 379–84, 422, 430, 444–5. The author may have been William Petyt.

[110] e.g. *Britannia languens*, p. 300, discussing the added value of labour in manufacturing.

[111] William Temple, *Miscellanea* (1680), pp. 99–101; William Temple, *Miscellanea: The Second Part* (1690), pp. 81–2, 89, 94–5, 194.

and the well-being of society as divine blessings.[112] They still sustained the belief that there were deadly sins meriting God's punishments, but improvements in material comfort were scarcely among the most serious of them, as Houghton's very qualified attempt at a retraction shows. At the end of the century the library of a wool-dealer and small farmer in Lancashire could comfortably shelve the works of Petty, Mun, and Locke, alongside another of Allestree's works, *The Gentleman's Calling* (1660), which emphasized the social role of emulation, 'the natural aspiring the lower sort have to approach the condition of their betters'.[113]

The arguments of Houghton and Barbon also attracted attention, whether sympathetic or critical, because there were similar debates, prompted by much the same economic circumstances, elsewhere in western Europe, which some of their readers would have known about. The Paris which Louis XIV was embellishing with monuments, and Colbert hoping to make richer and more populous, was applauded and sometimes condemned exactly as London was. In 1682 a French tract on *The Metropolis* explained how capital cities 'draw their life and glory from all parts of the state, and likewise give them back again' to the whole kingdom; and in the 1690s, when the policies of Colbert and Louis XIV seemed to have failed, the city was attacked by the political economist Boisguilbert as a parasite engaged in perpetual 'war with the provinces'.[114] In Vienna in the 1660s the Austrian Habsburgs thought they had found their own Colbert in one of the Cameralist writers, Joachim Becher, who had projects and interests as wide ranging as any member of the Hartlib circle, from minerals, alchemy, and language reform, to overseas plantations. Becher had visited Holland in order to learn the secrets of economic growth, and written about the importance of consumption and the need to encourage it.[115] Having earned some unpopularity in Germany by trying (somewhat inconsistently) to enforce an imperial ban on French imports, he was in England from 1679 until his death in 1682 and had close contact with the Royal Society.[116]

In the second half of the seventeenth century writers across Europe were debating the relationship between the public good on the one hand and private

---

[112] Alexandra Walsham, *Charitable Hatred: Tolerance and Intolerance in England, 1500–1700* (Manchester, 2006), p. 247; above, p. 113.

[113] H. R. French, *The Middle Sort of People in Provincial England, 1600–1750* (Oxford, 2007), pp. 215–16. Allestree's works were still being recommended as essential reading by a tradesman in Kendal in 1716: '*An Exact and Industrious Tradesman*': *The Letter Book of Joseph Symson of Kendal, 1711–1720*, ed. S. D. Smith (British Academy, Records of Social and Economic History NS 34, Oxford, 2002), p. 419.

[114] Alexandre Lemaître, *La Metropolitée ou De l'établissement des villes capitales* (Amsterdam, 1682), p. 5; Pierre de Boisguilbert, 'De la nécessité d'un traité de paix entre Paris et le reste du royaume', in *Pierre de Boisguilbert ou La Naissance de l'économie politique* (2 vols, Paris, 1966), ii. 799–818. On Boisguilbert, see below, pp. 246–7.

[115] Pamela H. Smith, *The Business of Alchemy: Science and Culture in the Holy Roman Empire* (Princeton, 1994); Hutchison, *Before Adam Smith*, pp. 90–3, 100–9; Carl Wennerlind, 'Credit-Money as the Philosopher's Stone', *History of Political Economy*, 35 (2003 Supplement), pp. 241–3. For Cameralism, see below, pp. 246, 251.

[116] Smith, *Business of Alchemy*, p. 18; Michael Hunter, *Boyle: Between God and Science* (New Haven, 2009), 187. Becher's *Magnalia naturae: or, The Truth of the Philosopher's Stone* (1680) was published in English, but not his other works. I know of no evidence that Becher met either Barbon or Houghton, but he might have done.

appetites, 'passions and interests', on the other;[117] and when they turned towards political economy, they could scarcely avoid discussing the motivation of consumers, and the economic and moral implications of their behaviour. An important ingredient in their intellectual orientation was the revival of Epicureanism in the middle of the century, which had rekindled the interest of scientists in atomism and of moral philosophers in pleasure and happiness and their compatibility with virtue.[118] Epicurus was a favourite author of Temple as well as Petty and other members of the Royal Society, and the Epicurean theme of worldly contentment prompted Hobbes to ponder the association between felicity and motion, and even the Dublin Philosophical Society to try to measure happiness in 1685.[119] It also had a powerful effect on contemporary thinking about commerce, towns, and the pursuit of material comfort, thanks largely to the works of a French Jansenist, Pierre Nicole, published in the 1670s.

Nicole's *Moral Essays* were translated into English between 1677 and 1680 and had already influenced Locke, prompting essays of his own on pleasure and morality which foreshadowed what he had to say about the quest for happiness in his *Essay Concerning Human Understanding* (1689).[120] Nicole's originality lay in his contention that in modern urban and market societies self-interest, *amour propre*, was fundamental to social cohesion and economic advance, directed as it generally was, and rationally should be, to achieving happiness, *bonheur*, and the ease and comfort of life, *les aises de la vie*. The vanity and avarice of fallen mankind extended to 'an infinity', he argued in an essay on 'Charity and Self-Love', but they were partly redeemed by their contribution to mutual cooperation, and hence to the general good.[121] The 'self-love' which a speaker on enclosure in the Commons had considered inimical to the common good in 1597, and which Lord North

---

[117] Cf. Albert O. Hirschman, *The Passions and the Interests: Political Arguments for Capitalism before its Triumph* (Princeton, 1977), pp. 9–66.

[118] Jerome B. Schneewind, *The Invention of Autonomy: A History of Modern Moral Philosophy* (Cambridge, 1998), pp. 264–9; Jon Parkin, *Science, Religion and Politics in Restoration England: Richard Cumberland's De legibus naturae* (1999), pp. 146–51; Reid Barbour, *English Epicures and Stoics: Ancient Legacies in Early Stuart Culture* (Amherst, Mass., 1998); Christopher Tilmouth, *Passion's Triumph over Reason* (Oxford, 2007), pp. 275–80. On the earlier rediscovery of Epicurus in the sixteenth century, see Stephen Greenblatt, *The Swerve: How the Renaissance Began* (2011).

[119] Michael Hunter, *Science and Society in Restoration England* (Cambridge, 1981), pp. 171–4; Jonathan I. Israel, *Radical Enlightenment: Philosophy and the Making of Modernity 1650–1750* (Oxford, 2001), p. 606; *Petty Papers*, i. 144; Tilmouth, *Passion's Triumph*, pp. 4–5; K. Theodore Hoppen, *The Common Scientist in the Seventeenth Century: A Study of the Dublin Philosophical Society, 1683–1708* (1970), pp. 122–3.

[120] Pierre Nicole, *Moral Essays* (2 vols, 1677–80); Locke, *Political Essays*, ed. Goldie, pp. xxiii, 251–2, 271; John Locke, *An Essay Concerning Human Understanding* (2 vols, 1715–16), Book II, ch. xxi 'Of Power'. Cf. Catherine Wilson, *Epicureanism at the Origins of Modernity* (Oxford, 2008), 207–16; below, pp. 205–6, 220. For Locke's translation of passages from Nicole on *bonheur* and *progrès*, which provoked what Goldie terms his 'hedonic turn', see Jean S. Yolton, *John Locke as Translator* (Studies on Voltaire and the Eighteenth Century, Voltaire Foundation, Oxford, 2000), pp. 31, 53, 121.

[121] Schneewind, *Invention of Autonomy*, pp. 275–9; Istvan Hont, *Jealousy of Trade: International Competition and the Nation-State in Historical Perspective* (Cambridge, Mass., 2005), pp. 47–50; Noel Malcolm, *Aspects of Hobbes* (Oxford, 2002), pp. 508–9.

thought led only to wasteful 'high living' in 1669, was now presented by Nicole as promoting social harmony and material progress.[122]

It is difficult to believe that these intellectual currents had not influenced Barbon's perception of the infinite wants of the mind being given free rein and fuller satisfaction in cities, or that Houghton had not picked up something from them when associating *England's Great Happiness* so closely with the consumer culture of a commercial society. Barbon and Houghton could not alone have created a climate of opinion in England more favourably disposed to consumer extravagance and urban growth than it had ever been, but they had contributed powerfully to it. Later observers were in no doubt that something remarkable had happened in the 1680s to English attitudes and modes of behaviour. After 1690 there were complaints that 'wanton emulation' had spread over the past century from the aristocracy down to 'the lower ranks of people' and recently given birth to 'a party' who argued 'that it is a benefit of the nation to be expensive in diet and apparel'. Looking back in 1724 Defoe was more precise about the chronology. 'This change or revolution in the manners and temper of the common people was in the height of its operation ... in the years 1684 to 88.'[123] High living seemed to have become a popular habit and the taste for it an acceptable, if not entirely respectable, frame of mind.

HIGH LIVING

Towards the end of his life in 1683, John Beale was prepared to allow that some things had changed for the better since 1660. He was composing a tract called 'From Utopia', apparently intended as a response to Houghton and a refutation of any notion that material progress brought England great happiness. What survives of it, as dictated to an amanuensis, shows him conceding more than he perhaps realized. He continued to think that Holland was 'our most dangerous neighbour', that Mun had established the crucial importance of a positive balance of trade, and that many of Roger Coke's complaints were fully justified. 'The great concernment of this mighty monarchy' was still its low population, and 'the honour of our nation' was at stake unless people recovered their 'pristine sobriety' in manners and especially in dress.[124] Yet Beale remained, in Houghton's words, a 'great favourer of anything that tends to the improvement' of the nation, and the more he thought about improvement, the more he found to admire.[125]

---

[122] Above, pp. 59–60, 131. For later interest in self-love, see Craig Muldrew, 'From Commonwealth to Public Opulence', in Steve Hindle et al., eds, *Remaking English Society* (Woodbridge, 2013), pp. 336–7; below, pp. 205, 210.

[123] H.J., *A Letter from a Gentleman in the Country to his Friend in the City* (1691), p. 17; Francis Brewster, *Essays on Trade and Navigation* (1695), p. 50; Daniel Defoe, *The Great Law of Subordination consider'd* (1724), p. 50.

[124] *ODNB*, sub Beale; Royal Society of London, MS 366/1/6, 'From Utopia' [dated 28 Jan. 1683], fos 2, 13ʳ.

[125] Houghton, *Collection*, iv. 2.

He distinguished again and again between the 'idle and profligate English', the 'degenerate' wastrels who were 'enemies to their native country', and the virtuous and diligent, perhaps a minority but a growing one, who were applying their energies in 'every kind of improvement' and setting an example which was already being copied in Scotland and Ireland. Sprat's response to Sorbière had persuaded him that there was 'more industry then stirring in England' than he had earlier appreciated. He now accepted that 'the numbers of the industrious' had 'wonderfully increased' since 1660. There had been a 'wondrous growth of industry amongst us', as evident in agriculture as in manufacturing, thanks to the 'silver trumpet' sounded by Hartlib and other writers on husbandry. There had even been an increase in population since John Graunt wrote about it, as Petty, 'our incomparable calculator', had proved. The English were improving, becoming more industrious and more populous, but were not yet improved enough.[126]

While the content of complaint about the condition of England had changed very little since 1660, the condition of the English had certainly improved. As indicated in the introduction to this book, there is now ample evidence that England was richer in the second half of the seventeenth century than it had ever been, and that national wealth per head was higher than at any time since 1500. GDP doubled in the seventeenth century, and average incomes increased by half between the 1650s and 1700.[127] Some of the things contemporaries complained about were among the causes of that improvement, like the low population and the expansion of trade with the East Indies. An appetite for more consumer goods was another, increasing and sustaining economic demand. 'High living' contributed to economic growth, and even its critics could scarcely deny that it was a symptom of material progress.

Although different historians have measured the increase in national wealth in different ways, there is no dispute that it was impressive and little about its causes or its broad consequences. As contemporaries knew when they referred to the customs accounts, foreign trade was expanding, with a particularly dramatic rise in trade beyond continental Europe. There was a fourfold increase in imports from Asia to England between the 1660s and the 1680s, and by 1700 America and Asia were providing a third of all imports and the re-export of a proportion of them made up nearly a third of all English exports.[128] Since population was no longer rising, but agricultural productivity per head increasing, there could be a notable rise in the proportion of the population engaged in secondary and tertiary occupations, industry, and services of all kinds; it may have reached at least 40 per cent by 1700. By the end of the century a much smaller agricultural labour force was providing enough grain to meet the country's needs and generate a surplus for export.[129] The industrial workforce was growing rapidly in new centres of textile production like Lancashire and the West Riding, and in coal mining, especially

---

[126] 'From Utopia', fos 4ʳ, 10ᵛ–11, 14.   [127] Above, pp. 12–13.
[128] Ralph Davis, *English Overseas Trade 1500–1700* (1973), p. 32; Ralph Davis, 'English Foreign Trade, 1660–1700', *EcHR* 7 (1954), pp. 150–66; Ogborn, *Indian Ink*, p. 67; Pincus, *1688*, pp. 85–6.
[129] Above, p. 10; below, p. 232.

around Newcastle, which was supplying by 1700 over half the nation's fuel requirements and having a multiplier effect on the economy similar to that which came from the growth of international trade.[130]

As Sprat and Beale suggested, agrarian and industrial improvements as conspicuous as these could scarcely be missed once they were drawn to people's attention; and neither could the increasing numbers employed in retailing and internal trade. By 1681 earlier hostility towards middlemen was being redirected towards the proliferation of shopkeepers, who could be found in almost all market towns by 1700, and of hawkers and pedlars, of whom there were at least 2,500 across the kingdom in 1698.[131] They would not have existed if there had not been more people buying more commodities, and an increase in their capacity to spend may well have been particularly palpable in the 1680s, when government could be funded from customs and excises, and direct taxation was particularly low, much lower than it had been in the 1660s.[132] Peace had in reality meant prosperity.

Generalizations about economic trends, the growth in national wealth, and its likely impact on the purchasing power of average incomes tell us little, however, about patterns of expenditure as incomes rose, and how they differed between different parts of the country and different status groups. In a society where there were gross inequalities in the distribution of wealth and disposable income, the ways in which income was spent depended as much on the opportunities available and the choices of individuals and peer groups, as on emulation of social superiors or metropolitan fashions. There is a distinction to be made too between different kinds of consumer goods. On the one hand there were those which most attracted the anger of social critics and proposers of sumptuary laws, chiefly items of apparel and food and drink which were supposed to reflect and display social rank, and most obviously signified 'excess' and 'debauchery' when they did not. At the opposite extreme were things enjoyed more privately, like improvements in domestic furniture and housing which might be regarded as among the comforts legitimately available to everyone. The two categories overlapped and there was a large middle ground between them, but there were different species of high living and the concept meant different things to different people.

It is not difficult to demonstrate that a growing metropolitan middle class was benefiting from higher standards of living of every kind, enjoying improvements in

---

[130] John Hatcher, *The British Coal Industry*, i: *Before 1700* (Oxford, 1993), pp. 45–6, 55. Work currently being undertaken by the Cambridge Group for the History of Population and Social Structure on the occupational structure of England in the very early eighteenth century suggests that the proportion of the male working population engaged in industry and manufacturing was already over 30 per cent: information kindly supplied by Dr Leigh Shaw-Taylor; see also John Dodgson, 'Gregory King and the Economic Structure of Early Modern England: An Input–Output Table for 1688', *EcHR* 66 (2013), p. 1013.

[131] *SCED*, pp. 390–4; Carole Shammas, *The Pre-Industrial Consumer in England and America* (Oxford, 1990), pp. 225–38; Margaret Spufford, *The Great Reclothing of Rural England* (1984), pp. 14–16; Jan de Vries, *The Industrious Revolution* (Cambridge, 2008), pp. 169–70. Cf. Nancy Cox, *The Complete Tradesman: A Study of Retailing, 1550–1820* (Aldershot, 2000), pp. 38–75.

[132] Chandaman, *English Public Revenue*, pp. 32–5, 330. Cf. David Ormrod, *The Rise of Commercial Empires* (Cambridge, 2003), p. 22 for comparisons and contrasts with the Dutch.

clothing and diet, and keeping warm even in the colder winters towards the end of the century thanks to better, if still overcrowded, housing, and more efficient chimneys and stoves.[133] There is also ample evidence of an improvement in domestic comfort in the provinces as the 'great amendment of lodging' which William Harrison had described in Elizabethan Essex spread to more distant parts of the country after 1660, and brought chimneys and glass windows, more rooms, parlours and bedchambers, and more opportunities to separate private from communal household activities.[134] There were still many labourers' cottages of only one room, especially in the north and west, but if they had no proper chimney and were 'open to the thatch [and] with no partitions', they were regarded by travellers like Celia Fiennes as distinctly primitive.[135]

Much more novel outside London, however, was the proliferation of more modest consumer items which became available as the country's import trades grew, its domestic industries expanded, and retailing mechanisms improved. In that context, small falls in the relative cost of consumables could have a marked effect on behaviour and make what had once been the luxuries of the few the indulgences of the many.[136] Since the price of linen fell by half across the seventeenth century while the wages of labourers doubled, for example, the linen cloth which maidservants had been able to afford only for a 'holiday apron' or pretty neck-cloth before 1650 became widely available. More people could have more and better shirts, smocks, and underclothes, and by the 1690s, when cheap Indian cottons were being imported in large quantity, even the poor were able to buy new clothes for their children.[137]

---

[133] Peter Earle, *The Making of the English Middle Class: Business, Society and Family Life in London, 1660–1730* (1989); Sara Pennell, '"Great quantities of gooseberry pye and baked clod of beef": Victualling and Eating out in Early Modern London', in Paul Griffiths and Mark S. R. Jenner, eds, *Londinopolis* (Manchester, 2000), pp. 228–49; William C. Baer, 'Stuart London's Standard of Living: Re-examining the Settlement of Tithes of 1638 for Rents, Income, and Poverty', *EcHR* 63 (2010), p. 635; William C. Baer, 'Landlords and Tenants in London 1550–1700', *Urban History*, 38 (2011), pp. 234–55; Hatcher, *British Coal Industry*, pp. 411–12, 415–17. Jeremy Boulton, 'Food Prices and the Standard of Living in London in the "Century of Revolution", 1580–1700', *EcHR* 53 (2000), pp. 455–92, shows that the cost of living in London was also increasing, in contrast to its cost elsewhere, which suggests that earnings, though probably not those of the labouring poor, must have been increasing to keep pace.

[134] Above, p. 17; R. Machin, 'The Great Rebuilding: A Reassessment', *P&P* 77 (Nov. 1977), pp. 33–56; Shammas, *Pre-Industrial Consumer*, pp. 158–64; Mark Overton et al., *Production and Consumption in English Households, 1600–1750* (2004), pp. 90, 99, 109–11; Lorna Weatherill, *Consumer Behaviour and Material Culture in Britain 1660–1760* (1988), pp. 6–8, 160.

[135] M. W. Barley, 'Rural Building in England', in Joan Thirsk, ed., *Agrarian History of England and Wales*, v.ii. (Cambridge, 1985), pp. 677–82; M. W. Barley, 'Rural Housing in England', in Thirsk, ed., *Agrarian History*, iv (Cambridge, 1967), pp. 762–6; *The Journeys of Celia Fiennes*, ed. Christopher Morris (1947), p. 207. Cf. Matthew Johnson, *Housing Culture: Traditional Architecture in an English Landscape* (1993), pp. 78, 128.

[136] John Styles, *The Dress of the People: Everyday Fashion in Eighteenth-Century England* (New Haven, 2007), p. 245. Cf. Donald Woodward, *Men at Work: Labourers and Building Craftsmen in the Towns of Northern England, 1450–1750* (Cambridge, 1995), pp. 247–8; John A. Chartres, 'The Marketing of Agricultural Produce', in Thirsk, ed., *Agrarian History*, v.ii. 406–8.

[137] John Hatcher, 'Labour, Leisure and Economic Thought before the Nineteenth Century', *P&P* 160 (Aug. 1998), pp. 94–5; *The Farming and Memoranda Books of Henry Best of Elmswell, 1642*, ed. Donald Woodward (British Academy, Records of Social and Economic History NS 8, Oxford, 1984),

There was a similar widening of consumer choice across the social range in the case of food and drink. The variety and quality of foodstuffs available was greatest among the social elite of town and country, but even 'the poorer sort' benefited as the real cost of staple items fell.[138] In the south of England at least, consumption of meat probably increased, and so did the quality of bread, since wheat was beginning to replace barley as the staple grain of the poor, and according to Aubrey had overtaken it as early as 1680.[139] Although consumption of dairy products was declining with enclosure and urbanization, per capita consumption of imported sugar more than doubled in the later seventeenth century, and consumption of alcohol, another source of calories, reached something of a temporary peak in the 1680s, when the English were drinking the equivalent of 400 pints of beer per head every year, as well as smoking nearly two pounds of tobacco in what was becoming another saturated market.[140] By the early eighteenth century England had an alehouse or inn for every 80 inhabitants—nearly double the provision in 1577—where the new consumer indulgences of the majority of the population could be displayed and shared outside the home.[141]

When it comes to assessing the diffusion of more durable consumer items enjoyed at home, the best evidence has proved to be that of probate inventories, which reveal how much depended on social status, occupation, and location. The 'middling sort', a diverse category perhaps amounting to nearly half the population by 1700, were naturally better placed than the poor to afford more investment in new kinds of domestic comfort. More of them had more and better furniture, cooking utensils of greater variety, and household linen of better quality and in greater quantity.[142] The distribution of such items was also skewed geographically, between town and country and between the south-east and counties further from London. By 1725, for example, 60 per cent of London inventories included

---

p. 111; Shammas, *Pre-Industrial Consumer*, pp. 96–7; Margaret Spufford, 'The Cost of Apparel in Seventeenth-Century England and the Accuracy of Gregory King', *EcHR* 53 (2000), pp. 691, 704.

[138] Adam Fox, 'Food, Drink and Social Distinction', in Hindle et al., eds, *Remaking English Society*, pp. 180–7.

[139] Craig Muldrew, *Food, Energy and the Creation of Industriousness: Work and Material Culture in Agrarian England, 1550–1780* (Cambridge, 2011), pp. 98–100; D. C. Coleman, *The Economy of England 1450–1750* (Oxford, 1977), p. 119; Chartres, 'Marketing of Agricultural Produce', p. 408. Aubrey may have exaggerated the extent of the change, however: de Vries, *Industrious Revolution*, pp. 167–8. On the difficulty of measuring meat production and consumption, see Muldrew, *Food, Energy and Industriousness*, pp. 83–102, 149–55.

[140] De Vries, *Industrious Revolution*, pp. 157, 181–2; Shammas, *Pre-Industrial Consumer*, pp. 141–6; John Chartres, 'No English Calvados? English Distillers and the Cider Industry in the Seventeenth and Eighteenth Centuries', in John Chartres and David Hey, eds, *English Rural Society 1500–1800: Essays in Honour of Joan Thirsk* (Cambridge, 1990), p. 319; Keith Thomas, *Religion and the Decline of Magic* (1971), p. 20.

[141] Hatcher, 'Labour, Leisure and Economic Thought', p. 93; Peter Clark, *The English Alehouse: A Social History 1200–1830* (1983), pp. 43–4.

[142] De Vries, *Industrious Revolution*, pp. 125–7, 178; Sara Pennell, 'The Material Culture of Food in Early Modern England, c.1650–1750', in Sarah Tarlow and Susie West, eds, *Familiar Pasts? Archaeologies of Later Historical Britain 1550–1860* (1999), pp. 37, 40. Cf. Paul Glennie and Ian Whyte, 'Towns in an Agrarian Economy 1540–1700', in Peter Clark, ed., *The Cambridge Urban History of Britain* (3 vols, Cambridge, 2000), ii. 188–9.

utensils, cups and mugs, for hot drinks such as coffee, compared with only 6 per cent in a sample of rural inventories; and books, pictures, and mirrors were much more common in towns than in villages.[143]

At the end of the seventeenth century there were similarly marked contrasts between widely separated counties. Upholstered furniture, feather beds, clocks, and mirrors were far more common in Kent than in Cornwall, and the amount of material wealth attributed to items like these had grown threefold since 1600 in Kent while remaining stationary in Cornwall. Since the contrasts were just as stark between householders of similar wealth in each of the two counties, they must have owed more to culture than to cost, and to spending habits more or less open to infection from London.[144] The same phenomenon is evident in contrasts between town and countryside. Urban tradesmen, especially retailers, shopkeepers, and others in service occupations, took the lead in adopting new consumer goods. They more often had window curtains than yeomen farmers of equivalent or greater wealth, for example, and they may have preceded the landed gentry in taking up new ways of cooking and eating, and new drinks like coffee and later tea. Virtually the only new consumer item which was more of a rural than an urban phenomenon among people of middling wealth was the clock, and that may have had more to do with its utility in the timing of farm activities than with competitive conspicuous consumption.[145]

There is certainly evidence here of consumer aspirations being determined by Barbon's insatiable 'wants of the mind', and a desire for 'the ease, pleasure and pomp of life', even if of a very modest kind. There is evidence too that new patterns of consumption were part of the gentrification of local elites in small towns and parishes, where the middling sort were competing with one another for social status.[146] But the purchase of new goods rested on more than competitive emulation between neighbours to see who could most quickly 'mend their former condition'. They were bought because they gave private satisfaction when they were used, worn, or consumed. It has been persuasively argued that this was not the old luxury of aristocracies advertising their distinctiveness by accumulating obviously costly expressions of wealth in buildings and plate, or coaches and apparel. It was a new luxury obtained from the purchase of cheaper items, quickly consumed or worn out, but nonetheless offering variety, comfort, and pleasure, and affordable and available once small gains in household incomes and new market mechanisms allowed it.[147]

The novel pleasures of consumer choice were even available to labourers, to judge by a large sample of their probate inventories drawn from six counties across

---

[143] Weatherill, *Consumer Behaviour*, pp. 31, 76.
[144] Overton et al., *Production and Consumption*, pp. 90–111, 140–5, 175–6. For comparable findings based on a different sample of inventories, see French, *Middle Sort of People*, 141–200.
[145] Overton et al., *Production and Consumption*, pp. 168–9 and *passim*. Cf. Weatherill, *Consumer Behaviour*, pp. 77–9, 165, 177, 185; Carl B. Estabrook, *Urbane and Rustic England: Cultural Ties and Social Spheres in the Provinces, 1660–1780* (Manchester, 1998), pp. 128–9; French, *Middle Sort of People*, pp. 27–8, 144–8.
[146] French, *Middle Sort of People*, pp. 185, 199.
[147] De Vries, *Industrious Revolution*, pp. 43–5, 55–8, 122–3.

the country. The median value of their household goods rose by almost a half in the later seventeenth century. They were not buying luxury items like mirrors, still less the imported carpets and carved furniture of wealthier tradesmen, but beds, bedding, and furnishings of better quality than they had before, and pewter or pottery rather than tableware made of wood. They were not aiming to emulate their betters but simply to enjoy a more comfortable standard of living.[148] They were also enjoying a better diet as food prices fell, and therefore had the energy to work harder in order to acquire more of all these things if they wished. In 1600 labourers had to be industrious in order to meet basic subsistence requirements; in the later seventeenth century labouring families, women and children as well as men, were working in order to accumulate a surplus.[149]

If there was a consumer revolution under way by the end of the seventeenth century, as many historians have argued, there was also an industrious revolution which paid for it by increasing productivity and delivering many of the goods, in industry as well as agriculture, which it consumed.[150] There was much debate among contemporaries about what would now be called leisure preferences, the propensity of workers to work less once earnings rose above the costs of an adequate standard of living. Voluntary leisure may well have been itself a consumer good chosen by some labourers, especially in heavy industries like mining, but there is little sign of it in agriculture where wages were lower and the rhythms of work varied with the seasons.[151] On the contrary, some agricultural labourers were choosing to keep their surplus earnings in cash. A small minority of those leaving probate inventories either possessed large amounts of money or had large debts owing to them, and the total sums recorded rose significantly after 1650. There was a widening gap between poor and wealthy labourers, but the latter were saving as well as spending and being industrious so that they could do both.[152]

The expectations engendered across the social spectrum by even modest increases in living standards were also related to the kinds of behaviour which kept fertility low, prevented population from rising, and thus sustained per capita incomes and economic growth. Improvements in the quality of housing and diet had no discernible effect on mortality rates, which continued to rise, but complaints like those in *Britannia languens* about luxury deterring people from marrying, and 'dispeopling' the nation, may have had something to be said for them.[153] Assumptions about present and future incomes had an effect on decisions about

---

[148] Muldrew, *Food, Energy and Industriousness*, pp. 163–200, 205.
[149] Muldrew, *Food, Energy and Industriousness*, pp. 162, 207, 258–9, 297. Cf. Robert C. Allen, *Enclosure and the Yeoman* (Oxford, 1992), pp. 250–1, on female employment in agriculture.
[150] De Vries, *Industrious Revolution*, pp. 85–7, 104, and *passim*.
[151] Hatcher, 'Labour, Leisure and Economic Thought', pp. 69–70, 82–92; Muldrew, *Food, Energy and Industriousness*, pp. 290–5. For a revealing case study, see Steve Hindle, 'Work, Reward and Labour Discipline in Late Seventeenth-Century England', in Hindle et al., eds, *Remaking English Society*, pp. 255–79.
[152] Muldrew, *Food, Energy and Industriousness*, pp. 201, 204, 319; Craig Muldrew. 'From Credit to Savings? An Examination of Debt and Credit in Relation to Increasing Consumption in England [c.1650–1770]', *Quaderni storici*, 137 (Aug. 2011), pp. 12–17.
[153] Above, p. 150; below, pp. 195–6, 221–2, 238–40.

marriage at every social level when people were free to make choices for themselves. We know that widows were remarrying much less often in the later seventeenth century than they had done before, for example, and the practice was becoming more acceptable. As early as 1660 Pepys heard 'a very good sermon' in praise of 'widowhood, and not as we do to marry two or three wives or husbands, one after another'; and in 1671 Edward Chamberlayne proposed a religious academy for young ladies which would also be a home for 'any devout widows or elder virgins, who intend not to marry'.[154]

It had been generally accepted for much longer that the young should not marry until they could afford to do so, well into their twenties, and it was more difficult to shift that fundamental assumption. When Petty tried to persuade Southwell that fertility could be doubled if more women married earlier, Southwell was horrified by the thought that it meant putting 'a girl at 16 and a lad at 18 to get children before they know how to maintain them, or have served out half their apprenticeships'. As Petty himself admitted, young people themselves were choosing not to marry because they were afraid they would 'not be able to maintain the children they shall beget'.[155] In a society acquiring new habits of spending and saving, young people may well have wanted to postpone marriage in order to avoid the kind of erosion in living standards which had occurred at the end of the sixteenth century.

It should be remembered also that average real incomes, although higher in 1700 than they had been in 1600, had still not reached the level they attained at the end of the fifteenth century, despite the fact that the national wealth was much greater.[156] We have no means of knowing whether there was some folk memory of late medieval halcyon days and a desire to return to them, but people knew they were better off than their grandfathers had been and must have been determined to hold on to and perhaps improve on what they had gained. When they described their 'worth' to the courts in terms of the value of their moveable goods between 1657 and 1681, almost all occupational and social groups reported sums larger in real terms than their equivalents a century before. The increases were necessarily modest in the case of labourers and servants, who did not have the valuable equipment and livestock which had pushed up the wealth of yeomen to remarkably high levels by 1650, but their gains were real.[157] When labour was scarce and rural industry offered increasing opportunities for women and children to contribute something to family earnings, there can have been few husbandmen like Edward

---

[154] Barbara J. Todd, 'Demographic Determinism and Female Agency: The Remarrying Widow Reconsidered...Again', *Continuity and Change*, 9 (1994), pp. 421–50; *Diary of Samuel Pepys*, i. 60; Edward Chamberlayne, *An Academy or Colledge* (1671), pp. 4–6; Bridget Hill, 'The Idea of a Protestant Nunnery', *P&P* 117 (Nov. 1987), p. 113.

[155] *The Petty–Southwell Correspondence*, ed. Marquis of Lansdowne (1928), pp. 145, 154; *Petty Papers*, i. 267.

[156] Above, p. 13. Cf. below, pp. 239–40.

[157] Alexandra Shepard and Judith Spicksley, 'Worth, Age, and Social Status in Early Modern England', *EcHR* 64 (2011), pp. 517, 527. The figures yield slightly divergent conclusions depending on whether one takes the median or mean as the better guide. On the yeomen, see above, p. 58.

Barlow's father in the 1640s, 'poor and in debt' and unable to provide his son with 'clothes fitting to go to church in... unless we would go in rags, which was not seemly'. In the 1680s emigration to the colonies was no longer being presented as a recipe for unemployment at home but as an opportunity for the 'industrious part of mankind' to enjoy even higher wages overseas.[158] Whatever pessimists may have thought in the debates of the 1660s and early 1670s, living standards had improved and might be improved further.

These were ideal conditions for the diffusion of an improvement culture being advertised, as Beale and Houghton observed, by the successors to Hartlib and his circle. Beale cited William Penn and his proposals for American plantations, the map-makers John Seller and William Morgan, and the two most prolific writers of improvement tracts, Richard Haines, a Baptist farmer in Sussex, and Andrew Yarranton, an engineer improving river navigation in Worcestershire, who filled two volumes with every conceivable means of ensuring *England's Improvement by Sea and Land* (1677, 1681) (Figure 5, overleaf).[159] All these authors were either based in London or had lodgings there, but they were aiming at improvement elsewhere. Other correspondents of Petty and Southwell were doing the same, like the Lowthers on their estates in Cumbria.[160] By the 1680s enthusiasm for improvement was well established among the English elite in Ireland and beginning to gain a hold in Scotland. In Dublin William Molyneux was the driving force behind debates in the Dublin Philosophical Society, including one on the relative populations of London and Paris, and reviving schemes for a further collaborative natural and civil survey of the kingdom and for maps like those being attempted in England.[161] In Edinburgh Robert Sibbald, another former student of medicine in Leiden and an admirer of the Royal Society of London, was doing much the same, and had turned a club for medical men into the Royal College of Physicians of Edinburgh in 1681.[162]

In all three kingdoms the culture of improvement was rooted in their capital cities, and it is an inescapable limitation of most of the evidence for its impact that it comes from the landed and professional classes and tells us little about its impact on economic behaviour. Improvements in farming were being encouraged in Scotland as well as Ireland, but as in England their adoption was conditioned

---

[158] Muldrew, *Food, Energy and Industriousness*, pp. 308, 315.

[159] Beale, 'From Utopia', fos 3ʳ, 4, 9ʳ, 12ᵛ; *ODNB, sub* Haines, Yarranton.

[160] Slack, *Reformation to Improvement*, p. 98; Paul Slack, 'Government and Information in Seventeenth-Century England', *P&P* 184 (Aug. 2004), p. 45; below, pp. 177, 231, 234. On Southwell's connections, see Ted McCormick, *William Petty and the Ambitions of Political Arithmetic* (Oxford, 2009), p. 264; *William Petty on the Order of Nature*, ed. Rhodri Lewis (Tempe, Ariz., 2012), pp. 19–22; Kate Loveman, 'Samuel Pepys and "Discourses touching Religion" under James II', *EHR* 127 (2012), p. 66.

[161] Barnard, *Improving Ireland?*, pp. 94–5; K. Theodore Hoppen, ed., *Papers of the Dublin Philosophical Society 1683–1709* (2 vols, Dublin, 2008), i. 92, 94.

[162] *ODNB sub* Sibbald; Charles W. J. Withers, 'Geography, Science and National Identity in Early Modern Britain: The Case of Scotland and the Work of Sir Robert Sibbald (1641-1722)', *Annals of Science*, 53 (1996), pp. 29–73; Adam Fox, 'Printed Questionnaires, Research Networks, and the Discovery of the British Isles, 1650-1800', *Historical Journal*, 53 (2010), p. 602.

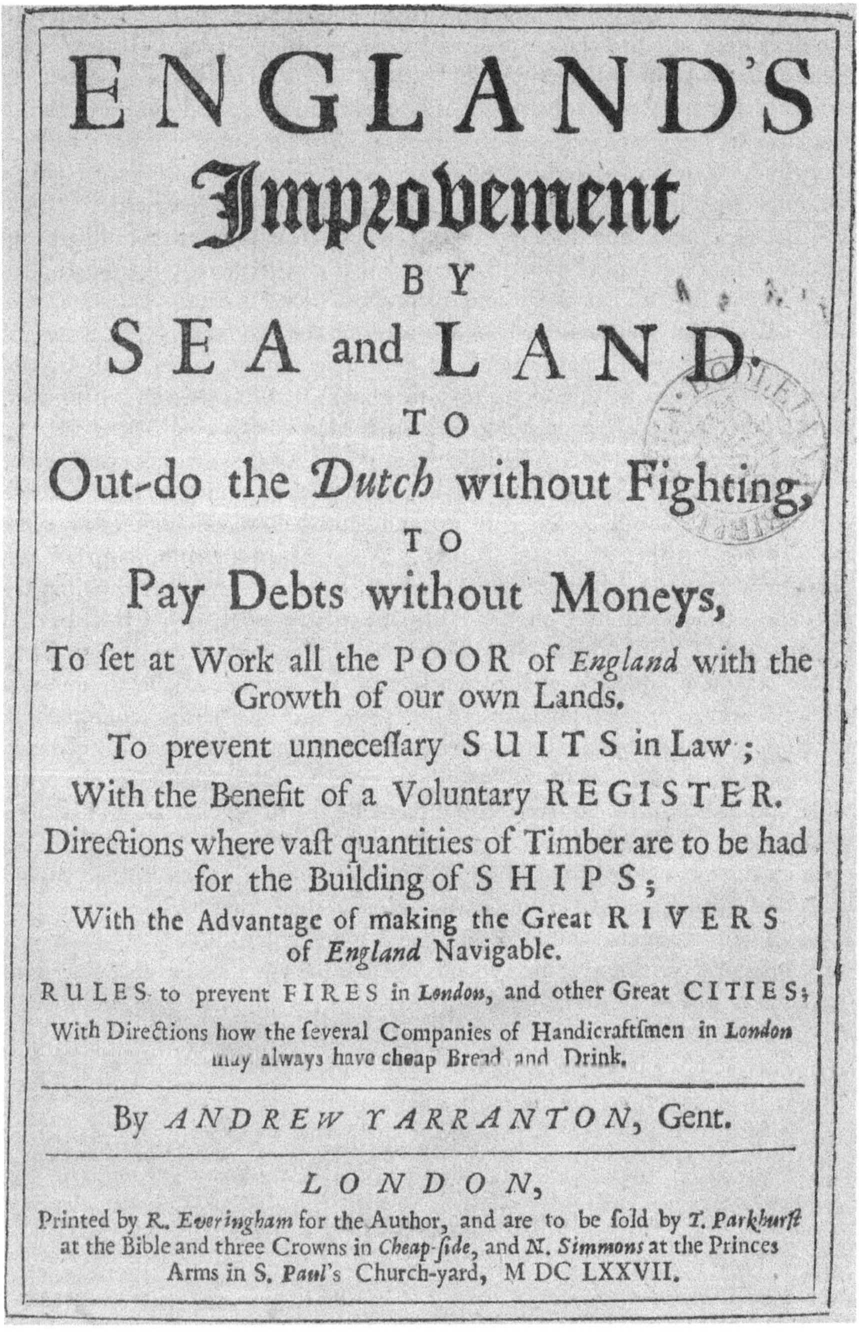

ENGLAND'S
Improvement
BY
SEA and LAND
TO
Out-do the *Dutch* without Fighting,
TO
Pay Debts without Moneys,

To set at Work all the POOR of *England* with the Growth of our own Lands.

To prevent unnecessary SUITS in Law;

With the Benefit of a Voluntary REGISTER.

Directions where vast quantities of Timber are to be had for the Building of SHIPS;

With the Advantage of making the Great RIVERS of *England* Navigable.

RULES to prevent FIRES in *London*, and other Great CITIES;

With Directions how the several Companies of Handicraftsmen in *London* may always have cheap Bread and Drink.

By *ANDREW YARRANTON*, Gent.

LONDON,

Printed by R. *Everingham* for the Author, and are to be sold by *T. Parkhurst* at the Bible and three Crowns in *Cheap-side*, and *N. Simmons* at the Princes Arms in S. *Paul*'s Church-yard, MDCLXXVII.

Figure 5. All England improved: Andrew Yarranton, *England's Improvement by Sea and Land* (1677).

by local circumstances and was always much less successful than its advocates imagined.[163] Nonetheless, once they were established among a metropolitan elite, such aspirations had the potential to shape educated opinion generally. That was not to happen in Scotland until the 1690s, and it was a slow process in Ireland, despite the fact that notions of improvement were sufficiently well rooted in Dublin in the 1680s for them to attract the mixture of mockery and admiration they had already encountered in London. Like the Royal Society, the Dublin Philosophical was 'sometimes called a society of useful learning, sometimes a shop of useless subtleties', and like earlier English projectors its members in their 'Petty-Mulleneuxian meeting' were accused of pursuing only private profit and their own 'particular interest'.[164]

In England, however, the process of diffusion was already well under way. One of its more striking manifestations in London was the growing influence in government and parliament of 'men of numbers', as Pepys called them in 1693,[165] members of the educated elite intrigued by measurements, calculations of proportions, and speculations about trends in population and the national wealth. Danby, whose interest in such things has already been mentioned, was probably the Lord Treasurer whom Peter Pett regarded as a 'master of the science of numbers', and not solely because of his instigation of the Compton Census of communicants in 1676.[166] William Blathwayt, who was to be as powerful a political figure as Danby in the 1690s, and who had been clerk to William Temple, was a contact of Petty's in the 1680s, receiving some of his papers and promising to read one of them out to James II.[167] Gregory King, one of Blathwayt's colleagues in the 1690s, had earlier been working on surveys and questionnaires for Ogilby's *Britannia*; and in the 1680s Charles Davenant was employed in managing the hearth and excise taxes.[168] The second generation of political arithmeticians belonged to intellectual networks and an intellectual environment inaugurated by the first.

When King and Davenant reworked Petty's calculation of the national wealth and copied his methodology, they departed very little from his conclusions despite their explicit reservations about them. They engaged once more in that vital comparison of England with France and Holland, and showed that income per head was highest in the Dutch Republic, rather lower in England, but much higher there than in France. King's estimate of England's total annual income was very little more than Petty's (£43.5 million as opposed to £40 million) and clearly

---

[163] J. H. Andrews, 'Land and People, c.1685', in T. W. Moody et al., eds, *A New History of Ireland*, iii: *Early Modern Ireland 1534–1691* (Oxford, 1976), pp. 467–8; Karin Bowie, 'New Perspectives on Pre-Union Scotland', in T. M. Devine and Jenny Wormald, eds, *The Oxford Handbook of Modern Scottish History* (Oxford, 2012), p. 314; below, pp. 188–90, 233–4.

[164] *Papers of the Dublin Philosophical Society*, ii. 913, 917.

[165] Colin Brooks, 'Projecting, Political Arithmetic and the Act of 1695', *EHR* 97 (1982), p. 45.

[166] Peter Pett, *The Happy Future State of England* (1688), p. 116; Slack, 'Government and Information', p. 46; above, p. 135.

[167] McCormick, *Petty*, pp. 273–4; Gertrude A. Jacobsen, *William Blathwayt: A Late Seventeenth Century English Administrator* (New Haven, 1932), pp. 100–4, 433.

[168] *ODNB sub* Davenant, King; Fox, 'Printed Questionnaires', pp. 597–8; above, p. 9.

influenced by him, though based on fuller and more careful calibration of its various elements.[169] The total was too low, and probably much too low, but there is still little agreement among economic historians on what a more reliable figure might be.[170]

Much more convincing than King's numbers, however, and a more accurate reflection of contemporary opinion, are his assumptions about the trends which lay behind them. He thought that the national income had been rising rapidly until war intervened in 1689. He put the surplus in 1688 as high as £2 million a year, an improbably high figure, but it supported his resounding general conclusion: 'We must allow this great fundamental truth, that the trade and wealth of England did mightily advance between the years 1600 and 1688.'[171] Whether they were interested in numbers or not, King's contemporaries, informed by their reading about the progress of commerce, would have endorsed that summary of national improvement across most of the century.

It did not follow that there was any educated consensus, even among improvers, about how general or secure the benefits of material progress and high living were. Although it exaggerated the number of the poor 'decreasing the wealth of the kingdom', and understated the numbers and incomes of those engaged in trade, King's famous table of the 'income and expense of the several families of England' for 1688 showed just how unequal the distribution of income was.[172] Writers on improvement themselves belonged at one time or another to different 'ranks' and 'degrees' in the table, and illustrate the risks involved in trying to move higher. The richest of them in the end, Child and Petty, knew from experience that only substantial holdings of land provided a safe form of investment with lasting returns, even in a period of agrarian depression, and that their transmission from one generation to the next depended on the right marriage choices and on demographic good fortune as much as good economic management.[173] As Graunt and

---

[169] Davenant, *PCW*, i. 128–30, 142–3, 249–52; King, 'Natural and Political Observations', pp. 31, 48–9, 63–9; Paul Slack, 'Measuring the National Wealth in Seventeenth-Century England', *EcHR* 57 (2004), pp. 619–28. On King's methodology, see Tom Arkell, 'Illuminations and Distortions: Gregory King's Scheme Calculated for the Year 1688 and the Social Structure of Later Stuart England', *EcHR* 59 (2006), pp. 32–69; Richard Stone, *Some British Empiricists* (Cambridge, 1997), pp. 77–100.

[170] Peter H. Lindert and Jeffrey G. Williamson, 'Revising England's Social Tables 1688–1812', *Explorations in Economic History*, 19 (1982), p. 393, suggests £54.4 million, close to an estimate indicated by Petty in 1687 (Slack, 'Measuring the National Wealth', p. 617). Different calculations by Steve Broadberry arrive at £53m (N. J. Mayhew, 'Prices in England, 1170–1750', *P&P* 219 (May 2013), p. 34), and by John Dodgson at between £51m and £57m (Dodgson, 'Gregory King and the Economic Structure', p. 1014). Muldrew, *Economy of Obligation*, pp. 90–2, argues for a much larger total. Recent work on economic growth in the eighteenth century, suggesting that it was less rapid than had previously been thought, would also point to a higher figure than King's: Peter Temin and Hans-Joachim Voth, *Prometheus Shackled* (Oxford, 2013), p. 150.

[171] Slack, 'Measuring the National Wealth', p. 623; *Two Tracts by Gregory King*, ed. G. E. Barnett (Baltimore, 1936), p. 61. King's £2m may also have been influenced by Petty: above, p. 123.

[172] *SCED*, pp. 780–1, discussed in Lindert and Williamson, 'Revising England's Social Tables', and Arkell, 'Illuminations and Distortions'.

[173] Cf. Richard Grassby, *The Business Community of Seventeenth-Century England* (Cambridge, 1995), pp. 370–80; *ODNB* sub Child, Petty.

Houghton discovered, retailing was never a foundation for riches, and even Barbon, the greatest entrepreneur and exemplar of high living among them, ended his life in debt.[174] All of these improvers were self-made men, aware for much of their lives of the obstacles in the way of their own material progress.

Another, much less successful, was Adam Martindale, who contributed articles on agrarian improvement to Houghton's periodical in 1682–3 and wrote tracts of his own on surveying, navigation, and mathematics. A Presbyterian minister ejected from his living in 1662, he became a schoolmaster and private chaplain, wholly dependent on sympathetic patrons in Lancashire and Cheshire.[175] Like many of the godly he wrote an autobiography searching for the meaning of afflictions and judgements which providence had visited upon himself and those he knew. He remembered a family story about his elder sister who had gone to London in the 1620s and survived an attack of plague, only to be ruined by her own extravagance; and fifty years later he saw his son, in a promising post in a London school, get into bad company, buy 'costly apparel' to keep up with his friends, and return home like the prodigal in debt and disgrace. Martindale's concluding reflections included one about judgements abounding in a country where even 'the meanest sort of servant maids' now dressed in a manner the daughters of 'the best sort of freeholders' would once have thought presumptuous; and another that 'a lower place with peace and comfort is to be preferred before an higher with trouble and vexation'.[176]

Martindale's intellectual world was very different from Barbon's, and closer to that of his father, Praisegod Barbon, who died in 1679 having lost his business premises in the Great Fire of 1666.[177] But there were many of that generation still alive in the 1680s, advocates of improvement among them, who were as hostile to prodigality as Martindale. After 1690, as we shall see, in the new world of stocks and shares, and the fire insurance companies pioneered by the younger Barbon, finance and trade still offered no security from afflictions, trouble, and vexation. In the 1680s, when the Crown's defaulting on its debts in 1672–4 was a recent memory, and specie still in short supply, success in trade depended all the more on credit and trust, and on self-discipline, prudence, and deliberate restraint. These were the virtues constantly urged on young tradesmen in sermons and advice literature throughout the period covered by this book, from Elizabethan tracts on diligence and thrift to Defoe's *Complete English Tradesman* in 1726.[178]

They were closely associated from the beginning with an ethic favouring self-advancement, in husbandry as well as trade,[179] and in the second half of the seventeenth century they could easily be incorporated into a discourse about industriousness and self-improvement without losing their relevance. In 1685 a

---

[174] On Graunt, see above, p. 118, and on Barbon, below, p. 211.
[175] *ODNB sub* Martindale; Houghton, *Collection*, iv. 58–62, 123–31, 212–37.
[176] *The Life of Adam Martindale*, ed. Richard Parkinson (Chetham Society 4, 1845), pp. 7–9, 43, 210–11.
[177] *ODNB sub* Praisegod Barbon.
[178] Laura Caroline Stevenson, *Praise and Paradox* (Cambridge, 1984), pp. 140–58, 196–7. Cf. Grassby, *Business Community*, pp. 286–8; below, p. 240.
[179] Keith Thomas, *The Ends of Life* (Oxford, 2009), pp. 32–3; above, pp. 18, 65.

popular broadside on *The New Art of Thriving* advertised the need for 'industry and frugality' as a 'good husbandry' essential to prosperity and the attainment of a 'comfortable subsistence'.[180] Among the parishioners of Myddle in Shropshire most admired by Richard Gough in 1701 were a couple whose happiness came from 'care and industry' and the practice of prudence, and others who were 'likely to live well' because they were 'provident and laborious'.[181] In 1684 the Nonconformist minister Richard Steele explained that the *Tradesman's Calling* demanded qualities of mind as well as manual skill. It might lead to 'vast progress...in the improvement of most trades if men did employ their brains within the sphere of their proper callings'. Only then would they attain that 'inward comfort and tranquillity of heart' which was 'the poor tradesman's riches' in the face of the manifold 'inconveniences and difficulties' of trade.[182]

These admonitions were often addressed to the young, and their lessons as often ignored in youth and then only learnt from bitter experience, as Lord North had discovered by the time he wrote about household economy in 1669.[183] The virtues of thrift, prudence, and industriousness were not inconsistent with improvement but inseparable from it. Yet they represented a culture conspicuously at odds with the one Barbon embodied and advocated, one of acquisitiveness and emulation of one's betters, driven by infinite passions of the mind. Inward comfort and tranquillity, and Martindale's willing acceptance of a lowly place with peace and comfort, were the only kinds of happiness available in this world. Contemporaries did not need modern sociological studies to tell them that, beyond a certain point, extra increments of income did not deliver additional felicity, only anxiety and frustration. A modest comfort of the kind that many were now enjoying was preferable to an appetite for ever higher living.[184]

Collective happiness, England's happiness, was similarly precarious, an aspiration only just beginning to be realized with an uncertain future ahead of it. In 1688 Peter Pett, a friend of Houghton's as well as Petty's, published *The Happy Future State of England*, his enormous compilation of information about population, national finance, scientific progress, and rational religion—the whole apparatus from which improvement was constructed and whose rationale and purpose he set out to defend. Like Houghton and other Tories writing about England's happiness when James II came to the throne, he assumed that rising national wealth would bring universal contentment with a properly funded government.

---

[180] Brodie Waddell, *God, Duty and Community in English Economic Life, 1660–1720* (Woodbridge, 2012), pp. 98–101. The broadsheet was a summary of Henry Peacham's popular *Worth of a Peny* (1641 and many later editions).

[181] Muldrew, *Food, Energy and Industriousness*, p. 307.

[182] Richard Steele, *The Trades-man's Calling* (1684), pp. 51–5, 164. Steele had earlier written *The Husbandmans Calling* (1668) with the same message about the need for 'prudence and diligence' (p. 241), but less emphasis on brain-work.

[183] Karen Harvey, *The Little Republic* (Oxford, 2012), p. 70.

[184] Cf. below, pp. 215–17. On the sociological and economic literature on happiness, see Richard Layard, *Happiness: Lessons from a New Science* (2006), pp. 29–53; Luigino Bruni and Pier Luigi Porta, *Economics and Happiness: Framing the Analysis* (Oxford, 2006); Wilfred Beckerman, *Economics as Applied Ethics: Value Judgements in Welfare Economics* (2010).

Like Petty he hoped that religious divisions could be reconciled as quickly as the squabbles of political parties. He had talked to Danby, and Danby's census of communicants and his own political arithmetic showed that the number of Nonconformists was declining and the number of papists insignificant.[185] Even a Catholic monarch, he implied, could scarcely reverse the tide.

Pett took a similar view about the undoubted problems impeding economic advance. He accepted most of the diagnosis in *Britannia languens*. Population was too low and trade in decay, rampant luxury created an adverse balance of trade, and there seemed to be no 'improving manufacture in England' except that of periwigs. Yet all this was being remedied. He was as certain as Petty that in the long run nature would have its course, and its course was determined by past history. 'The political energy of the Reformation' had vastly 'contributed to the increase of the value of our land, and the number of the people, and the extent of our commerce, and indeed of commerce itself'. It was 'very probable' that the population was now much larger than anyone supposed, and although trade had been depressed since 1648, 'nature' would necessarily 'hasten its improvement'. He had observed that 'after a long age of dissoluteness and luxury', there was generally 'a contrary humour' reigning, and it was 'a humour of which I think we now see the tide coming in'.[186] If there had been little growth in manufactures, he saw considerable promise in novelties like 'crape' and linen, and the recent bankruptcies in London which had infected 'so many of our country traders' must 'by natural necessity oblige them to countenance the improvement of the realm by new commodities'.[187]

Pett presented a digest of all the evidence and arguments which could now be assembled to show that improvement must inevitably deliver a happy, even Utopian, future state. On the very eve of the Revolution of 1688 he found it easy to condemn writers of almanacs for wasting their readers' time with talk of 'the lottery of fate' and future 'revolutions and events' whose outcomes were unpredictable. Even the almanacs had referred to improvements in agriculture which justified 'rational expectations of the future state of the earth meliorated by its culture'. There was already 'a pleasant and profitable prospect of such improvement near our metropolis and other great cities' which would make England 'the garden of the world'.[188] Given 'this state of improvement that the world is arrived at', the time could not be far distant when all men would 'improve their fortunes by the improvement and culture of the earth', and by their industry and diligence 'fill all hands with profit and eyes with pleasure'. Nothing was ever wholly certain, but a paradise, luciferous more than lucriferous 'as my Lord Bacon's phrase is', seemed to beckon, 'a new heaven and a new earth that perhaps we may see shortly in old England'.[189]

---

[185] Slack, 'Politics of Consumption', pp. 622–3; Mark Goldie, 'Sir Peter Pett, Sceptical Toryism and the Science of Toleration in the 1680s', in W. J. Sheils, ed., *Persecution and Toleration* (Studies in Church History 21, Oxford, 1984), pp. 201–2, 251–2, 269–72.
[186] Pett, *Happy Future State*, pp. 66, 107, 116, 184–5, 250.
[187] Pett, *Happy Future State*, pp. 251, 257. [188] Pett, *Happy Future State*, pp. 100, 277.
[189] Pett, *Happy Future State*, pp. 91, 275. On 'luciferous', see above, p. 112.

Few of Pett's readers would have disagreed with that significant 'perhaps', or thought his Utopia quite so close at hand as he did. But they would all have understood his idiom and shared his interests. Many would have been among the growing number of the 'curious' whom he described as now 'inquiring into the totals of the numbers of people in states and kingdoms and their chief cities' as avidly as they had once talked about 'the number and strength of their ships of war'.[190] They would also have been talking about contrasts between the wealth and commerce of different countries, and the character of their peoples. In 1685 one of them wrote 'an Essay on the Interest of the Crown in American Plantations' which circulated in manuscript, perhaps through Southwell's connections.[191] Its author has not been identified, and he concentrated chiefly on the opportunities presented by the colonies, especially in the West Indies. But he described how overseas plantations had within living memory changed the character of 'our people at home', and he presented a picture of a new commercial England strikingly different from Pett's.

'The very genius of the people is altered,' he asserted. Now there were 'few of the middle gentry of the nation' or even of the nobility who did not have 'some branch of their family transplanted and thriving' overseas. In consequence a passion for trade and navigation had 'infected the whole kingdom, and no man disdains to marry or mix with it'. The writer manipulated and misinterpreted Petty's conclusions about the distribution of national income in order to prove that three-quarters of it came solely from trade; and he drew on Harrington when arguing that commerce was now 'much the overbalance of the wealth of the nation' which must necessarily have political consequences. Trade was 'a natural emanation of the mind's freedom', filling people's heads 'by degrees... with popular and republican notions'. There was 'eternal noise and wild manner of talk' in every coffee-house about whether Holland or even Spain or Turkey was 'the happiest place in the world', and the whole people were encouraged 'to think felicity is to be had in every country but this; and no wonder then if they love the government accordingly'.[192]

Here was an England less contented than Pett's, disrupted not united by improvement, but it was a sign of the times that other authors similarly overstated the importance of commerce and thought it a threat to the political establishment.[193] The essay on American Plantations offered remedies which might rebalance the interest of the Crown relative to that of trade, chiefly through use of the royal prerogative to give the king a share in it; and James II was only too ready to do that, by favouring commercial monopolies like the Royal African and East India Companies. He was equally keen to promote economic policies, including freedom of conscience and a general naturalization, which he had neither the time nor

---

[190] Pett, *Happy Future State*, sig. N2^v.
[191] BL, Add. MS 47131, fos 22–8; Pincus, *1688*, pp. 89–90. This is among the Egmont Papers, some of which certainly came from Southwell, who is referred to in Add. MS 47131, fo. 4^v.
[192] BL, Add. MS 47131, fos 24^v–27.   [193] Pincus, *1688*, pp. 88–90.

perhaps the authority to implement.[194] In the political and intellectual climate of the 1680s, however, it was no longer necessary to labour the point which Graunt had put to Pepys in the 1660s about trade being as great as ever and in many more hands.[195] England had changed.

There were therefore multiple ironies in the revolutions and events of 1688. In conditions of peace and relative plenty, and with a monarch pursuing commercial advance, England was to all appearances becoming ever more like Holland just at the moment when William of Orange invaded at the invitation of its landed aristocracy. He brought with him war, heavy taxation, a reformation of manners, and even Charles II's three-piece suit.[196] That was a turn of the lottery of fate unpredicted by almanac-writers, and one which required careful negotiation if wealth and happiness, comfort and content, were to continue, and a culture of improvement survive.

---

[194] Pincus, *1688*, pp. 372–82; J. P. Cooper, *Land, Men, and Beliefs: Studies in Early-Modern History*, ed. G. E. Aylmer and J. S. Morrill (1983), p. 217; Beinecke Library, Yale University, OSB MSS 1, Poley Papers, Box 2, folder 68; Sowerby, *Making Toleration*, pp. 62–3.

[195] Above, p. 138.

[196] Tony Claydon, *William III and the Godly Revolution* (Cambridge, 1996); Kuchta, *Three-Piece Suit*, p. 90.

6
Challenges to Affluence 1690–1730

In 1697 the author of *A Letter to a Member* of parliament reflected on what the events of 1688–9 had meant. The English were 'already a people bred up in affluence and plenty', but now there were multiple challenges to their continuing enjoyment of 'happiness above any nation in the world'. The challenges included the 'public calamity' of war, competition from other aggressive trading nations, especially France, and their own propensity to 'luxury and effeminate expense'.[1] After 1688 wars and the need to finance and win them exacerbated old anxieties about the economic and moral capacity of the English to fight them; and they brought into clearer focus what had earlier been suspected, that there might be a finite limit to the nation's economic growth beyond which stagnation if not national decline must inevitably follow. Only after peace returned in 1713 could there be a restoration of the general confidence of the 1680s about England's 'present affluence',[2] and it is the purpose of this chapter to show how that came about.

The author of the *Letter* put greatest emphasis, however, not on war, or economic uncertainty, or luxury, but on the problems of domestic politics, on the absence of the 'great veneration due to parliaments' by the public who elected them. Without that, no government could effectively restrain private excess in the public interest, fund a mounting national debt, or ensure that English commerce and industry remained, in practice as well as principle, 'as improvable as the mind and understanding of man'.[3] With a parliamentary regime and a freer press, the instabilities of the Interregnum seemed to have returned, with all the consequent difficulties in achieving consensus about improvements for the public good; and we should begin there, with the character of the 'public politics' which again offered opportunities and hazards in equal measure.[4]

PUBLIC POLITICS

The opportunities and hazards were both foreseen by John Evelyn towards the end of 1688, when he tried with a mixture of expectation and foreboding to imagine

---

[1] *A Letter to a Member of the Honourable House of Commons, In Answer to Three Queries* (1697), pp. 5–6, 9–10, 20–3.
[2] Below, p. 201. For an earlier use of affluence in the modern sense, see above, pp. 130–1.
[3] *Letter to a Member*, pp. 5–6, 28.
[4] On the character of 'public politics' at this time, see Mark Knights, *Representation and Misrepresentation in Later Stuart Britain* (Oxford, 2005), pp. 66–108.

what the 'approaching revolution' might bring. He had a long list of the national improvements which he hoped might at last be achieved: a census of the population, a new poor law, a standing committee to promote 'all projects convertible to the public benefit', public libraries in every county, and a reformation of manners. There was now an opportunity to 'render this nation as happy as human endeavour were, with the blessing of God, capable to make it', but only if there was a parliament not 'influenced by faction' or 'corrupted by self-interest', and determined to use its powers 'with moderation, justice, piety, and for the public good'. It was asking a lot, and he was far from sanguine, wondering whether things might not 'dissolve [in]to chaos again', as they had in the 1640s, and whether there could be any 'improvement of mankind' at all in 'this declining age'.[5]

Other advocates of improvement who had recently looked to Charles II and James II for support understandably had similar misgivings. In 1689 Peter Pett abandoned any prediction of England's happy future state when he republished his large volume, and gave it a new title, promising only a historical discourse on the *Growth of England in Populousness and Trade since the Reformation*.[6] He had purged it of its most obvious Tory characteristics, but it was not only Whig dominance of the new regime which prevented optimism about the future in a world where political influence depended less on access to the court and more on engagement in a complex public dialogue between interest groups, parliaments in almost permanent session, and a ministry trying to maintain some sense of direction.[7] In such conditions, it seemed doubtful whether proposals for national improvement would receive anything like a favourable hearing, and if they were listened to, whether they would produce results any more concrete than they had in the past.

The answer to the first question was more positive and more immediate than the answer to the second. Publishers rushed to print works old and new about the national wealth, how it was acquired and distributed, and how it could be improved. They included the essays which Petty had been reluctant to publish, and which now gave him posthumous renown. An authorized version of *Political Arithmetick* at last appeared in 1690, after a new edition of the *Treatise of Taxes* published in 1689, and one of his slighter papers, now styled a 'treatise of naval philosophy', was included in an *Account of Several New Inventions and Improvements now necessary for England* in 1691.[8] In the same year *Verbum sapienti* was published as a supplement to his *Political Anatomy of Ireland*, and immediately received the attention it deserved. Petty's calculation of the national income was

---

[5] Steve Pincus, 'John Evelyn: Revolutionary', in Frances Harris and Michael Hunter, eds, *John Evelyn and his Milieu* (2003), pp. 185, 193–5; Tim Harris, *Revolution: The Great Crisis of the British Monarchy, 1685–1720* (2007), p. 306. Cf. Pincus, *1688: The First Modern Revolution* (New Haven, 2009), pp. 215–17.
[6] Mark Goldie, 'Sir Peter Pett', in W. J. Sheils, ed., *Persecution and Toleration* (Studies in Church History 21, Oxford, 1984), p. 252. In 1691, however, Pett was able to tell Anthony Wood that his prediction about the growth of the linen trade (above, p. 167) had been fulfilled: Bodleian Library, MS Wood F 43, fo. 211.
[7] Knights, *Representation and Misrepresentation*, pp. 67–8.
[8] Thomas Hale, *An Account of Several New Inventions* (1691), pp. 130–2.

welcomed as 'the first considerable essay of this nature... treading in an unbeaten path', and used by both Whig and Tory writers as an estimate to be taken 'for granted till there appears to be a better'.[9]

There was similar demand for tracts on trade, to judge by the number of competing titles. Josiah Child's *Discourse about Trade* appeared in 1690, and was reprinted as *A New Discourse of Trade* in 1693, although it was a work almost entirely written in the later 1660s.[10] Dudley North's *Discourses upon Trade*, largely written with an eye to the parliament of 1685, were printed posthumously in 1692. They included arguments about 'the exorbitant appetites of men' being 'the main spur' to trade, industry, and ingenuity, which owed something to Barbon, and Barbon published his own *Discourse of Trade* in 1690 to drive his point home.[11] Not to be outdone, and aiming to capture a far larger audience, Houghton resurrected his newspaper in 1692, after calling for subscribers and correspondents in *A Proposal for Improvement of Husbandry and Trade* endorsed by 28 Fellows of the Royal Society, Pett, Evelyn, and Robert Southwell among them. It announced his intention to 'inform' landlords and tenants, merchants and artisans, physicians and political arithmeticians, about markets, prices, new agricultural improvements, the health of London, and 'the trade, strength and policy of other nations'—'in short all useful things fit for the understanding of a plain man' and of 'public benefit'.[12]

Houghton's new *Collection*, published weekly at two pence an issue and often just a single broadsheet, delivered most of what was promised. Its short articles were designed to provoke private curiosity and public discussion, to stimulate as well as inform. Alongside quantitative information, about weights and measures for example, there were brief essays with a particular point. The utility of political arithmetic was illustrated by the probability that customs records might show the English to be consuming a pound of tobacco per head every year. It was if anything an underestimate, but a neat demonstration of Barbon's argument that new desires contributed to infinite economic growth.[13] London's booming population was said to be increasing demand for Newcastle coal, Norfolk textiles, Suffolk and Cambridgeshire butter, and Cheshire cheese. The success of England's glass industry in removing dependence on foreign imports was cited as an instance of the familiar

---

[9] H.J., *A Letter from a Gentleman in the Country* (1691), p. 3; Dalby Thomas, *An Historical Account of the Rise and Growth of the West-India Collonies* (1690), p. 2, and compare [Dalby Thomas], *Some Thoughts Concerning the Better Security of our Trade and Navigation* (1695), p. 4, for some computations modelled on Petty's original. For similar references to Petty, see Daniel Defoe, *An Essay upon Projects* (1697), pp. 139–40; John Cary, *An Essay on the Coyn and Credit of England* (Bristol, 1696), p. 2.

[10] William Letwin, *The Origins of Scientific Economics* (New York, 1964), pp. 42–4, 251.

[11] Dudley North, *Discourses upon Trade* (1691 [Old Style]), p. 297; Richard Grassby, *The English Gentleman in Trade: The Life and Works of Sir Dudley North 1641–1691* (Oxford, 1994), pp. 231 n. 5, 287.

[12] John Houghton, *A Proposal for Improvement of Husbandry and Trade* (1691).

[13] John Houghton, *A Collection for the Improvement of Husbandry and Trade*, ed. Richard Bradley (4 vols, 1727–8), i. 19–20; above, p. 157.

adage about the English being 'very backward in invention' but excellent at 'improving of arts'.[14]

Houghton was also quick to respond to demand for new kinds of information, especially about the financial markets and investment opportunities created since the 1680s by more joint-stock companies and new public funds. His newspaper was the first to give regular lists of stock prices, sometimes in as many as fifty companies, as well as about the trading conditions which might influence them, like the amount of traffic in the port of London.[15] When the *Collection* folded in 1703, with Houghton still confident that improvement would make England 'the richest and happiest nation the sun sees',[16] other trade papers took its place, some of them published by insurance companies like Charles Povey's Sun-Fire Office. Povey's *General Remark on Trade*, which began in 1705, was initially distributed free to shops, taverns, and coffee-houses and sent out in bundles to the provinces, and it soon laid claim to 3,500 subscribers. Houghton had kept indexed volumes of useful improvements in his shop for interested enquirers, and Povey similarly claimed to have a 'Traders Exchange-House' at home, an information exchange echoing Henry Robinson's venture in the 1650s.[17] They were no longer so necessary when appetites for useful information could be met by the proliferation of printed media, and especially news-sheets, after parliament's failure to renew the Licensing Act in 1695.[18]

Print and speculative financial enterprises were also necessary conditions for the 'projecting age' whose recent arrival was described by Defoe in 1697. Although he referred to some earlier projects mentioned in this book, like the New River of 1609, Defoe dated the birth of this new 'monster' to 1680, citing as an early example William Dockwra's penny-post in London in that year, which Evelyn welcomed as a 'useful, cheap, certain, and expeditious' service and which was already delivering 700,000 letters a year in 1688.[19] He might also have mentioned the first insurance companies, one of them set up by Barbon and insuring 4,000 London houses against fire by 1684, and the banking schemes of the same decade in which Houghton and Hugh Chamberlen (son of the Interregnum projector) had an interest.[20] In the 1690s there was a host of similar projects for the public good as

---

[14] Houghton, *Collection*, i. 441–3, ii. 48. Cf. above, p. 29; below, p. 230. For the success of English glassmaking, see Eleanor S. Godfrey, *The Development of English Glassmaking 1560–1640* (Oxford, 1975), pp. 210–11; Pincus, *1688*, p. 58.
[15] Anne L. Murphy, *The Origins of English Financial Markets: Investment and Speculation before the South Sea Bubble* (Cambridge, 2009), pp. 37–8, 98–9; P. G. M. Dickson, *The Financial Revolution in England: A Study in the Development of Public Credit, 1688–1756* (1967), pp. 486–7; Natasha Glaisyer, *The Culture of Commerce in England, 1660–1720* (Woodbridge, 2006), pp. 145–55.
[16] Houghton, *Collection*, iii. 378.
[17] Glaisyer, *Culture of Commerce*, pp. 156–72; *ODNB* sub Povey; Perry Gauci, *The Politics of Trade: The Overseas Merchant in State and Society, 1660–1720* (Oxford, 2001), p. 165; P. G. M. Dickson, *The Sun Insurance Office 1710–1960* (Oxford, 1960), p. 20.
[18] Knights, *Representation and Misrepresentation*, pp. 16–18.
[19] Defoe, *Essay upon Projects*, pp. 1–2, 10–11, 24, 27; Pincus, *1688*, p. 72.
[20] Pincus, *1688*, p. 72; Nicholas Barbon, *A Letter to a Gentleman in the Country* (1684), p. 2; Dickson, *Sun Insurance*, pp. 6–16; J. Keith Horsefield, *British Monetary Experiments, 1650–1710* (1960), p. 282. For Peter Chamberlen, Hugh's father, see above, p. 101.

rival promoters scrambled for investors in charitable funds, annuities, lotteries, and waterworks, or some mixture of them. In 1690 Chamberlen's new *Proposal to make England Rich and Happy* proposed a fund which would relieve the poor, finance highway improvements, profit its subscribers, and do 'universal good without hurt to any'.[21] That failed to get off the ground, and other enterprises, including some of Povey's, collapsed amid accusations of malpractice. It proved no easier than it had been in the 1650s to reconcile private profit with the public good when, as Defoe implied, every project could claim to be 'built on the honest basis of ingenuity and improvement' and to add public benefit to private advantage.[22]

The message of improvement was being delivered in multiple ways through different genres of printed literature and to different, if overlapping, audiences. Neither the volume of literature nor the size of the audience is easily measured. Some of the formal discourses and manuals about trade joined older publications like those of Mun and Malynes in the libraries of gentlemen as well as merchants, and their number was increasing in the 1690s, but they remained a tiny proportion of the two thousand titles being published every year. Public and parliamentary debate on particular issues boosted their number for a time. More than a hundred pamphlets about the Recoinage appeared in 1695–6, for example, but most of them were ephemeral, sometimes anonymous, and likely to be forgotten afterwards.[23] If we add newspapers like Houghton's to the list, economic information must have reached a large sector of the reading public, especially in London, though probably not the plain men he hoped to find among his readers. Houghton himself thought that the sociability of coffee-houses was an important means of exchanging information which people might have read about in one form or another. There 'an inquisitive man' could learn more about 'arts, merchandize, and all other knowledge' in an evening than he could get from books in a month. Coffee-houses were as productive as his newspaper in achieving his purpose, 'that trade may be better understood, and the whole kingdom made as one trading city'.[24]

One indication of increasing familiarity with economic arguments, at least in London, is the way in which they were picked up, copied, and deployed for different purposes, by a variety of authors, all of them expecting to find an interested audience. One of them was Thomas Tryon, famous as an early advocate of vegetarianism and the author of popular handbooks on health. Born in 1634 among 'an industrious sort of people' in Gloucestershire, he was a self-taught polymath who once worked as a shepherd, spent some time in Holland and

---

[21] *Dr Hugh Chamberlen's Proposal to make England Rich and Happy* (1690), title page; Horsefield, *Monetary Experiments*, pp. 104, 156–7.

[22] Defoe, *Essay upon Projects*, pp. 11, 14. For a well-documented example of a serious and even 'godly' project which caused 'financial havoc' see the discussion of Humphrey Mackworth's enterprises, in Koji Yamamoto, 'Piety, Profit and Public Service in the Financial Revolution', *EHR* 126 (2011), pp. 806–34.

[23] Julian Hoppit, 'The Contexts and Contours of British Economic Literature', *Historical Journal*, 49 (2006), pp. 85–8 and *passim*; Knights, *Representation and Misrepresentation*, pp. 16–17; Pincus, *1688*, p. 497 n. 6; Murphy, *Origins of Financial Markets*, pp. 111–12; Horsefield, *Monetary Experiments*, pp. 289–311.

[24] Miles Ogborn, *Indian Ink* (Chicago, 2007), pp. 182, 197.

Barbados, and finally prospered as a hatter in London. In *England's Grandeur and the Way to Get Wealth*, and a history of the country's 'unheard of' growth in trade which followed it, he described the great benefits brought by 'improvement of arts and sciences' in his lifetime; and he used some political arithmetic to show that 'mightily improved' farming had allowed a large slice of the population to move from agriculture into services and industries. He was as familiar with the advantages of excise taxes, 'paid by the working people and tradesmen' and 'not so much the countryman nor shepherd', as he was with the arguments for and against popular consumption and the trade with France, both of which he favoured. Sometimes unconventional and inconsistent, he was in many ways exceptional, but he was far from unusual in using the language of improvement and political economy in the expectation of influencing public and political opinion. He assumed that everyone, 'in these latter times especially', knew that trade's 'improvement tends not only to the support of personal but national interests'.[25]

Political arithmeticians themselves had to recognize that a free press and regular parliaments had brought discussion of the economy firmly into the public realm. Gregory King had qualms about it as late as 1710, objecting when John Chamberlayne's *Magnae Britanniae notitia* printed some of his calculations because they might inform the country's enemies about its wealth and military strength.[26] Yet King clearly intended that his 'Natural and Political Observations and Conclusions upon the State and Condition of England', written in 1696, should be published, and Davenant printed some details from it in order to correct the errors of 'wretched projectors and contrivers of deficient funds, who are always buzzing about the ministers'.[27] In an age of projects and a popular political press, there could be no prospect of restricting new kinds of information to manuscript circulation between friends or to the king and his chief ministers, as Graunt once supposed. The question now was whether 'clear knowledge' would, as Graunt had imagined, serve to balance parties and factions in a common endeavour for national improvement.[28]

The difficulty turned out to be that knowledge was rarely clear, never complete, and could always be manipulated to suit the case of contending parties and factions claiming a parliamentary hearing. There was never any lack of information, from petitioners for particular interests and from merchants like John Cary, the 'great projector' of Bristol and a friend of Locke, who published tracts on commerce in 1695 for a House of Commons he thought 'much in love with trade' and only in

---

[25] *ODNB* sub Tryon; Thomas Tryon, *Some Memoirs of the Life of Mr. Tho. Tryon* (1705), p. 7; Thomas Tryon, *England's Grandeur and the Way to Get Wealth* (1699), p. 21; Thomas Tryon, *A Brief History of Trade in England* (1702), pp. 15, 19, 81–2, 97, 116, 121, 153–4; Thomas Tryon, *Some General Considerations Offered, Relating to our present Trade* (1698), pp. 1, 7–8, 10, 13. Its content and style suggest that the *Brief History* was written by Tryon.
[26] John A. Taylor, *British Empiricism and Early Political Economy: Gregory King's 1696 Estimates of National Wealth and Population* (Westport, Conn., 2005), pp. 59–60.
[27] Ted McCormick, *William Petty and the Ambitions of Political Arithmetic* (Oxford, 2009), p. 293; Davenant, *PCW*, i. 128–30, 138. King's great work was not in fact published until 1802.
[28] Petty, *EW*, ii. 397; above, pp. 119–20.

need of expert tuition if they were to see it 'truly' and in 'a better light'.[29] Cary was one among many who thought a new Council of Trade, more representative than those of the past, would be able to arbitrate between different interest groups, but the new Board of Trade of 1696 was not designed for that; neither did it have the time to be the clearing house for improvement projects which Evelyn had envisaged. Left to its own devices, the House of Commons was therefore a reactive body, responding to pressure from outside when a majority could be found for a proposal, but rarely taking the initiative. The lapse of the Licensing Act was a conspicuous example of failure to agree when there were alternative ways forward: it was not a deliberate move towards a freer press. In such circumstances piecemeal legislation was easier than general reform. The Commons hesitated to agree on wholesale reform of weights and measures in 1697–8, for example, but happily legislated for national standards for particular commodities.[30]

Efforts to reform the poor law, which had been high on the improvement agenda ever since the 1640s, similarly had only partial effect. The Board of Trade gave the matter unusually sustained attention when bad harvests and economic recession after the recoinage threatened to raise the cost of poor relief to an unsustainable level. In a genuine search for meaningful data in 1696, the Board instituted a thorough inquiry into poor rates in every parish, which concluded that the total cost 'may amount to £400,000 p.a.' The figure was much lower than the £1 million or more commonly guessed at, and seems to have reassured no one, perhaps because it was little advertised. In 1697 members of the Board each wrote down their own version of a 'scheme for the employment of the poor', and pooled their ideas in a report which received more parliamentary attention, since it might reduce the cost, whatever its level. The Commons debated a general reform of the law on at least thirteen occasions between 1694 and 1704, but the poor law of Elizabeth was never replaced by the poor law of Anne.[31]

Instead parliament passed a series of statutes between 1696 and 1712 which authorized fourteen specific towns to implement local reforms of their own, by erecting Corporations of the Poor to centralize parochial relief and support workhouses. They were based on the London Corporation for the Poor of 1647 which Hartlib and his friends had supported, and like that model they were the product of local projectors and their contacts in the Commons able to command sufficient civic and parliamentary support. In Bristol, the first successful town, the sponsors

---

[29] John Cary, *An Essay on the State of England, in Relation to its Trade* (Bristol, 1695), sig. A6ʳ.
[30] William Pettigrew, 'Regulatory Inertia and National Economic Growth: An African Trade Case Study, 1660–1714', in Perry Gauci, ed., *Regulating the British Economy, 1660–1850* (Farnham, 2011), pp. 25–38; Julian Hoppit, *Land of Liberty?* (Oxford, 2000), pp. 245–6, 341–2.
[31] Slack, 'Government and Information in Seventeenth-Century England', *P&P* 184 (Aug. 2004), pp. 56–7; Paul Slack, *From Reformation to Improvement* (Oxford, 1999), pp. 109, 117; Locke, *Political Essays*, ed. Goldie, pp. 182, 189; Beinecke Library, Yale University, OSB MSS 2, Blathwayt Papers, Series I, Box 7, folder 143, 1 Oct. 1697. James Puckle, *England's Path to Wealth and Honour* (2nd edn, 1700), sig. A3ᵛ, estimated the annual cost of the poor rates at £1m a year. In 1680, however, another author put the figure at £400,000: *Britannia languens*, in J. R. McCulloch, ed., *Early English Tracts on Commerce* (Cambridge, 1952), p. 378.

were Cary and the city's Whig MPs, backed by local Dissenters and Quakers, who were excluded from the town council. There were similar alliances in Hull, Colchester, and Exeter which secured statutes with Whig backing in the Commons in the 1690s, and in London, which revived its Interregnum Corporation in 1698. In the next decade, when there were parliaments with larger numbers of Tories, towns controlled by them, like King's Lynn, Gloucester, Worcester, and Norwich, were able to join them. Whether Tory or Whig in political complexion, however, most of these towns had been experimenting with workhouses for their poor and promoting urban improvements of all kinds since the 1660s.[32]

Henry Bell of King's Lynn, for example, whose father had been involved in Interregnum projects for the poor, had erected a new customs house there in 1683 as an exchange for 'conference in trade and commerce'. He was active in schemes to improve navigation and set up a charity school, and he got a clause for lighting the streets tacked to the Corporation Act of 1701 in order to 'enlighten' Lynn. In Gloucester, councillors shared some of the interests of the founder of their charity school, Timothy Nourse, a local landowner and Catholic Tory, who had written a tract on the 'benefits and improvements of husbandry' and advocated model villages, 'little towns' such as the Lowthers erected at Lowther New Town in Westmorland, as well as uniformity of measures, pawnshops, and 'colleges' to employ the poor.[33] Groups of improvers were similarly active in towns which did not need, or failed to obtain, Corporations of the Poor. Bath obtained parliamentary sanction for a street-lighting scheme along with Lynn and Norwich. Liverpool joined Bristol and London in getting statutes for new waterworks. The citizens of Taunton expected that a navigation scheme for the river Tone, backed by another of Locke's friends, Edward Clarke, would provide funds for a town workhouse and school, although in the event all the profits went to local investors.[34]

The number of statutes for local improvements of one kind and another after 1689 testifies to pent-up local demand for them, after a decade in which parliament had scarcely met at all. It testifies also to the capacity of parliaments meeting in regular and predictable sessions to facilitate successful lobbying for them. In the case of expensive improvement schemes like those for rivers and roads, parliament had the further advantage of offering greater protection to undertakers and investors than the royal patents and charters which sometimes sanctioned them in the past had in practice provided. The undisputed power of parliamentary statute to interfere with property rights significantly reduced the risks involved.[35] It has been shown that investment in transport improvements increased substantially in

---

[32] Above, p. 105; Slack, *Reformation to Improvement*, pp. 103–9.
[33] Slack, *Reformation to Improvement*, pp. 104–6. On Nourse, see Paul Warde, 'The Idea of Improvement, c.1520–1700', in Richard W. Hoyle, ed., *Custom, Improvement and the Landscape in Early Modern Britain* (Farnham, 2011), pp. 139, 148.
[34] Slack, *Reformation to Improvement*, p. 102; Mark Knights, 'Regulation and Rival Interests in the 1690s', in Gauci, ed., *Regulating the British Economy*, pp. 74–5, 80.
[35] Cf. Julian Hoppit, 'Patterns of Parliamentary Legislation, 1600–1800', *Historical Journal*, 39 (1996), pp. 125–8.

the middle of the 1690s, and the same applies to urban improvements of other kinds.[36] For all its limitations, a parliamentary regime was showing that it could provide more fertile ground for the improvement agenda than a would-be absolute monarchy.

The effects were to become fully apparent only in the later eighteenth century. By 1730, however, there had been 81 Turnpike Acts and 34 statutes for river improvement since 1688, which lowered transport costs and encouraged regional specialization and integration. More than half the mileage of the thirteen major roads out of London had been turnpiked, and 500 miles of river made navigable. A 'transport revolution' was well under way.[37] The same can be said about the 'urban renaissance' which transformed the appearance, cleanliness, and amenities of the larger English towns between 1660 and 1770.[38] A great fire sometimes prompted large-scale redevelopment, as it had in London in 1666, in Northampton in 1675, where Henry Bell of King's Lynn first learnt from Robert Hooke about modern urban planning, and then in Warwick in 1694. After that, however, regular parliaments enabled towns to copy piecemeal improvements from one another and engage in a process of competitive emulation which led many of them in the end to acquire their own statutory 'improvement commissions'. They were better equipped to act for the more energetic and publicly spirited citizens than narrow town councils which one contemporary thought generally hostile to 'ingenuity and improvements'. The first provincial example, in Salisbury in 1737, had a more broadly representative board of directors and trustees closely modelled on those of the Guardians of the Corporations of the Poor. There were to be 160 such commissions by 1799.[39]

The political revolution of 1688 had established what was genuinely a 'reactive state',[40] but also a powerful one, not directing but responding effectively to a culture of improvement so firmly established and so widely dispersed that there was no lack of projects for the purpose or willing investors to fund them. It was no mean achievement when there was a war on and the state had to act to mobilize the resources necessary to pay for it. That also required parties and factions to be held in some kind of balance, and a contribution from political arithmetic and political economy, to achieve it.

---

[36] Dan Bogart, 'Did the Glorious Revolution Contribute to the Transport Revolution? Evidence from Investment in Roads and Rivers', *EcHR* 64 (2011), pp. 1075, 1100; Knights, 'Regulation and Rival Interests', pp. 66–7.

[37] Hoppit, *Land of Liberty?*, p. 329; Bogart, 'Glorious Revolution', p. 1073.

[38] Peter Borsay, *The English Urban Renaissance: Culture and Society in the Provincial Town, 1660–1770* (Oxford, 1989).

[39] Slack, *Reformation to Improvement*, pp. 98, 104, 132–3; David Eastwood, *Government and Community in the English Provinces, 1700–1870* (Basingstoke, 1997), p. 68; E. L. Jones and M. E. Falkus, 'Urban Improvement and the English Economy in the Seventeenth and Eighteenth Centuries', in Peter Borsay, ed., *The Eighteenth-Century Town: A Reader in English Urban History 1688–1820* (1990), pp. 142–3; Paul Langford, *Public Life and the Propertied Englishman, 1689–1798* (Oxford, 1991), pp. 222–32.

[40] Lee Davison et al., eds, *Stilling the Grumbling Hive: The Response to Social and Economic Problems in England, 1689–1750* (Stroud, 1992), 'Introduction'.

WAR AND EMPIRE

Among the *Strange and Wonderful Prophecies and Predictions* on an illustrated broadsheet published as wartime propaganda in 1691 was the expectation that the war recently begun would 'in all likelihood and probability' ensure 'the glory of the British empire' and 'the ruin and decay of the French Greatness'. There would be battles on land and at sea, but husbandmen could safely go to war while women tended the fields, and victory against Louis XIV and popery was assured. England would gain new lands and more trade, and everyone 'sit at ease and in plenty'. 'A long prosperity is promised', and it was expected to begin soon.[41] The two wars against France, the Nine Years War (1689–97) and War of the Spanish Succession (1702–13), in fact lasted for twenty years, and might have ended any prospect of national improvement and material progress for a generation.

The military commitments involved, on the continent of Europe as well as at sea, were unprecedented in the country's history, and so were the costs, which could only be met through a financial revolution which entailed new institutions such as the Bank of England and the National Debt, and a recoinage which had savage deflationary consequences. In the event England had sufficient credit, accumulated capital, and ready cash to survive in the 1690s; and after 1702 the country was even able to enjoy a trade surplus thanks to fortuitous events in the Indies and northern Europe which encouraged English exports and re-exports. But neither the country's survival in the 1690s nor its recovery in the 1700s could safely have been predicted in 1690, and the first was a close-run thing since the government came very close to bankruptcy in 1696.[42]

It was obvious that the outcome of such costly wars must also depend upon features of the economy which political economists had been discussing for the best part of a century, like the relationship between the wealth of a state and its power, methods of taxation, the money supply, international currency flows, and the balance of trade. As in the crisis of the 1620s, therefore, their advice was sometimes called for and often volunteered, and it was now published and disseminated to larger audiences than ever before. That is not to say that their prescriptions always determined the actions of governments, who were necessarily reacting to rapidly changing events. But the language of political economy was used to define policy options and to justify decisions once they had been taken. The pressure of events also had an impact on economic writers themselves. The Nine Years War was as educative for political economists as it was for politicians and the wider political nation. It drove home the lessons they had been learning since the first Dutch War, that economic and commercial policies must be framed with regard to a world

---

[41] Derek Hirst, *Dominion: England and its Island Neighbours 1500–1707* (Oxford, 2012), pp. 3–6. At a more local and popular level, there were ghost stories in Westmorland in the 1690s about an 'apparition in the shape of soldier' making predictions about the outcome of the war: Adam Fox, 'Vernacular Culture and Popular Customs in Early Modern England: Evidence from Thomas Machell's Westmorland', *Cultural and Social History*, 9 (2012), p. 343.

[42] D. W. Jones, *War and Economy in the Age of William III and Marlborough* (Oxford, 1988), pp. 1–23. Jones comments (p. 308) that 'Albion was extremely lucky as well as perfidious'.

where there were several European powers contending for dominance by pursuing trade and empire.

According to a tract on trade in 1695, these conditions had created something wholly new, nothing less than a 'modern system of politics' operating on a global scale. Political power now depended on money more than military valour—'no money, no soldiers'—and money depended on trade and dominion of the seas. The Dutch had shown the way, but 'in this age...not only republics but even absolute monarchies' like France, Sweden, Florence, and the Papacy were pursuing 'improvement of their wealth by trade'. Now that France was engaged in a war for supremacy on land and sea, the outcome was bound to reorder 'the balance of trade and naval power in Europe', not necessarily to England's advantage.[43] The author was describing a modern states-system characterized by what Hume called 'jealousy of trade', and one which was to influence Enlightenment thinking after the Peace of Utrecht in 1713.[44]

Political arithmeticians drew their own measured conclusions. Davenant remarked that 'the whole art of war' was 'quite changed from what it was in the time of our forefathers' and had been wholly 'reduced to money'. Gregory King's conviction that a war against 'so potent a monarch' as Louis XIV would inevitably be 'long and very expensive' led him to rework Petty's national accounts, and conclude in 1695 that it could not be afforded at all if it lasted another two years. His pessimism may well have influenced some of his calculations, but he was right to warn that the national income was now declining, though probably not by the £3 million a year he estimated. At that rate, according to King, there would be famine by 1698 and no possibility of war continuing. Both of them dismissed as absurd the opinion of many in authority, as well as 'the people in general', that there would be a rapid and easy victory; and Davenant laid the blame for such 'vanity' squarely at Petty's door. 'Great genius' though he was, in *Political Arithmetick* he had vastly overstated the strength of England as compared with that of France, and 'rather made his court than spoke his mind' in order to please Charles II.[45]

Davenant nevertheless followed Petty when advising the government on how wartime expenditure could be met. The public charge, which now fell almost wholly on landowners and active traders (through land taxes and customs duties), could only be made 'easy and supportable' if it was 'put upon all degrees of men alike, with geometrical proportion'. Excises on consumption were therefore essential, and might more 'equally rate all sorts of wealth and substance', but they

---

[43] *Considerations Requiring greater Care for Trade* (1695), pp. 1–2, 10–12.

[44] Istvan Hont, *Jealousy of Trade: International Competition and the Nation-State in Historical Perspective* (Cambridge, Mass., 2005), 'Introduction'; Craig Muldrew, 'From Commonwealth to Public Opulence', in Steve Hindle et al., eds, *Remaking English Society* (Woodbridge, 2013), pp. 323–4; J. G. A. Pocock, *Barbarism and Religion*, ii: *Narratives of Civil Government* (Cambridge, 1999), pp. 2–3.

[45] Davenant, *PCW*, i. 3, 16, 129–30; King, 'Natural and Political Observations', pp. 31, 61–3, 69; Gregory King, 'Burns Journal', in Peter Laslett, ed., *The Earliest Classics* (Farnborough, 1973), p. 167; Jones, *War and Economy*, p. 15. In 1695, however, Davenant happily followed Petty in his own calculations of the national wealth: Davenant, *PCW*, i. 62.

were likely to miss the capital wealth of many financiers and merchants.[46] When he tried to put a plausible case for new taxes in 1690, and even argued that they would stimulate 'an improvement of trade in time of war', Defoe also recognized that they would be 'an insupportable grievance' unless they were levied on consumption, and, somehow or other, on the funds and stocks of those who did not spend.[47]

Political economy and good sense could not solve what was fundamentally a political problem. Parliament had not helped itself by abolishing the unpopular hearth tax in 1689, and new excises were introduced only slowly since they were regarded as 'dangerous to liberty' and characteristic only of absolute and military governments.[48] There were other new taxes, in some of which Gregory King or his friends had a hand. One was a Duty Act of 1695 on births, marriages, and burials, which contained clauses imposing annual taxes on bachelors over the age of 15 and childless widowers. It was an attempt at pro-natalist policies like those introduced by Colbert in France in the 1660s, and a small victory for political arithmetic, but it was of little financial, let along demographic, consequence. So complicated that King had to produce a printed table to explain it, and open to criticism because the duties payable were graduated more by social rank than by wealth, it collapsed within a decade.[49] Welcoming its demise, William Paterson acidly commented that people must expect heavy taxes, but it was 'a great deal too much to find impositions upon their very coming into and going out of the world'.[50]

That left the land tax as the largest source of direct taxation. Established in 1692, and remodelled in various ways between then and 1698, it was levied at a rate of 4 shillings in the pound on the value of property, which meant that many landowners were paying a fifth of their income for nearly twenty years. The tax was scarcely equitable, however, let alone geometrically proportionable. Houghton displayed its disproportions in a table he published in 1693 comparing its distribution with the number of acres and houses in every county; and Davenant published further numbers to the same effect in 1695. Both of them were writing for parliament and may have influenced a final revision of the tax in 1698, but the northern counties and Cornwall remained under-assessed and the Home Counties visibly overrated.[51] Representatives of the Home Counties found themselves outvoted whenever the matter was discussed in the Commons because, as another published table demonstrated in 1698, the distribution of parliamentary seats bore

---

[46] Davenant, *PCW*, i. 16–17, 60–4, 62, 142–3. Cf. above, p. 80, on geometrical proportion.

[47] Daniel Defoe, *Taxes no Charge* (1690), title page, sig. A2^r, pp. 10, 11, 15. The argument about an increase in trade turned out to have some validity after 1702: Jones, *War and Economy*, pp. 195–209.

[48] Davenant, *PCW*, i. 80; Richard Temple, *An Essay upon Taxes* (1693), pp. 10, 13–14. Cf. Colin Brooks, 'Projecting, Political Arithmetic and the Act of 1695', *EHR* 97 (1982), p. 47.

[49] Brooks, 'Projecting', pp. 31–53; Gregory King, *A Scheme of the Rates and Duties*... (1695). For proposals for tax incentives of this kind, see *An Essay or Modest Proposal of a Way to encrease the Number of People* (1693), p. 3; above, p. 137.

[50] William Paterson, *An Inquiry into the Reasonableness... of an Union with Scotland* (1706), p. 121.

[51] C. G. A. Clay, *Economic Expansion and Social Change* (2 vols, Cambridge, 1984), ii. 267; Houghton, *Collection*, i. 71–8; John Houghton, *An Account of the Acres & Houses with the Proportional Tax &c* (1693); Davenant, *PCW*, i. 38–45; Philip Loft, 'Political Arithmetic and the English Land Tax in the Reign of William III', *Historical Journal*, 56 (2013), pp. 321–43.

no relationship to the distribution of taxation.[52] Davenant was forced to conclude that an unequal tax (but one administered by landowners themselves) was the price 'the gentlemen of England' were ready to pay for the security of a government wholly dependent on regular parliaments.[53]

The larger inequity, of taxes falling much more heavily on land than trade, was less easy for gentlemen to tolerate, despite the constant attempts of writers on the economy to demonstrate that, if properly understood, the interests of land and trade must be the same.[54] It also left governments needing to fill a mounting gap between current revenue and expenditure. Annual government expenditure in the Nine Years War was around £5 million a year, three times its level in the 1680s, and £8 million a year in the Spanish Succession War, and although revenue from taxation rose to around 9 per cent of the national income in 1710, it was never enough. The national debt, which had not existed at all in 1688, rose to £17 million in 1698 and £36 million in 1714.[55] Governments had to borrow what they could not extract through taxation, and projectors and political economists produced schemes for lotteries, annuities, and especially banks, for the purpose.[56]

The great pamphlet war in 1694–5 between advocates of the Bank of England and its alternatives, most of them land banks, set Whigs against Tories and exacerbated perceived conflicts of interest between trade and land, but it involved much the same rhetoric about the national interest on both sides.[57] William Paterson, chief advocate of the successful Bank and vociferous critic of the 'banks beyond the moon' which others were proposing, insisted again and again on its benefits 'for trade and improvements'.[58] Hugh Chamberlen, his Tory opponent, author of that earlier project to make England 'rich and happy', readily admitted that the nation's trade should be 'the glory of the landed man'.[59] There was also at

---

[52] Slack, 'Government and Information', pp. 50–1; Patrick K. O'Brien, 'The Political Economy of British Taxation, 1660–1815', *EcHR* 41 (1988), p. 19; John Smart, *A Scheme of the Proportions the Several Counties in England paid to the land tax . . . compared with the Number of Members they send to Parliament* [1698].

[53] Davenant, *PCW*, i. 75.

[54] William Paterson, *A Brief Account of the Intended Bank of England* (1694), p. 13; *A Discourse of the Nature, Use, and Advantages of Trade* (1694), p. 9; [Thomas], *Some Thoughts Concerning the Better Security of our Trade*, p. 4.

[55] Hoppit, *Land of Liberty?*, p. 124. Davenant, following King, thought the tax burden in the 1690s already over 10 per cent of the national income: *PCW*, i. 142. For some recent calculations for the 1690s, see Sowerby, *Making Toleration*, pp. 51–2.

[56] Murphy, *Financial Markets*, pp. 45–8.

[57] On the foundation of the Bank and the large contemporary literature about it, see Dickson, *Financial Revolution*, and Horsefield, *British Monetary Experiments*. The party-political issues involved are discussed in Pincus, *1688*, pp. 388–90; and in Steve Pincus and Alice Wolfram, 'A Proactive State? The Land Bank, Investment and Party Politics in the 1690s', in Gauci, ed., *Regulating the British Economy*, pp. 45–55. I am more impressed than Professor Pincus is by the rhetoric about trade and economic growth common to both Whigs and Tories.

[58] *Some Observations upon the Bank of England* (1695), pp. 1, 3, 11, 15, 17, 25. Paterson may not have been the author of this tract, but it seems to me likely given similarities of style with other works by him, e.g. *Brief Account of the Intended Bank of England*, p. 4, and *The Occasion of Scotland's Decay in Trade* (n.p., 1705), p. 8.

[59] Hugh Chamberlen, *Some Useful Reflections upon a Pamphlet called A Brief Account of The Intended Bank of England* (1694), p. 13; above, p. 174.

least one Tory energetically engaged on Paterson's side. In *England's Glory* (1694), Humphrey Mackworth applauded the Bank's success in settling 'the great question' of how to raise 'a fund that shall be credited by all'. It would 'make the king great, the gentry rich, the farmers flourish, the merchant trade, ships increase, seamen to be employed, set up new manufactures, and encourage the old'. Mackworth took much of his language about infinite economic growth and England being 'ten times richer' in a generation from Barbon and Houghton, and from Carew Reynell's *True English Interest*, written twenty years before.[60]

Yet no amount of familiar promises about the country's future affluence can explain why the Bank was able to raise £1.2 million within a fortnight of its launch, and the government to fund one-third of its wartime expenditure by borrowing. The immediate cash came from the accumulated funds of financiers and the redundant trading capital of merchants formerly trading with France and trying, so far without success, to break into the East India trade.[61] The longer-term confidence, in regular parliaments and an initially Whig ministry, on which 'the public credit' depended, proved much more precarious; and the need to sustain that, when the coinage was being clipped and losing value, necessitated the Recoinage of 1696, the most difficult economic decision of an English government in the seventeenth century, perhaps the most important in the longer run, and certainly the most disruptive.[62]

Given their longstanding worries about a shortage of coin, political economists might have been expected to offer concerted advice in favour of the modest devaluation which was advised by William Lowndes, Secretary at the Treasury, but they were divided. The government consulted some supposed experts, Davenant, Child, Wren, Newton, and Locke among them. They came to no conclusion, partly because, as Wren said, they lacked reliable information and no one knew 'what the present cash of England is'. The many writers on the issue split between those, like Barbon, who favoured devaluation because it would boost the domestic economy, and others like Locke and Cary who opposed it on a variety of grounds, the most persuasive being the damage it would do to government credit at a time of massive borrowing. The king, determined to maintain the value of the large remittances funding his armies abroad, may have had the last word. The coinage was restored to its old standard, with consequences at home which Barbon had

---

[60] H[umphrey] M[ackworth], *England's Glory: or, The Great Improvement of Trade in general by a Royal Bank* (1694); Carl Wennerlind, *Casualties of Credit* (Cambridge, Mass., 2011), pp. 112–13; Paul Slack, 'The Politics of Consumption', *EHR* 122 (2007), pp. 626–7; Pincus, *1688*, pp. 391–2. Professor Pincus argues that H.M. was a Whig and thinks him unlikely to have been Mackworth, but I see no reason to doubt the usual attribution. On Mackworth, see Yamamoto, 'Piety, Profit and Public Service'.

[61] Jones, *War and Economy*, pp. 12–13, 249–50; Murphy, *Origins of English Financial Markets*, pp. 56–9; Hoppit, *Land of Liberty?*, pp. 124–7.

[62] Nicholas Mayhew, *Sterling: The History of a Currency* (1999), pp. 98–102. On the contemporary debate, see Joyce Oldham Appleby, *Economic Thought and Ideology* (Princeton, 1978), pp. 199–241; Wennerlind, *Casualties of Credit*, pp. 128–35. There was further alarm over 'public credit' in 1710, under a Tory ministry: Wennerlind, *Casualties of Credit*, pp. 162–95.

accurately predicted: 'want of... money, a stop of trade, and a general complaint and poverty all over the nation.'[63]

Like the successful launch of the Bank of England, the Recoinage was the product of hard political and fiscal realities in wartime, and commercial policies had to be tailored to suit the same circumstances. As Davenant reluctantly acknowledged, the scarcity of money in a long war inevitably made 'any exportation of bullion... a great grievance of which in quiet times, we should not be sensible'. It focused attention on the balance of trade, despite all the difficulties involved in its measurement.[64] In 1697 the Board of Trade summarized conventional opinion about which areas of commerce should be 'improved and extended', and which were wholly out of balance and needed somehow to be restrained. In a long report on 'The State of Trade' which was widely circulated, it identified the usual suspects: the trade with France, where there was an estimated imbalance of £1 million a year before 1689, and which would need to be tackled in any future peace negotiations; and trade with the East Indies, where the imbalance was impossible to compute for lack of 'certain information', but generally assumed to be large.[65] In neither case was there any easy solution, and efforts to find one taught parliament and the public something more about constraints on policy-making in the 'modern system' of international politics.

In 1697 the East India trade and the company which controlled it were already political hot potatoes of long standing. The East India Company was attacked from all sides, by Whigs because it was governed by Tories like Child, and by would-be interlopers who exploited a revival of demands for free trade for their own purposes. The trade faced opposition from textile manufacturers threatened by imports of cheap Indian cottons and was predictably defended by Houghton and Barbon because it promoted consumer expenditure. But the issues raised by a peculiarly complex area of commerce were intricate and intractable, as a pamphlet debate between Davenant and John Pollexfen about the merits of both trade and Company showed.[66] Some kind of untidy compromise was inevitable. The Company survived, but at the price of tolerating a rival company, funded in exchange for a loan of £2 million to the government, until the two amalgamated in 1702. It also had to face the consequences of the Silk Act, passed after much petitioning in 1700, which restricted the import of a wide range of silk and cotton goods.[67]

---

[63] Slack, 'Government and Information', pp. 54–5; Jones, *War and Economy*, pp. 244–5; William Lowndes, *A Report Containing an Essay for the Amendment of the Silver Coins* (1695), p. 3; Nicholas Barbon, *A Discourse Concerning Coining the New Money lighter* (1696), sig. A4^r; Cary, *Essay on the Coyn and Credit of England*, pp. 2, 17.

[64] Davenant, *PCW*, i. 94. Barbon noted that an accurate balance was difficult to calculate in any instance and impossible to interpret even if it could be: Barbon, *Discourse Concerning Coining*, p. 36.

[65] *SCED*, pp. 568–70, 574. The report was presented to the Commons in 1699 and considered again by the Lords in 1713.

[66] Murphy, *Financial Markets*, pp. 72–7; John Blanch, *The Interest of England Considered* (1694), sig. A5^r, p. 55; Houghton, *Collection*, ii. 6–8, 133–5; Barbon, *Discourse Concerning Coining*, p. 44; Hont, *Jealousy of Trade*, pp. 240–6.

[67] Hoppit, *Land of Liberty?*, pp. 274–5; Murphy, *Financial Markets*, p. 48; Gauci, ed., *Regulating the British Economy*, p. 1.

The French trade produced even more public clamour when peace returned in 1713 and the government had to conclude a commercial treaty with France. The terms proposed by the Tory ministry amounted to a free trade agreement of the kind Thomas Tryon would have welcomed, encouraging trade but also benefiting English manufacturers threatened by French competitors.[68] They were defeated in parliament, but they had produced a press campaign which had a lasting effect on political opinion. In 1712 there had seemed to be unanimous agreement about the benefits peace would bring to commerce, reflected in public addresses to the Crown from several counties, even those from Merionethshire in Wales and from 'the loyal clans of Scotland'. In 1713, however, two new weekly newspapers set out to persuade their readers that the terms negotiated were either the best or the worst that could have been achieved.[69]

Mercator, edited for the government by Davenant and Defoe, and Henry Martin's *British Merchant*, for the opposition, presented rival interpretations and calculations of the current balance of trade. Yet both of them provided histories of trade to set their arguments in context which were similar stories of commercial growth, differing only in detail. The *British Merchant* in particular quoted from a range of authorities on the wealth and trade of the nation, Petty, Temple, Fortrey, and Davenant himself among them; and it carried correspondence from readers sometimes critical of these authorities.[70] It was engaged in a deliberate exercise in public education. At the peak of the controversy, over 10,000 copies of trade-related journals were being produced a week, and the *British Merchant*, collected in three volumes in 1721, became a handbook to English political economy. It was translated into French in 1753 and recommended by Lord Shelburne in 1786 as 'a book which has formed the principles of nine tenths of the public since it was first written'.[71]

Economic principles were not the same as economic policies, which had to be negotiated between contending parties and involved uncomfortable compromises, as the examples of the East India and French trades show. But principles were guides to policy, and the economic legislation of the years around 1700, taken as a whole, points to the establishment of a coherent policy which had been in the making since at least the 1650s, if not the 1620s, and which, wherever practicable, protected the country's manufacturing and commercial interests against threats from outside. While the remaining powers of chartered monopolies like the Merchant Adventurers and Russia Company were gradually whittled away, there

---

[68] Tryon, *General Considerations*, p. 10; D. C. Coleman, 'Politics and Economics in the Age of Anne', in D. C. Coleman and A. H. John, eds, *Trade, Government and Economy* (1976), pp. 190–2; Ahn Doohwan, 'The Anglo-French Treaty of Commerce of 1713: Tory Trade Politics and the Question of Dutch Decline', *History of European Ideas*, 36 (2010), pp. 167–80.

[69] Perry Gauci, *The Politics of Trade: The Overseas Merchant in State and Society, 1660–1720* (Oxford, 2001), pp. 238–71. On similar press debate in 1711, occasioned by the credit crisis, see Wennerlind, *Casualties of Credit*, pp. 200–15.

[70] *Mercator: Or, Commerce Retrieved*, nos 1 and 2, May 1713; no. 63, 15–17 Oct. 1713; *The British Merchant: or Commerce Preserv'd*, no. 93, 22–5 June 1714, pp. 134–5; Charles King, ed., *The British Merchant* (3 vols, 1721), i. xix, 165–89.

[71] Gauci, *Politics of Trade*, pp. 165, 270; below, p. 250.

were protectionist tariffs against imports, and exports were encouraged by a reduction in duties. At the same time the landed interest was protected and agricultural improvement and productivity encouraged by corn bounties which had been established permanently in 1689.[72]

Neither economic principles nor policies for economic improvement could be applied consistently to the economic problems of empire, however. That was partly because there was no consensus about what kind of empire it should be. Was it, as Barbon asked in 1690, an empire 'for trade' based on command of the sea, or 'an empire upon the land' of ever greater size?[73] In 1668 Worsley had thought it was both. It was 'an absolute necessity' to improve trade with the American plantations because emigration to them was essentially an export of English labour to places where it could be most productive, furnishing commodities which were import substitutes in England, and benefiting the balance of trade. At the same time, plantations increased 'the limits of our dwelling' and the king's 'territories and dominions', and made 'the Empire of England...more august, formidable and considerable abroad'.[74]

Barbon, however, was convinced that in the modern system of international politics naval power and the protection of trade were fundamental. Past empires might have been built on territory; now command of the seas gave England an opportunity to create 'an empire not less glorious and of a much larger extent, than either Alexander's or Caesar's'.[75] By 1720 most metropolitan analyses of the benefits of plantations agreed with Barbon that theirs was an empire whose essential purpose was greater trade not territory, but they took Worsley's points on board to the extent of recognizing that the colonies must flourish in order to promote manufactures, shipping, and employment at home. Their empire should therefore be one dependent on 'hands' and not 'land'. That was what made it, as Montesquieu recognized, a 'thing without precedent', 'combining trade with empire'.[76]

It was always an uneasy combination, nevertheless, because the 'hands' in the other parts of the empire had ambitions and interests of their own which they were unwilling to sacrifice to those of the mother country. Most of the problematic consequences were identified by the Board of Trade in its 1697 report. The East India Company was developing its own multilateral trades in the Indian Ocean and exporting ever larger quantities of cheap Indian textiles to Europe. In America the northern plantations of New England, 'settled by an industrious people', were producing manufactures for export, growing in population perhaps at the demographic expense of the home country, and not even capable of organizing their own

---

[72] David Ormrod, *The Rise of Commercial Empires* (Cambridge, 2003), pp. 43–51, 141, 168–73, 216–17, 343–5; Hoppit, *Land of Liberty?*, pp. 321–2; Julian Hoppit, 'Bounties, the Economy and the State in Britain, 1689–1800', in Gauci, ed., *Regulating the British Economy*, pp. 142, 156.
[73] Armitage, *The Ideological Origins of the British Empire* (Cambridge, 2000), p. 143. The issues and debates are analysed in Armitage, *Ideological Origins*, pp. 146–69.
[74] Thomas Leng, *Benjamin Worsley* (Woodbridge, 2008), p. 152; *SCED*, pp. 535–6.
[75] Armitage, *Ideological Origins*, p. 143.
[76] Armitage, *Ideological Origins*, pp. 166–7; Montesquieu, *Persian Letters*, trans. C. J. Betts (1977), p. 242 (Letter 130).

defence.⁷⁷ Ireland had been increasing its production of woollen textiles when it ought to have been concentrating on linen, and Scotland presented particular 'dangers which threaten our trade', having recently set up its own company trading to Africa and the Indies, and being about to found the Darien settlement on the isthmus of Panama.⁷⁸

Every part of this empire was ready to press its claims to special treatment from London, with Ireland denying it was merely a 'colony for trade', and even the West Indies plantations aspiring to be as much 'a part of England' and entitled to trade as freely with it as the Isle of Wight or Anglesey or the fenlands reclaimed from the sea.⁷⁹ Total constitutional reconstruction such as Petty had envisaged in one of his projects for a 'Union' of all English dominions in an 'Improved Empire' might have helped towards a solution, but was never likely. Neither was that lesser union of the three kingdoms of England, Scotland, and Ireland, which was sometimes projected in the years around 1700.⁸⁰ Only union with Scotland in 1707, driven by the urgent need to guarantee a Protestant Succession in both kingdoms, proved to be practical politics. English and then British parliaments were left managing imperial commerce as best they could, responding to pressure groups at home and in the colonies who had agents in London, and favouring broadly mercantilist policies as they had encouraged domestic improvement, by piecemeal legislation for such items as West Indian sugar, Irish and American textiles, and Indian silks.

The fundamental economic problem of empire was beyond easy political control, however, and well understood by Davenant and Defoe.⁸¹ It lay in the difference between a rich country like England and poorer ones able to produce manufactured goods more cheaply. Hence the threat to English manufacturers presented by the North American colonies, Ireland and Scotland, and especially India, the source of the cheapest textiles in the world. Writing in 1701, with greater originality than he was later to display in the *British Merchant*, and arguing in the 'manner of political arithmetic', Henry Martin positively welcomed the competition. It was an incentive to improvements, to the increase of 'arts, and mills, and

---

⁷⁷ *SCED*, pp. 571, 574; Stephen Foster and Evan Haefeli, 'British North America in the Empire', in Stephen Foster, ed., *British North America in the Seventeenth and Eighteenth Centuries* (Oxford, 2013), pp. 19–21; Cary, *State of England*, p. 69; Beinecke Library, Yale University, Osborn MS fb. 237, paper of 1696 on defence of the American plantations. Houghton took the demographic issue seriously but concluded that 'plantations do not depopulate, but rather increase or improve our people': Houghton, *Collection*, iv. 36–40 (1681). The 'or' is perhaps significant.
⁷⁸ *SCED*, pp. 575–6, 578.
⁷⁹ Armitage, *Ideological Origins*, pp. 154–5, 163–5; Simon Clement, *The Interest of England, as it Stands, with Relation to the Trade of Ireland* (1698), p. 18; Hont, *Jealousy of Trade*, pp. 222–33; Edward Littleton, *The Groans of the Plantations* (1689), pp. 1, 29.
⁸⁰ Armitage, *Ideological Origins*, pp. 152, 162–3; McCormick, *Petty*, p. 233; *The Queen an Empress: And her Three Kingdoms one Empire* (Dublin, 1706), pp. 24–7. Cf. the proposal for a 'United Grand-Council' of the three kingdoms in Clement, *Interest of England*, p. 21.
⁸¹ Hont, *Jealousy of Trade*, pp. 215–17, 229–31, 246–58. For Defoe's discussion of the same issue with respect to Union with Scotland, see Laurence Dickey, 'Power, Commerce and Natural Law in Defoe's Political Writings, 1698–1707', in John Robertson, ed., *A Union for Empire: Political Thought and the British Union of 1707* (Cambridge, 1995), pp. 82–96.

engines which save the labour of hands' at home. The argument was indisputable, and scarcely new. Locke, for example, thought that 'arts and inventions, engines and utensils' were useful because they served to 'shorten the labour and improve several things'. But advocates of improvement, Locke included, were still reluctant to move to the logical conclusion that high wages must be a good thing. Martin was aware that his argument ran 'directly contrary to... received opinions'.[82]

All the problems of empire were necessarily exacerbated by a common economic and political culture. Improvement and political economy had been transplanted to England's neighbouring kingdoms and plantations, and even to the ports of India. In 1682 the East India Company claimed to have developed Madras from 'a small village... to a great city of above 100,000 inhabitants' in less than thirty years. The city soon had its own insurance facilities as well as a corporation with a mayor and alderman. Bombay copied London's building regulations after its own great fire in 1671, and developed policies to encourage 'all manner of workmanship as well as trade' so as to 'increase our inhabitants'. They were based, as in political arithmetic, on considerations of 'number, weight, measure, and place', but place had been added to the conventional trio.[83] In India, just as in New England, the English were developing their own identities by adapting English culture to new circumstances and opportunities.[84]

Improvement and political economy seemed especially threatening when they infected Ireland and Scotland. Now famously attached to the cause of 'improvement' and 'the advance of trade', as Paterson remarked, the Protestant establishment in Dublin and their most articulate spokesman, William Molyneux, naturally agreed with Houghton that it was 'better for England to have Ireland rich and populous than poor and thin'.[85] That cut no ice with a London parliament determined to protect the English woollen industry against Irish competition in 1699.[86] In the eighteenth century, however, the Dublin parliament was promoting road and river improvements as enthusiastically as that in Westminster, and London was supporting the efforts of the Irish Linen Board founded in 1711 to encourage at least one industry. In 1738 a Church of Ireland bishop thought few

---

[82] Henry Martin, *Considerations upon the East-India Trade* (1701), sig. A2ʳ, pp. 65–70; Locke, *Political Essays*, ed. Goldie, pp. 261–2, 323. Cf. Richard C. Wiles, 'The Theory of Wages in Later English Mercantilism', *EcHR* 2nd ser. 21 (1968), pp. 113–26. Contemporaries were often uncertain about these issues. Puckle (*England's Path*, pp. 26–7) thought high wages would encourage idleness and 'stop improvements', while Tryon was in favour of high wages but against labour-saving inventions: Tryon, *General Considerations*, pp. 7, 8; *Brief History*, pp. 38–41.

[83] Philip J. Stern, *The Company-State: Corporate Sovereignty and the Early Modern Foundation of the British Empire in India* (Oxford, 2011), pp. 30, 38–40, 84–5, 93.

[84] Cf. John H. Elliott, 'Introduction', in Nicholas Canny and Anthony Pagden, eds, *Colonial Identity in the Atlantic World, 1500–1800* (Princeton, 1987), pp. 8–11; T. H. Breen, 'Creative Adaptations: Peoples and Cultures', in Jack P. Greene and J. R. Pole, eds, *Colonial British America* (Baltimore, 1984), pp. 195–232; below, pp. 254–6.

[85] Paterson, *Inquiry*, p 75; Houghton, *Collection*, iv. 91–2.

[86] Armitage, *Ideological Origins*, pp. 164–5. The Act of 1699 prohibited the export of woollens from Ireland and the American plantations, but was never enforced in the latter: Ormrod, *Commercial Empires*, pp. 168–9; Nuala Zahedieh, *The Capital and the Colonies: London and the Atlantic Economy, 1600–1700* (Cambridge, 2010), p. 265.

places in the world 'more improved' in so short a space of time than Ireland had been by an 'era of labour and industry' since 1689. Yet it was still 'less cultivated and improved than any other country in Europe'.[87] Ireland remained poor, though increasingly populous, and there was no alteration in its ambiguous status, partly a colony, partly a kingdom with its own parliament.

Scotland was different, as Davenant pointed out. Although a 'kingdom confederated' with England since 1603, it was a 'distinct state' with its own laws and free to trade as it wished.[88] When improvement and political economy arrived there in full force in the 1690s Scotland tried to do precisely that and forge its own economic identity. There were proposals for 'improvements... for the wealth of the kingdom' in agriculture, the mechanical arts, commerce, and navigation, in order to show that Scotland was 'as improvable for national advantages' as any country in the world.[89] The Darien adventure of 1698 was one result, supported by Paterson with arguments drawn from English political economy: 'Trade will increase trade... money will beget money,' and since trade was 'capable of making greater alterations in the world than the sword' the Scots would be able to found an empire more quickly and easily than Alexander or Caesar and become 'arbitrators of the commercial world'.[90]

The failure at Darien and the famine of 1695–9 between them consumed a fifth of the country's liquid capital and close to 15 per cent of the population, and left Scottish improvers with no option but to seek access to England's empire and markets on the best terms that could be agreed.[91] That did not guarantee a majority in the Scottish parliament for the Union which emerged, but the political management and public propaganda which delivered it came from practised advocates of improvement. Paterson, speaking to the educated, now supported Union as enthusiastically as he had the Bank of England because it was 'a way to further improvement'. Defoe, addressing a wider public, cleverly linked the issue of Protestant Succession with access to English trade, arguing that the two together would ensure 'security and quiet possession of... wealth and improvements' on both sides of the Tweed.[92]

The Union was about more than political economy, but it is impossible to conceive of it happening on the terms that it did without public consciousness of its

---

[87] Julian Hoppit, 'The Nation, the State, and the First Industrial Revolution', *Journal of British Studies*, 50 (2011), pp. 322, 327–8; Barnard, *Improving Ireland?*, pp. 35–6.

[88] Davenant, *PCW*, ii. 248.

[89] Charles W. J. Withers, 'Geography, Science and National Identity... the Work of Sir Robert Sibbald', *Annals of Science*, 53 (1996), pp. 47, 62; Ryan K. Frace, 'Religious Toleration in the Wake of Revolution: Scotland on the Eve of the Enlightenment (1688–1710s)', *History*, 93 (2008), p. 365; Christopher A. Whatley, *The Scots and the Union* (Edinburgh, 2006), p. 43.

[90] Armitage, *Ideological Origins*, pp. 159–60; William Paterson, *Proposals and Reasons for Constituting a Council of Trade* (Edinburgh, 1701), sig. 3ʳ, p. 195.

[91] Karen J. Cullen, *Famine in Scotland: The 'Ill Years' of the 1690s* (Edinburgh, 2010), pp. 27, 188–90.

[92] Whatley, *Scots and the Union*, pp. 274–315; Paterson, *Inquiry*, p. 123; Katherine R. Penovich, 'From "Revolution Principles" to Union: Daniel Defoe's Intervention in the Scottish Debate', in Robertson, ed., *Union for Empire*, p. 233.

economic benefits for both sides.[93] It has been described as the 'first instance of a modern state formation where considerations of competitive trade played a major part', and it created a 'united kingdom' (without Ireland) which was now the largest free trade area in Europe.[94] It left Scotland still a poorer relation, lacking the widely diffused private capital which had enabled England to invest in projects for improvement as well as finance its wars, and more dependent, like Ireland, on public funds to promote economic enterprise. Unlike the Irish, however, Scots now had equal access to English markets and the empire, and in the 1720s Scotland had an 'Honourable Society of Improvers in the Knowledge of Agriculture', and a 'Board of Trustees for Fisheries and Manufactures' intended to inspire 'our country to industry'. By then improvement was a feature of economic perceptions as fundamental in Scotland as it was in England.[95]

The Act of Union was the most conspicuous example of England's ability to adapt to the demands of a modern system of power-politics based on economic self-interest. The parliamentary regime had managed to fashion financial, commercial, and imperial policies which allowed the country to sustain two long wars, and arguments drawn from political economy had given them public legitimacy. Political economy itself had become better attuned to the realities of modern politics in the process. After 1690 it was difficult to pretend that international commerce meant 'mutual accommodation' between different interests, and united all peoples 'into one common society'.[96] It was a tool to be used by one nation to exploit the resources of others. Modern politics, as Andrew Fletcher the Scottish critic of the Union complained, were 'framed . . . with respect only to particular nations . . . without any regard to the rest of mankind'.[97]

Britain had emerged from that contest in 1713 as a major European and imperial power, with all the trappings of a military-fiscal state.[98] It had shown that it could preserve its affluence and even increase it in the short term. Whether it had the capacity to ensure economic growth and material progress in the longer run was another and larger question, and one which political economists were already beginning to ponder.

---

[93] John Robertson, 'An Elusive Sovereignty: The Course of the Union Debate in Scotland 1698–1707', in Robertson, ed., *Union for Empire*, p. 227; Whatley, *Scots and the Union*, pp. 252–6, 306–11. For a metropolitan view of potential improvements in Scotland, see Guy Miège, *The Present State of Great Britain* (2 vols, 1707), ii. 8–12, 24–6.

[94] Hont, *Jealousy of Trade*, p. 63.

[95] Hoppit, 'Nation, State and Industrial Revolution', pp. 326–7; Whatley, *Scots and the Union*, pp. 365–6.

[96] For efforts to do so, see *The Character and Qualifications of an Honest Loyal Merchant* (1686), pp. 2, 13–14; and (less successfully) *A Discourse of the Nature, Use, and Advantages of Trade* (1694), pp. 2–3, 5–6, 11.

[97] Hont, *Jealousy of Trade*, p. 65; below, pp. 212–13.

[98] Patrick K. O'Brien, *Power with Profit: The State and the Economy, 1688–1815* (Inaugural Lecture, University of London, 1991), pp. 12–14.

A STATIONARY STATE?

A stationary state was Adam Smith's term in *The Wealth of Nations* (1776) for the position which any society with a 'full complement of riches' must inevitably approach, a point at which it could advance no further.[99] I borrow it here because it serves to summarize the anxieties about potential checks to economic growth which had been aroused by circumstances earlier in the century, as we have seen, and which were sharpened by the events of the 1690s. They were the result also of modes of historical and economic enquiry which seemed to indicate that, whatever the immediate circumstances, material progress could never be endlessly sustainable, and must soon come to a halt if it had not already done so.[100] If improvement was to survive, in aspiration or reality, that intellectual challenge had to be met, and since it could never be wholly silenced, alternative arguments and assumptions had to be articulated and broadcast more loudly.

Initial doubts in the 1690s about whether war itself would bring economic growth to an end proved in the event to be short-lived. In 1698 Davenant published many of King's calculations, but since the war was now over and had been afforded he was able to use King's findings, not as warnings of imminent economic decline, but as plain evidence of the country's continuing wealth. He was worried about the National Debt, 'the huge engine of credit' that war had left behind and which might threaten England in the future as it had already ruined the great empire of Spain. But King's demonstration that England's national wealth had fallen less than that of France (although the Dutch had increased theirs) had led Davenant to revise his earlier opinion that the war 'did every year more impair and prejudice the condition of England than that of France'. Although it was impossible for a political arithmetician, even 'the best computer in the world', to calculate the losses accurately, his own computations suggested that England had lost far less than France, and was now richer than it had been in the 1660s. By 1706 it would be back where it was in 1688 and producing an annual surplus in the national income of £2.4 million.[101]

There were other signs of an early restoration of confidence. Defoe's claim in 1697 that England was 'not at all diminished or impoverished by this long, this chargeable war, but, on the contrary, was never richer since it was inhabited', may be dismissed as an example of his characteristic hyperbole. There may have been no more than hindsight in the rumour reported later that Louis XIV was ready for peace as early as 1696 because 'if England could maintain a war, and at the same time remedy the ill state of their coin, it was in vain to contend with them longer'.[102] By the time war against France resumed in 1702, however, there was more optimism on

---

[99] Adam Smith, *The Wealth of Nations*, ed. Andrew Skinner (2 vols, 1970), i. 184. I owe the reference to E. A. Wrigley, *Energy and the English Industrial Revolution* (Cambridge, 2010), pp. 197–8.
[100] See above, pp. 38–9, 109–10, 135–7.
[101] Davenant, *PCW*, i. 163, 168, 262–6, 380; ii. 277; King, 'Natural and Political Observations', pp. 68–9.
[102] Defoe, *Essay upon Projects*, p. 2; *The Autobiography of William Stout of Lancaster 1665–1752*, ed. J. D. Marshall (Manchester, 1967), p. 117.

the English side, and much less on the French, that it could be afforded. In 1697 a French observer was astonished that England, 'not worth a quarter of France' in population and wealth, had funded William of Orange's war, 'without reducing the people to beggary or obliging them to give up cultivating the land', which was what was happening across the Channel.[103] George Stepney, a diplomat who corresponded with King and knew something of his work, concluded in 1701 that France would have great difficulty maintaining another war since its wealth and population had clearly fallen since 1666;[104] and some of the calculations by Vauban and Boisguilbert, who reached similar conclusions when they copied Petty's and Davenant's political arithmetic, were already becoming known in England.[105]

There were persistent doubts, however, about whether England's wealth had increased in the longer term, since the middle of the seventeenth century, as both King and Davenant supposed. They were resurrected most forcefully by John Pollexfen in his debate with Davenant about the East India trade in 1697–9, where Pollexfen tried to distinguish between arguments about 'principles', about how a nation's wealth should be defined, and about 'matter of fact', about whether it had increased. The arguments of principle were lengthy but largely predictable. For Pollexfen, a nation's wealth resided in the labour of its people and the manufactures they made for export, which brought in treasure, and since the balance of England's trade since the 1670s had been consistently negative, the national wealth had declined. Davenant responded with the standard arguments that bullion and money were simply commodities for exchange like any others, and that imported goods contributed substantially to the riches of the kingdom.[106] The argument about the facts was equally predictable but more interesting. It was the argument about the visible affluence of England, and whether that signified genuine wealth, a dispute which had begun in exchanges between Josiah Child and his critics in the 1660s and been revived in 1694 in an attack on this 'great monarch' of the East India trade by John Blanch, a Gloucestershire clothier.[107]

Blanch had seized on Child's contention, repeated when his works were republished in 1690, that the splendours of London, including its contentious coaches, demonstrated that the whole nation 'was at present in a very thriving condition'. On his travels in the provinces Blanch had observed quite the reverse. There were 'many outhouses unthatched, the brewhouses ... untiled, and the timber rotten',

---

[103] François Crouzet, 'The Sources of England's Wealth: Some French Views in the Eighteenth Century', in P. L. Cottrell and Derek H. Aldcroft, eds, *Shipping, Trade and Commerce: Essays in Memory of Ralph Davis* (Leicester, 1981), p. 61.

[104] George Stepney, *An Essay upon the Present Interest of England* (2nd edn, 1701), p. 39; Gregory King, 'Burns Journal', p. 171.

[105] Paul Studenski, *The Income of Nations: Theory, Measurement, and Analysis* (New York, 1958), pp. 53–60. 'De Souligné', *The Desolation of France Demonstrated* (1697), pp. 2–52, used some of Boisguilbert's findings; an English edition of Vauban was published in Vauban, *A Project for a Royal Tythe, or, General Tax* (1708). Cf. below, pp. 246–7.

[106] John Pollexfen, *England and East-India Inconsistent in their Manufactures* (1697), pp. 5–8; Davenant, *PCW*, ii. 347–69; Hont, *Jealousy of Trade*, pp. 240–5.

[107] Above, pp. 133, 139; John Blanch, *An Abstract of the Grievances of Trade which Oppress our Poor* (1694), p. 15.

Challenges to Affluence 1690–1730 193

all of this a manifest 'emblem of poverty'. Like Tryon and others, he considered that England's apparent affluence was narrowly concentrated in London and the commercial sector. 'The wealth of some particular men' such as Child was 'no proof of our riches in the general... We may be poorer in the general tho' rich in particulars.'[108] When Davenant responded to Pollexfen in 1698 he took the opportunity to answer Blanch as well. In 1666 there was already 'a general face of plenty upon all England, all the different ranks of men were at their ease, the common people were well fed and clothed, and the farmhouses were in good repair, which is the truest sign of wealth increasing in a kingdom'. Between 1666 and 1688, moreover, even more land had been improved than in the previous half-century, and farmhouses remained in good repair. The kingdom had 'accumulated more wealth of all kinds than any other part of Europe' and there was now 'no country in the world where the inferior rank of men were better clothed and fed, and more at their ease'.[109]

Pollexfen remained unpersuaded. 'All sorts of improvements' were indeed essential to economic growth, but they were attributable 'to our labour and industry, and not... gotten by foreign traffic'. 'Costly furniture of houses, sumptuous apparel and equipage' purchased with English coin and bullion made no lasting addition to the national wealth. They were the product of a taste for 'luxury and the expensive way of living', which had the 'usual consequences'—idleness and 'neglect of manufacturing'—and it had started with the rebuilding of London in 1666 and increased ever since. Despite all the superficial evidence to the contrary, Pollexfen was insistent that 'this anno 1666 is fix't for the year of our declension'.[110]

Much of the case for national declension was based, as it had been for decades, on the observation that poverty was increasing and poor rates rising, that the 'the whole mass' of the people preferred 'pleasure to industry', and that luxury was the enemy of all improvement, a prospect which worried Davenant himself, as we shall see in the next section of this chapter.[111] The case was reinforced by the suspicion that all of these were symptoms of the decline to which all states and kingdoms in history had been subject, some, like Athens and Rome, starting their progress earlier than others, and all perhaps ending, as Bacon once supposed, with the improvement of 'mechanical arts and merchandise' which came only in their final 'declining age'.[112]

At the end of the 1690s the problem of the Spanish Succession and the imminent dismemberment of what had once been the greatest empire in the world was a reminder of those historical cycles of growth and collapse which all

---

[108] Blanch, *An Abstract*, p. 15; John Blanch, *The Interest of England Considered* (1694), sig. A3ᵛ, pp. 3, 60–2.
[109] Davenant, *PCW*, i. 360–1, 370–2.
[110] John Pollexfen, *A Vindication of Some Assertions relating to Coin and Trade* (1699), pp. 38–41, 54, 71; Pollexfen, *England and East-India*, pp. 47, 51.
[111] Peter Paxton, *A Discourse Concerning the Nature, Advantage, and Improvement of Trade* (1704), pp. 72–4.
[112] Vickers, *Bacon*, p. 454; above, p. 39–40. Cf. Dickey, 'Power, Commerce and Natural Law', in Robertson, ed., *Union for Empire*, pp. 82–5.

societies must experience. Cyclical models were implicit in contemporary histories of trade, which showed the 'commercial ball' being tossed from one place to another, as it had once been from Flanders to Holland, and in tracts on English industry and commerce which warned that many countries had 'degenerated from the great achievements of virtue and industry' and that all must sooner or later 'pass their zenith'.[113]

Writing about the Anglo-Saxons but with an implicit contemporary moral in 1695, William Temple pointed to 'the usual circle of human affairs': 'war ended in peace, peace in plenty and luxury, these in pride; and pride in contention, till the circle ended in new wars.'[114] In the following year Edmund Bohun wrote to Cary about the cycles visible in the rise and fall of trading and landed empires from ancient Tyre and Rome to modern Spain, and warned that the dominions of the English might be similarly threatened and prospering American plantations, even perhaps Scotland and Ireland, soon determine their own future 'in the weak condition we are entering into'.[115] Even Houghton, in one of his more reflective moments in 1696, was worried that war might be the final blow to England's greatness, so that 'like flies when we have crept to the top of the glass, we shall fall down again'.[116]

The literary battles between Ancients and Moderns in the same decade might have reassured the Moderns that history was not a story of cycles but one of consistent progress, but much of the historical evidence seemed to be pointing towards ancient cities and kingdoms that were richer and more populous than any now known. Classical Rome, for example, was said to have contained at least seven and perhaps even fourteen million people, and certainly been far bigger than modern London and Paris put together.[117] Defenders of modernity had to respond, one Londoner arguing in 1706 that 'old Rome' had lacked modern London's shops, busy markets, services, and low taxes. It could never have been so prosperous or—he suggested—so populous.[118] The great weakness in the case made by the Moderns, however, and not only in England, was their inability to

---

[113] King, ed., *British Merchant*, i. xxix; *A Discourse of the Necessity of Encouraging Mechanick Industry* (1690), sig. A2ʳ, p. 2; *Considerations Requiring greater Care for Trade* (1695), pp. 12–13.

[114] William Temple, *An Introduction to the History of England* (1695), p. 60. Cf. Edmund Bohun's opinion, after reading Milton's *History of Britain*, that 'the general corruption of morals' was a cause of the Norman Conquest: *The Diary and Autobiography of Edmund Bohun Esq.*, ed. S. Wilton Rix (Beccles, 1853), p. 33; and Temple to similar effect in 1690, above, p. 38.

[115] BL, Add. MS 5540, fos 59, 62ᵛ–65ʳ. For similar views about the future of the plantations, see Sir Thomas Browne, *Certain Miscellany Tracts* (1683), pp. 181–3; *Nehemiah Grew and England's Economic Development*, ed. Julian Hoppit (British Academy Records of Social and Economic History NS 47, Oxford, 2012), p. 107.

[116] Houghton, *Collection*, ii. 54.

[117] James C. Riley, *Population Thought in the Age of the Demographic Revolution* (Durham, NC, 1985), p. 42; Isaac Vossius, *Variarum observationum liber* (1685), p. 34.

[118] 'De Souligné', *A Comparison Between Old Rome in its Glory, as to the Extent and Populousness, and London as it is at present* (1706), pp. 33, 47–8. Thomas Templeman took a similar view: *A New Survey of the Globe* (1729), p. iii. In the seventeenth century, however, large estimates of the population and area of ancient Rome had been accepted: above, p. 38; Petty, *EW*, ii. 532. Classical Rome had a population of one million and a thriving retail trade: Claire Holleran, *Shopping in Ancient Rome: The Retail Trade in the Late Republic and the Principate* (Oxford, 2012), p. 1 and *passim*.

demonstrate that the populations of modern commercial societies were in fact growing, and their fears that they might be declining.

In England the labours of political arithmeticians might have been expected to have clarified the issue. They had not. Since there was still no census, there was no more certain knowledge about the size of England's population than about the size of its money supply, although several numbers were suggested, the more reliable of them based on calculations from the hearth tax. In 1688 Pett noted that previous guesses had ranged from 6 to over 16 million, and called for 'some such accurate survey' as Petty 'that mathematical statesman' had hoped for. On 6 January 1692 Thomas Neale, Master of the Mint and a famous projector, told the House of Commons: 'I compute in England there are about eight millions of people.' Six days later he spoke again: 'I compute there are in this kingdom about six millions of people.' Someone had spoken to him in the interval, perhaps Paul Foley who later in the month said the total was as low as 4.6 million.[119] King's laborious calculations in 1695 produced a figure of 5.5 million, now thought to be reasonably accurate, and it ought perhaps to have settled the matter. But many others still preferred 7 million or 8 million, figures taken from Petty and used by Davenant before he was corrected by King. In 1700 *Angliae notitia* said 7 million, and that remained the conventional estimate into the early eighteenth century.[120] Since Graunt had put the total population at 6.4 million in 1662, there was room for argument about whether or not it had grown afterwards, but if it had grown it had not grown by very much.

In consequence there was still general agreement at the end of the century on the need to 'multiply our people' since the country was far from fully peopled: only 'one third peopled, if compared with other parts of the universe', according to an author who must have had China or India in mind, only two-thirds, according to Davenant, only one-half, according to John Bellers. Thomas Tryon even thought there were 'much fewer natives in England than we had in the reign of Queen Elizabeth'.[121] There was agreement about the causes too. Emigration was still complained about, although it had passed its peak;[122] plague had disappeared, but it was evident that life-expectation was kept low by new 'sicknesses and distempers', especially in towns and cities;[123] and there was particular stress once

---

[119] Pett, *Happy Future State*, pp. 113–18; *The Parliamentary Diary of Narcissus Luttrell, 1691–3*, ed. Henry Horwitz (Oxford, 1972), pp. 112, 123, 144.

[120] King, 'Natural and Political Observations', p. 36; above, p. 140; *Nehemiah Grew*, ed. Hoppit, pp. 94–5; Davenant, *PCW*, i. 19, 62, 197; ii. 221. For other references to 7 or 8 million, see Barbon, *Discourse of Trade*, p. 47; Cary, *State of England*, p. 163; Chamberlayne, *Angliae notitia* (1700), p. 46; Tryon, *Brief History*, p. 50; King, ed., *British Merchant*, i. 165; Erasmus Philips, *The State of the Nation* (1725), p. 47.

[121] *Discourse of the Necessity of Encouraging Mechanick Industry*, p. 4; *Discourse of the Nature, Use, and Advantages of Trade*, p. 26; Davenant, *PCW*, i. 74; *John Bellers: His Life, Times and Writings*, ed. George Clarke (1987), p. 76; Tryon, *Brief History*, p. 64. For earlier discussion of population increase, see above, pp. 136–8, 159–60; and cf. below, pp. 238–40.

[122] e.g. Tryon, *Brief History*, p. 46; Cary, *State of England*, p. 66; Thomas, *Historical Account*, p. 1.

[123] Tryon, *Brief History*, pp. 69–70. Expectation of life at birth fell throughout the seventeenth century, and was particularly low in the 1680s: E. A. Wrigley et al., *English Population History from*

more on low fertility and the decline of marriage, although marriage rates were probably now recovering.[124]

In 1702 Tryon believed that the number of marriages in London had declined by a third since the 1680s, and an attempt to stem the apparent tide, *Marriage Promoted* of 1690, asserted that 'near one half of the people of England' died single, and that a third of those who married did not marry as soon as they might.[125] These were guesses, for want of precise information. In 1693, however, Edmond Halley was able to use the bills of mortality of Breslau to calculate actual and potential fertility rates there; and his conclusion presented to the Royal Society, that 'there might well be four times as many births as we now find' if people married earlier, must have been intended to strike a chord at home.[126] It seemed plain that England's demography had changed at some point in the middle of the seventeenth century, and that the trend must somehow be corrected.

The rationale for that conviction was the persistent assumption that economic growth could not occur without demographic growth because 'people are the wealth of a nation', the old axiom much repeated into the 1720s and beyond.[127] Faced with mounting evidence to the contrary, economic writers could have followed the hints of Evelyn and others in the 1670s that national income could grow without an increase in population, and wealth increase per capita; and they might have added Henry Martin's point about the value of labour-saving inventions. Few of them followed that escape route from their dilemma. Houghton recognized that a shortage of labour stimulated invention and improvement in Holland, but England was not nearly so densely populated and he hoped it would never have to face 'such necessity'.[128] Old assumptions that there must be ever more producers and consumers if there was to be infinite economic growth remained undisturbed. Arguing in 1694 that the Bank of England 'would beget trade and people' and 'make us infinitely rich to eternity', Henry Mackworth had to answer the objection that the English might in the end 'make more goods than we can consume, or the world will utter'. He responded that 'people will increase, for trade will bring in people as well as riches to the nation', and 'where people are,

---

Family Reconstitution (Cambridge, 1997), pp. 282–4, 348–9; John Landers, *Death and the Metropolis: Studies in the Demographic History of London 1670–1830* (Cambridge, 1993), pp. 130–9, 192–3.

[124] Wrigley et al., *English Population History from Family Reconstitution*, p. 195. Clandestine marriages may have concealed the recovery: Jeremy Boulton, 'Clandestine Marriages in London: An Examination of the Neglected Urban Variable', *Urban History*, 20 (1993), pp. 208–10.

[125] Tryon, *Brief History*, p. 168; *Marriage Promoted: In a Discourse of its Ancient and Modern Practice* (1690), p. 27. For similar concerns, see King, 'Natural and Political Observations', pp. 44–6; Daniel Defoe, *Some Considerations upon Street-Walkers* (1726), pp. 6–7; Nehemiah Grew, ed. Hoppit, pp. 95–9.

[126] Edmond Halley, 'Some Further Considerations on the Breslaw Bills of Mortality', *Philosophical Transactions of the Royal Society*, 17.198 (1693), pp. 655–6.

[127] Thomas, *Historical Account*, p. 1. Cf. Grassby, *English Gentleman in Trade*, p. 319; Nehemiah Grew, ed. Hoppit, p. 77; John Trenchard and Thomas Gordon, *Cato's Letters* (4 vols, 1723–4), iv. 1, 63–4.

[128] Above, pp. 136, 187–8; Christine McLeod, *Inventing the Industrial Revolution: The English Patent System, 1660–1800* (Cambridge, 1988), p. 208.

there will be consumption of all commodities'.[129] When similarly confronting the possibility that there might be a finite amount of trade in the world, Houghton and William Paterson both borrowed Mackworth's answer. In a successful commercial nation, trade would multiply people and economic growth be self-sustaining.[130]

Preconceptions about the inevitability of population growth were also shaped by the common presupposition, referred to earlier, that it was inherent in the providential course of history, part of the divine plan for the multiplication of mankind which would continue until the Last Judgement.[131] That was the case put by natural theologians like Matthew Hale, and it was often repeated, not only because it answered claims by some Deists and free-thinkers that the world was eternal, but because it showed that the Moderns were superior to the Ancients in this as in other respects. According to William Whiston and William Nicholls writing in the 1690s, a gradual increase of mankind over the centuries, with only short intervals caused by famines and plagues, was as much a part of God's benevolent design as the gradual 'progress' in 'invention of arts' and sciences, which had advanced more rapidly than ever before in the 'present age'.[132]

Since the world was not eternal, natural theologians knew that it must have an end as well as a beginning, though they declined to put a date to it as certain as their date for the Creation.[133] Economists were similarly aware that there must be limits to local or global increases of numbers dictated by limited food supply, but since much of the world was thinly populated, and agriculture could everywhere be improved, demographic increase would long continue. Intellectually fertile as ever, political arithmeticians tried to calculate when a ceiling might be reached, using estimates of the number of acres needed per person, and of present populations and their likely rate of growth.[134] Although none of these was known with any certainty, Petty took his projections of global population forward another 2,000 years, to the year 3680, by which time he thought 'the whole world will be fully peopled' and then 'according to the prediction of the Scriptures' there would be 'wars and great slaughter &c' and the Last Judgement would occur.[135] King took similar projections to 3600 or even further, 'in case the world should last so long'.[136] They pushed the end of demographic growth for the world as a whole into a conveniently far distance.

---

[129] Mackworth, *England's Glory*, sig. A3ʳ, pp. 18–19, 21.

[130] Houghton, *Collection*, ii. 52–3; Paterson, *Inquiry*, p. 140. Cf. *Bellers*, ed. Clarke, pp. 75–6.

[131] Above, pp. 125–7, 140.

[132] William Whiston, *A New Theory of the Earth from its Original to the Consummation of all Things* (1696), pp. 174, 385–8; William Nicholls, *A Conference with a Theist* (1698), part I, pp. 66–75, 80. At least one political economist later referred respectfully to Nicholls: Jacob Vanderlint, *Money answers all Things* (1734), pp. 14, 28.

[133] William Poole, *The World Makers: Scientists of the Restoration and the Search for the Origins of the Earth* (Oxford, 2010), pp. 155, 163–4. For some contemporary estimates, see Jed Z. Buchwald and Mordechai Feingold, *Newton and the Origin of Civilization* (Princeton, 2013), pp. 336–7.

[134] For their various calculations, see Slack, 'Plenty of People', pp. 8, 20 nn. 33–4.

[135] Petty, *EW*, ii. 464. Cf. Hobbes's 'last remedy of all', when 'all the world is overcharged with inhabitants': Thomas Hobbes, *Leviathan*, ed. Noel Malcolm (3 vols, Oxford, 2012), ii. 540.

[136] King, 'Natural and Political Observations', p. 41; King, 'Burns Journal', pp. 1–2.

They recognized nonetheless that in the case of England the ceiling might be reached much sooner, and Graunt seems to have assumed that it had already been reached in 1662. Only Petty, however, looked closely at such a possibility and its implications for economic growth, and even he found it impossible to emancipate himself entirely from assumptions about providential history. As aware as any of his contemporaries of rising mortality in urbanized societies, subject to 'the mischiefs of plagues and contagions', he thought England's population might still continue to grow but not for very long.[137] In *Another Essay in Political Arithmetick* (1683), chiefly about the growth of London, he concluded that the end to demographic growth in England would come in little more than a century, in or around the year 1800. London was already the largest city 'in the universe', and by then it would contain 5 million people out of a national total of nearly ten million. Country would no longer be able to support town, and national demographic growth must soon cease.[138] A stationary state would have arrived.

Yet what was the alternative? In another of his revealing thought experiments, Petty put forward two 'suppositions', two scenarios, and left his readers to choose between them. One was a London which continued to grow towards its maximum. It would be able to employ and feed its population, expand commerce, multiply manufactures through the division of labour, and propagate the 'improvement of useful learning' and of every art and science through competition and emulation. It would enjoy everything that brought a 'convenient, commodious and comfortable' standard of living—apart from improvements in health. The alternative was that London's growth should somehow be brought to a premature halt, along with the manifold benefits that came with it. A choice had to be made: between the healthy and lasting simplicities of less dense populations, and the manifold advantages of an urbanized economy destined in the end to overwhelm itself.[139]

Petty's preference was evident, and he left the matter there in 1683. He probably intended to elaborate his thinking in the book he was writing on 'the growth, increase and multiplication of mankind', and he might there have referred to his view of the world's end which he discussed in his letter to Southwell in 1685.[140] The prophecies of Scripture showed that the end would arrive with the perfection of all knowledge and learning, and that was only possible in great conurbations like London and not in thinly populated agrarian societies whose only conceivable improvements might be in husbandry.[141] It would seem that dense populations, with all their hazards, were necessary for wisdom as well as economic growth. Divine providence and the logic of political arithmetic, Petty might have argued, were wholly consistent with one another.

---

[137] Petty, *EW*, ii. 371–2, 471.
[138] Petty, *EW*, ii. 464–5, 505–9. In 1801 London's population was 900,000 out of a total for England and Wales of 9 million, but other towns had grown faster than the capital: P. J. Corfield, *The Impact of English Towns, 1700–1800* (Oxford, 1982), pp. 66, 69.
[139] Petty, *EW*, ii. 470–6. Cf. above, p. 146.
[140] Petty, *EW*, ii. 453; above, p. 127.
[141] *Petty–Southwell Correspondence*, pp. 153–7; Petty, *EW*, ii. 474. For Southwell's objections and Petty's further response, see *Correspondence*, pp. 158–74.

For the modern historian, however, it is the economic reasoning behind Petty's two scenarios which resonates. In a few speculative pages, he put his finger on the paradoxes of the advanced organic economy enjoyed by England and Holland and described in the Introduction to this book.[142] In the conditions which obtained in north-western Europe in the century after the 1650s, high mortality, growing towns, and rising living standards were indeed positively related. The diseases spread by trade and aggravated by urbanization had led indirectly to higher incomes per head.[143] Moreover, notwithstanding high mortality and rising living standards, preventive checks to increases in fertility, in the shape of late marriage and prolonged celibacy, continued to operate, at least until the 1720s. And the counterfactual argument also holds: if the population had grown much more quickly than it did after 1650, per capita incomes would have fallen and economic growth been restrained.[144] Between 1650 and 1750, in a country such as England with growing towns and a growing international commerce, not only did affluence impede population growth, but an increase in numbers would have imperilled economic progress.

That apparent antithesis between increasing population and increasing prosperity lay at the centre of much of the European debate between Ancients and Moderns when it revived in the middle decades of the eighteenth century, and apologists for modern commercial societies again found themselves pondering why their populations seemed no longer to be growing.[145] Only then was there some reiteration of Petty's point, though without any reference to him. When David Hume contributed to the debate in 1752, for example, he drew attention to the unique character of the advanced economies of modern north-western Europe. Nothing like their 'great and populous cities... so stocked with riches and inhabitants' had been seen in antiquity; and he was forced to consider the possibility that their wealth and population density were unlikely ever to be exceeded in the future. In a later letter Hume questioned whether rich commercial nations could 'go on increasing trade *in infinitum*, or whether they do not at last come to a *ne plus ultra*', a point at which they must 'check themselves' and 'finally stop their progress'.[146] But he made no reference to Petty, whose essay, when cited at all by other authors, was used to show how the multiplication of mankind would continue, not that it must soon stop.[147]

---

[142] Above, pp. 10–12. Cf. below, pp. 238–40.

[143] Cf. Richard Smith, 'Periods, Structures and Regimes in Early Modern Demographic History', *History Workshop Journal*, 63 (2007), pp. 212–16; Nico Voigtländer and Hans-Joachim Voth, 'Malthusian Dynamics and the Rise of Europe: Make War, Not Love', *American Economic Review*, 99 (2009), pp. 248–54.

[144] Wrigley, *Energy and the English Industrial Revolution*, pp. 142–3, 147. Cf. Robert C. Allen, *The British Industrial Revolution in Global Perspective* (Cambridge, 2009), pp. 14, 128–9.

[145] Riley, *Population Thought in the Age of the Demographic Revolution*; Andrea A. Rusnock, *Vital Accounts* (Cambridge, 2002), pp. 179–209; David V. Glass, *Numbering the People: The Eighteenth-Century Population Controversy and the Development of Census and Vital Statistics in Britain* (Farnborough, 1973), pp. 11–89.

[146] David Hume, 'Of the Populousness of Antient Nations', in *Essays and Treatises on Several Subjects* (2nd edn, 1753), iv. 216–17 (first published in 1752); Hont, *Jealousy of Trade*, pp. 269–79.

[147] Petty's pamphlet, *Another Essay in Political Arithmetick* (1683), was reprinted in 1686, 1699, and 1755, and cited by Whiston and Nicholls and in *Journal oeconomique, ou Mémoires, notes, et avis*

Until the middle of the eighteenth century, the presumption of growth remained, despite increasing awareness in the 1720s of heavy urban mortality especially from smallpox, and despite debates about the population of London whose bills of mortality could be manipulated to show either growth or decline.[148] The presumption was sustained by the arguments of natural historians and natural philosophers who joined natural theologians in an enthusiasm for modern commerce and improvement which either took population growth for granted or predicted that it must soon be resumed. In 1691 the naturalist John Ray had no doubt that God preferred a country 'improved to the height of all manner of culture for the support and sustenance... of innumerable multitudes of people' to 'a barbarous and inhospitable Scythia... or a rude and unpolished America, peopled with slothful and naked Indians'.[149] William Derham's influential Boyle lectures of 1711 and 1712 expounded on a divine plan which included 'the great improvements made in the last and present age, in arts and sciences, in navigation and commerce', and assumed without particular emphasis that population would continue to grow for at least a thousand years.[150] That style of intellectual speculation left no room for thought of a stationary state.

The works of Nehemiah Grew, botanist, physician, and Fellow of the Royal Society, provide a particularly extravagant example of the same improving frame of mind. He had read Petty's essays, including his manuscript about the great scale of creation, and imbibed their message. His book on natural theology, *Cosmologia sacra* (1701), pictured 'the improvement of language and trade, and of all other arts and sciences' as divinely ordained, 'the better to show that man, by nature, is an improvable creature'; and he had no doubt that England above all other nations had the potential to produce 'inventions new and infinite to the end of the world'. In his manuscript treatise on 'the means of a most ample increase of the wealth and strength of England in a few years', presented to Queen Anne in 1707, he set out a blueprint for rapid economic growth. A country with lands, manufactures, and commerce, all 'improved' as he proposed, would increase its national income fivefold to £190 million a year.[151]

There were obstacles to be overcome. Like Tryon he thought too many people were engaged in unproductive 'selling trades', perhaps three or four million; but

---

sur les arts, l'agriculture, le commerce (Paris, 1756), p. 189. The earliest reference to Petty's prediction about an end to London's growth, however, seems to be in Thomas Short, *A Comparative History of the Increase and Decrease of Mankind in England, and Several Countries Abroad* (1767), p. 20. Joseph Massie, *A Plan For the Establishment of Charity-Houses* (1758), p. 92, referred to Petty's projection for London but not to an end point.

[148] Rusnock, *Vital Accounts*, pp. 46–9; Joanna Innes, *Inferior Politics: Social Problems and Social Policies in Eighteenth-Century Britain* (Oxford, 2009), pp. 127–31; Charles Maitland, *Mr Maitland's Account of Inoculating the Small Pox* (1722), p. 3.

[149] Alexandra Walsham, *The Reformation of the Landscape* (Oxford, 2011), pp. 389–40. Cf. John Haynes, *Great Britain's Glory* (1715), p. 1.

[150] Larry Stewart, *The Rise of Public Science: Rhetoric, Technology, and Natural Philosophy in Newtonian Britain, 1660–1750* (Cambridge, 1992), pp. 48–54; William Derham, *Physico-Theology* (1714), pp. 177–8.

[151] Nehemiah Grew, *Cosmologia sacra* (1701), pp. 103–4; Stewart, *Rise of Public Science*, p. 52; *Nehemiah Grew*, ed. Hoppit, 'Introduction', p. 94; above, pp. 127–8.

many of them could be moved into more productive employments. Like others, he wanted restrictions on emigration and encouragements to marriage, since 'fewer are to be seen in Holland unmarried at 25 years of age than in England at 40', but determined action would double the population in a generation. His voluminous reading, which included Bacon, Cotton, and Child, as well as Petty, taught him that England's recent history was one of constant improvement. In former ages English manufactures were 'very contemptible'; 'we knew not how to pave our streets'; most houses were of wood and thatch and had no glass windows, most clothes simple woollens, most plates made of pewter or wood. In the present age all that had changed, and the process would continue.[152]

Much of this was wishful thinking, but similar sentiments, without Grew's exaggerations, could reasonably be voiced once peace returned in 1713 because they bore some relationship to the economic facts. In 1718 William Wood's *Survey of Trade*, a digest of much of the economic writing and thinking of the previous two decades, highlighted the 'prodigious' growth in trade which had occurred since the 1690s. The wartime prohibition on imports from France and, still more, a boom in exports since 1700, chiefly of textiles and grain to northern Europe, had produced a healthy balance of trade.[153] The interests of land and trade had both benefited and 'greater improvement' still could be expected. There were some clouds on the economic horizon. One might be a fall in population, which was always the mark of a 'declining country', but Wood insisted that 'the numbers of our people' had 'not at all . . . or very little diminished' since 1688. Any unfavourable commercial treaty with France like that recently proposed would similarly put a brake on the growth of trade. Nevertheless, the 'rise and progress of the English greatness' seemed set to continue. Wood followed Davenant and not Pollexfen in his description of England's present wealth. All the evidence of conspicuous consumption, great quantities of plate and costly furniture, and those farmhouses still in excellent repair, indicated a 'prosperous people' and a 'thriving nation'.[154]

As Wood was well aware, the size of the National Debt was an even larger cloud dampening economic optimism, but writers continued into the 1720s to try to minimize it. Erasmus Philips's *State of the Nation* (1725), a tract like Wood's which was twice reprinted, sought to combat those 'so gloomy that they thought us in a worse condition than we really are'. Although he wished the French commercial treaty had been ratified, while Wood opposed it, he argued that England's 'present affluence', visible in its trade, buildings, gardens, and plate, showed that 'she is become richer than she was' in 1688. He too brushed aside fears about population, since it had 'without doubt' increased over the past century and now reached 8 million. The money supply had increased too and at last allowed a reduction in the rate of interest (to 5 per cent) in 1714 which boded well for the public credit.

---

[152] *Nehemiah Grew*, ed. Hoppit, pp. 35, 83–6, 95–100.
[153] William Wood, *A Survey of Trade in Four Parts* (1718), pp. 45–9; Jones, *War and Economy*, pp. 169–209.
[154] Wood, *Survey of Trade*, pp. 34, 45, 54–6, 74–6, 198.

The 'progress of trade' had continued even in wartime and its proper management would ensure that the commercial sun continued to shine.[155]

For the moment, therefore, until it was revisited in the 1750s, the prospect of a stationary state had been pushed into the background, by political economists, natural theologians, and all who joined the Moderns against the Ancients in embracing the alternative prospect of endless improvement. The economic crisis created by war in the 1690s had been seen off. The inconvenient fact of demographic stagnation could be minimized because, in the absence of a census, there was no hard evidence for it, and if it was real it must be merely a temporary hiccup. A priori assumptions about the progressive course of history had been shaken, but only temporarily. All that said, however, there was another inconvenient fact produced by affluence which could not so easily be dismissed, and which Philips was compelled to acknowledge, if only briefly. While there must be some 'foundation of real substance' to England's wealth for it to have lasted 'so long a time', he had to admit that 'luxury' had a part to play in the 'great consumption of commodities' which 'generally attends affluence'.[156] That consequence of evident prosperity had to be faced by advocates of modernity and natural theology alike because it was incompatible with public and private virtue. Luxury was the challenge to affluence which attracted most attention after 1688 and created most anxiety and controversy after peace returned in 1713.

LUXURY

The moral backlash against the extravagance and infinite indulgence of consumer appetites which Barbon and Houghton had sought to demoralize, and which seemed to have spread from the Stuart court to all manner of people by the 1680s, won unanimous support in the 1690s. Patriotic calls for restraint and self-denial were inevitable in wartime, and spearheaded from the top. William III opened his reign, much as Charles II had done his but with greater enthusiasm, sponsoring 'a general reformation of the lives and manners of all our subjects', which included sumptuary restraint.[157] Something like it was 'indispensably required', according to an author in 1690, because the kingdom had recently 'degenerated ... into all the soft indulgences of sloth', and had adopted the 'effeminate and luxurious course of life' which had destroyed Assyrian, Persian, Greek, and Roman empires before it. In 1704, James Whiston found evidence all around of 'idleness, luxury, debauchery, profaneness and deism', and the same 'decay of religion, virtue and common justice' which had been responsible for 'all the

---

[155] Philips, *State of the Nation*, sigs A4–5ʳ, pp. 5, 12, 15, 42–7. Cf. Erasmus Philips, *An Appeal to Common Sense* (1720), p. 13. On the effects of the reduction in interest rate, see Temin and Voth, *Prometheus Shackled*, p. 94.
[156] Philips, *State of the Nation*, sig. A5, p. 44.
[157] Faramerz Dabhoiwala, *The Origins of Sex* (2012), pp. 52–3; Tony Claydon, *William III and the Godly Revolution* (Cambridge, 1996); David Hayton, 'Moral Reform and Country Politics in the Late Seventeenth-Century House of Commons', *P&P* 128 (Aug. 1990), pp. 48–91.

revolutions of empires, kingdoms, or states that have degenerated from better to worse'.[158] The whole history of domestic troubles in England and Scotland in the seventeenth century was attributed by Paterson in 1701 to a people 'inclinable to gratify their passions and appetites'; now there was an opportunity for them to 'pursue the public good' they had always claimed to be aiming at.[159]

That they might at last be able to do so was the promise held out by the new Societies for Reformation of Manners prosecuting all forms of vice, from swearing, gambling, and drunkenness, to breaches of the Sabbath, adultery, and fornication. By 1700 there were ten of them in major English cities, the first in London in 1689, and others followed or were planned in smaller towns, several counties, in the plantations as well as Dublin and Edinburgh, and in other parts of Europe, even including frugal Holland.[160] Intended, as one of their publicists said, 'to discountenance strife and restore unity', their appeal extended across political and religious parties. Their members included supporters of the lay religious societies like the Society for Promoting Christian Knowledge and Dissenters as well as Anglicans, and although High Church and High Tory Anglicans tended to hold aloof, they nonetheless gave general support to a moral reformation which had become a political as well as religious endeavour.[161]

These alliances did not hold together for very long. They were broken by the renewed polarization of denominational and party divisions in the latter part of Anne's reign which opened the activities of the religious societies to criticism from one sectional group or another. Even the Reformation Societies lost their initial impetus, and in London turned increasingly to the policing of sexual offences. They were not helped by the absence of statutory support. Bills in parliament against immorality fared no better than proposals for sumptuary legislation, one MP commenting that people who 'would not take the Old and New Testament for a rule of life would never be reformed by an act of parliament'.[162] What persisted, in England as elsewhere in western Europe, was a more widely shared anxiety about the general spectre of luxury which loomed ever larger as living standards improved. Material progress presented prospects of moral and social damage almost as intractable as the challenges of affluence which have been identified in modern commercial societies.[163] Insatiable consumer appetites seemed to have dissolved all established values and social relationships and undermined distinctions of class

---

[158] *Discourse of the Necessity of Encouraging Mechanick Industry*, sig. A2, pp. 2, 32–3; James Whiston, *England's State Distempers* (1704), pp. 3, 13. Cf. James Whiston, *The Mismanagements in Trade Discovered* (1704), p. 3.

[159] Paterson, *Proposals and Reasons for Constituting a Council of Trade*, pp. 192–3.

[160] Dabhoiwala, *Origins of Sex*, pp. 55–6.

[161] Shelley Burtt, *Virtue Transformed: Political Argument in England 1688–1740* (Cambridge, 1992), pp. 39–48; Slack, *Reformation to Improvement*, pp. 113–14.

[162] Slack, *Reformation to Improvement*, p. 115; Dabhoiwala, *Origins of Sex*, pp. 54, 57–61.

[163] Istvan Hont, 'The Early Enlightenment Debate on Commerce and Luxury', in Mark Goldie and Robert Wokler, eds, *The Cambridge History of Eighteenth-Century Political Thought* (Cambridge, 2006), pp. 377–418. On modern discontents about affluence, see Avner Offer, *The Challenge of Affluence: Self-Control and Well-Being in the United States and Britain since 1950* (Oxford, 2006).

and occupation, without delivering any of that general happiness which Houghton and others had promised.

The consequences were easy to identify but impossible to eradicate. In 1701 Charles Povey lamented *The Unhappiness of England* in what he regarded as 'this degenerate Iron Age'. He recognized 'the outward conveniences of life' which 'the meaner sort of people usually measure their felicity by', but the lack of 'due care for the distinction of persons', and the crime and depravity of the metropolis, made it impossible for him 'to walk the streets in safety' and enjoy 'the comforts of life with an uninterrupted tranquillity of mind'. Increasing wealth had brought no increase in general happiness, and men and women were now 'in a much more miserable condition than the people that lived in former ages'.[164] Yet Povey had been a coal merchant, dependent on the growing consumer demand in the metropolis, and he was ready to admit that the whole nation needed more trade to 'make us completely happy'.[165]

Other critics of luxury were equally in two minds, and far from wishing to put a stop to economic progress. James Whiston was a commodity broker convinced that commerce was in principle 'capable of increase *ad infinitum*'.[166] Francis Brewster, a Dublin merchant and the MP who identified a new 'party' defending expenditure on diet and apparel, thought the aristocracy and landed elite should be encouraged to spend rather than hoard their money because it was economically productive. It was the duty of merchants to be parsimonious and not waste their stock in 'unnecessary expense', and of the poor to make manufactures not buy them.[167] John Cary could welcome 'emulation' and the 'variety of fashions' which offered 'wings to men's inventions', but at the same time attack the 'fancy' for 'fashion' which led even maidservants to dress above their station.[168] All of them wanted to have their cake and eat it, to enjoy the benefits of modern affluence without its disadvantages.

Thomas Tryon was a predictable exception. Although himself dedicated to an 'abstemious clean way of living' in his diet, he was in favour of 'eating, drinking, and genteel living' on the part of anyone who could afford it. The trouble was that there were not enough of them. Despite 'the contrary mistaken notions' which had 'got a footing' with men like Brewster, people generally were not living 'more luxuriously' in the 1690s than they had ever done. Most of them were still too 'poor and miserable' to join the better-off, women as well as men, who bought 'cloths, stuffs, silks and a hundred other things', which boosted manufacturing and commerce. 'Of what value would the trade of this nation be, were it not for extravagant fools, drunkards, smokers, gluttons and madmen?'[169] Dalby Thomas, a colonial merchant, had been more restrained but equally robust in 1690. He

---

[164] Charles Povey, *The Unhappiness of England* (1701), sig. A7ʳ, pp. 51, 72–3, 118, 130.
[165] Povey, *Unhappiness*, 'Preface', p. 23.
[166] Whiston, *Mismanagements*, p. 18. On their careers, see *ODNB*.
[167] Francis Brewster, *Essays on Trade and Navigation* (1695), pp. 40–53; above, p. 153.
[168] Cary, *State of England*, pp. 53, 58–9, 150, 162, 186.
[169] Tryon, *Some Memoirs*, p. 30; Tryon, *General Considerations*, p. 8; Tryon, *Brief History*, pp. 43, 116.

denied that consumer goods 'fit for the necessities, ease, and ornaments of life' such as drink, clothes, houses, and even coaches were 'baits to vice and occasions of effeminacy'. The truth of the matter was that they were 'spurs to virtue, valour, and elevation of the mind, as well as the just rewards of industry'.[170]

The most earnest advocates of moral reformation might have been expected to take a more principled view of virtue than Thomas or Tryon, but some of them were equally enmeshed in the new commercial world and had somehow to come to terms with it. They included members of the new Anglican religious societies inspired by Anthony Horneck, a fashionable London preacher who himself had works on economic improvement and political arithmetic in his library.[171] One of the founders of the SPCK, for example, was Humphrey Mackworth, author of *England's Glory* in praise of the Bank of England, and the promoter of a mining enterprise in Glamorgan which advertised itself as 'for the public good, charitable to the poor and profitable to every person who shall be concerned therein'. Among the investors, who expected some profit, were High Tories and clergy like Robert Nelson who waxed eloquent in public about the vices of 'intemperance and luxury', 'the vain diversions of a wicked age'. He must have shared Mackworth's aim to 'adjust a due care of temporal affairs and of spiritual together', but that was easier said than done. Mackworth's diary reveals him worrying privately about the amount of time he spent on 'a crowd of worldly projects', some of them of dubious probity, 'without the least concern for my poor soul'.[172]

One popular tactic, though Nelson might not have readily agreed with it, was to admit that luxury was a sin, and accept it as the moral price which had to be paid for affluence in a commercial society. That was the case made by Nicole and French Jansenists in the 1670s, and it was sufficiently well known by 1691 for it to be attacked by Sir George Mackenzie, an eminent Scottish lawyer, turning in retirement to the study of moral philosophy in the Bodleian Library. In a *Moral History of Frugality* (1691) he took a hard line against 'some very devout men who would persuade us that it is not fit to decry luxury too much in this age'. They had argued that it was not only good for commerce, and an incentive to industry, but a means to 'diffuse riches among the indigent'. Any such charitable consequences, according to Mackenzie, were the result of a disinterested 'prodigality'. Luxury was a different thing altogether. 'In this age', as he rightly said, it was often given the 'bewitching' name of 'convenience', but it was always the product of 'self-love' and avarice.[173]

Such fine verbal distinctions provided no refuge for Locke, who had been persuaded by Nicole that there was much more to be said about self-love and appetites for comfort and convenience than that. When writing about interest rates in the 1690s, he recognized the economic incentives of emulation and 'vanity' while equally praising frugality; and in an essay on labour, he wanted both to see all

---

[170] Thomas, *Historical Account*, p. 5.   [171] Slack, *Reformation to Improvement*, p. 111.
[172] Yamamoto, 'Piety, Profit and Public Service', pp. 813, 822–4, 827–8; Robert Nelson, *A Companion for the Feastivals and Fasts of the Church of England* (3rd edn, 1705), p. 5; Robert Nelson, *An Address to Persons of Quality and Estate* (1715), p. 9.
[173] Above, p. 152; George Mackenzie, *The Moral History of Frugality with its opposite Vices* (1691), 'Dedication', pp. 4, 40–1, 66–7, 71.

mankind supplied with 'the real necessities and conveniency of life... in greater plenty than they have now', and for 'luxury and vanity' to be suppressed at the same time. Since he was conscious that sumptuary laws were unlikely to have an effect in an age that 'inclines to luxury and excess', he must have known that both had to be lived with.[174]

That was the conclusion which Davenant and Defoe arrived at when struggling with the same issue. Davenant had no wish to deny that trade was 'in its nature a pernicious thing', creating private extravagance and public corruption, but he saw no easy way of making wealth and virtue 'coexist together'. Any effort by public authority to make people suppress their 'appetites and passions' was bound to be compromised from the start. Laws had to be 'accommodated' to popular expectations and 'the inclinations of the people', and in any case trade was essential to finance modern warfare. Pernicious or not, recent history had made it 'a necessary evil'.[175] Defoe took a similar historical view. The extravagance and debauchery which had first gained a footing in the court of James I and then infected everyone ought if possible to be moderated. It might be done by taxes on consumption, or by the Societies for the Reformation of Manners which he supported until he was repelled by their blatant hypocrisy in not reforming their own manners first.[176] But Defoe was as realistic as Davenant in thinking that a people who had become accustomed to 'ease and wealth' were unlikely to be quickly won over to frugal virtues, and that governments needed the profits of commerce since wars were now won by whichever nation had the longest purse.[177]

Like reformers of manners, preachers were sometimes left exposed when they publicly rode two horses at once. Sermons to the Levant Company, for example, welcomed England's 'extraordinary advancements of wealth and reputation abroad this last century' while struggling to explain how the acquisition and display of wealth could be reconciled with the conscience and self-respect of 'the happy man'.[178] Nelson was convinced that they could not. The pursuit of great riches led only to anxiety, trouble, and vexation, and true happiness lay in being satisfied with 'the necessaries and conveniences' of life 'within a small compass'.[179] But the conveniences of life were themselves purchasable goods and the happiness they delivered attractive to ordinary consumers in the circumstances of the 1690s. In 1694 an Essex woolcomber, Joseph Bufton, copied down some local verses for a town crier, which must have appealed to him in current circumstances:

---

[174] John Locke, *Locke on Money*, ed. Patrick Hyde Kelly (2 vols, Oxford, 1991), i. 230–1, 276; ii. 494–5; Locke, *Political Essays*, ed. Goldie, p. 255.
[175] Hont, *Jealousy of Trade*, pp. 206–16; Davenant, *PCW*, i. 390–2.
[176] Daniel Defoe, *The Poor Man's Plea* (1698), pp. 4–5; Daniel Defoe, *Giving Alms no Charity* (1704), pp. 27–8; Daniel Defoe, *Taxes no Charge* (1690), p. 10; Daniel Defoe, *The Great Law of Subordination consider'd* (1724), pp. 50, 54–5; Daniel Defoe, *Reformation of Manners: A Satyr* (1702); Dabhoiwala, *Origins of Sex*, p. 67; Maximilian E. Novak, *Daniel Defoe: Master of Fictions* (Oxford, 2001), pp. 129–30, 171, 317.
[177] Daniel Defoe, *An Argument shewing, that a Standing Army... Is not Inconsistent with a Free Government* (1698), pp. 5, 15, 20; Daniel Defoe, *The Present State of the Parties* (1712), p. 327. Cf. Novak, *Defoe*, pp. 622–3; J. G. A. Pocock, *The Machiavellian Moment* (Princeton, 1975), pp. 432–5.
[178] Glaisyer, *Culture of Commerce*, pp. 87–90, 94.   [179] Nelson, *Companion*, pp. 302–3.

> Thrice happy's he who in a middle state,
> Feels neither want nor studies to be great,
> He eats and drinks and lives at home at ease,
> Whilst warlike monarchs cross the raging seas.[180]

Happiness remained as useful a concept as it had been in the 1650s and 1680s in covering a multitude of sins and public and private appetites.[181] Although Locke among others complained that governments were now so 'intent upon the care of aggrandising themselves' that they neglected 'the happiness of the people', the concept of national happiness continued to have resonance. It was often repeated in Anne's reign, in addresses congratulating her on the victory of Blenheim in 1704, for example.[182] The virtues of moderation in a small compass and the middle state could similarly still be employed by advocates of improvement who wished to avoid luxurious excess, and who looked to achieve the modest sufficiency for everyone which seemed closer to realization now than it had ever been. Alongside the several references to 'ease' in the literature I have been citing were recurrent statements that people in general either had a natural right to the comforts or conveniences of this life, a comfortable life or comfortable subsistence, or were in the process of acquiring them.[183] In his 1707 guide to *The Present State of Great Britain*, Guy Miège shared the qualms about luxury of his collaborator Defoe, but he left his readers in no doubt that the English were living 'at ease and in plenty'.[184] Plenty could be thought to have delivered a broadly based comfort, and the process of disarming much of the spectre of luxury by translating it into this moderate, modest, and above all respectable, image of itself had begun.[185]

The language of ease clearly did little to paper over the cracks in contentious territory. Yet the assumption that there was an essential harmony between individual betterment and an increase in the national wealth, and that they could be held in harness with virtue by moral suasion, and a private if not public reformation of manners, retained its potency. Such optimism was essential to sustain the legitimacy of a new regime, defending the public good and the national wealth in wartime; and wealth, as Davenant explained, was whatever contributed to keeping not only the prince but 'the general body of his people in plenty, ease and safety'.[186] The conventional image for that collective sense of purpose also remained: the beehive which Cary incorporated into the seal of his Bristol workhouse in 1696 because it was a model of industry, and which Henry Mackworth used as a 'perfect emblem' of England's consumers and tradesmen prospering and proliferating after

---

[180] French, *Middle Sort of People*, pp. 247–8.   [181] Above, pp. 111–14, 166–7.
[182] *Locke on Money*, ii. 495; Knights, *Representation and Misrepresentation*, pp. 150–1.
[183] In addition to Locke (above, p. 206), see, for example, Defoe, *Essay upon Projects*, p. 143; Pett, *Happy Future State*, p. 85; Josiah Child, *A Discourse about Trade* (1690), p. 29; Povey, *Unhappiness*, pp. 51, 97, 130; Hont, *Jealousy of Trade*, p. 218 (Davenant); below, p. 217. According to the *OED*, 'conveniences', usually in the plural, and denoting 'material arrangements conducive to personal comfort', seems first to have come into use in the later seventeenth century.
[184] Guy Miège, *The Present State of Great Britain* (2 vols, 1707), i. 21, 133, 219–21; Frank Bastian, *Defoe's Early Life* (1981), pp. 156–7.
[185] See below, pp. 215–16.   [186] Davenant, *PCW*, i. 381.

the foundation of the Bank, 'and the more the merrier, like bees in a hive, and better cheer too'.[187]

In 1705, however, it was also the emblem satirized in Bernard Mandeville's *The Grumbling Hive*, a hive 'feared in wars' and famous for its 'sciences and industry', with vast numbers of bees living 'in luxury and ease' and enjoying all life's 'conveniences' and 'comforts' because their private vices, unrestrained, delivered public benefits.[188] Mandeville compelled improvers to do what they had been so anxious to avoid, and recognize the fragility of their supposition that either the beehive or the national wealth had anything at all to do with virtue. It took some time for Mandeville's message to be appreciated, partly perhaps because in 1705 he was contributing to a European debate initiated by Fénelon, bishop of Cambrai, critic of the policies of Louis XIV and author in 1699 of his own fable about bees which envisaged a Utopia without luxury.[189] When the original verses were republished in 1714, however, in *The Fable of the Bees, or Private Vices, Publick Benefits*, with long prose 'remarks' as well a subtitle pointing their moral, they were plainly intended for an English audience.

Even then they received little more attention. Sustained and hostile reaction came only with the second edition in 1723, because recent events had created a susceptible public. It included an essay attacking charity schools, the most successful enterprise of the Anglican societies, and they now had their Whig critics who argued that they did more harm than good. It appeared in the wake of the collapse of the South Sea Bubble in 1720, which had shaken any remaining public confidence in the infinite improvement promised by projects for banks and the multiplication of paper money, and aroused suspicion that the pursuit of ease might too readily be transformed into an appetite for large and sudden riches. Public scandal about the malpractices of the Charitable Corporation, set up to provide pawnshops for the poor, was adding further fuel to the suspicion that all enterprises for the public good, even 'the most virtuous schemes', were founded 'upon principles of corruption'.[190] That was the raw nerve which Mandeville struck with clinical accuracy in the 1720s.

He brought to the exercise not only wit but a ruthless logic which enabled him to puncture the commonplaces of conventional morality and political economy alike. Like Barbon, he had been trained in the medical school at Leiden, and he had absorbed even more of the Epicurean and Augustinian strains in French moral philosophy which stressed the role of appetites, passions, and self-love in driving

---

[187] Jonathan Barry and Kenneth Morgan, eds, *Reformation and Revival in Eighteenth-Century Bristol* (Bristol Record Society 45, 1994), p. 51; Mackworth, *England's Glory*, pp. 20, 23. Cf. above, pp. 113–14, and for Cary's flexible use of the bee metaphor, BL, Add. MS 5540, fo. 75. Mackworth as usual had borrowed a good phrase from another author: above, p. 141.

[188] Bernard Mandeville, *The Fable of the Bees or Private Vices, Publick Benefits*, ed. F. B. Kaye (2 vols, Oxford, 1924), i. 17–18, 24, 26, 36.

[189] Hont, 'Enlightenment Debate on Commerce and Luxury', pp. 382–3, 387–9.

[190] Thomas A. Horne, *The Social Thought of Bernard Mandeville: Virtue and Commerce in Early Eighteenth-Century England* (1978), p. 68; Trenchard and Gordon, *Cato's Letters*, iv. 205–9 (no. 133, June 1723); Slack, *Reformation to Improvement*, pp. 119–21; Wennerlind, *Casualties of Credit*, pp. 237, 240.

human behaviour.¹⁹¹ He had no time for the fashionable new philosophy of Lord Shaftesbury, which pictured mankind as naturally sociable, benevolent, and virtuous. Virtue and vice were not permanent realities but social constructs, and he had learnt from Bayle that men commonly acted against their principles.¹⁹² He seems also to have read many of the texts about the wealth of England and its improvement which I have been citing, and he was able to use their own language to point to conclusions they had failed, perhaps deliberately, to reach.

Central to the impact of Mandeville's satire was the charge of hypocrisy. Vice was unavoidable in commercial societies. Trade brought in 'riches, and where they are, arts and sciences will soon follow', but so would avarice and luxury, their 'inseparable companions'. 'While man advances in knowledge, and his manners are polished, we must expect to see at the same time his desires enlarged, his appetites refined, and his vices increased.'¹⁹³ Charitable works like the hospitals of Thomas Guy and John Radcliffe were no argument to the contrary; they were simply public demonstrations of past avarice, and pride and vanity had 'built more hospitals than all the virtues together'.¹⁹⁴ Far from 'enervating and effeminating people', vice was fundamental to the security of a wealthy nation which aimed to 'live in all the ease and plenty imaginable'. Part II of the *Fable*, published in 1729, reiterated the message. 'The national happiness which the generality wish and pray for' was 'to live in ease, in affluence and splendour at home, and to be feared' abroad. None of that could be attained without 'avarice, profuseness, pride, envy, ambition and other vices'.¹⁹⁵

Mandeville allowed one exception to his general rule. In the case of the Dutch Republic, circumstances had dictated frugality. A country 'so small and so populous' had to save to pay for defence against Spain and for the import of foodstuffs and manufactures from abroad. The case of Britain was wholly different. Here envy and emulation were essential to 'set up a variety of manufactures, ... leave no ground uncultivated', and boost exports.¹⁹⁶ It was therefore futile to attempt to regulate either lavishness or frugality, because they depended on the circumstances of the people. In modern Britain luxury was both natural and necessary. Following 'the road that leads to virtue' would only have the effect of returning artificers to the plough, turning merchants back into farmers, and emptying London, 'sinful over-grown Jerusalem', of 'the covetous, the discontented, the restless and ambitious'. It would produce 'an harmless, innocent and well-meaning people'

---

¹⁹¹ Harold J. Cook, *Matters of Exchange* (New Haven, 2007), pp. 397–406; E. G. Hundert, *The Enlightenment's Fable: Bernard Mandeville and the Discovery of Society* (Cambridge, 1994), pp. 36–8, 45–9; Dabhoiwala, *Origins of Sex*, pp. 112–13.
¹⁹² John Robertson, *The Case for the Enlightenment: Scotland and Naples 1680–1760* (Cambridge, 2005), pp. 261–70.
¹⁹³ Mandeville, *Fable*, i. 184–5.
¹⁹⁴ Mandeville, *Fable*, i. 260–1; Slack, *Reformation to Improvement*, p. 133.
¹⁹⁵ Mandeville, *Fable*, i. 122–3; ii. 106.
¹⁹⁶ Mandeville, *Fable*, i. 184–9. Cf. Hans Blom, 'Decay and the Political Gestalt of Decline in Bernard Mandeville and his Dutch Contemporaries', *History of European Ideas*, 36 (2010), pp. 153–66.

without 'fraud and luxury', without 'wealth and power', and with no incentives to improvement.[197]

It would be possible to annotate most of the statements I have quoted with references to English authors from Temple to Tryon and Dalby Thomas. Mandeville's remark about a population of ploughmen is reminiscent of Houghton's case against sumptuary laws and Petty's thinking about London, and his view of emulation, that it 'adds spurs to industry, and encourages the skilful artificer to search after further improvements', might have been penned by Barbon.[198] Yet none of them, with the possible exception of Barbon, had abandoned talk of morality and virtue altogether. When Mandeville showed that to be the logical consequence of their position, if they wanted to avoid the charge of hypocrisy, he presented all advocates of improvement with a formidable challenge.

Determined as most of them were to maintain the uneasy alliance between morality and commerce, they were unable to accept the logic. To have done so would have been to shatter that consensus about the importance of material progress for the public good on which the culture of improvement depended. They could only keep silent or argue about the terms of the debate, about the language not the logic. Many of Mandeville's critics, especially the clergy among them, naturally engaged in argument about Christian revelation, natural religion and sociability, defended the charities he attacked, and condemned the extremes of vicious behaviour like prostitution and polygamy which he defended.[199] The future bishop, Joseph Butler, the ablest moral philosopher among them, distinguished between benevolence and self-love much more effectively than Mackenzie had done, and emphasized the importance of conscience and 'moral consideration' at every point of the argument.[200] Those who considered the *Fable*'s economic argument, however, concentrated chiefly on the definition of luxury.

George Blewitt, for example, author of one of the first responses, readily accepted that much of what Mandeville said was consistent with received wisdom. All the fruits of the earth had been providentially designed for the service of man, and 'his skill and capacity in the improvement of them were given him by nature, to make his present being easy and agreeable'. Since 'men's wants are their desires' and never satisfied, and since 'political writers' were all agreed that the national wealth depended on improving the soil and extending commerce, private appetites were essential to economic growth.[201] Frugality was a virtue but so were the liberality and generosity ('especially when assisted by an elegance of taste') which came with 'the possession or enjoyment of abundance'. Luxury for Blewitt consisted only in 'the excess of ease and pleasure, or in the abuse of plenty'. When expenditure was disproportionate to people's means, then, but only then, was it 'both a private vice

---

[197] Mandeville, *Fable*, i. 197, 231–3. [198] Mandeville, *Fable*, i. 130; above, pp. 144–5, 198.
[199] Replies to Mandeville are discussed in Horne, *Social Thought of Mandeville*, pp. 76–95, and several of them printed in J. Martin Stafford, ed., *Private Vices, Publick Benefits? The Contemporary Reception of Bernard Mandeville* (Solihull, 1997).
[200] Joseph Butler, *Fifteen Sermons Preached at the Rolls Chapel* (1726), pp. 1–13, 237–8.
[201] Stafford, ed., *Private Vices, Publick Benefits*, pp. 265, 347, 365.

and a public prejudice', tending 'to debauch and corrupt a people'. In all other circumstances it was essential to improvement.[202]

That attempt at a narrow definition of luxury left plenty of room for argument about what was disproportionate expenditure and what not. Blewitt wholly failed to counter Mandeville's point that luxury was either everything or nothing. Once one accepted that men's desires were limitless, then what was superfluous for some people was a necessity for others. As Voltaire put it in 1738, luxury was either everywhere or nowhere.[203] Nonetheless, Blewitt's was the usual response to Mandeville adopted by his English critics. Francis Hutcheson concluded, on 'the question of fact in this matter' that 'it is but a small part of our consumptions which is owing to our vices'; and Bishop Berkeley, who thought paper money would stir up industry and urged Irish landowners to spend more lavishly in order to encourage employment, would have agreed.[204] No one anticipated Dr Johnson and said plainly that 'you cannot spend money in luxury without doing good to the poor', but that is what philanthropists and advocates of benevolence and liberality were expecting people to do.[205]

Writers in Ireland as well as England continued to disagree about whether luxury was detrimental to the national interest, and about whether it could ever be morally benign. Whenever wars were going badly, there were again calls for a 'speedy reformation' of manners, and fears that effeminacy, luxury, and self-interest would ruin the nation.[206] But the intellectual ground was plainly shifting when a growing sector of respectable opinion made subtle distinctions between different forms of extravagance and looked at their consequences for good or ill in the terms set by political economy. Mandeville's satire had not cleansed English political economy from all taint of hypocrisy. Still less, however, had it provoked a moral backlash sufficient to discredit the pursuit of affluence. Instead political economy had emerged fortified. Mandeville had pushed the demoralization of luxury still further, and made moral ambivalence more tolerable for those persuaded, like him, that 'religion is one thing and trade is another'.[207]

That seems to have been Mandeville's effect on the elderly Defoe. Still looking for some means of reconciling economic appetites with virtue in 1727, and referring once again to the projects and financial speculations which had brought Barbon himself to bankruptcy in 1699, he was compelled to ponder at some length the message of the *Fable*. In the long run, 'the sins of the people' might indeed be 'the greatest blessing to trade', and he confronted the stark possibility that

---

[202] Stafford, ed., *Private Vices, Publick Benefits*, pp. 265–6.
[203] Mandeville, *Fable*, i. 108, 123; Voltaire, 'Observations sur MM Jean Law, Melon et Dutot sur le commerce, le luxe, les monnaies et les impôts', in Louis Moland, ed., *Œuvres complètes de Voltaire* (Paris, 52 vols, 1877–85), xxii. 363.
[204] Stafford, ed., *Private Vices, Publick Benefits*, p. 400; Horne, *Social Thought of Mandeville*, pp. 83–4; Wennerlind, *Casualties of Credit*, p. 239.
[205] Berry, *Idea of Luxury*, p. 98.
[206] Martyn J. Powell, *The Politics of Consumption in Eighteenth-Century Ireland* (Basingstoke, 2005), pp. 44–7; *The Diary of Thomas Turner 1754–1765*, ed. David Vaisey (Oxford, 1985), p. 125 (1757). Cf. Innes, *Inferior Politics*, pp. 181–3.
[207] Berry, *Idea of Luxury*, p. 138; Mandeville, *Fable*, i. 356.

'reforming our vices' would 'ruin the nation'. In the following year he was content to push such moral qualms about economic progress to one side, applauding the 'due and daily progression' of British commerce and resolutely predicting that it was bound to 'complete the glory and prosperity of the whole nation'—if not its virtue—in the end.[208]

Luxury was being accommodated. It could be confined to a narrow domain when much of it was rechristened as comfort and ease, and even the remnant could be defended, or more or less quietly accepted, because it contributed to material progress. The bees of the *Fable*, living in luxury and ease, presented a different image from that of the busy bees improving each shining hour in Isaac Watts's popular poem for children of 1715.[209] Like contemporary moralists and economists they nevertheless managed to live more or less happily side by side.

The culture of improvement had survived the challenges which events after 1688 presented to confidence in affluence and its future growth. The rhetoric of improvement proved indispensable in holding together the parties and interest groups on which political stability depended, and to that extent facilitated the emergence of a fiscal-military state able to harness the resources of the kingdom and its dominions in war. The language of improvement had also proved resilient enough to withstand the arguments of those who thought material progress must in the end come to a full stop, and flexible enough to answer those who continued to argue that material progress necessarily meant moral decline and decay. The reality of improvement had similarly survived, and was visible in economic growth after 1700, and in local investments in social welfare and the economic infrastructure even before that. A parliamentary regime and the public press which reflected and influenced its deliberations had not prevented the pursuit of improvement, but rather given its purposes and rationale greater publicity. Improvement was more firmly rooted in 1730 than it had ever been.

In some respects it had become a word of such general application that, like luxury, it was impossible to pin down because it was everywhere and embraced everything. But improvement had also, under pressure of events, become more circumscribed in its ambition, concerned (as Mandeville appreciated) with delivering plenty at home in order that the country could compete successfully abroad in the modern system of international politics. There were alternative voices. In 1696 Edmund Bohun clung to aspirations close to Hartlib's, insisting that it was 'the great design of God Almighty' to 'civilize the whole race of mankind, to spread trade, commerce, arts, manufactures, and by them Christianity, from people to people round the whole globe of the earth'. In 1703 Andrew Fletcher, the Scottish opponent of union, foresaw London becoming ever more dominant and

---

[208] Daniel Defoe, *The Complete English Tradesman* (2 vols, 2nd edn, 1727), ii. 162, 165; Peter Earle, 'The Economics of Stability: The Views of Daniel Defoe', in Coleman and John, eds, *Trade, Government and Economy*, pp. 286–8; Daniel Defoe, *A Plan of the English Commerce* (1728), p. xvi.

[209] Isaac Watts, 'Against Idleness and Mischief', in *Divine Songs...for the Use of Children* (2nd edn, 1716), p. 29.

argued that several large cities in the three kingdoms would better serve 'to the improvement of all arts and sciences' for everyone. He was anticipating David Hume's vision of a European republic of states, connected rather than divided by commerce, and bringing peace and enlightenment to the Continent. In England, however, wars for trade, Fletcher's 'golden ball for which all nations of the world are contending', brought narrower horizons.[210]

In that constricted sphere, improvement and the intellectual tools which propelled it had also demonstrated their limitations. Political economy provided a language and lens through which immediate and difficult issues could be observed and brought into focus. But it had not provided solutions when there were several options on offer with respect to taxation, banking, the commerce of the East India Company, or that of Ireland. There the outcomes depended on political horse-trading and the pressure of events. Neither had information been of much help, either because it was open to dispute, as with figures for the balance of trade, or wholly deficient, as it was with respect to the country's money supply and population. Political arithmetic, now used in a host of contexts, had come to seem—as one critic said—'a perfect Popish nose of wax', to be moulded to any purpose. Though not wholly discredited, it was often mocked, as in Swift's *Modest Proposal*, with its calculation of the 'public good' to be gained from converting 100,000 Irish children into a dish for gentlemen, and in Joseph Addison's 'Trial' of the hooped skirt, whose multiple petticoats were defended because they would bring 'a prodigious improvement of the woollen trade and . . . sink the power of France in a few years'. Both of them were satirizing the kind of thinking which Petty thought should be characteristic of 'a well-governed state'.[211]

By Petty's standards England was scarcely that after 1688. Although the character of a parliamentary regime would scarcely have surprised him or anyone else who had lived through the Interregnum, it held out little prospect of a clear-minded sovereign power using the tools of number, weight, and measure to engage in social and economic engineering. That helps to explain why political arithmetic lost much of its sense of purpose in the early eighteenth century.[212] The few clauses which its devotees managed to insert in the Duty Act of 1695 were scarcely evidence of a sovereign or his chief ministers pursuing improvement with the determination that Colbert and Louis XIV were reputed to have shown in France.[213] The initiative, as with the Corporations of the Poor and statutes for local improvements, came from an informed public applying pressure on Whitehall and Westminster from outside.

---

[210] Thomas, *Ends of Life*, p. 146; Andrew Fletcher, *Political Works*, ed. John Robertson (Cambridge, 1997), pp. 193, 214; Pocock, *Barbarism and Religion*, ii: *Narratives of Civil Government*, pp. 169–70, 189.

[211] Nicholls, *Conference with a Theist*, p. 78; Jonathan Swift, *A Modest Proposal* (Dublin, 1729); Erin Mackie, *The Commerce of Everyday Life: Selections from The Tatler and The Spectator* (1998), pp. 482–5 (1710). (I owe the latter reference to Keith Wrightson.)

[212] Jessica Warner, 'Faith in Numbers: Quantifying Gin and Sin in Eighteenth-Century England', *Journal of British Studies*, 50 (2011), pp. 77–9; Innes, *Inferior Politics*, pp. 127–30; Julian Hoppit, 'Political Arithmetic in Eighteenth-Century England', *EcHR* 49 (1996), pp. 516–40.

[213] See, for example, *Britannia languens*, pp. 461–2; below, p. 224.

Yet public pressure was irrefutable evidence of improvement's success. It had lost all novelty and much of its cutting edge. It had become a conventional and inescapable part of political and public discourse, sometimes papering over divisions of opinion, and never far from the centre of economic discussion when it did not. *England's Happiness Improved* was the familiar title of a 1697 handbook on 'good housewifery and management' which included an exhortation not to buy foreign produce because it consumed the nation's 'treasure' and hindered 'the circle of inland trade'. When popular agitation for a complete ban on imports of calicoes from India revived in 1719, a piece of doggerel verse could similarly appeal to 'the rules for the common improvement of nations' as if they were familiar to everyone, as indeed they now were.[214] National improvement had become a commonplace aspiration in early eighteenth-century Britain. The next chapter must ask what the consequences were, and set English and now British improvement in a European context in order to determine what was distinctive about it.

---

[214] Harvey, *Little Republic*, p. 31; Brodie Waddell, *God, Duty and Community in English Economic Life, 1660–1720* (Woodbridge, 2012), p. 169. On popular agitation in 1719–20, see Beverly Lemire, *Fashion's Favourite: The Cotton Trade and the Consumer in Britain, 1660–1800* (Oxford, 1991), pp. 35–42.

7

England's Improvement

Improvement has appeared in a number of guises in the course of the period covered by this book, and before coming to a conclusion we should try to assess what it meant by the end of it, in early Georgian England. Improvement still had a long future ahead of it. It was fundamental to its character that the word and the culture could be used, as circumstances changed, to interpret and respond to new intellectual or economic challenges, and it could survive and adapt as creatively in the industrializing Britain of reform, classical economics, and an evangelical revival, as it had in the trading nation of glorious revolution, mercantilism, and moral reformation a century earlier. But improvement had amply demonstrated that capacity to evolve by the 1730s. In that sense its creative period was over, and it is an appropriate point at which to judge its impact. It has been suggested more than once in earlier chapters that improvement was becoming a distinctive culture of considerable potential. This chapter will try to substantiate the claim by examining what the culture signified in the early eighteenth century, how much it had contributed to English (and after 1707 British) economic development, and finally how both the culture and the outcomes differed from those to be found in other countries on the eve of the European Enlightenment.

COMFORT AND PROGRESS

It was one sign of England's distinctiveness in the eighteenth century that it was reputed to have invented the concept of material comfort.[1] It is open to argument whether the French aristocracy or the Dutch bourgeoisie were the first to create the domestic lifestyle necessary for its achievement in reality,[2] but the word 'comfort', in the material sense of the term, was—like improvement—an English invention. It had no obvious synonym in other languages, and had to be exported to Germany and even to France. In 1688 in his *Great French Dictionary*, Guy Miège needed to translate 'the comforts of this life' as *les plaisirs (les douceurs, les aises) de la vie*, and

---

[1] Paul Langford, *Englishness Identified: Manners and Character, 1650–1850* (Oxford, 2000), pp. 117–19; Daniel Roche, *A History of Everyday Things: The Birth of Consumption in France, 1600–1800* (Cambridge, 2000), pp. 107–8. Cf. John E. Crowley, *The Invention of Comfort: Sensibilities and Design in Early Modern Britain and Early America* (Baltimore, 2001), pp. 149–59.
[2] Jan de Vries, *The Industrious Revolution* (Cambridge, 2008), pp. 126–8.

'to live a comfortable life' as *vivre agréablement, content ou à son aise*.³ More than a century later, Jean-Baptiste Say thought 'comfortable' was a word that ought to be added to every French dictionary, because the vast majority of houses, even in the richest areas of France, still lacked 'the things that the English call comfortable'.⁴ Comfort was the mark of England's exceptional material well-being.

In seventeenth-century French 'confort' still meant consolation, and comfort retained that meaning in English. Like improvement again, however, it was a word employed in different senses depending on contexts, and open to changes of emphasis over time. Like 'happiness' and 'ease' it moved in a decidedly material direction when it joined 'peace and plenty' among the acknowledged goals of good government in England in the later seventeenth century. The purposes of improvement, including better relief for the poor, the further growth of London, and commercial expansion, had all been presented as bringing greater comfort—'comfortable maintenance', 'comfortable livings', 'comfortable subsistences'—for more people.⁵ When Davenant observed in 1698 that the English unlike the French lived 'in plenty and at their ease', he was commenting on a contrast which was growing ever greater.⁶

Comfort nevertheless retained some of its non-material dimensions and resonance, for part of the utility of the word, as with happiness, was its capacity to extend to new goals when old ones had been attained and to offer compensating satisfactions if they had not been fulfilled. Hence a promoter of fisheries as a remedy for unemployment in 1720 thought that constant labour would be 'a continual ease and comfort' to the poor 'by amusing and diverting them from thinking on their poverty and other misery': cold comfort, one might think.⁷ More often, however, pleasure and even delight were regarded as the natural companions of material comfort and inseparable from it. They were symptomatic of the civility and politeness which writers like 'Cato' in the 1720s thought elevated Augustan England above the 'wild empires of the east' with their 'gothic governments', and above Poland and the highlands of Scotland, still languishing in an early barbarous stage of civilization. The modern 'inventions of arts and sciences' made life easy and pleasant, and persuaded people—if persuasion was required—that 'other things are necessary to their happiness besides those which nature has made necessary. Thus the luxury of the rich becomes the bread of the poor.'⁸

---

³ Guy Miège, *The Great French Dictionary: In Two Parts* (1688), Part II, *s.v.* 'comfort', 'comfortable'. Voltaire used the word ease, 'aisance', for the comforts which he thought had allowed the English population to grow: *Œuvres complètes de Voltaire*, ed. Louis Moland (52 vols, Paris, 1877–85), xxiv. 512.

⁴ Michael Sonenscher, 'The Emergence of Fashion and the Rise of Capitalism in Eighteenth-Century France', *P&P* 216 (Aug. 2012), p. 256.

⁵ Richard Haines, *Provision for the Poor* (1678), title page; Petty, *EW*, ii. 470; Josiah Child, *A Discourse about Trade* (1690), p. 29.

⁶ Davenant, *PCW*, i. '242' (*recte* 252).

⁷ *The Importance and Management of the British Fishery Consider'd* (1720), p. 14. Richard Steele had made the same point in 1684: above, p. 166.

⁸ John Trenchard and Thomas Gordon, *Cato's Letters* (4 vols, 1723–4), ii. 136.

That was naturally the historical vision Mandeville incorporated into the notes to his *Fable*. He pictured even the richest men in the earliest stages of society as lacking 'many comforts of life that are now enjoyed by the meanest and most humble wretches'; 'whatever has contributed since to make life more comfortable' had become essential, although it 'more or less deserves the name of luxury'.[9] The same point was later to be a commonplace among philosophers and historians of the Scottish school, like David Hume and Adam Ferguson, who wrote about 'that complicated apparatus which mankind devise for the ease and convenience of life: their buildings, furniture, equipage, clothing... all that assemblage which is rather intended to please the fancy than to obviate real wants, and which is rather ornamental than useful'.[10] Hume avoided controversial territory when he changed the title of one of his essays on such things from 'Of Luxury' to 'Of Refinement in the Arts'; and English writers had earlier done the same. Picking up an image from Barbon in 1701, Henry Martin remarked that the true riches of a modern nation were not only 'meat and bread and clothes and houses', 'the conveniences as well as the necessaries of life', but also the 'several refinements and improvements of them'.[11]

Historians in the Scottish school presented the advance of societies through stages from barbarism to modern civility as a story of progress, and progress was also a notion already familiar to an English audience: it was not a concept freshly minted in the middle of the eighteenth century.[12] The word was first used in the sense of movement forward and upward in descriptions of the advance of religion and reformation, in America or England, and of knowledge, learning, and science, especially after the founding of the Royal Society; but by the 1670s it was being applied to the past and present advance of commerce and trade in Holland and England, and even Spain.[13] After that it could be extended, especially by Davenant, to apply to England's growing wealth and comfort, and extended even further by Walter Moyle and those who borrowed his phrase about the 'the rise and progress of the English greatness'.[14]

Thanks to the political arithmeticians, England's economic progress was also being given some historical depth and specificity. In 1725 Erasmus Philips dated 'the progress of trade in this nation', which had created its present affluence, to the

---

[9] Bernard Mandeville, *The Fable of the Bees*, ed. F. B. Kaye (2 vols, Oxford, 1924), i. 169.
[10] Adam Ferguson, *An Essay on the History of Civil Society* (Edinburgh, 1767), p. 375; David Hume, *An Enquiry concerning the Principles of Morals* (1751), p. 138; Adam Smith, *The Theory of Moral Sentiments* (1759), pp. 108–9.
[11] Christopher J. Berry, *The Idea of Luxury* (Cambridge, 1994), pp. 142–4, 150–2; Henry Martin, *Considerations upon the East-India Trade* (1701), p. 16. Cf. Nicholas Barbon, *A Discourse of Trade* (1690), p. 15.
[12] David Spadafora, *The Idea of Progress in Eighteenth-Century Britain* (New Haven, 1990), surveys the eighteenth-century literature. The suggestion (p. 211) that notions of 'improvement in human affairs *generally*—in culture or civilization or even happiness' were rarely more than implicit or cursory before 1760 seems to me open to challenge.
[13] Paul Slack, 'Material Progress and the Challenge of Affluence in Seventeenth-Century England', *EcHR* 62 (2009), pp. 579–80; above, pp. 136, 197.
[14] Davenant, *PCW*, i. 305–6 (Moyle); above, p. 201.

century and a half before 1688, a chronology which Adam Smith copied,[15] and David Hume's *History* followed Davenant in asserting that 'the commerce and riches of England did never, during any period, increase so fast as from the restoration to the revolution' of 1688.[16] Although it was not until Hume's day that the 'progress of society' in general was widely debated,[17] that is what English writers on the growth of commerce and refinement were aspiring to describe, and no one more often than Defoe. Having finished his *Tour* in 1724, which surveyed the present improvements of England and explained why they were absent from most of Scotland, he embarked upon a *General History of Discoveries and Improvements in useful Arts* (1725–6) which set out to chart their 'growth and progress' ever since the Flood. His purpose was to show 'who were the inventors, who the improvers, who the patrons of them, through all the ages of their improvement in the world'.[18]

Though written in the new language of improvement, Defoe's history was like Ralegh's a century before, a story of providential design. It was similarly unfinished, since Defoe stopped in the fifteenth century, but no less intended to end with the triumphs of the English in commerce, navigation, and overseas plantations, the great historical developments on which Defoe concentrated.[19] Other nations could have anticipated them, but they had not done so, and their failures were object lessons for the present. The 'progress of the Carthaginians' in trade might have been 'improved upon and increased in the world', had they had not been defeated in war. Once they had an empire, the Romans failed to do enough to encourage commerce and navigation. There was a danger too that modern Europeans might waste time and resources on futile endeavours, and Defoe (in provocative mode) counted the search for a North-West Passage and for a reliable means for measuring longitude at sea among them. But the English had at least been farsighted enough to realize the commercial potential of their wool before 1500, and their recent plantations on other continents gave them a springboard for the diffusion of their technologies across the globe.[20] Like Bohun before him, Defoe foresaw the possibility that Europeans overseas might one day pick up the baton and run even further, but for the moment English improvement placed Britain at the head of the field.[21]

None of this had come about by historical accident. As Defoe made clear in 1704, 'God almighty, who never acts in vain', had brought the wealth and the power of Holland into the world from the 'ruin of the Flemish liberty', made

---

[15] Erasmus Philips, *The State of the Nation* (1725), pp. 5, 15; Joel Mokyr, *The Enlightened Economy* (New Haven, 2009), p. 13.

[16] David Hume, *The History of England* (8 vols, 1763), viii. 317.

[17] The index to Shaftesbury's *Characteristics* (first published in 1711) had used 'progress of society' when referring to a passage on the development of arts and manners: Anthony Ashley Cooper, 3rd Earl of Shaftesbury, *Characteristicks of Men, Manners, Opinions, Times*, ed. Philip Ayres (Oxford, 1999), p. 391.

[18] Daniel Defoe, *A General History of Discoveries and Improvements* (1725–6), pp. v, 215.

[19] Above, p. 39.

[20] Defoe, *General History of Discoveries*, title page, pp. vii–viii, 100, and *passim*. On longitude, see below, pp. 237–8.

[21] Defoe, *A General History of Trade: For the month of July* (1713), pp. 18–19; above, p. 194.

England 'a prodigy of trade' since Elizabeth's reign, and raised her now to 'a pitch of glory, superior to all people in the world', 'wealth flowing, blessings probable... and the people happy'.[22] Defoe's view of history was always more providential than Davenant's, but it led him to the same point. The Almighty, 'wisely and foreknowingly no doubt', had given England modern comfort along with wealth, a 'new way of living' which had altered the diet, housing, and 'the habits and the furniture of the people'.[23] In 1709 he remarked that—in contrast with Scotland a decade earlier—'we know not in England what belongs to famine... it amounts to no more than this, that... your wheat is dear'; and in 1726 he expanded further on the unprecedented 'well-living of our people here'. Even the 'poorer part of the people clothe better, and furnish better, and thus increase the consumption of the very manufactures they make'.[24]

Defoe was referring to the changes in standards and styles of living during his lifetime which have been described in earlier chapters, and which were continuing. In the early eighteenth century they were most obvious in the centre of towns, with new public buildings, and their streets being paved, cleaned, better lit, and in every respect 'beautified'.[25] In some of them fresh water was available, thanks to the pipes and water-raising machines which had begun with London's New River project in 1609 and now spread to the provinces, where the celebrated engineer George Sorocold showed what could be done in Macclesfield in 1685 and a dozen other towns later.[26] The great urban fires which had led to rebuilding in London and elsewhere were declining in number as brick and tile replaced wood and thatch, and when they occurred some modest protection or at least reassurance was provided by new fire engines, improved by the German and Dutch inventions which Richard Newsham claimed to have perfected in 1721, and by the fire insurance companies whose services were multiplying.[27] Thanks to emulation and the demonstration effect of innovations in the capital, the improvements of London envisaged by Petty were being realized across the kingdom.[28]

Bricks and water also brought improvement to the domestic environment, raising standards of cleanliness in the home, although scarcely to the heights attained by the

---

[22] Defoe, *Giving Alms no Charity* (1704), pp. 5–6, 8. Cf. Laurence Dickey, 'Power, Commerce and Natural Law in Daniel Defoe's Political Writings, 1698–1707', in John Robertson, ed., *A Union for Empire* (Cambridge, 1995), pp. 92–3.

[23] Defoe, *General History of Trade: For the month of July*, pp. 6, 27, 47–8.

[24] Mokyr, *Enlightened Economy*, pp. 16–17 (citing Defoe's *Review*, 20 Oct. 1709); Daniel Defoe, *The Complete English Tradesman* (1726), pp. 386–7.

[25] Peter Borsay, *The English Urban Renaissance* (Oxford, 1989), pp. 70–2; Malcolm Falkus, 'Lighting in the Dark Ages of English Economic History: Town Streets before the Industrial Revolution', in D. C. Coleman and A. H. John, eds, *Trade, Government and Economy in Pre-Industrial England* (1976), pp. 248–73.

[26] E. L. Jones and M. E. Falkus, 'Urban Improvement and the English Economy in the Seventeenth and Eighteenth Centuries', in Peter Borsay, ed., *The Eighteenth-Century Town* (1990), p. 128; Paul Slack, 'Great and Good Towns 1549–1700', in Peter Clark, ed., *The Cambridge Urban History of Britain*, ii: *1540–1840* (Cambridge, 2000), p. 311.

[27] Jones and Falkus, 'Urban Improvement', pp. 120–5; G. V. Blackstone, *A History of the British Fire Service* (1957), pp. 57–68.

[28] Jones and Falkus, 'Urban Improvement', p. 142.

Dutch, since English towns were still growing and parts of them were increasingly overcrowded. The habits of the people changed as much as their furniture as diets improved even among the labouring poor, and cheaper and lighter clothing permitted more regular changes of linen. Personal cleanliness had begun to improve, partly because washing and even bathing were being recommended as good for health. In the seventeenth century the English felt themselves superior to the 'stinking people' of Scotland,[29] and by 1726 a French visitor found it remarkable that 'English women and men are very clean: not a day passes without their washing their hands, arms, faces, necks and throats in cold water, and that in winter as well as in summer'.[30] Cleanliness, however, was in the eye of the beholder, and standards of acceptable public and private hygiene were always rising. One historian of Georgian towns comments that their citizens 'do not seem to have become more contented as their environment improved' because 'previously overlooked nuisances were noted as the threshold of decency changed'.[31] In this as in other respects, once they had been aroused, expectations of improvement were self-sustaining and never satisfied.

It followed that improvement could never produce the happiness, national or individual, its advocates had always promised. John Locke was as insistent on the importance of the 'conveniences of life' as Hartlib had been,[32] and even more on mankind's potential for self-improvement, that development of mental capacities which he thought could be achieved as readily as material improvement through the application of labour and industry.[33] But he was more realistic in recognizing that neither of them produced felicity. People were 'seldom at ease' in reality: the quest for happiness rested on 'uneasiness', whether caused by hunger or cold or weariness or by the 'fantastical uneasiness', for honour, power, and riches, inculcated by 'fashion, example and education'.[34] The old observation that appetites for self-advancement and consumer satisfaction only led to anxiety and discontent was often reiterated. 'Is not both the possession and the pursuit of wealth, to those who really love it, ever anxious?', asked James Harris in a treatise on happiness in 1744.[35]

---

[29] Keith Thomas, 'Cleanliness and Godliness in Early Modern England', in Anthony Fletcher and Peter Roberts, eds, *Religion, Culture and Society in Early Modern Britain: Essays in Honour of Patrick Collinson* (Cambridge, 1994), pp. 59–60, 67, 74–5; Anthony Weldon, *A Perfect Description of the People and Country of Scotland* (1649), p. 1 (first published in 1626). 'The ground might be fruitful had they the wit to manure it', Weldon added.

[30] Mokyr, *Enlightened Economy*, p. 292.

[31] Emily Cockayne, *Hubbub: Filth, Noise and Stench in England 1600–1770* (New Haven, 2007), p. 231.

[32] According to Locke, children had a 'right not only to bare subsistence, but to the conveniences and comforts of life' so far as their parents could afford them: *Two Treatises of Government* (1690), p. 116 (ch. IX, para. 89).

[33] David Armitage, *Foundations of Modern International Thought* (Cambridge, 2013), p. 122. Hartlib advocated the improvement of the 'intellectual abilities' of children: *Samuel Hartlib and the Advancement of Learning*, ed. Charles Webster (Cambridge, 1970), p. 149. Cf. Bacon, cited above, p. 5.

[34] John Locke, *An Essay Concerning Human Understanding* (2 vols, 1715–16), i. 213 (Book II, ch. 21, para. 45).

[35] James Harris, *Three Treatises* (2nd edn, 1766), p. 130 (first published 1744). Cf. above, pp. 86, 112, 165, 206, and below, p. 263. Julian Hoppit, *Land of Liberty?* (Oxford, 2000), p. 2, characterizes the period from 1689 to 1727 in Britain as an 'anxious age'.

Harris was examining human emotions with the cool sensibilities recently made fashionable by his uncle, the 3rd Earl of Shaftesbury, but the same anxieties were more sharply articulated by writers inspecting their consciences in the older tradition of Puritan diarists and autobiographers. The voluminous diaries of Joseph Ryder, a successful Leeds clothier, were as full of references to 'troubles' in the 1730s and 1740s as those of his seventeenth-century predecessors, and he was equally determined to avoid poverty and achieve economic security while trying to escape the moral hazards which worldly 'excess' might bring with it. Torn between a temptation to indulge in too much 'ease' and an inclination to devote 'too much care and labour' to his business, he wanted to find happiness in a 'middle way consistent with... real practical holiness'.[36] He could never achieve it, but such moral qualms did not prevent him being industrious, and he would have agreed with the conventional advice which John Cannon gave to all young people 'in profitable places', that they should be 'careful, frugal, and take the example of the industrious ant or laborious bee' because that was for the general good.[37] Morality did not inhibit private enterprise, but it did not make people wholly comfortable or contented with it either.

That conclusion would scarcely have surprised Defoe who had struggled with the same issues a decade earlier.[38] More remarkable to Defoe's generation was the failure of improvement to deliver two other much publicized and equally unattainable purposes. Despite high wages and plentiful employment opportunities, neither poverty nor idleness had been eliminated, nor even to all appearances much reduced. Defoe saw some of the reasons more clearly than most of his contemporaries. Rising wages and increasing demand for labour had given workers greater freedom of choice and the ability to exercise their leisure preferences. He recognized that genuine poverty must have declined, since if it had not, the English would have enlisted in the army as willingly as the French did. Increasing expenditure on poor relief and subsidies to employment through workhouses was therefore wasteful and counterproductive. A century later Tocqueville commented on the paradox that England, the richest country in Europe, was the one with the largest proportion of its population dependent upon organized charity because it could afford it. Defoe simply concluded that prosperity, 'mighty wealth', had made England 'the most lazy, diligent nation in the world'.[39]

There was a much more fundamental obstacle to the full delivery of improvement's agenda, however. It lay in another paradox, discussed in the previous chapter. England's economic growth between 1650 and 1750 only allowed an increase in income per head because the population was no longer rising, and it was no longer rising partly because mortality, and especially infant and child mortality,

---

[36] Matthew Kadane, *The Watchful Clothier: The Life of an Eighteenth-Century Protestant Capitalist* (New Haven, 2013), pp. 10, 55, 102–3, 184–6.
[37] Craig Muldrew, *Food, Energy and the Creation of Industriousness* (Cambridge, 2011), p. 307.
[38] Above, pp. 211–12.
[39] Defoe, *Giving Alms no Charity*, pp. 24, 26; Muldrew, *Food, Energy and Industriousness*, pp. 310–15; Paul Slack, *Poverty and Policy in Tudor and Stuart England* (1988), p. 5; above, pp. 154, 159.

had increased.[40] Urbanization and commercialization accelerated the transmission of infections by increasing migration between countryside and towns and between continents, and in that sense increases in morbidity and mortality were the price which had to be paid for prosperity.[41] Their effect on individuals was to increase anxiety and insecurity, and suggest that moral reformers were right to insist that true happiness could never be found in this world. When the infant daughter of a Manchester wig-maker died in 1713, he noted in his diary that 'devouring death' had now 'taken two dear wives and five sweet infants from me' and left 'no refuge to flee to but God almighty'; and he vowed, not for the first time, 'to come to an universal reformation of my life for the time to come'. In the 1680s Essex woolcombers like Joseph Bufton, happy in their middle state, had set up an early friendly society which supported any of its members in sickness and other 'calamities, afflictions and troubles, that do attend us'.[42]

'Health's improvement', an ambition since the 1650s, was therefore long in coming. It was not for want of effort or investment. In the 1680s Thomas Sydenham was leading the way towards an 'improvement of physic' by his Baconian attention to 'matters of fact', examining what were apparently new fevers and in the process introducing the notion of diseases as separate entities for the first time; but that had little immediate impact in terms of prevention or therapy.[43] In the later seventeenth century professional medical help, from doctors and nurses, was more widely available and affordable, at least in south-eastern England, but that offered more in the way of comfort than cure to the patients.[44] Urban building in brick, which was thought likely to reduce the risk of epidemic disease in the seventeenth century, and improvements in hygiene and diet, must have had an impact, but it came only slowly.[45] In the eighteenth century London's West End was growing steadily healthier than the rest of the metropolis, and the contrasts in mortality rates between the aristocracy and the rest of the population, and between England and France, were much greater in 1800 than they had been in 1700.

---

[40] Above, p. 199; E. A. Wrigley et al., *English Population History from Family Reconstitution 1580–1837* (Cambridge, 1997), pp. 282–4.

[41] Mary Dobson, 'The Last Hiccup of the Old Demographic Regime: Population Stagnation and Decline in Late Seventeenth and Early Eighteenth-Century South-East England', *Continuity and Change*, 4 (1989), pp. 395–428; John Landers, *Death and the Metropolis* (Cambridge, 1993), pp. 86–7.

[42] *The Diary of Edmund Harrold, Wigmaker of Manchester 1712–15*, ed. Craig Horner (Aldershot, 2008), p. 67; H. R. French, *The Middle Sort of People in Provincial England, 1600–1750* (Oxford, 2007), pp. 245–6; Peter Clark, *British Clubs and Societies 1580–1800* (Oxford, 2000), pp. 353–4; above, pp. 206–7.

[43] Above, p. 108; Thomas Sydenham, *The Whole Works*, trans. John Pechey (1696), sigs A2^r, a1^v, a4^r; William Simpson, *A Short Essay towards the History and Cure of Fevers* (1678), title page, p. 1; Richard Neve, *Arts Improvement* (1715), sigs A3^v–4^r. Cf. Mokyr, *Enlightened Economy*, pp. 242–5; Harold J. Cook, 'Markets and Cultures: Medical Specifics and the Reconfiguration of the Body in Early Modern England', *TRHS* 6th ser. 21 (2011), pp. 123–4, 131, 140.

[44] Ian Mortimer, 'The Triumph of the Doctors: Medical Assistance to the Dying, c.1570–1720', *TRHS* 6th ser. 15 (2005), pp. 97–116; Ian Mortimer, *The Dying and the Doctors: The Medical Revolution in Seventeenth-Century England* (Woodbridge, 2009).

[45] Jones and Falkus, 'Urban Improvement', p. 127; Paul Slack, *From Reformation to Improvement* (Oxford, 1999), p. 98; Mokyr, *Enlightened Economy*, pp. 293–5.

While adult mortality in England began to improve early in the eighteenth century, however, infant and child mortality did not begin to decline until the 1730s.[46]

Against that background, it is easy to understand why the disappearance of plague after 1666 was not immediately recognized as a triumph for deliberate collective action to improve public health, which is arguably what it was. The disease was kept at bay by the quarantine precautions against ship-borne infection adopted in England and other European ports, and the great panic caused by the threat of its reintroduction from Marseilles in 1720 was a stark reminder that they were sometimes fallible. The danger was highlighted by a new Quarantine Act in 1721, and by the public controversy which forced the repeal of some of its clauses which would have imposed rigid quarantine around London if the disease returned. Defoe and other writers rehearsed the arguments of the 1660s about whether plague was contagious, and whether it would need to be attacked by improved methods of isolating the sick such as Petty had proposed.[47] Once the imminent threat from Marseilles was over, however, medical improvers were forced to turn their attention to the other epidemic diseases responsible for England's worsening mortality regime, and there they faced a long and initially unrewarding battle.

Smallpox was the most obvious target, newly virulent since the 1680s, plainly contagious, and now the 'destroying angel' plague had once been.[48] Old and new weapons were vigorously employed against it. The new one was inoculation, practised first in China in the later seventeenth century and then in the Ottoman Empire, discussed in the *Philosophical Transactions* of the Royal Society in 1714, and advocated and defended by its Fellows when it was introduced into England in the 1720s. John Arbuthnot and John Jurin gathered relevant data and used political arithmetic to calculate the risks involved, concluding that while at least 1 in 10 of those who contracted smallpox died of it, no more than 1 in 100 died after inoculation.[49] The odds in favour of inoculation weighed more heavily than the arguments, often of theological principle, against it. In the 1760s the technique was much improved by Robert Sutton and was being successfully practised in towns and parishes where there were general inoculations.[50] By the 1780s improvers and physicians had good reason to claim that Britain had taken the lead in the fight against smallpox, and some could even predict that, if parliament legislated for it, inoculation would wholly 'exterminate the smallpox from Great Britain'. That,

---

[46] Wrigley et al., *English Population History from Family Reconstitution*, pp. 215, 206, 263–8, 285–6, 292–3; Landers, *Death and the Metropolis*, pp. 161, 239–40, 354; T. H. Hollingsworth, *The Demography of the British Peerage* (1964), pp. 53–61.
[47] Above, p. 63; Paul Slack, *The Impact of Plague in Tudor and Stuart England* (1985), pp. 311–37.
[48] Charles Maitland, *Mr Maitland's Account of Inoculating the Small Pox* (1722), p. 3.
[49] Geoffrey Parker, *Global Crisis* (New Haven, 2013), pp. 630–1; Andrea A. Rusnock, *Vital Accounts: Quantifying Health and Population in Eighteenth-Century England and France* (Cambridge, 2002), pp. 45–55; Joanna Innes, *Inferior Politics* (Oxford, 2009), p. 131.
[50] Mary J. Dobson, *Contours of Death and Disease in Early Modern England* (Cambridge, 1997), pp. 278–9, 481; Peter Razzell, *The Conquest of Smallpox* (Firle, 1977), p. 44; Rusnock, *Vital Accounts*, pp. 92–4. On opposition to inoculation, see Edmund Massey, *A Letter to Mr Maitland, in Vindication of the Sermon against Inoculation* (1722); Philip Rose, *An Essay on the Small-pox, Whether Natural or Inoculated* (1724).

however, had to wait for Jenner's discovery of vaccination in 1796, and for its compulsory use, which only began in 1853.[51]

The ultimate conquest of smallpox stands as one of the great medical achievements of the modern world. It deserves a place in any history of improvement, and a place in this one because, although inoculation was only just beginning in 1730 and had been invented elsewhere, it was improved, experimented with, and finally transformed in England. In its eighteenth-century context, however, it was only one part of a wholesale campaign to avoid disease which was being waged across Europe and which took much of its inspiration from English authors, especially Sydenham. It focused particularly on the environmental causes of disease, like poor ventilation and contaminated water supplies in crowded towns, and the bad air of low-lying rural marshlands. It must have contributed to the slow general decline of mortality, but it is difficult to separate its effects from those of other improvements in living standards, and equally difficult to tie them to the decline of particular diseases.[52] A fall in the number of deaths from dysentery in London after 1740, which contemporaries commented on, may be an exception.[53] A more obvious one was a decline in morbidity and mortality which followed the drainage of fens and marshes, and in England that too came after 1740.[54]

The deliberate effort to improve public health in the early eighteenth century was therefore remarkable more for its ambition and persistence than for any rapid victory. It was remarkable also in England because, in contrast to other countries then and later, it was not promoted by the state. As with other forms of improvement, it depended on local initiative, sometimes by local government, more often by energetic projectors and philanthropists who could mobilize local support. After the great outcry against the domestic clauses in the 1721 Quarantine Act, the central government learnt its lesson and intervened only when it could be sure of public backing, as with the re-imposition in 1745 of the strict controls it had first imposed against cattle plague in 1714.[55] In the 1720s English writers were full of admiration for the 'improvements' and 'many projects in favour of arts and trade' introduced by Peter the Great in Russia after his visit to western Europe, just as they had earlier acknowledged how much the power and wealth of France under Louis XIV owed to Colbert, but there was no suggestion that either country was able to match what England had achieved under a free constitution.[56] The last

---

[51] J. R. Smith, *The Speckled Monster: Smallpox in England 1670–1970, with Particular Reference to Essex* (Chelmsford, 1987), pp. 55, 121; Mokyr, *Enlightened Economy*, pp. 291–2.

[52] James C. Riley, *The Eighteenth Century Campaign to Avoid Disease* (Basingstoke, 1987), pp. 115–38 and *passim*.

[53] Alan Macfarlane, *The Savage Wars of Peace: England, Japan and the Malthusian Trap* (Oxford, 1997), p. 109; Landers, *Death and the Metropolis*, pp. 238–41.

[54] Dobson, *Contours of Death and Disease*, pp. 110–12, 343, 521–3.

[55] John Broad, 'Cattle Plague in Eighteenth-Century England', *Agricultural History Review*, 31 (1983), pp. 104–15; Slack, *Reformation to Improvement*, pp. 145–6. For greater government support later in the eighteenth century, see John V. Pickstone, 'Dearth, Dirt and Fever Epidemics: Rewriting the History of British "Public Health", 1780–1850', in Terence Ranger and Paul Slack, eds, *Epidemics and Ideas* (Cambridge, 1992), pp. 125–48.

[56] Trenchard and Gordon, *Cato's Letters*, ii. 281–2 (1722); *The Autobiography of William Stout of Lancaster 1665–1752*, ed. J. D. Marshall (Manchester, 1967), pp. 192–3; Joshua Gee, *The Trade and*

thing English improvers wanted or needed was an absolute monarch. Improvement could blossom and flourish fruitfully on its own.

The ways in which it flourished later in the eighteenth century lie outside the chronological limits of this book, but they merit some brief discussion because they show how much was owed to improvement's inventors. Political arithmetic, for example, gained a new lease of life in the middle of the century, when governments again faced major wars and there was public interest in the measurement of national resources and the collection of relevant data, some of it generated by the office of the Inspector General of Imports and Exports, a post first filled by Davenant in 1703.[57] When Andrew Hooke and Joseph Massie tried to measure national income and population in the 1750s, they looked back to Petty, Davenant, and King and repeated their calculations. Massie produced lower figures than Hooke, and probably much more accurate ones. His updated version of King's social analysis showed that the national income had risen to £61 million a year, up by a third since 1688, but the population (6 million) had scarcely risen at all.[58] Since his population estimate was no more firmly based than those of Petty and King, Massie was scarcely fair when he condemned them for preferring 'hypothesis before useful facts', but he had a point when he observed, albeit with the benefit of hindsight, that England had entered a period of demographic stagnation. Petty and King had both 'taken it for granted that the people of England did and would increase, without inquiring whether the causes of increase still existed, or even regarding the great depopulation that there was in their own times'.[59]

The irony is that in Massie's own time the population was starting to increase once more, and that no one knew it because there was still no national census, despite the obvious need and pressure for it. Thomas Potter's proposed bill, which had some ministerial backing, narrowly failed to get through parliament in 1753, chiefly because its provisions were too ambitious and complicated and too easily opposed as threats to English liberties. It was not a proposal for a single national census, but for the annual registration of the number of people, births, marriages, deaths, and paupers in every parish, supervised by the Board of Trade. Petty could have asked for little more. Even in the new climate of opinion which produced a radical revision of the calendar in 1752, and Hardwicke's Act against clandestine marriages which competed with Potter's bill for parliamentary time

---

Navigation of Great-Britain Considered (1729), Preface; Davenant, *PCW*, i. 7; *Nehemiah Grew and England's Economic Development*, ed. Julian Hoppit (British Academy, Records of Social and Economic History NS 47, Oxford, 2012), p. xxvi.

[57] Innes, *Inferior Politics*, pp. 133–41; Julian Hoppit, 'Political Arithmetic in Eighteenth-Century England', *EcHR*, 49 (1996), pp. 519–33. For political arithmetic later in the century, see S. J. Thompson, 'The First Income Tax, Political Arithmetic, and the Measurement of Economic Growth', *EcHR* 66 (2013), pp. 873–94.

[58] Andrew Hooke, *An Essay on the National Debt and National Capital* (1750), pp. 9–10, 26–8; Joseph Massie, *Calculations of Taxes for a Family of each Rank*... (1756), p. 7; Peter Mathias, *The Transformation of England: Essays in the Economic and Social History of England in the Eighteenth Century* (1979), pp. 178, 189; Innes, *Inferior Politics*, p. 134; Hoppit, 'Political Arithmetic', p. 520.

[59] Massie, *A Plan for the Establishment of Charity-Houses* (1758), pp. 91–7.

in 1753, however, there was a limit to the amount of centralization parliament could stomach.[60]

The search for reliable information, useful facts, about population therefore continued, and interest in its early investigators increased still further. Statistics from bills of mortality extending over the previous century were collected, published, and analysed, there was a new edition of Graunt's *Observations*, and some of Petty's essays were reprinted in 1755 because they were 'very scarce and much sought after'.[61] Earlier in the century both authors had often been referred to in definitions of political arithmetic, but now they were no longer simply plagiarized, or mocked by the likes of Swift and Addison.[62] Their works had become classics to be given the respect they deserved, not as the last word but as texts with lessons for the future. In a biographical dictionary published in 1761–2 Graunt and Petty were acclaimed as pioneers in the progress of useful knowledge. Although his conversion to Catholicism was regrettable, Graunt must be hailed as the 'first founder' of a new 'science', a man comparable to Newton in his genius, and the man who had shown how a government could become 'powerful and the people who live under it happy'. Petty had risen from humble beginnings to become 'a singular instance of an universal practical genius', fertile in invention, and the author of essays and a final testament which pointed to the improvements 'of most use to mankind' which remained to be accomplished.[63] There was work still to do.

Eighteenth-century improvers had already embarked on some of the unfinished business of the pioneers, seeking to alleviate if not eliminate poverty and idleness as well as disease, and trying to calculate not only the growing wealth of the nation but how its benefits were distributed. Massie's elaborate revision of King's social table confirmed what earlier commentators had only been able to speculate about, that there was a continuing redistribution of employment and income away from agriculture into trade and manufacturing. It led him to the same conclusions as theirs. 'Hard-working people' whose labour made 'this nation rich and happy' enjoyed only 'a slender proportion of the national benefits', and they had a right to the 'necessaries and conveniences which belong to their station, not only as industrious and hardworking people, but as Englishmen'.[64] Jacob Vanderlint had

---

[60] Innes, *Inferior Politics*, pp. 138–9; Hoppit, 'Political Arithmetic', pp. 526–7; David V. Glass, *Numbering the People* (Farnborough, 1973), pp. 17–20; S. J. Thompson, 'Census-Taking, Political Economy and State Formation in Britain, c.1790–1840', Ph.D. thesis, University of Cambridge, 2010, pp. 21–8. I am grateful to Dr Thompson for letting me see a copy of his thesis. On calendar reform, see Robert Poole, *Time's Alteration: Calendar Reform in Early Modern England* (1998), pp. 111–20.

[61] Innes, *Inferior Politics*, pp. 134, 139; Thomas Short, *New Observations on City, Town and Country Bills of Mortality* (1750); Thomas Short, *A Comparative History of the Increase and Decrease of Mankind in England, and Several Countries Abroad* (1767).

[62] John Harris, *Lexicon technicum: or, An Universal English Dictionary of Arts and Sciences* (2nd edn, 2 vols, 1704–10), ii *sub* 'Political Arithmetick'; Ephraim Chambers, *Cyclopaedia* (2 vols, 1728), ii *sub* 'Political Arithmetick'; above, p. 213.

[63] *A New and General Biographical Dictionary* (11 vols, 1761–2), vi. 45–50; ix. 291–304.

[64] Joseph Massie, *Reasons Humbly Offered Against laying any farther Tax upon Malt or Beer* (1760), pp. 1, 11; Innes, *Inferior Politics*, p. 134. For similar points earlier, see John Pollexfen, *A Discourse of Trade, Coyn, and Paper Credit* (1697), p. 44; above, p. 204.

used much the same language when he constructed household budgets for labourers and tradesmen in 1735 and used them to argue that there was 'too great an inequality of property'.⁶⁵

What was new in the middle of the eighteenth century was not a fresh agenda for improvement but the energy, exuberance, and cash which enabled new forms of philanthropy, remodelled institutions, and even the most Utopian enterprises to get off the ground. There were successive schemes claiming to be a 'new improved method of relieving the poor' after 1723, each more ambitious than the last. None of them delivered all that was hoped for, but they contributed, along with flexible settlement laws and private philanthropy, to a structure of welfare provision responding more effectively to social and economic need. There were hospitals and infirmaries, the first of them in Westminster in 1720, and some of them were specialized institutions like Thomas Coram's Foundling Hospital erected in 1739, which had been planned in 1722 and looked back to Hartlib's anxieties about the plight of poor children.⁶⁶ In the 1720s Laurence Braddon similarly picked up John Bellers's proposal of 1695 to solve the problem of poverty by means of 'colleges of industry', and expanded it into a scheme for large 'collegiate cities' along the Thames housing in total as many as 600,000 people. Neither of them received the financial backing they needed. A generation later, however, schemes scarcely less wild and visionary, like William Hanbury's projects for colleges, libraries, and new forests, evoked a generous philanthropic response, partly because they were effectively promoted as offering pleasure to the charitably minded. The most successful philanthropic entrepreneur of the century, Jonas Hanway, was adept at advertising charities which were 'most amusing and pleasant' and designed to make Britain 'a kind of new creation'.⁶⁷

As the country's affluence and comforts increased, the attractions of sociability enabled philanthropy to become more expansive and once again universal in its ambitions. In 1751 Malachy Postlethwayt's *Universal Dictionary of Trade and Commerce* had a frontispiece showing Britannia receiving and dispensing the wealth of the world to the benefit of all, along with some lines from John Gay's ode 'to his native country': 'That Benefit is unconfin'd | Diffusing Good among Mankind.'⁶⁸ The benefits and pleasures of useful knowledge were also being diffused, partly by philanthropists, partly by the increasing number of clubs and societies dedicated to it. In 1675, when this new form of sociability was only just beginning, with

---

⁶⁵ Jacob Vanderlint, *Money answers all Things* (1734), pp. 24, 27–8, 75–6, 102–4.
⁶⁶ Slack, *Reformation to Improvement*, pp. 137–44; Joanna Innes, 'The "Mixed Economy of Welfare" in Early Modern England', in Martin Daunton, ed., *Charity, Self-Interest and Welfare in the English Past* (1996), pp. 139–80; Anne Winter and Thijs Lambrecht, 'Migration, Poor Relief and Local Autonomy: Settlement Policies in England and the Low Countries in the Eighteenth Century', *P&P* 218 (Feb. 2013), p. 125.
⁶⁷ J. C. Davis, *Utopia and the Ideal Society* (Cambridge, 1981), pp. 339–50; Sarah Lloyd, *Charity and Poverty in England, c.1680–1820: Wild and Visionary Schemes* (Manchester, 2009), pp. 79–80, 117–18, 130–65.
⁶⁸ Stanley H. Palmer, *Economic Arithmetic: A Guide to the Statistical Sources for English Commerce, Industry, and Finance, 1700–1850* (New York, 1977), p. 1.

meetings in inns and coffee-houses, the author of *The Art of Good Husbandry, or the Improvement of Time* complained that 'these clubs and societies (how civil soever they appear to be)' encouraged idleness and other vices when time would be better spent in 'continual labour'. With civility, however, came the exchange of information and the opportunity, as Houghton observed, to improve all kinds of arts and knowledge.[69]

That was what clubs and societies claimed to be doing as they multiplied through the eighteenth century, from the societies for 'the improvement of medical knowledge' in Edinburgh and for 'improving husbandry and manufactures' in Dublin set up in 1731 to the learned and scientific societies which followed the model of the Lunar Society of Birmingham and were meeting in Manchester, Derby, Leeds, and Newcastle by 1793. They met for the pleasure of agreeable company as much as for instruction or scientific enquiry, but the Bath Agricultural Society had its own experimental farm and the Kent Society for Useful Knowledge was active in trying to control an outbreak of gaol fever in Maidstone, while Chester had a Smallpox Society, founded in 1778.[70] Their historian concludes that they 'contributed to making British society not just sympathetic to economic innovation but increasingly obsessed with it' by giving 'improvement its fashionable status'.[71]

Yet they scarcely altered its fundamental character. In 1715, in what was even then a conventional account of *Arts Improvement*, Richard Neve advertised the benefits which had been brought to 'the whole race of mankind' by 'diffused knowledge', and by 'experiments and observations' of the kind advocated by Bacon and Boyle. Many of them were useful, as in medicine, building, husbandry, gardening, and mechanics, and others 'ludicrous', by which he meant playful and pleasurable—like a method for playing tunes on drinking glasses. All of them had created new products and trades, from the fine glass made in London to the sugar coming from Jamaica to all of Europe, and there was a long list of such 'inventions and noble improvements': 'silk-stockings, Mortlake tapestry, earthenware of Fulham, speaking trumpets, dipping of cloth to keep out the wet, air-pumps, making of lutestring, musical automata . . . by clockwork.'[72]

An equivalent list compiled in 1750 or 1780 would have differed only in its details. For most of the eighteenth century a culture of improvement expressed and established national pride in England's enjoyment of comfort and material progress. But the culture was also designed to add further to them, and we must ask whether inventions and noble improvements had contributed to economic growth or merely been symptoms of it.

---

[69] R.T., *The Art of Good Husbandry, Or, the Improvement of Time* (1675), pp. 2–4; above, p. 174.
[70] Clark, *British Clubs and Societies*, pp. 86, 110, 271–2; Smith, *Speckled Monster*, p. 55. Some cities from the 1760s had Chambers of Commerce to represent business interests based on French models: Robert J. Bennett, *Local Business Voice: The History of Chambers of Commerce in Britain, Ireland and Revolutionary America 1760–2011* (Oxford, 2011), pp. 13–16; below, p. 247.
[71] Clark, *British Clubs and Societies*, pp. 438–9.
[72] Neve, *Arts Improvement*, sigs A2–B2.

A KNOWLEDGE ECONOMY

The connections between culture and economic development have been much debated by historians, and particularly by historians discussing the reasons for Britain's early industrialization. They have looked for the preconditions which made England so precocious, and in recent years have agreed that many of them were already in place and having an impact before 1700. The most obvious of them were economic. By that date England had a comparatively urbanized and high-wage economy, which produced incentives to labour-saving, and to investment in new technologies for the purpose; and it was exploiting the coal resources and opportunities for international trade which ultimately prevented any prospect of a stationary state.[73] Some of the institutional preconditions for economic growth were also already established. Although public confidence in the financial institutions created by the 1688 revolution was much more fragile than has sometimes been suggested, there was considerable legal protection for property rights, and there were generally agreed mechanisms for overriding them when necessary, in the shape of a powerful parliament equipped with statutes for the purpose.[74] Parliaments in the 1650s and again in the 1690s had proved themselves powerful enough to fund armies and navies which gave them the potential to take command of the oceans and prevent other powers, notably France, from doing the same.

A third set of contributing factors, interacting with the other two, are acknowledged to have been cultural ones, although there is disagreement about their relative importance. People and peoples are not always driven to maximize personal and collective profit, and may be inhibited by the cultural influence of assumptions and habits based on religion or custom, as well as by institutional constraints or their economic condition. In the English case, however, much has been made of the role of culture in creating what Joel Mokyr has called a 'knowledge economy'. It would be generally agreed that 'inventiveness' was as important as economic incentives.[75] The propensity to distribute and develop new knowledge and technology, and to cooperate to innovate, gave England the entrepreneurs and skilled workforce, the 'social capital', essential for industrialization.

Some of the social capital was already being created before a culture of improvement was invented, and ought to be regarded as a source of that culture and not its

---

[73] Robert C. Allen, *The British Industrial Revolution in Global Perspective* (Cambridge, 2009), pp. 1–22 and *passim*; E. A. Wrigley, *Energy and the English Industrial Revolution* (Cambridge, 2010), pp. 22–5; Ronald Findlay and Kevin H. O'Rourke, *Power and Plenty: Trade, War, and the World Economy in the Second Millennium* (Princeton, 2007), pp. xxiii, 240, 254, 351–2.

[74] Douglass C. North and Barry R. Weingast, 'Constitutions and Commitment: The Evolution of Institutions Governing Public Choice in Seventeenth-Century England', *Journal of Economic History*, 49 (1989), pp. 803–32; Douglass C. North, *Institutions, Institutional Change and Economic Performance* (Cambridge, 1990), pp. 131–40. On the fragility of financial institutions and their dependence on political management after 1689, see Anne L. Murphy, 'Demanding "Credible Commitment": Public Reactions to the Failures of the Early Financial Revolution', *EcHR* 66 (2013), pp. 178–80; Aaron Graham, 'Auditing Leviathan: Corruption and State Formation in Early Eighteenth-Century Britain', *EHR* 128 (2013), pp. 806–8.

[75] Joel Mokyr, *The Gifts of Athena: Historical Origins of the Knowledge Economy* (Princeton, 2002); Findlay and O'Rourke, *Power and Plenty*, p. xx.

product. High levels of literacy and numeracy, for example, depended on educational facilities which had been established largely for religious purposes, and which were being exploited by parents and children as much for the pleasure and personal fulfilment to be found in reading, and to a lesser extent counting, as for their economic utility.[76] Nonetheless, a large part of the knowledge element in England's knowledge economy was incorporated within Bacon's experimental philosophy and channelled by the conscious improvers he inspired into a collective search for useful innovations and inventions. It is no accident that the skills of the English in improving rather than inventing novelties became proverbial by 1700.[77] In 1693 even the popular advice columns of John Dunton's *Athenian Gazette* were distinguishing between invention and improvement. Already engaged in correspondence about whether women were 'capable of making as great improvement' through study as men, and arguing with an eye to his female readers that they were, the editor raised the question of whether the telescope was improvable. The predictable answer was that its first invention had been vastly more improbable than the several subsequent improvements to it.[78] According to Mandeville in 1720, the British were 'excellent artificers in most handicrafts, but more noted for improvements than invention', and de Saussure commented acidly in 1717 that 'though not inventive', they were 'capable of improving and finishing most admirably what the French and Germans have invented'.[79]

There was nothing very new in England's developing inventions and innovations borrowed from other countries, but that was also a practice which gathered pace after 1650. It was part of the process of international technology-transfer which occurred across Europe between the sixteenth and eighteenth centuries, and which William Cecil had encouraged in the 1570s, not only through patents offered to improvers, but by importing the skills of Dutch and Flemish refugees who brought new technologies with them. As one of their English supporters remarked, they were needed because 'we are not so good devisers as followers of others'. A prominent example was the production of the new draperies, which diversified England's cloth industry and its overseas trade at the end of the sixteenth century, employing surplus labour in town and country and taking English textiles to southern as well as northern Europe. The necessary skills were learnt from Protestant refugees from Flanders who were welcomed into English towns precisely because they offered an alternative to a declining broadcloth industry.[80] English governments and landowners would not have known how to drain the fens without Bradley and Vermuyden, or which new forms of farming to try without reading

---

[76] Allen, *British Industrial Revolution*, pp. 52–4; above, p. 65.   [77] Above, pp. 29, 172–3.
[78] *Athenian Gazette*, 1 Aug. 1693, 4 Nov. 1693. Women's improvement was a controversial issue: *Athenian Gazette*, 8 Dec. 1691; Mary Astell, *The First English Feminist: Reflections upon Marriage and Other Writings*, ed. Bridget Hill (Aldershot, 1986), p. 129.
[79] Nigel Goose, 'Immigrants and English Economic Development in the Sixteenth and Early Seventeenth Centuries', in Goose and Lien Luu, eds, *Immigrants in Tudor and Early Stuart England* (Brighton, 2005), p. 154. Cf. Mandeville, *Fable of the Bees*, ed. Kaye, ii. 299–300.
[80] Above, p. 57; Goose, 'Immigrants', p. 154; Robert C. Allen, 'Why the Industrial Revolution was British: Commerce, Induced Invention, and the Scientific Revolution', *EcHR* 64 (2011), p. 364.

German and French handbooks or going to see practice in the Low Countries for themselves. German mineworkers were employed in the Lake District at the end of the sixteenth century, and at the end of the seventeenth Huguenot refugees brought their skills to watch- and clock-making, textile-printing, and paper-making, in London. Italian glassmakers were responsible for the invention of lead 'flint' glass there in the 1670s, and in 1713 John Lombe stole Italian machinery designs for his Derby silk factory.[81]

By then technology was beginning to be exported as well as imported. It is significant that the first Act of parliament to prevent British artisans emigrating came in 1719, and there was further French complaint soon afterwards that the English were wholly unwilling to acknowledge how much they owed to industrial innovations in countries like France which they had emulated.[82] The British were now claiming what they had as their own, and what they had was the regular practice as well as the idea of technological improvement. The ways in which new skills could be improved, locally as much as internationally, were explained in the 1690s by Sir John Lowther, a Fellow of the Royal Society who took a close interest in his Cumberland estates and especially his growing new port of Whitehaven. He wanted to encourage a local textile industry making 'cottons' and Norwich 'stuffs' in order to employ idle hands there. 'If we can but get an entrance into any manufacture', he wrote, 'I doubt not but it will prosper and enrich all that side of the country.' To get it started, someone would have to identify suitable employees, spotting talent like a schoolmaster watching 'schoolboys with the copybooks'; and a 'master workman', 'a southern artist' imported from somewhere like Norwich, would no doubt be required to bring the work to 'perfection' and 'profit'.[83]

That was how the transfer of skills worked in much more specialized industries like paper-making, which the Board of Trade in its 1697 Report thought could be improved to 'make as good as what comes from abroad'. It was also the process which Plattes and Blith had predicted would bring plenty if new inventions and improvements were applied in practice. 'This very nation may be made the paradise of the world', Blith asserted with reference to improvements in husbandry, 'if we can but bring ingenuity into fashion.'[84] It was already sufficiently in fashion in England by 1650 for Defoe to suppose that Cromwell's soldiers had taken agrarian

---

[81] Joan Thirsk, 'Making a Fresh Start: Sixteenth-Century Agriculture and the Classical Inspiration', in Michael Leslie and Timothy Raylor, *Culture and Cultivation in Early Modern England: Writing and the Land* (Leicester, 1992), pp. 15–34; Mokyr, *Enlightened Economy*, p. 106; Christine MacLeod, 'The European Origins of British Technological Predominance', in Leandro Prados de la Escosura, ed., *Exceptionalism and Industrialisation: Britain and its European Rivals, 1688–1815* (Cambridge, 2004), p. 113.

[82] MacLeod, 'The European Origins of British Technological Predominance', p. 114; Antoine Augustin Bruzen de la Martinière, *Le Grand Dictionnaire géographique et critique* (The Hague, Amsterdam, Rotterdam, 9 vols, 1726–39), i. 400.

[83] *The Correspondence of Sir John Lowther of Whitehaven 1693–1698*, ed. D. R. Hainsworth (British Academy, Records of Social and Economic History NS 7, Oxford, 1983), pp. 416–17, 606, 661–2. The project seems to have taken off, but was brought to an abrupt end when the import of Irish wool on which it depended was banned by the Irish woollens Act of 1699.

[84] *SCED*, pp. 576–7; Christine MacLeod, *Inventing the Industrial Revolution* (Cambridge, 1988), pp. 20–11; above, pp. 99–100.

innovation to Scotland, and Josiah Child to assume the same 'industrious' soldiers had begun 'the vast improvement of Ireland'.[85] When trying to ascertain the contribution which a culture of ingenuity and improvement made to economic growth, therefore, it makes sense to begin where practical improvement began, on the land.

Over the two centuries between 1560 and 1760 English agriculture became more efficient and more productive. At the end of that period three-quarters of England was enclosed when less than half had been at the beginning.[86] Drainage of marshes and fens, including 95,000 acres in the Bedford Level alone, had extended the arable acreage by perhaps 10 per cent.[87] Farm output doubled over the two centuries, and labour productivity in agriculture almost doubled also, which allowed towns to grow even when (after 1650) the total population ceased to do so. In 1600 farm output per worker was roughly the same in England as in France; by 1750 it was twice as big, and output per acre was 50 per cent higher than it was in France.[88] The gains had come less from expansion of the cultivated area than from changes in the way it was used, with a shift from pasture to arable in the later sixteenth century and back again in the later seventeenth, so that the number of people working on the land had first increased and then declined. By 1760 the most obvious features of the agrarian landscape were no longer open fields but enclosed pasture and meadow populated by livestock.[89] In the early eighteenth century English agriculture was feeding a population double that of the early sixteenth, without any great increase in the number of workers on the land.

The various improvements responsible for that outcome had changed over time, from engrossing and enclosing in the sixteenth century to the introduction in the seventeenth of new practices and techniques like convertible husbandry, the floating of water meadows, and the cultivation of crops like clover and turnips, which were grown for fodder after 1650 but had the unintended consequence of improving yields.[90] By the early eighteenth century their cumulative effect astounded foreign observers. Reports to the French Foreign Office in the 1730s described the 'high degree of perfection' reached by English agriculture in terms worthy of Davenant: the land was better cultivated than in other countries, livestock were more numerous and farm buildings better maintained, and England was exporting a large surplus of grain. Agriculture had become for England what the silver mines

---

[85] Above, p. 97; Josiah Child, *A Discourse about Trade* (1690), p. 13 (written in the later 1660s).

[86] James Simpson, 'European Farmers and the British "Agricultural Revolution"', in Prados de la Escosura, *Exceptionalism and Industrialisation*, p. 81; Mark Overton, *Agricultural Revolution in England* (Cambridge, 1996), pp. 147–8.

[87] Overton, *Agricultural Revolution*, pp. 89–90; Frances Willmoth, *Sir Jonas Moore* (Woodbridge, 1993), p. 91.

[88] Overton, *Agricultural Revolution*, pp. 75, 88; Allen, *British Industrial Revolution*, pp. 59–60, 66; Mokyr, *Enlightened Economy*, p. 172.

[89] Overton, *Agricultural Revolution*, pp. 88, 93, 110–13.

[90] Simpson, 'European Farmers', p. 72; Overton, *Agricultural Revolution*, pp. 99–100, 110–21; Joan Thirsk, *Alternative Agriculture: A History from the Black Death to the Present Day* (Oxford, 1997), pp. 72–4.

of Peru had once been for Spain, Voltaire famously remarked in 1756, the source of its wealth and a much more lasting one.[91]

It could not have become that if there had not been institutional support. It might come from the Crown, exercising prerogative power, or more often from parliament, in legislation from the 'Statute of Approvement' of 1549 to statutes protecting investment in the fens in the seventeenth century and authorizing enclosure in the eighteenth; and the law courts provided formal and agreed mechanisms for adjudicating when property rights conflicted or tenants disputed the actions of landlords.[92] Economic incentives were equally important, and decisive in determining the kinds of improvement undertaken at different points in time. They account most obviously for the lull in enclosing activity from the 1670s, when grain prices were low and profit to be found in alternative forms of husbandry. Corn bounties after 1688 helped to keep grain prices from falling as fast in England as they did in other countries in similar circumstances, but the more powerful incentive to increasing productivity came now from growing urban demand and the market integration it stimulated.[93] None of this could have been achieved, finally, without available capital and the propensity of landlords and landowners to invest in schemes where returns might be delayed and sometimes never happen at all. The scrupulously kept estate records of families like the Lowthers in the north-west and the Le Stranges in the opposite corner of England, at Hunstanton on the Wash, show them consciously taking risks and sometimes losing money, as Sir Hamon and Sir Nicholas Le Strange did on their investments in fen drainage in the mid-seventeenth century.[94]

The ways in which contemporaries thought institutions and economics might inhibit improvement and hinder its culture are nicely illustrated by comments from a Scottish improver in 1697 on why innovations were 'so universally' condemned there. According to James Donaldson, landlords and tenants lacked any 'skill to calculate or forecast' the profit to be made from innovation, 'most people thinking it better to take a scant crop of corn than leave their land grass' and enjoy the benefits of crop-rotation, for example. The tenant took a short-term view because he had less security of tenure than his English equivalent, and if he made 'any improvement of his ground' he was faced with a higher rent or forced out. Hence the Scottish proverb 'Bouch [i.e. botch] and sit; improve and flit'. Landlords were no better, in their case because of their poverty, 'that great enemy to virtue'.

---

[91] François Crouzet, 'The Sources of England's Wealth: Some French Views in the Eighteenth Century', in P. L. Cottrell and Derek H. Aldcroft, eds, *Shipping, Trade and Commerce* (Leicester, 1981), p. 67.

[92] Richard W. Hoyle, 'Introduction: Custom, Improvement and Anti-Improvement', in Richard W. Hoyle, ed., *Custom, Improvement and the Landscape in Early Modern Britain* (Farnham, 2011), pp. 12–14, 37; Paul Warde, 'The Idea of Improvement c.1520–1700', in Hoyle, ed., *Custom, Improvement and the Landscape*, p. 132; Bill Shannon, 'Approvement and Improvement in the Lowland Wastes of Early Modern Lancashire', in Hoyle, ed., *Custom, Improvement and the Landscape*, pp. 176–8.

[93] Hoyle, 'Introduction', p. 25; Simpson, 'European Farmers', pp. 72–3, 81–2.

[94] Jane Whittle and Elizabeth Griffiths, *Consumption and Gender in the Early Seventeenth-Century Household: The World of Alice Le Strange* (Oxford, 2012), p. 208.

Without spare cash to invest, they were 'glad to accept anything that first offers rather than wait for future great things'.[95] Donaldson's argument was exaggerated, and itself a product of an improvement culture whose diffusion accelerated after the Union, but eighteenth-century Scotland unlike England still needed public funding to help make improvement a reality.[96]

In England, however, where there was relative security and affluence for both landlords and many tenants, improvement had become habitual. In the 1630s Sir Hamon Le Strange could afford to build up a library with all kinds of improving manuals, including books on mathematics and geometry, along with two somewhat dated handbooks on husbandry, translations from Latin and French of works by Conrad von Heresbach and Claude Estienne. Like some of their neighbours, such as the Bacons of Stiffkey, relatives of Sir Francis, the Le Stranges learnt about fashionable innovations as much from talking to one another as from reading about them, and they looked down on other local families who did neither. They also took evident pride and pleasure in planning and paying for improvement. Even when drainage of their marsh lands produced a financial loss, it improved opportunities for shooting, hawking, and fishing, and left a visibly well-managed landscape.[97] The Lowthers thought much the same. After the civil wars, Sir John Lowther, uncle of the industrial entrepreneur already cited, confessed that he 'naturally loved the improvement and meliorating of grounds, holding it no less good husbandry to improve the wastes and barren grounds as to purchase new'; and although he 'never found much profit' in growing corn, 'it was a pleasure to see things mended, the poor set on work and the ground thereby amended'.[98]

The persistence of investors and projectors in the great enterprise of draining the fens tells a similar story about the power of a fashionable improving frame of mind. What was being achieved was worth all the early failures and the problems caused later when the level of the fenland fell.[99] According to a versifier in 1685, it was something of greater importance than the discoveries in America—not some 'late discovered Isle, nor old Plantation new christened, but a kind of New Creation'. The vogue for improvement fortified contempt for those who opposed it. The

---

[95] James Donaldson, *Husbandry Anatomized, or An Enquiry into the Present Manner of teiling and manuring the ground in Scotland* (Edinburgh, 1697), pp. 124–5. A similar proverb was current in England in the early seventeenth century: 'Botch and sit, build and flit': Morris Palmer Tilley, *A Dictionary of the Proverbs in England in the Sixteenth and Seventeenth Centuries* (Ann Arbor, 1950), p. 60. It would be interesting to know whether it was falling out of use south of the border by 1700.

[96] Above, p. 190. For similar comments later, see Daniel Defoe, *A Tour Through the Whole Island of Great Britain*, ed. G. D. H. Cole and D. C. Browning (2 vols, Everyman, 1962), ii. 280; Hoyle, 'Introduction', pp. 34–5. On contrasts between English and Scottish tenants, see Rab Houston, 'Custom in Context: Medieval and Early Modern Scotland and England', *P&P* 211 (May 2011), pp. 45–51.

[97] Elizabeth Griffiths, '"A Country Life": Sir Hamon Le Strange of Hunstanton in Norfolk, 1583–1654', in Hoyle, ed., *Custom, Improvement and Landscape*, pp. 211–15, 236; Whittle and Griffiths, *Consumption and Gender*, p. 208.

[98] Hoyle, 'Introduction', p. 28.

[99] For similar persistence in draining marshes in Lancashire, see *The Account Book of Richard Latham 1724–1767*, ed. Lorna Weatherill (British Academy, Records of Social and Economic History NS 15, Oxford, 1990), p. xviii.

fenlanders, according to Dugdale, were 'a rude and almost barbarous sort of lazy and beggarly people', just like the rogues Norden had found in forests and on wastes which ought to be improved. The Yorkshire landlord in the 1720s who thought tenants resistant to 'experiments' were 'old cart horses—one can't thrust 'em out of their beaten tracks', was repeating what Aubrey and Hartlib had said half a century earlier.[100]

Those who lost out when improvers moved in had their own ways of life and habits of husbandry to protect, and they had a serious case to make, which they had defended in the law courts as well as by riots in open-field counties and the fens in the seventeenth century. They soon appreciated, however, that defence of custom was less effective than the defence of property rights. It is significant that no counter-culture to the culture of agrarian improvement ever developed, or at least not one coherent enough to have lasting momentum.[101] An economic case has been made that yeomen on open fields were farming just as efficiently as landowners with large estates, but that put them on the improving side. When the economic incentives were right, as they were around 1700, copyholders keen to maintain or improve their standards of living were ready to introduce new crops like turnips and clover by common agreement, a process of collective innovation as productive as that of landowning families emulating one another elsewhere.[102]

There was also what might be called an institutional case against agrarian improvement when it seemed to threaten the balanced distribution of property which Harrington and classical republicans had thought essential to liberty. In the later eighteenth century, when the distribution of wealth was of general concern and large estates seemed to be growing ever larger, the argument was wielded again, and critics of parliamentary enclosure now cited Francis Bacon as well as Harrington in their support.[103] They had misread their Bacon and were behind the times. In his *History of Henry VII* Bacon had praised the first legislation against engrossing and the decay of tillage in 1488–9 because it preserved the yeomanry, small farmers, and freeholders, and hence the military forces of the kingdom; but he was not against all enclosures in principle since that, he said, would be 'to forbid the improvement of the paternity of the kingdom', that is to say its natural resources.[104] In any case, it was wishful thinking in the later eighteenth century to suppose that the nation's military potential still depended on sturdy yeomen who were now improving land themselves. Whatever people thought about further enclosure, no one wanted to forbid agrarian improvement.

---

[100] Willmoth, *Jonas Moore*, pp. 103–4; Hoyle, 'Introduction', p. 34; above, p. 108.
[101] Hoyle, 'Introduction', pp. 36–8. Even the Diggers were not an exception: above, p. 94.
[102] Allen, *British Industrial Revolution*, pp. 68, 77; Robert C. Allen, *Enclosure and the Yeoman: The Agricultural Development of the South Midlands 1450–1850* (Oxford, 1992), pp. 200–7.
[103] Allen, *Enclosure and the Yeoman*, pp. 303–7; S. J. Thompson, 'Parliamentary Enclosure, Property, Population, and the Decline of Classical Republicanism in Eighteenth-Century Britain', *Historical Journal*, 51 (2008), pp. 621–42.
[104] Francis Bacon, *The Historie of the raigne of King Henry the seventh*, ed. Michael Kiernan (*OFB* viii, Oxford, 2012), pp. 54–5; Warde, 'Idea of Improvement', p. 137; Joan Thirsk, ed., *The Agrarian History of England and Wales*, iv: *1500–1640* (Cambridge, 1967), p. 214.

By then England's manufactures and industry had changed just as much as its agriculture since the sixteenth century, and a Baconian culture had contributed to their improvement in similar ways. There were connections between the two economic sectors. A family like the Lowthers was as interested in coal, textiles, and shipping as in husbandry, and the technologies involved in draining the fens, improving river navigation, bringing fresh water to towns, and draining mines were closely related. The timing of change in the two sectors was different, however. While the formative periods for practical agrarian improvement came with enclosure in the later sixteenth century and the alternative agriculture of the later seventeenth, the deliberate and concerted application of new technology outside agriculture began at the very end of the seventeenth century and flourished with industrialization in the later eighteenth. The annual number of patents for new inventions predictably rose rapidly after 1760, but there had been a temporary boom in the 1690s, when several of them were for waterworks of one kind or another, and when many writers were beginning to advertise the economic utility of public science.[105]

The science and technology advocated by the Hartlib group took longer to have practical effect than its improved husbandry. In 1666 Robert Hooke noted that gains in scientific understanding had so far 'ended only in some small inconsiderable product hardly worth naming' because they had not been 'united, improved, or regulated by art'; crafts and skills still needed to be found and developed in order to 'conquer the difficulties of natural knowledge'.[106] That required sustained application and not the eclectic interests of someone like Petty, who put his energies into the eye-catching double-bottom boat in whose further development no one was willing to invest, or Charles Povey, the coal merchant, journalist, and manager of an insurance company, who found time to invent an improved coal hoist, fire engine, and even a self-playing organ.[107] By 1700, however, England had more than a handful of experienced engineers, and at least one of national renown in the shape of the 'ingenious' George Sorocold, who designed not only river improvements and waterworks, but dockyards and factories like the one at Derby later taken over by Lombe for the production of silk.[108]

In one of the letters published in 1734, after his visit to England, Voltaire praised the revolution in science which Bacon had inspired, but judged 'useful inventions' in technology to be distinctly different. "Tis to a mechanical instinct, which is found in many men, and not to true philosophy, that most arts owe their origin.'[109] That might have been true about their origin, but not, as Voltaire no doubt knew, about their fruitful development. Eighteenth-century advances in the use of water power depended on the skills of men like Sorocold, for example, but also on knowledge of hydraulics and mathematics. They brought some hard science into

---

[105] MacLeod, *Inventing the Industrial Revolution*, pp. 81, 146, 150.
[106] Mokyr, *Gifts of Athena*, p. vii.   [107] Above, p. 116; *ODNB sub* Povey.
[108] *ODNB sub* Sorocold; above, p. 219. For his reputation, see John Houghton, *A Collection for the Improvement of Husbandry and Trade*, ed. Richard Bradley (4 vols, 1727–8), i. 106–7; Thomas Savery, *The Miners Friend* (1702), p. 31; Ralph Thoresby, *Ducatus Leodiensis* (1715), p. 80.
[109] Voltaire, *Letters Concerning the English Nation* (Dublin, 1733), p. 77.

contemporary technology, even if English disciples of Bacon were inclined to magnify its role. In 1736 a river improvement proposed for Chester was said to rest on 'the knowledge of natural philosophy and mathematics in all its branches'. When looking for examples of the utility of Newtonian science in 1701, John Arbuthnot included hydraulic machines for raising water among those obviously of 'great use and comfort'.[110]

There are two other examples of the scientific revolution of the seventeenth century feeding into the industrial revolution of the eighteenth, and perhaps the only two which led to technologies of general application that proved to be genuinely transformative. One was steam power, with its origins in investigations of atmospheric pressure from Galileo to Hooke. The other was 'clock-work' or gearing, which revolutionized the design of machinery, and sprang from the need for precise time-keepers in order to achieve what Defoe thought an unachievable goal, the precise measurement of longitude at sea.[111] Both of them rested on scientific research undertaken across western Europe. Yet both owed their development and application to English improvement. Better clocks, watches, and air-pumps than those to be found in other countries were among the many 'noble inventions and improvements' which Chamberlayne's handbook on the state of England was able to celebrate in 1700.[112]

A writer in 1661 on the potential of human industry had cited watches as obvious candidates for improvement, and Hooke placed them alongside water pumps and telescopes as things not yet perfected but which soon would be: 'Let us see what the improvement of instruments can produce.' In the course of his competition with Huygens, he was himself to design the first machine to produce cheap and accurate gears, an early example of the standardization of machine parts.[113] As for steam power, the steam pump perfected by Newcomen in 1712 was notorious, as Desaguliers remarked, for having been 'a long time in an improving condition'. Its 'progress and improvement' went back to Savery and Papin, both of them public advocates of improvements of all kinds, and before that to the Marquis of Worcester, whose *Century of Inventions* Savery had read, and whose laboratory and ordnance factory at Vauxhall Hartlib had hoped to turn into a 'College of Invention'.[114]

---

[110] Mark Knights, 'Regulation and Rival Interests in the 1690s', in Perry Gauci, ed., *Regulating the British Economy, 1660–1850* (Farnham, 2011), p. 71; John Arbuthnot, *An Essay on the Usefulness of Mathematical Learning* (Oxford, 1701), p. 32. Cf. Mokyr, *Gifts of Athena*, pp. 46–7.

[111] Robert C. Allen, 'The British Industrial Revolution in Global Perspective', *Proceedings of the British Academy*, 167, *2009 Lectures* (Oxford, 2011), pp. 216–21; Allen, *British Industrial Revolution*, pp. 7, 204–5; above, p. 218.

[112] Edward Chamberlayne, *Angliae notitia* (1700), p. 48. These had not been mentioned in editions of this work in the 1670s and 1680s.

[113] Above, p. 115; Lisa Jardine, *The Curious Life of Robert Hooke: The Man Who Measured London* (2003), p. 47; James A. Bennett et al., *London's Leonardo: The Life and Work of Robert Hooke* (Oxford, 2003), p. 72; Allen, 'Why the Industrial Revolution was British', pp. 375–6.

[114] John Theophilus Desaguliers, *A Course of Experimental Philosophy* (2 vols, 1734, 1763), ii. 465–7; Savery, *Miners Friend*, pp. 3–4; Thomas Savery, *Navigation Improv'd* (1698), Preface; Larry Stewart, *The Rise of Public Science* (Cambridge, 1992), pp. 24–7. On Worcester, see Edward Somerset, Marquis of Worcester, *A Century of the Names and Scantlings of such Inventions, as at present I can call to mind to have tried and perfected* (1663); Charles Webster, *The Great Instauration* (1975), pp. 347–8.

As with agrarian improvement, the development and application of such technologies, and investment in them, owed something to the security delivered by institutions, including the protection of patents. But that did not inhibit competition, and competition was more important than any encouragements to invention and improvement provided by the state. The parliamentary statute giving Worcester monopoly rights to his 'water-commanding engine' did not prevent more effective versions of the same thing, any more than the financial incentives provided by the Board of Longitude, set up in 1714 with the backing of Newtonians like William Whiston, produced quick success in the search for it.[115] In these cases, as in the long history of fen drainage, a culture of improvement helps to explain why the quest for viable solutions survived multiple disappointments.

The diffusion and practice of new technologies as opposed to their mere survival depended, however, on economic conditions prevailing after 1650 and not before then. Two of them were particularly important. One was the stimulus to inventiveness and industry at home which came with overseas plantations and empire. In 1700 England had yet to produce successful substitutes for Indian cottons, but their import boosted a printing and dyeing industry already reaching 'perfection', according to the Board of Trade; and the expansion of sugar production in the West Indies was transforming the English copper industry, where output rose rapidly and required new technologies in metal smelting in order to meet demand for copper cauldrons. Bacon and Worsley would have expected nothing less.[116]

The second new circumstance conducive to technological enterprise was a shortage of labour. When population was no longer rising and labour costs increasing, there were incentives to find and use labour-saving devices; and occupational mobility along with rising real incomes per head provided a workforce able to rise to the challenge. William Lee invented the stocking frame in 1589, but failed to find backers either in London or in Rouen when he took it there in 1612, presumably because labour was plentiful and cheap; his 'engines' only caught on in England after the 1650s.[117] By contrast, at the end of the century England had the first machine for putting heads on nails, called the 'oliver', and the first factory for making pins built by the Dockwra Copper Company in 1692.[118] In such instances, the technologies which contributed to economic growth depended on a population growing richer but not larger.

Yet even here, in the realm of demography, culture may well have contributed to the result. If one reason for a stationary population was high mortality, and another emigration, the third and more decisive one, at least until the 1720s, was low

---

[115] MacLeod, *Inventing the Industrial Revolution*, pp. 81–8; H. W. Dickinson, 'The Steam Engine to 1830', in Charles Singer and Richard Raper, eds, *A History of Technology* (7 vols, Oxford, 1954–78), iv. 170–3; Stewart, *Rise of Public Science*, p. 187; Mokyr, *Enlightened Economy*, p. 135.

[116] Beverly Lemire, *Fashion's Favourite: The Cotton Trade and the Consumer in Britain 1660–1800* (Oxford, 1991), pp. 31–3; Nuala Zahediah, 'Colonies, Copper, and the Market for Inventive Activity in England and Wales, 1680–1730', *EcHR* 66 (2013), pp. 805–25.

[117] *ODNB sub* Lee; Joan Thirsk, *Economic Policy and Projects* (Oxford, 1978), p. 99; *SCED*, p. 267.

[118] Allen, 'Industrial Revolution in Global Perspective', pp. 211–12; Allen, 'Why the Industrial Revolution was British', p. 371.

fertility because of late ages at marriage and large numbers of people never marrying at all. The reasons for changes in marital behaviour have been much debated by historians and were much discussed by contemporaries, as we have seen. Some of them were probably institutional, like the activity of parish officials in preventing the poor from marrying and the growth of an increasingly well-funded system of social welfare which meant that the elderly did not need to remarry at all.[119] There is some empirical evidence for both phenomena, and for the likely impact of changes in occupational structure and in the ways in which better-off fathers provided for their children.[120]

Expectations were also changing, however. When Edmond Halley remarked in 1693 that most people hesitated 'to adventure on the state of marriage from the prospect of the trouble and charge of providing for a family', and Petty referred to the poor being deterred from marriage by the likely cost of maintaining their children, they were pointing to the importance of assumptions about an adequate maintenance; and those assumptions must have risen with rising real incomes.[121] A tract addressed to women in 1695 which argued that even the 'advantages of celibacy' were subject to 'improvement' may carry no more weight than the comments of those who attributed low fertility to degenerate and effeminate luxury, but where we have the opinions of women themselves, it seems clear that some of them had freedom to choose and chose to stay single.[122] There is increasing evidence that people were able to make choices about the size of their families, and to postpone or defer marriage when it was economically inconvenient.[123]

Much remains uncertain and speculative about the reasons for changes in behaviour which raised or depressed fertility, especially given the fact that the numbers marrying and marrying earlier rose after 1730. That suggests that rising incomes did not have the effect of keeping fertility low for very long. When we remember that in 1700 average real incomes per head had still not caught up with their level in 1500, however, a lag of perhaps a generation between incomes beginning to rise in the mid-seventeenth century and people marrying earlier in

---

[119] Above, pp. 63, 136–7, 140, 159–61, 195–9. Cf. Mokyr, *Enlightened Economy*, pp. 283–8.

[120] Craig Muldrew, '"Th'ancient Distaff" and "Whirling Spindle": Measuring the Contribution of Spinning to Household Earnings and the National Economy in England, 1550–1770', *EcHR* 65 (2012), pp. 519–20; Pamela Sharpe, *Population and Society in an East Devon Parish: Reproducing Colyton, 1540–1840* (Exeter, 2002), pp. 164–75; Judith Spicksley, 'Usury Legislation, Cash, and Credit: The Development of the Female Investor in the Late Tudor and Stuart Periods', *EcHR* 61 (2008), pp. 277–301.

[121] Above, p. 160; Edmond Halley, 'Some Further Considerations on the Breslaw Bills of Mortality', *Philosophical Transactions of the Royal Society*, 17.198 (1693), p. 655.

[122] Edward Stephens, *A Letter to a Lady, Concerning the due Improvement of her Advantages of Celibacie, Portion, and Maturing of Age and Judgment* (n.p., ?1695), p. 8; Sara Mendelson and Patricia Crawford, *Women in Early Modern England* (Oxford, 1998), pp. 165–74. Stephens published for a 'Religious Society of Single Women' striving for spiritual improvement. For similar groups see Astell, *The First English Feminist*, ed. Hill, pp. 24–8. On the dangers of effeminacy, see above, pp. 38, 150, 170, 202, 205, 209, 211.

[123] Emma Griffin, 'A Conundrum Resolved? Rethinking Courtship, Marriage and Population Growth in Eighteenth-Century England', *P&P* 215 (May 2012), pp. 142–5; Andrew Hinde, *England's Population: A History since the Domesday Survey* (2003), pp. 146–8.

consequence seems wholly understandable.[124] It may have been old rather than new expectations which depressed fertility, but in either case it is probable that for a time people were making marriage choices in order to maximize family incomes, and that for half a century many of them were doing as Defoe advised, not taking on too soon 'the necessary charge of a wife and a family' and so being able eventually to buy the consumer goods they made.[125] If that was so, then a culture of improvement which encouraged consumerism and higher standards of living contributed significantly to economic growth. It postponed demographic recovery for long enough to raise incomes per head, which in turn raised economic expectations. Economy and culture evolved together, each reinforcing the other.[126]

Fortunately perhaps, a persuasive argument for the 'positive-feedback' effect of an improvement culture on economic growth well before the industrial revolution does not depend upon its potential demographic impact. It rests on the ways in which a pervasive attachment to improvement encouraged the accumulation of skills and appetites necessary for innovations and investment in them. By the beginning of the eighteenth century, it reinforced aspects of English and British culture to which historians of economic development have drawn attention: the ways in which 'useful knowledge increased by feeding on itself' and came 'wrapped in an ideology that encouraged material prosperity', for example, the emergence of 'the new idea of the economy as a separate thing', and 'the frenetic pursuit of income to buy novel consumer goods' which was a necessary condition for 'economic progress'.[127] All these were part and parcel of improvement, and it had two other elements, essential components, which served to mitigate what might have been negative feedback effects of economic and cultural change.

Improvement's two central claims were that it delivered benefits both to the nation and to every citizen, and that economic progress would continue, was indeed potentially infinite. Neither proposition could have been persuasively sustained if there had not been the per capita economic growth evident between 1650 and 1750. They were culturally important, however, because they gave improvement a moral dimension and a momentum sufficiently powerful to counteract arguments, not only about the unequal distribution of wealth or taxation or property rights, but about luxury and the immorality of private profit. The repeated assertion, against much of the evidence, that the interests of land and trade, landowners and merchants, were always identical is one example of that.[128] Another is the continued insistence on the need for restraint and moderation in the pursuit of profit and ease, and the avoidance of 'excess', because these virtues

---

[124] Above, pp. 159–60. On the time lag of 15–20 years between turning points in real wages and marriage behaviour, see E. A. Wrigley and R. S. Schofield, *The Population History of England 1541–1871: A Reconstruction* (2nd edn, Cambridge, 1989), pp. xxiv, 471.
[125] Defoe, *The Complete English Tradesman* (1726), pp. 159–60, 164, 386–7.
[126] Allen, *British Industrial Revolution*, p. 11.
[127] Mokyr, *Gifts of Athena*, p. 33; Margaret C. Jacob, *The Cultural Meaning of the Scientific Revolution* (Philadelphia, 1988), p. 4; Deirdre N. McCloskey, *Bourgeois Dignity* (Chicago, 2010), p. 349; Allen, *British Industrial Revolution*, p. 13.
[128] Above, pp. 139, 182–3. Cf. McCloskey, *Bourgeois Dignity*, pp. 348–9.

were essential to private self-esteem and collective enterprise alike.[129] Moreover, arguments of this kind helped to overcome what some historians of economic development term the 'free-rider' problem. They persuaded people to cooperate and engage in collective action for the general good even when it ran against the individual self-interest of many of them. Improvement provided a powerful 'morality of cooperation'.[130] It was patriotic, part of the national character and national purpose.

'The Englishman is never satisfied with what he has obtained,' one Frenchman remarked in the middle of the century. 'His mind gets bored when in rest' and 'the desire to increase always his property by continuous speculations destroys in him the love of tranquillity' which (the author assumed) usually inclined 'all well-to-do men towards idleness'. Hume had said much the same about the 'spirit of the age' in any commercial society, where 'the minds of men, being once roused from their lethargy and put into a fermentation... carry improvement into every art and science', but it was a frame of mind especially potent in England.[131] It explains why improvers persisted for so long when they often failed, and when the environment seemed wholly against them, as in the fens and in the face of epidemic disease, and why most of them continued to think that the growth of commerce must soon produce a growth of population.[132] All would come right in the end.

By the 1750s many things were coming right, even the population, although population growth brought problems of its own by postponing further increases in average real wages. Improvement had not delivered all it promised, but it had done enough to allow writers of the Scottish school like Hume to take material progress for granted as a mark of the superiority of modern commercial societies above any that preceded them, and to force other countries to look to Britain for lessons in how comfort and progress could be achieved. Economic advance was only one of the purposes of the European Enlightenment of the later eighteenth century, but it was a necessary condition for all the others, and Britain already had what Mokyr terms 'an enlightened economy' which accounted for its 'precocious modernity'.[133] To say that Britain was first with something, however, is not necessarily to say that there were not other paths that could be taken towards a similar result, in this case comfort and material progress.[134] If we look at other countries and the economies and cultures which evolved from their different histories, institutions, and resource

---

[129] Above, pp. 210–11, 221.
[130] North, *Institutions*, pp. 42, 90, 132–3; Patrick K. O'Brien, 'The Nature and Historical Evolution of an Exceptional Fiscal State and its Possible Significance for the Precocious Commercialization and Industrialization of the British Economy from Cromwell to Nelson', *EcHR* 64 (2011), p. 410. I owe this point to discussion with Steve Pincus.
[131] Crouzet, 'Sources of England's Wealth', p. 68; David Hume, *Political Discourses* (Edinburgh, 1752), pp. 26–7.
[132] Joel Mokyr, 'Knowledge, Enlightenment and the Industrial Revolution: Reflections on *The Gifts of Athena*', *History of Science*, 45 (2007), p. 188.
[133] Mokyr, *Enlightened Economy*, pp. 31–2.
[134] The point is made with respect to British industrialization by Peter Mathias: *Transformation of England*, p. 14.

endowments, we may see more clearly what made England precocious and what, if anything, was distinctive about its improvement culture.

THE EUROPEAN CONTEXT

England was part of a common European culture, and for most of the period covered by this book it was not at the centre of it. The culture was fashioned and held together by books and translations, and by personal correspondence and temporary or permanent migration. It was important for England's future share in it that Hartlib had come and stayed there, and that he and the Royal Society had an international network of correspondents and attracted visitors like German Cameralist writers to London in the 1660s.[135] At the beginning of the seventeenth century Montchrétien spent a few years in England before writing the first work explicitly about political economy, but his visit had no noticeable effect on his book.[136] Until the very end of the century, Malynes and Mun were the only English authors on trade and commerce to be widely read elsewhere.[137] Holland, not England, was the great commercial success story, attracting visitors, as it attracted Petty, Worsley, and Barbon, and the information exchange of Europe in the later seventeenth century.

In the later sixteenth century, the books which governed attitudes towards political economy came from Italy and France. The works of Bodin and Botero on reason of state, and on how economic resources should be harnessed to make states powerful, were read and translated across the Continent, in Germany and Spain as well as England.[138] Most European countries consequently had their 'Machiavellian moment' which impelled them to think in new ways about how their economies functioned.[139] Later on, though at different times in different places, they all had what might be called an Epicurean moment, when Nicole and the French Augustinians directly or indirectly forced them to confront the realities

---

[135] Pamela H. Smith, *The Business of Alchemy: Science and Culture in the Holy Roman Empire* (Princeton, 1994), p. 38; Richard Bonney, 'Early Modern Theories of State Finance', in Bonney, ed., *Economic Systems and State Finance* (Oxford, 1995), pp. 184–5.

[136] Jean-Yves Grenier, *Histoire de la pensée économique et politique de la France d'ancien régime* (Paris, 2007), pp. 109–11.

[137] Kenneth H. Carpenter, *Dialogue in Political Economy: Translations from and into German in the 18th Century* (Baker Library, Boston, 1977), p. 7; Terence W. Hutchison, *Before Adam Smith: The Emergence of Political Economy, 1662–1776* (Oxford, 1988), pp. 90, 95–6; Pieter de la Court, *The True Interest and Political Maxims of the Republick of Holland and West-Friesland* (1702), pp. 40–1; Jacob Soll, *The Information Master: Jean-Baptiste Colbert's Secret State Intelligence System* (Ann Arbor, 2009), p. 116.

[138] Cf. above, p. 45.

[139] I borrow the term from John Pocock's influential analysis of classical republican thought (J. G. A. Pocock, *The Machiavellian Moment* (Princeton, 1975)); but I have given it a more narrowly economic focus here, influenced partly by Steve Pincus, 'Neither Machiavellian Moment nor Possessive Individualism: Commercial Society and the Defenders of the English Commonwealth', *American Historical Review*, 103 (1998), pp. 706–7; and Steve Pincus, 'From Holy Cause to Economic Interest', in Alan Houston and Steve Pincus, eds, *A Nation Transformed: England after the Restoration* (Cambridge, 2001), pp. 274–8.

of private interests and appetites; and later still Mandeville made all of them grapple with the issue of luxury.[140] They likewise faced similar economic problems which gave these intellectual challenges immediate relevance: inflation and population growth before 1650, the absence of them after that date, and the rising costs of war, including wars for global trade. They had a shared culture and shared predicaments.

That did not mean that there were shared outcomes, and least of all a common view of improvement, since different countries started with different economies, institutions, and histories. France was three times the size of England in 1700, with a population four times as big, a much larger agrarian sector similar to England's only in the north, and a monarchy aspiring to the absolute power which it needed to defend its frontiers and control privileged corporations and social elites. With its poor soils and a smaller but dense population wholly dependent on commerce, the Dutch Republic was at another extreme, a decentralized economy based on the fragile cooperation between towns and provinces which had brought it into existence. Germany and the Habsburg dominions had an even larger array of institutions, which needed to cooperate in managing mines, local industries and trades, and vast forests, so that appeals to the 'common weal'—*gemeine Wesen*—were essential to overcome formidable obstacles to collective innovation.[141]

Spain was different again. When the silver from exploitation of its American empire began to run out at the beginning of the seventeenth century, its intellectuals, projectors, and politicians were among the first to turn to Botero and Bodin for remedies for incipient economic decline; but they found their prescriptions for demographic and commercial growth—including a census, banks, and trading companies—hampered at every turn by an aristocratic culture resistant to 'novelties' of any kind. Their successors in the early eighteenth century were writing comprehensive economic tracts recommending agrarian, industrial, and mercantile politics like those adopted in France and England, but they were similarly frustrated by the political constraints of what has been called a 'polycentric state'.[142]

In comparison England had been endowed by a benevolent providence with most of what was necessary for economic growth in a competitive commercial world. Quite apart from accidents of geography, like an island location astride Atlantic trade routes and plentiful and accessible coal deposits, it had for centuries

---

[140] John Robertson, *The Case for the Enlightenment: Scotland and Naples 1680–1760* (Cambridge, 2005), pp. 8–9, 287 and *passim*.

[141] Paul Warde, *Ecology, Economy and State Formation in Early Modern Germany* (Cambridge, 2005), pp. 175–83; Keith Tribe, *Governing Economy: The Reformation of German Economic Discourse 1750–1840* (Cambridge, 1988), p. 52; Andre Wakefield, *The Disordered Police State: German Cameralism as Science and Practice* (Chicago, 2009), pp. 17–18; Mack Walker, *German Home Towns: Community, State and General Estate, 1648–1871* (Ithaca, NY, 1971), pp. 145–50. On corporate restraints on growth, see Sheilagh Ogilvie, 'Consumption, Social Capital, and the "Industrious Revolution" in Early Modern Germany', *Journal of Economic History*, 70 (2010), pp. 287–325; Joachim Whaley, *Germany and the Holy Roman Empire* (2 vols, Oxford, 2012), ii. 286; Parker, *Global Crisis*, pp. 638–9.

[142] J. H. Elliott, 'Self-Perception and Decline in Early Seventeenth-Century Spain', *P&P* 74 (Feb. 1977), pp. 43–57; Regina Grafe, 'Polycentric States: The Spanish Reigns and the "Failure" of Mercantilism', in Philip J. Stern and Carl Wennerlind, eds, *Mercantilism Reimagined: Political Economy in Early Modern Britain and its Empire* (Oxford, 2014), pp. 243–7.

enjoyed a monarchy strong enough to keep subsidiary corporate bodies under strict central control. It was also strong enough to seize the property of the church, whose transfer to the landed classes had the unintended consequence of preventing royal absolutism and substituting rule by a large social elite eventually as interested in commerce as in land. It was fortunate even in its misfortunes, since the collapse in the later sixteenth century of its chief export trade in textiles to Antwerp provided the incentive for merchants to look for new markets elsewhere, and hence for the production of manufactures for export, like the new draperies, whose growth was largely unrestrained by guild or other restrictions. Its political revolutions in the seventeenth century had among other things left it with a political elite increasingly attached to free trade at home but also protective of its own landed base, so that bounties and corn laws helped to keep grain prices higher in England than in other parts of Europe, and to sustain agrarian improvements which only became profitable and therefore fashionable elsewhere when prices rose again after 1750.[143]

The point scarcely needs labouring, but such contrasts meant that there was no incentive, let alone any necessity, for other countries to follow the English road to an enlightened economy.[144] It was a consequence of their shared culture that they nonetheless engaged in competitive emulation of one another. They were conscious also of their institutional and economic differences; and they could learn more about them at the end of the seventeenth century from histories and works of descriptive geography published for a European audience which compared the resources, institutions, and manners of different states, and sometimes their weaknesses, including England's fondness for 'alterations' which Pufendorf thought accounted for its persistent 'civil dissensions'.[145] Comparisons between England, France, and the Dutch Republic, like those made by Petty and Davenant, were common from the 1660s once the three engaged in what Colbert called 'a perpetual struggle' for commerce, and obvious contrasts were drawn between the religious tolerance of the Dutch which encouraged immigration and trade and French absolutism which hindered them. In the Dutch Republic itself, the de la Court brothers pointed to the virtues of frugality and thrift, diligence and industriousness, which went hand in hand with civic virtue and the pursuit of self-interest under a free government, and distinguished it from the conspicuous prodigality and corruption of the English under a monarch and courtier aristocracy.[146]

None of these perceptions prevented states borrowing from one another, although they might be used as reasons to reject unwelcome proposals. According to his early biographer, when John Law offered his scheme for a land bank to Lord

---

[143] Simpson, 'European Farmers', pp. 72–3.
[144] Cf. Jeff Horn, *The Path not Taken: French Industrialization in the Age of Revolution, 1750–1830* (Cambridge, Mass., 2006), pp. 4–8; Hoppit, 'Nation, State, and the First Industrial Revolution', pp. 330–1.
[145] Samuel Pufendorf, *An Introduction to the History of the Principal Kingdoms and States of Europe* (1695), p. 171.
[146] Hutchison, *Before Adam Smith*, pp. 87–8; de Vries, *Industrious Revolution*, pp. 59, 62–3; Arthur Weststeijn, *Commercial Republicanism in the Dutch Golden Age: The Political Thought of Johan and Pieter de la Court* (Leiden, 2012), pp. 181, 193, 200–4; de la Court, *True Interest*, pp. 278, 285, 366.

Treasurer Godolphin in 1704, it was rejected because it 'could never be put into execution under a limited government' which lacked 'the authority of an absolute prince to carry it through'. When he later offered it in Paris he met with the opposite argument that banks, which had been invented in republics, could only sustain their credit under a representative government.[147] In effect, since London already had a central bank and scarcely needed another, the incident shows that an English parliament was by then more powerful than an absolute prince.

A century earlier, however, England had looked to France for lessons in what could be done to increase wealth and power. There, as in England, 'oeconomy' was coming to mean the good management of land and the national estate, resources were being measured, new manufactures encouraged as import substitutes, and a start made on draining coastal marshes.[148] Projects for the country's economic recovery and revival after the French Religious Wars, like Henry IV's plans for a rebuilt Paris and Sully's for an information archive in the Louvre, were admired in England, and French handbooks on husbandry read and translated there. In 1607 part of a tract on new crops by Olivier de Serres was published as *The Perfect Use of silk-wormes*, with a dedication to James I and a prefatory poem by Drayton hailing the introduction of mulberry trees as a new discovery worthy of Columbus.[149] Montchrétien's treatise on political economy in 1615 may have had little influence inside or outside France, but he was comparable to Malynes and Mun because he at least had some practical commercial experience, a rare distinction among French political economists then and later.[150]

In the 1660s and 1670s Colbert tried, vainly as it turned out, to remove some of the obstacles which prevented France from being like England, by edicts to raise the status of merchants and pressure to remove internal tolls, both of them issues still unresolved a century later, and by giving royal privileges and backing to new manufactures, which, except in the case of luxury industries, often failed in the absence of economic incentives to their diffusion.[151] Nevertheless, he was as assiduous in collecting information as Sully had been; and his successful projects, and especially the Canal du Midi linking the Atlantic to the Mediterranean which

---

[147] Richard Bonney, 'Towards the Comparative Fiscal History of Britain and France in the "Long" Eighteenth Century', in Prados de la Escosura, ed., *Exceptionalism and Industrialisation*, pp. 203–4; *ODNB sub* Law. For the problems of war finance in France, see Guy Rowlands, *The Financial Decline of a Great Power: War, Influence, and Money in Louis XIV's France* (Oxford, 2012), pp. 236–9.

[148] Simon Schaffer, 'Introduction', in Lissa Roberts, Simon Schaffer, and Peter Dear, eds, *The Mindful Hand: Inquiry and Invention from the Late Renaissance to Early Industrialisation* (Amsterdam, 2007), pp. 96–9; Grenier, *Histoire de la pensée économique*, pp. 53–7, 99, 111–12; Raphaël Morera, *L'Assèchement des marais en France au XVIIe siècle* (Rennes, 2011).

[149] Olivier de Serres, *Le Théâtre d'agriculture et mesnage des champs* (Paris, 2nd edn, 1603); Olivier de Serres, *The Perfect Use of silk-wormes, and their benefit* (1607), trans. Nicholas Geffe, sigs A2–3.

[150] Jean-Claude Perrot, *Une histoire intellectuelle de l'économie politique: XVIIe–XVIIIe siècle* (Paris, 1992), pp. 63–7; Grenier, *Histoire de la pensée économique*, pp. 109–11.

[151] Charles Woolsey Cole, *Colbert and a Century of French Mercantilism* (2 vols, New York, 1939), i. 363–4; James Van Horn Melton, *The Rise of the Public in Enlightenment Europe* (Cambridge, 2001), p. 47; Christine MacLeod, 'The European Origins of British Industrial Predominance', in Prados de la Escosura, ed., *Exceptionalism and Industrialisation*, pp. 121, 123; Robert C. Allen, *Global Economic History: A Very Short Introduction* (Oxford, 2011), pp. 28–9.

was pushed through against much local opposition, made France once again, by 1690, a model of an apparently successful political economy which all Europe admired.[152]

The other model was still the Dutch Republic, though England was now catching up. German Cameralist writers, who were generally court officials, scientists, or academics, visited both countries, but went first to Holland when they were looking for the secrets of commerce which might assist recovery from their own wars of religion after 1648. Even so, they had to devise remedies for common ailments which suited German circumstances. They could not say, as the de la Courts did when meeting the Epicurean challenge, that private passions were already and necessarily being turned to the public good and the modest accumulation of riches in free republics;[153] and Joachim Becher, for all his advocacy of consumption, could not visualize it as something to be encouraged by freer trade, as Barbon did. Their antidote to dangerous private appetites was a benevolent prince uniting privileged and corporate bodies in common pursuit of the wealth, welfare, and happiness of all. The idiom was the same as that employed in England, where Becher and von Schröder (who had read Mun and Child) were welcomed by the Royal Society and the latter elected a Fellow, but in Germany it had its roots in much earlier imperial aspirations.[154]

At the end of the century England was at last exporting some of its own political economy, along with the science of Newton and the philosophy of Locke, and especially to France, where it was employed by critics of the policies of Colbert and Louis XIV seeking to explain why they had ruined the country. Vauban may have owed less to English political arithmetic than to a French tradition of numerical enquiry going back to Bodin when he calculated the size of the population in 1698 and found it smaller than generally supposed; but his use of 'supputations', suppositions, and of periods of 'doubling' when discussing demography, is distinctly reminiscent of Petty.[155] Although Boisguilbert's economic thinking was influenced more by Nicole and the Augustinians than by English authors, he seems similarly to have borrowed Petty's methodology and some of his arguments in his *Détail de la France* of 1695. He defined wealth as the product of land and labour, for example, and 'consumption and income' as 'one and the same thing', and he argued that the economy should be left to take its own course with as little

---

[152] Soll, *Information Master*, pp. 68–78, 97, 116; Chandra Mukerjii, 'Demonstration and Verification in Engineering: Ascertaining Truth and Telling Fictions along the Canal du Midi', in Roberts et al., eds, *Mindful Hand*, pp. 169–86; above, p. 213.

[153] Harold J. Cook, *Matters of Exchange* (New Haven, 2007), pp. 262–3; Weststeijn, *Commercial Republicanism*, pp. 168–84, 353; Pieter and Johan de la Court, *Fables Moral and Political* (2 vols, 1703), i. 81; ii. 42, 176.

[154] Smith, *Business of Alchemy*, pp. 27–31, 210–12; Walker, *German Home Towns*, pp. 145–50; Whaley, *Germany and the Holy Roman Empire*, ii. 261–2; R. J. W. Evans, *The Making of the Habsburg Monarchy, 1550–1700* (Oxford, 1979), pp. 163–4, 296–8; above, p. 151.

[155] Perrot, *Histoire intellectuelle de l'économie politique*, pp. 26, 400; N. Meusnier, 'Vauban: arithmétique politique, Ragot et autre *cochonnerie*', in Thierry Martin, ed., *Arithmétique politique dans la France du XVIIIe siècle* (Paris, 2003), p. 92; Geoffrey Parker, *Global Crisis* (New Haven, 2013), p. 628; above, p. 117.

government interference as possible: *laissez faire la nature*. He sounded very like Petty or Barbon also when he remarked that 'true wealth consists of a full enjoyment, not only of the necessaries of life, but even of all the superfluities and all that which can give pleasure to the senses', but his purpose, unlike theirs, was to demonstrate that the peasantry who paid for Louis's wars enjoyed none of those things.[156]

Only the misery of the peasantry and the damage inflicted on the country's agrarian economy could have brought together an opposition which included aristocratic reactionaries keen to strengthen social distinctions and restore power to the nobility, and writers as different as Boisguilbert and Fénelon, whose attack on luxury and the 'superfluities' now regarded as necessaries provoked Mandeville's *Fable*.[157] Such an unholy alliance was inconceivable in England, with its commercial agriculture and commercial culture. It showed some of the structural features which made France so different, the rigid distinctions of status which separated land from trade, left the peasantry paying most of the taxes, and created privileged enclaves, like the provincial merchants who opposed local 'chambers of commerce' when they were proposed by a new Council of Commerce set up in 1700, and the faculties of medicine which successfully obstructed the practice of inoculation against smallpox for a generation after 1720.[158]

The contrasts with England could scarcely be missed when French observers tried to explain why, despite France's superior wealth, its campaigns had ended in defeat. In 1699 one of them drew attention to the remarkable fact that in England 'many noblemen of the highest rank and standing are openly interested in trade, and several of their children are actually merchants ... without losing their rank in any way'.[159] In the reaction after the death of Louis XIV, in the Paris of the Regency and the Entresol club, there was an understandable fascination with England which led Montesquieu and Voltaire to go and see such differences for themselves, and it injected a much stronger dose of English political economy into French thinking.[160] Some of it was derived from English publications, but it was also communicated in the works of four authors who published or wrote in Paris in the 1720s, though only one of them was French: John Law from Scotland, Law's secretary, Jean-François Melon, Richard Cantillon, an Irishman, and Ernst Ludwig Carl, who had studied Cameralism in Halle. Between them they conveyed

---

[156] Gilbert Faccarello, *The Foundations of Laissez-faire: The Economics of Pierre de Boisguilbert* (1999), pp. 13, 21–9, 138; Perrot, *Histoire intellectuelle*, pp. 89, 149–54; Grenier, *Histoire de la pensée économique*, pp. 138–42; Hutchison, *Before Adam Smith*, pp. 108–14; Lionel Rothkrug, *Opposition to Louis XIV: The Political and Social Origins of the French Enlightenment* (Princeton, 1965), pp. 359–60. In some of its editions, Boisguilbert's *Détail* was subtitled 'La France ruinée sous le règne de Louis XIV'.

[157] Hont, *Jealousy of Trade*, pp. 25–6; above, p. 208.

[158] Thomas J. Schaeper, *The French Council of Commerce 1700–1715* (Columbus, Oh., 1983), pp. 257–8; Rusnock, *Vital Accounts*, pp. 73–5.

[159] Crouzet, 'The Sources of England's Wealth', p. 71. Cf. above, p. 34.

[160] Grenier, *Histoire de la pensée économique*, pp. 152–5; Perrot, *Histoire intellectuelle*, pp. 43–4. Cf. Nick Childs, *A Political Academy in Paris, 1724–1731: The Entresol and its Members* (Studies in Voltaire and the Eighteenth Century 10, Oxford, 2000).

something of the character of English improvement, and their publications merit attention here because they illustrate the problems involved in cultural transmission, how much had to be added by way of critical gloss or constructive elaboration for an international audience, and how much could be lost, sometimes literally in translation.

Law's short tract, *Money and Trade*, was translated into French and German in 1720, largely because of his brief and disastrous career in Paris as chief finance minister, but it had been published first in Edinburgh in 1705, as a contribution to the debate about economic improvement there. Like Paterson and Donaldson, he argued that the Scots might grow rich if there was a will 'to improve the country' as there was in England. They were inhibited only by their poverty, and Law's antidote was a massive expansion of the money supply in order to boost consumption and investment. England itself could benefit, he added, if it had an extra £50 million in cash, and even then it 'would not improve... so far as it is capable for improvement'.[161] Here was the vision of endless improvement, but its impact was diminished when, in both French and German translations, the improvement of a country became its general 'enrichment' and the improvement of land merely its 'cultivation'.[162]

Carl's much longer treatise, 'on the wealth of princes and their states', published in French in three volumes in 1722 and 1723, was more conventional, but he took some pains with his vocabulary and its likely effect. He talked about perfection. The 'infinite' diversity of human appetites and the encouragement of consumption, 'the soul of all commerce', would induce the perfection of husbandry, *la perfection de l'agriculture*, and of industry, *la perfection des arts*. He might have found some of this in Boisguilbert and his French contemporaries, and he owed much of what he said about the obligations of the Prince, about schools and hospitals, and about the improvement of knowledge, to his Cameralist education. When he talked about the duty of a Prince to ensure 'the comfortable enjoyment (*jouissance aisée*) of the necessities and conveniences of life' for all his subjects, however, his exemplar of near-perfection was England. The universal enjoyment of comfort was what the English boasted about so much 'under the name of liberty and property', and what they valued 'more than any other nation in the world as the principal supports of the happiness of a state'.[163]

Melon's *Essai politique sur le commerce*, first published in 1734 and translated into English in 1738, was wholly novel in its methodology, though less so in its conclusions. It deployed economic models of general application in order to derive an argument which must 'hold universally true in all countries'. Melon imagined

---

[161] John Law, *Money and Trade considered, with a Proposal for supplying the Nation with Money* (Edinburgh, 1705), pp. 71, 111, 117, 121; above, pp. 189, 233–4.
[162] John Law, *Considerations sur le commerce et sur l'argent par Mr Law, Controlleur Général des Finances* (The Hague, 1720), pp. 174, 183, 187; John Law, *Gedancken von Waaren- und Geld-Handel* (Leipzig, 1720), pp. 126, 135. Cf. above, pp. 4–6.
[163] Ernst Ludwig Carl, *Traité de la richesse des princes, et de leurs états* (3 vols, Paris, 1722–3), i. 8, 226–7, 403; ii. 461; Hutchison, *Before Adam Smith*, pp. 156–9. Cf. Voltaire on English comforts: *Letters Concerning the English Nation*, pp. 58–9.

four islands, each with a different economy, and how they might fare in competition with one another. One of them was 'Corn', representing France; 'Wool' was England.[164] He also made great play with political arithmetic, and added a chapter on the topic to a new edition in 1736.[165] Although he thought Petty had grossly understated the population and wealth of France, he had no doubt that in order to compete successfully with England, it must add the advantages of 'Wool' to 'Corn'. The French must encourage manufactures, commerce, and technological innovations like water-raising engines, as the English did. Melon's mantra was not perfection but gradual 'progress'. In France the progress of trading companies and colonies had been too slow; and the progress of industry must be promoted, as it was in England, because it would 'always present new needs' and had no limit.[166]

The fourth author, Cantillon, was by some distance the most original, the least deferential to English achievements and English authors, and in the end the most influential. His essay 'on the nature of commerce in general' was published in full and in French only in 1755, long after it was written, perhaps in English and partly in London, where he died in 1734, apparently murdered by his French cook.[167] He was sharply critical of Law, having made a fortune from speculating against his financial schemes, and he was no less critical of the political arithmeticians, whose works he had read and found 'purely imaginary', throwing no light on the realities of human behaviour. Like Melon, he was a model-builder in Cartesian style, and his model was one covering the whole course of economic development from primitive barter to modern, monetized markets, analysed in a manner which looked forward to the Physiocrats and Adam Smith. Cantillon's economic history was not the story of constant improvement, perfection, or progress, which his contemporaries thought England exemplified, but one of economic cycles punctuated by periods of decline, from which even England could not be immune. The cyclical concept was scarcely new, but Cantillon showed more persuasively than anyone else before Smith that when a state 'arrived at the highest point of wealth' it must gradually 'fall into poverty by the ordinary course of things'.[168]

When it finally appeared in 1755 Cantillon's essay was a major contribution to a new phase in international discussion of political economy, more theoretical than that of the 1720s, and much more heavily influenced by French authors. There was another wave of interest in the works of English writers, but they were more than matched now, in quantity and quality, by French publications. Before 1750 there

---

[164] Grenier, *Histoire de la pensée économique*, p. 159; Robertson, *Case for the Enlightenment*, pp. 340–7; Hont, *Jealousy of Trade*, pp. 30–3.
[165] Jean-François Melon, *Essai politique sur le commerce* (new edn, n.p., 1736), pp. 318–57.
[166] Jean-François Melon, *Essai politique sur le commerce* (n.p., 1734), pp. 42–6, 58, 76, 109–12.
[167] *ODNB sub* Cantillon; Antoin E. Murphy, *Richard Cantillon: Entrepreneur and Economist* (Oxford, 1989). Richard Cantillon, *Essai sur la nature du commerce en général*, ed. Henry Higgs (1931), contains the first English translation of the full text.
[168] Hutchison, *Before Adam Smith*, pp. 163–75; Richard Cantillon, *Essai sur la nature du commerce en général* (London [i.e. Paris?], 1755), pp. 54, 108, 244, 312. For English authors reaching a similar conclusion, see above, pp. 193–4. By the 1770s, French observers were thinking England had now reached the highest point of its wealth: Crouzet, 'The Sources of England's Wealth', p. 63.

were many more works on economic matters published annually in England than in France, perhaps more than five times as many, depending on one's definition of economic.[169] After 1750, as Voltaire and others noticed, French publications on political economy suddenly exploded, and by the 1760s were appearing at a faster rate than new novels for an eager public. Between 1750 and 1769 there were five times as many as there had been in the previous two decades.[170] They included translations of Cary and Child, and even the *British Merchant* of 1713–14, which appeared as *Le Négotiant anglois* in 1753.[171] As in England, appropriate tribute was being paid to the political arithmeticians, in a long essay in the first volume of the *Encyclopédie*, for example, and their importance was acknowledged in other countries besides France, but they were figures from the past.[172]

English political economy was beginning to look dated partly because there was little sign of any reciprocal sense in England, as opposed to Scotland, that French publications warranted close attention. Part of Cantillon's book was published in London by Malachy Postlethwayt in 1749, probably from the original manuscript, and most of it was reproduced in a version edited by Cantillon's cousin in 1759, but neither editor drew attention to its originality. The preface to the first simply hoped that the 'emulous spirit' which Cantillon had encouraged in France would promote further 'improvement of trade' in England, and the introduction to the second had a history of commerce and manufactures ending with praise for England's achievements.[173] Yet Cantillon's text now belonged to a pan-European Enlightenment debate about political economy in which the Scotland of David Hume was fully involved. In 1742 Hume had welcomed the current exchange of ideas between different countries because 'the emulation which naturally arises ... is an obvious source of improvement'.[174] In the 1750s that international republic of learning had arrived, and several countries were seeking their own version of an enlightened economy, except England where people thought they already had one.

The others borrowed from one another only what suited their circumstances, so that writers in Naples, for example, were interested in what the French *économistes* had to say about land, and fully conscious that their kingdom lacked the clearly

---

[169] I owe the point to Jonathan Hoppit. For the number of British publications, see Julian Hoppit, 'The Contexts and Contours of British Economic Literature, 1660–1760', *Historical Journal*, 49 (2006), pp. 84–9.

[170] C. Théré, 'Economic Publishing and Authors, 1566–1789', in Gilbert Faccarello, ed., *Studies in the History of French Political Economy: From Bodin to Walras* (1998), pp. 11–12, 18, 28; John Shovlin, *The Political Economy of Virtue: Luxury, Patriotism, and the Origins of the French Revolution* (Ithaca, NY, 2006), p. 2; William H. Sewell Jr, 'The Empire of Fashion and the Rise of Capitalism in Eighteenth-Century France', *P&P* 206 (Feb. 2010), pp. 83–4; Grenier, *Histoire de la pensée économique*, p. 180.

[171] Murphy, *Cantillon*, p. 308; Robertson, *Case for the Enlightenment*, pp. 354–5; Charles King, ed., *Le Négotiant anglois*, trans. François Véron de Forbonnais (2 vols, Dresden, 1753); above, p. 185.

[172] Denis Diderot and Jean d'Alembert, eds, *Encyclopédie, ou Dictionnaire raisonné des sciences, des arts et des métiers* (17 vols, Paris, 1751–65), i. 678–80; *Journal œconomique* (Paris, 1756), pp. 187–9. Cf. P. Crépel, 'Arithmétique politique et population dans les métamorphoses de l'*Encyclopédie*', in Martin, ed., *Arithmétique politique*, pp. 47–69.

[173] Malachy Postlethwayt, *A Dissertation on the Plan, Use, and Importance, of The Universal Dictionary of Trade and Commerce* (1749), p. 23; Philip Cantillon, *The Analysis of Trade, Commerce, Coin, Bullion ...* (1759), pp. xvii–xix; Cantillon, *Essai*, ed. Higgs, pp. 376–7, 383.

[174] Mokyr, *Gifts of Athena*, p. 278.

discernible rights to private property which promoted agrarian improvement in England.[175] Different countries also made their own distinctive contributions. In Germany Süssmilch referred to Graunt as the Columbus of demography but took that science to a new level of excellence in 1742,[176] and in Italy economists were applying mathematics of a sophistication far beyond that of political arithmetic soon afterwards.[177] At the end of the century German Cameralists' distinctive contribution to political economy was a political model for the protection of public health, a 'medical police', which other countries admired and emulated.[178] Even the Bourbon monarchy of Spain was able, after several experiments, to conduct the first census of a large European state in 1787.[179]

The Dutch Republic, which had once been the model of a small country building power and an embarrassment of riches on commerce, was the conspicuous exception. Though still the place where books written elsewhere were published and bought, it had no political economy of its own worthy of the name during the eighteenth century.[180] The Anglo-mania of the 1720s took the literary form there of 'spectatorial' periodicals copying Addison in their focus on rational sociability, and the Dutch reaction to recent and rapid economic decline, and to the challenge posed by Mandeville, was a renewed emphasis on the civic virtues of frugality and polite restraint which reinforced the corporate particularism of guilds and towns.[181] By the 1770s the Dutch were copying France and England in founding literary and scientific clubs, and there were more than fifty 'economic societies', but the latter had few members by 1795.[182] Their equivalents were much more popular in Germany, where there was even one devoted to bee-keeping, a 'Physico-Oeconomical Bee Society', in Lautern in 1769, and Cameralists were busily developing an economic 'science' which owed more to French than to English writers.[183]

---

[175] Robertson, *Case for the Enlightenment*, pp. 382–5.
[176] Sabine Reungoat, *William Petty: Observateur des Îles Britanniques* (Paris, 2004), pp. 279–87; Hutchison, *Before Adam Smith*, pp. 248–9.
[177] Hutchison, *Before Adam Smith*, pp. 178–9, 255–6.
[178] G. Rosen, 'Cameralism and the Concept of Medical Police', *Bulletin of the History of Medicine*, 27 (1953), pp. 21–42; Laurence Brockliss and Colin Jones, *The Medical World of Early Modern France* (Oxford, 1997), pp. 735–6.
[179] Daniel R. Headrick, *When Information Came of Age: Technologies of Knowledge in the Age of Reason and Revolution, 1700–1850* (Oxford, 2000), pp. 76–7.
[180] Bonney, 'Early Modern Theories of State Finance', pp. 223–4; Jan A. F. de Jongste, 'The Restoration of the Orangist Regime in 1747: The Modernity of a "Glorious Revolution"', in Margaret C. Jacob and Wijnand W. Mijnhardt, eds, *The Dutch Republic in the Eighteenth Century: Decline, Enlightenment, and Revolution* (Ithaca, NY, 1992), p. 53; Erik S. Reinart, 'Emulating Success: Contemporary Views of the Dutch Economy before 1800', in Oscar Gelderblom, ed., *The Political Economy of the Dutch Republic* (Farnham, 2009), pp. 30–1.
[181] Wijnand Mijnhardt, 'The Dutch Enlightenment: Humanism, Nationalism, and Decline', in Jacob and Mijnhardt, *Dutch Republic*, pp. 206–8; Jacob and Mijnhardt, 'Introduction', in Jacob and Mijnhardt, *Dutch Republic*, pp. 12–13; de Jongste, 'Restoration of Orangist Regime', pp. 53–5; Wyger R. E. Velema, 'Ancient and Modern Virtue Compared: De Beaufort and Van Effen on Republican Citizenship', *Eighteenth-Century Studies*, 30 (1997), pp. 437–40.
[182] Mijnhardt, 'Dutch Enlightenment', pp. 218–20.
[183] Whaley, *Germany and the Holy Roman Empire*, ii. 494–8; Tribe, *Governing Economy*, pp. 55, 99–100, 135; Bonney, 'Early Modern Theories of State Finance', pp. 186–8.

In short, there were different paths to an enlightened knowledge economy, and none of them was the same as that taken much earlier in England. One historian of the different versions of Enlightenment adopted in different countries remarks that the English were unusual after 1740 in not having at the forefront of their intellectual life 'commitments which were central to the Enlightenment itself'. They included 'the development of the sciences of man and of political economy, the historical investigation of the progress of society, and the critical application of ideas for human betterment to the existing social and political order'. These were still being discussed, sometimes critically, in England, but any novel commitment and sense of urgency had disappeared, because they had already been debated at length since the 1640s.[184] They had been incorporated into a culture of improvement which helped to make England economically precocious and served English interests well enough. The useful knowledge it had produced could also be exported, especially when grain prices rose after 1750 and there was an 'agricultural enlightenment' across Europe, drawing on writers like Arthur Young and accepting, as English agrarian improvers had done long before, that a finite amount of land was no barrier to an increase in productivity over time.[185] In 1750, however, the English culture of improvement remained different from anything to be found elsewhere, and it is important to try to identify where the difference lay.

In the middle of the eighteenth century other countries plainly had, or were soon to acquire, many of the ingredients which we have seen contributing to improvement in England. With the notable exception of the Dutch, most of them had some version of political economy,[186] and in France there were political economists of particular theoretical sophistication. There was no one in Britain of the stature of Melon or Cantillon and the Physiocrats until Adam Smith. The empirical pursuit of useful knowledge was also common ground by 1750, and its utility in promoting economic innovation and projects for social reform was widely acknowledged. Happiness, for states and citizens alike, had been a goal of princes and governors everywhere since at least the middle of the seventeenth century, sometimes longer, and by the eighteenth material wealth and its increase was regarded as one of its necessary conditions.[187] In consequence, economists, moralists, and politicians sooner or later recognized the public utility of consumer appetites and profit motives for economic growth and communal well-being. Although they responded to Mandeville and the problem of luxury in ways which depended upon how material wealth was distributed and displayed in their own country, they were all engaged in debate about the distribution of wealth and welfare.

What other countries did not have, at least in any coherent or consensual form, was the collective conviction that material progress for a state and its citizens

---

[184] Robertson, *Case for the Enlightenment*, p. 42; Mokyr, *Enlightened Economy*, pp. 31–2. Cf. Innes, *Inferior Politics*, p. 141.
[185] Mokyr, *Enlightened Economy*, pp. 184–6.
[186] Robertson, *Case for the Enlightenment*, pp. 29, 325–6.
[187] Richard Drayton, *Nature's Government: Science, Imperial Britain, and the 'Improvement' of the World* (New Haven, 2000), pp. 69–10; Perrot, *Histoire intellectuelle*, p. 94.

could be infinitely prolonged, the boundless optimism which drove the whole improvement agenda and sustained it against its critics into the 1740s. There is no shortage of writers referring to the 'infinite' economic potential of human appetites, as Carl did, or to the 'limitless' possibilities of industrial expansion, as in Melon's *Essai*; but there is little evidence of a general view that the main purpose of economic management was endless growth, and only in England was it powerful enough to generate a collective morality of cooperation.[188]

Where there were collective moralities they were conservative, not forward-looking, like the Dutch attachment to frugality and civic virtue and the German Cameralists' faith in the 'common weal' as the ideological cement linking ambitious princes and corporate bodies which had power on the ground. In France many of the various candidates for a new collective morality similarly looked backwards, to some earlier better state, before Colbert and Louis XIV, as with Fénelon, or to an economic and social equilibrium, as with Boisguilbert and some of the Physiocrats. Despite his vision of nature's fecundity delivering incremental improvement, Quesnay's purpose was to ensure stability, and his overall conception has been described as 'essentially static'.[189]

The peculiarity of English improvement was most obvious in the century between 1650 and 1750 when the shadow of economic and demographic depression hung across the whole of western Europe, and not least over France. After the famines of Louis XIV's reign, the fear of depopulation was much stronger than in England and persisted into the 1750s when the population was in fact increasing. We have seen the same perception in Britain, clearly articulated by Massie and hinted at by Petty and others earlier, but there the expectation and observation of improvement delayed and obscured such perceptions long enough for improvement to retain its vigour.[190] It seems to me arguable too that the view of the contest for international commerce as a 'zero-sum game' was more powerful in France than in England, especially after 1713 when Britain had the largest share;[191] and that fears of a stationary state were similarly stronger in France because it seemed between 1720 and 1750 to have arrived there.

England was more fortunate in its military and economic achievements after 1688, but it was already different in its culture. French historians have noted that French political economy during the *ancien régime* placed much more emphasis on happiness than on the idea of progress, and that even the notion of progress rarely extended to the idea of economic growth as 'a process not limited in time'. They suggest that the primary ambition of French political economists was political and social conservation, 'not the continuous amelioration of the material condition of

---

[188] Above, p. 241.
[189] Faccarello, *Foundations of* Laissez-faire, pp. 58–68; Liana Vardi, *The Physiocrats and the World of the Enlightenment* (Cambridge, 2012), p. 275.
[190] Perrot, *Histoire intellectuelle*, pp. 159–62; above, pp. 200, 225.
[191] Steve Pincus points to the lack of any British consensus about the 'zero-sum' concept in 'Rethinking Mercantilism: Political Economy, the British Empire, and the Atlantic World in the Seventeenth and Eighteenth Centuries', *William and Mary Quarterly*, 69 (2012), pp. 12–29. Cf. Mokyr, *Enlightened Economy*, p. 64; Hutchison, *Before Adam Smith*, p. 87; above, pp. 109–11, 120.

individuals'.[192] Continuous amelioration was improvement's alpha and omega. In its European context, especially in the century after 1650, English improvement was different because it was by definition of endless potential.

England's improvement was therefore *sui generis*. Its culture was precocious in the sense that parts of it were copied or independently conceived later in other countries, but neither the culture nor its practical results were replicable *in toto* in different environments. They could be transplanted most completely where the English governed, close to home, in the British archipelago, in Scotland and Ireland. Improvement could also be rooted more or less effectively where the British settled and held sway on other continents, in parts of India, the West Indies, and especially North America, despite the intention of some of the founders of godly societies there to break away entirely from the private and public corruptions of old England. In Pennsylvania, for example, Penn and his colleagues in the 1680s failed to create their ideal of an improved commonwealth, one which avoided 'emulation' by having a common trading stock, making 'everyone... interested in everyone's prosperity', all of them 'improving in a united way'. They were soon divided by appetites for private profit and property and improving like other plantations in the English way.[193]

The cultural interactions between the American colonies and the mother country were always as contested as their political and economic relationships, marred by an inferiority complex on one side and arrogance on the other, but they testified to what were at root common assumptions and modes of behaviour.[194] In New England in the 1640s ideals of commutative and distributive justice in market relations and the monopoly privileges of corporations and guilds were giving way in the face of economic realities and demands for free trade; and there were laws protecting the intellectual property of inventors, and encouraging manufactures and industry, and 'committees for furthering of trade', just as there were at home.[195] By 1700 there was an educated public in the larger cities reading works on political economy, and reflecting on what they read. In Boston in 1719 Cotton Mather, a disciple of Boyle and early advocate of smallpox inoculation, was perhaps referring indirectly to Mandeville's *Fable* when he wondered why there should be

---

[192] Perrot, *Histoire intellectuelle*, p. 94; Grenier, *Histoire de la pensée économique*, p. 260. Turgot may have been an exception: Jonathan Israel, *A Revolution of the Mind: Radical Enlightenment and the Intellectual Origins of Modern Democracy* (Princeton, 2010), pp. 7–8; Hutchison, *Before Adam Smith*, p. 320.

[193] Nicholas More, *The Articles, Settlement and Offices of the Free Society of Traders in Pennsilvania... for the better Improvement and Government of Trade in that Province* (1682), sig. B1ᵛ; J. H. Elliott, *Empires of the Atlantic World* (New Haven, 2006), pp. 212–13.

[194] See, for example, Sarah Yeh, 'Colonial Identity and Revolutionary Loyalty: The Case of the West Indies', in Stephen Foster, ed., *British North America in the Seventeenth and Eighteenth Centuries* (Oxford, 2013), pp. 195–209; and more generally, T. H. Breen, *The Marketplace of Revolution: How Consumer Politics Shaped American Independence* (Oxford, 2004); Kariann Akemi Yokota, *Unbecoming British: How Revolutionary America became a Postcolonial Nation* (Oxford, 2011).

[195] Stephen Innes, *Creating the Commonwealth: The Economic Culture of Puritan New England* (New York, 1995), pp. 175, 220–35; Barry Levy, *Town Born: The Political Economy of New England from its Founding to the Revolution* (Philadelphia, 2009), pp. 129–30. Cf. above, pp. 80–1, 87–8.

any idle poor in a colony where work and land were so plentiful: 'What a pity it is that such a hive should have any drones in it.'[196]

Benjamin Franklin of Philadelphia, who had met Mandeville in London in 1725 and found him 'a most facetious entertaining companion', seems to have read everything, including some of the works of Thomas Tryon and Nicole; and he similarly used his reading to view New England through European spectacles. He found difference as well as similarity.[197] In 1755 he noted that the North American colonies were blessed by a total absence of luxury, which explained why more people married and married earlier than they did in fully settled and urbanized countries, and why their populations were rising so quickly. They were also places where economic advance could rapidly accelerate because they had recently moved through all the early stages of civilization and were now ready, at least in Philadelphia in 1731, for 'every kind of improvement', including 'the refinements of life'.[198]

Thomas Jefferson in Virginia also visualized the colonies as being in one of the later stages of human development, where rising populations multiplied 'the chances of improvement' and where 'one improvement begets another'. Like many other writers, he interpreted the colonies in classical republican terms as societies of small farmers, almost wholly engaged, as Jefferson was himself, in the 'improvement' of land. The spectre of luxury and the civic corruption which came with urbanization and industries loomed larger for him than for Franklin.[199] Yet despite continuing differences of culture between North and South, at the end of the eighteenth century there was a general expectation across the new nation of a 'greater perfection and happiness than mankind has yet seen'.[200] In the 1780s, having observed the march of progress over half a century, Franklin would not have been alone in wishing, as Petty might have done, that he could live another two hundred years and see America then, since 'inventions of improvement are prolific, and beget more of their kind'.[201]

The founding fathers were exceptional historical figures, but colonial North America was an exceptional continent with limitless natural resources broadly distributed among its colonizers, and Franklin was right to think it fertile territory

---

[196] Gary B. Nash, *The Urban Crucible: Social Change, Political Consciousness, and the Origins of the American Revolution* (Cambridge, Mass., 1979), pp. 129–39, 188; Stephen Foster, *Their Solitary Way: The Puritan Social Ethic in the First Century of Settlement in New England* (New Haven, 1971), p. 137.

[197] Alan Houston, *Benjamin Franklin and the Politics of Improvement* (New Haven, 2008), pp. 8, 13, 44, 114.

[198] Benjamin Franklin, 'Observations on the Increase of Mankind', in *Political, Miscellaneous, and Philosophical Pieces* (1779), pp. 1–2, 8 (first published 1755); Houston, *Franklin and the Politics of Improvement*, p. 14. For Franklin's personal dislike of luxury, see Yokota, *Unbecoming British*, p. 73.

[199] Thomas Jefferson, *Notes on the State of Virginia* (2nd American edn, Philadelphia, 1794), pp. 92, 240–1; P. J. Marshall, *Remaking the British Atlantic: The United States and the British Empire after American Independence* (Oxford, 2012), p. 91. On Jefferson's interest in inventions and agrarian improvement, see Robert Friedel, *A Culture of Improvement: Technology and the Western Millennium* (Cambridge, Mass., 2007), pp. 273, 324–6, 432.

[200] Gordon S. Wood, *Empire of Liberty: A History of the Early Republic, 1789–1815* (Oxford, 2009), p. 13; Elliott, *Empires of the Atlantic*, pp. 242–4, 401–2.

[201] Houston, *Franklin and the Politics of Improvement*, p. 15.

for endless improvement. On the continent of Europe, where most countries already had dense populations, established institutions, and economies shaped by long histories, inventions of improvement found stonier ground. The sharp contrasts between the two continents suggest that we need to look back over the invention of improvement in England, and ask, in the Conclusion, what historical, institutional, and economic conditions had led to its emergence there in the first place.

8
Conclusion

The prolonged and peculiar history of English improvement began with an accident of language, the chance discovery of an English word which could be applied to the whole domain of human experience, which implied endless progress, and which could be understood wherever the language took root, as in North America. We have seen, however, that its long development and its success depended upon much more than that. It emerged and took hold because of a conjunction of cultural, institutional, and economic conditions interacting in creative ways in the middle of the seventeenth century, and producing outcomes no more replicable elsewhere than the word itself.

Circumstances of political revolution, when all kinds of alternative futures could be conceived, were not only crucial in joining the Baconian pursuit of useful knowledge to millenarian expectations and giving publicity to the results. The abolition of the monarchy and advent of a republic also produced institutions vital for England's subsequent economic performance and the character of its political economy. The parliaments and councils of the Interregnum created a fiscal and military state capable of funding a standing navy as well as army, passing a Navigation Act, confiscating large parts of Ireland, conquering Scotland and Jamaica, and setting a precedent and a model for the effective protection of the country's commercial interests in Europe and across the oceans. The model remained in place after 1660 because it was in the interest of both the landed and the commercial elite that it should do so. It has been persuasively argued that the second revolution, in and after 1688, did not introduce the new institutional foundations for economic growth envisaged by some economic historians but protected and built upon what had already been largely achieved.[1]

These political events were important to the character and plausibility of an improvement culture because they showed that there could be no rapid route to perfection, either in revolution with its expectation of an imminent millennium, or in total restoration of some earlier golden age. That left improvement—gradual, piecemeal, and cumulative progress—in command of the intellectual field. A radical change in economic circumstances had a similar effect. Its coincidence

---

[1] Patrick O'Brien, 'The Nature and Historical Evolution of an Exceptional Fiscal State and its Possible Significance for the Precocious Commercialization and Industrialization of the British Economy from Cromwell to Nelson', *EcHR* 64 (2011), pp. 425–36; Ronald Findlay and Kevin H. O'Rourke, *Power and Plenty: Trade, War, and the World Economy in the Second Millennium* (Princeton, 2007), pp. 240–2.

with political and cultural change was another accident, and the consequences of the coincidence were only gradually appreciated, but they were profound. The demographic watershed of the mid-seventeenth century put an end to rapid inflation and population growth for a century, and the immediate reaction to it, in the 1660s and 1670s, was complaint about economic decline and decay. Yet that itself proved to be a stimulus to economic enquiry; and its economic effects, given that the country's foreign and internal trade continued to expand, proved wholly positive, as urbanization accelerated and a shrinking agricultural workforce boosted agrarian productivity. A growing national income meant that investment in improvements could be afforded and an increase in incomes per head offered some justification to the claims of improvers that economic growth was the product of a thriving consumer culture and contributed to the well-being of everyone.

The coincidences involved in the emergence of improvement both as an ideology and a reality make it tempting to ask counterfactual questions about what the outcome would have been if one feature or another in the conjuncture had been absent. Such questions are easier to ask, and much easier to answer, with respect to particular political or military events than about longer-term changes in culture, institutions, or economies. If Charles I had defeated parliament in 1643, or James II defeated William of Orange in 1688, or if France had won the War of the Spanish Succession after 1701, things would certainly have been different in the short term. Institutional change would have been delayed, the advocates of improvement been more defensive, and economic growth perhaps retarded, especially if Louis XIV had won. But it is difficult to believe that ultimate outcomes would not have been broadly the same because they depended on the distribution of wealth and other resources, on assumptions about property, and on norms of behaviour normally untouched by single events.[2] Such things changed only slowly over time, and that suggests that we need to look back beyond 1640, to the sixteenth century if not earlier, in order to discern what made England ready for improvement when it finally emerged.

One of the things was London, a city in the 1580s just embarking on its extraordinary growth, with a level of real wages already high by European standards and developing techniques and skills, in mathematics and navigation for example, which encouraged the production of useful knowledge. Another was the Elizabethan beginning of colonization in North America which rested explicitly on 'use', on planting and cultivating the land, as a justification for possession of thinly peopled territories quite different from the empires and cities which the Spanish had encountered in the south.[3] There was clearly no lack of entrepreneurial ambition among the Spanish conquistadors, as some of their plans for sugar plantations and Pacific trade show; and sixteenth-century Spanish authors had

---

[2] Cf. Douglass North, *Institutions, Institutional Change and Economic Performance* (Cambridge, 1990), pp. 139–40; Douglass C. North and Barry R. Weingast, 'Constitutions and Commitment: The Evolution of Institutions Governing Public Choice in Seventeenth-Century England', *Journal of Economic History*, 49 (1989), pp. 831–2.

[3] Robert C. Allen, *The British Industrial Revolution in Global Perspective* (Cambridge, 2009), pp. 25–56; above, pp. 67–75.

much to say about how Hispaniola and New Spain were being 'ameliorated', or, as the English would have said, improved. As John Elliott has speculated, when raising another counterfactual question, English and Spanish improvement might well have taken a different course, at least in the New World, if the Spanish had gone north and the English south. English agrarian improvement would have had less opportunity to demonstrate its potential there.[4]

The conquistadors, however, were not farmers, and the English were already practising agrarian improvement at home. It is arguable that English colonizers, wherever they went, took with them a culture which distinguished them, not only from the Spanish but, according to Josiah Child, from the Dutch when they too ventured overseas. In a passage published in 1690, but probably written in the 1660s, Child commented that the Dutch in the West Indies 'did never much thrive in planting', which hindered the 'improvement' of Tobago and Curaçao in contrast to that of English Jamaica. 'In other parts of the world' similarly, the Dutch were only interested in war, trade, and building fortified ports for commerce, and not in 'clearing, breaking up of the ground, and planting as the English have done'.[5] Whatever the truth of the observation, and the realities changed over time, the English thought themselves different.

The third element which accounts for the emergence of English improvement, therefore, in addition to urbanization and colonization, was agricultural innovation. The earliest, and perhaps the most important, it was the work of sixteenth-century landowners, yeomen, and their surveyors, whose activities as engrossers and enclosers helped to feed a growing London, provided a rationale for American plantation, and gave other kinds of improvement the same name. Agrarian improvement was itself assisted by institutional support and change, by concepts of private property, by statutes going back into the Middle Ages, and—after the dissolution of the monasteries—by an active land market which left a large and growing propertied elite with ample opportunity to respond to the incentives created by rising grain prices. By the end of the sixteenth century members of the aristocracy were also beginning to invest in iron and coal mines on their estates, and to see the attractions of developing their property in and around a London already benefiting from the diversification of England's overseas trade.[6] Later commentators on the country's recent economic growth were right to trace its origins back to the middle of the sixteenth century.

There cannot be any doubt that England in 1600 already had an elite which was both fully engaged in improvement of land and far from hostile to the progress of

---

[4] John H. Elliott, 'The Seizure of Overseas Territories by the European Powers', in David Armitage, ed., *Theories of Empire 1450–1800* (Aldershot, 1998), pp. 148–9; J. H. Elliott, *The Old World and the New 1492–1650* (Cambridge, 1970), p. 78.

[5] Josiah Child, *A Discourse about Trade* (1690), pp. 185–6. On the date of composition, see William Letwin, *The Origins of Scientific Economics* (New York, 1964), pp. 251–2. I am grateful to John Elliott for drawing Child's comment to my attention, and discussing it with me.

[6] Lawrence Stone, *The Crisis of the Aristocracy, 1558–1641* (Oxford, 1965), pp. 335–84; Theodore K. Rabb, *Enterprise and Empire: Merchant and Gentry Investment in the Expansion of England, 1575–1630* (Cambridge, Mass., 1967).

industry and commerce. In Elizabethan and Jacobean England, not only were landed gentry investing in colonial and commercial enterprises, but merchants were claiming the status of gentlemen and even apprentices defending the honour of their crafts. In 1741 Hume remarked that 'commerce in my opinion is apt to decay in absolute governments, not because it is there less secure, but because it is less honourable' than under a freer regime.[7] Long before then, the English gentry were placing their younger sons in trade, and although there was much aristocratic prejudice against 'mere merchants', there were no institutional or status barriers of privilege such as inhibited social mobility in France.[8] In England more than any other country, the interests of land and trade, though never identical, were acknowledged to be closely allied.

If we want to find economic conditions in England broadly similar to those of the century after 1650, however, we have to look back beyond the sixteenth century, to the later Middle Ages, and that also proves to be an instructive comparative exercise. The long delay before the English population recovered from the heavy mortality of the Black Death and the plagues which immediately followed it in the later fourteenth century meant that for more than a century large sections of the population enjoyed higher incomes, new opportunities to own land, and access to consumer markets. Living standards for the majority rose and remained high.[9] Parishes and small towns could afford to invest in corporate activities like the rebuilding of parish churches and other public buildings, and in projects to improve urban services, like water supplies, which would later be called improvements.[10] Material progress could be observed. It was in the fifteenth century that the English first congratulated themselves on having living standards superior to those of the French peasantry, and, as suggested earlier, that expectations about an adequate standard of living may have been set which persisted for another two centuries.[11]

Yet there was no new culture which explicitly conceived that material progress might continue ad infinitum. The fifteenth century produced an equivalent to the notion of improvement, which appeared in a large number of different contexts, and combined an appeal to harmony and prosperity with insistence on the obligations of rulers to promote them, and that was the 'common weal'. Like improvement, the term was a linguistic and rhetorical device, a flexible and loosely defined concept, and when it was put to new uses in the later fifteenth century it shaped ways of thinking about politics and society and how they might be changed. But the common weal, and its equivalents elsewhere in late medieval Europe,

---

[7] Deirdre N. McCloskey, *Bourgeois Dignity* (Chicago, 2010), p. 346; above, pp. 69–72, 75. Professor McCloskey is mistaken in implying that the honour of trade was something new in England after 1688.
[8] Above, pp. 34, 247. [9] Allen, *British Industrial Revolution*, pp. 14, 21, 238; above, p. 11.
[10] Peter Clark, 'Improvement, Policy and Tudor Towns', in G. W. Bernard and S. J. Gunn, eds, *Authority and Consent in Tudor England: Essays Presented to C. S. L. Davies* (Aldershot, 2002), pp. 233–5; Paul Slack, *From Reformation to Improvement* (Oxford, 1999), pp. 151–3.
[11] Christopher Dyer, *Standards of Living in the Later Middle Ages: Social Change in England c.1200–1520* (Cambridge, 1989), pp. 271–2; above, pp. 160–1.

pointed backwards rather than forwards, to the restoration of 'the lost perfection of the commonwealth'.[12]

The character of common-weal thinking may have been influenced by other features of the fifteenth-century economy which were different from those in the later seventeenth. In the period after the Black Death international commerce and England's share in it were depressed, the real incomes of the social elite declined, and there was no rapid urbanization to stimulate technological innovation.[13] The cultural resources available were also different, and it would be absurd to expect the fifteenth century to have had the intellectual equipment of the seventeenth. But the fifteenth-century case is relevant because it shows that economic conditions of rising real wages were not in themselves sufficient to account for a belief in endless progress. They might sometimes produce a conservative culture as formidable as that based on the common weal, and they did precisely that in some parts of Europe after 1650.[14]

It may therefore have been important for the persistence and resilience of an improvement culture in England in the later seventeenth and eighteenth centuries that it had been born in the very different circumstances of the later sixteenth, and that it was fully developed, thanks to Hartlib and his colleagues, before the 1660s. By then it was strong enough to repudiate complaints about economic decline and decay, despite all the evidence to the contrary, and to accommodate new elements, particularly the Epicurean preoccupation with private appetites which ran counter to respectable opinion and alienated the moral majority. Equally fundamental to improvement's success was the new seventeenth-century science of political economy, emerging in England in the economic crisis of the 1620s and shaped by Petty and others after 1660 into a tool capable of measuring the wealth and resources of the kingdom. The economy came to be conceived as a separate thing, something as open to dissection and diagnosis as the human body; and it is no accident that several of its investigators had studied medicine. Petty and Mandeville, and even Barbon and Locke, might all have been said, like Quesnay the French physician and Physiocrat, to have turned from prescribing 'medicine for the individual' to prescribing 'medicine for government and the whole human race'.[15] They were trying to formulate theoretical principles for a new discipline and looking for empirical evidence to substantiate them. They were creating knowledge and organizing information.

In the course of this book we have seen several examples of the use of information, and the deployment of data and statistics from old archives or new enquiries.

---

[12] John Watts, '*The Policie in Christen Remes*: Bishop Russell's Parliamentary Sermons of 1483–84', in Bernard and Gunn, *Authority and Consent*, pp. 43, 50; David Rundle, 'Was There a Renaissance Style of Politics in Fifteenth-Century England?', in Bernard and Gunn, *Authority and Consent*, pp. 23–4. On the European context and comparisons, see John Watts, *The Making of Polities: Europe, 1300–1500* (Cambridge, 2009), pp. 8–9, 287, 384–7. For later uses of the concept, see above, pp. 53–65.

[13] Dyer, *Standards of Living*, pp. 26, 188–9, 194–5.   [14] Above, pp. 243, 253.

[15] Richard Drayton, *Nature's Government: Science, Imperial Britain, and the* 'Improvement' *of the World* (New Haven, 2000), p. 70. Worsley, who had served as a surgeon in Ireland and claimed an MD in the 1660s, might be added to the list: *ODNB sub* Worsley.

Numbers on their own rarely determined opinions, let alone political decisions. They were rhetorical devices supporting standpoints already held, and conveying the illusion of certainty. The data was always incomplete, most notably about the country's population; it was easily massaged for propaganda purposes, as in the case of figures for the balance of trade between England and France; and it was often ignored when inconvenient, as with the Board of Trade's investigation into the cost of poor relief.[16] England was not yet an 'information state' routinely collecting ordered information for every conceivable purpose,[17] although statesmen from Burghley onwards were pushing it in that direction. What was important and innovative in political arithmetic, as Petty and Davenant emphasized, was the construction of hypotheses: Petty's 'suppositions' showing the way to knowledge,[18] which stimulated the search for new kinds of information in order to confirm, refute, or refine them. In that sense, 'invention and improvement', intellectual research and development, were as fundamental to political economy as they were to industry and technology.

The political arithmeticians who gave English political economy some renown in early eighteenth-century Europe also made a further contribution to the intellectual stature of England's improvement culture. The hypotheses they constructed from the limited information available to them were designed to illuminate change over the very long run, beginning in the distant past and projected into the future.[19] Scattered historical data from early customs accounts or subsidies led them to suppositions about changes in England's wealth since the sixteenth century, and their interest in biblical history prompted them to plot the course of population change ever since the Flood. They knew that there were short-term changes, caused particularly by wars or plagues or famines, and sometimes found data to illuminate them, as with Graunt writing about plague in London, and Davenant and King about the probable effects of William III's war. But they were predisposed by the character of their evidence and their reading to see a history of growth and to assume that, albeit with temporary hiccups, it would continue. Material progress seemed to be historically predetermined.

The notion of endless improvement owed its strength, finally, to its diffusion after the Interregnum, and especially after 1688, when information about land and trade, population, wealth, and well-being, and speculation about their history and future, circulated through parliamentary debate and a popular press, as well as in correspondence between friends and conversations in coffee-houses. That was what enabled improvement to become a morality of collective cooperation in a national purpose. One question hovering over the argument earlier in this book has been whether talk about improvement was simply a metropolitan phenomenon, communicated to a London audience and reflecting only their conditions and concerns. Improvement was certainly defined in the metropolis, but the country's culture was

---

[16] Above, pp. 131, 176, 184–5.
[17] Edward Higgs, *The Information State in England: The Central Collection of Information on Citizens since 1500* (Basingstoke, 2004), pp. vii–viii.
[18] Above, p. 117.   [19] Above, pp. 125, 197.

as integrated as its economy by 1700, and information circulated almost as easily and productively as money, or, as contemporaries would have said, as blood nourishing the whole body politic.

When improvement was exported, its diffusion depended on local circumstances. In Ireland and Scotland it was initially the property of an intellectual elite and rooted in their capital cities, and then communicated more widely, especially in eighteenth-century Scotland thanks to its larger literate and reading public.[20] Across the Atlantic, even more than in Ireland, improvement was the creed of conquerors and colonizers, and its supposed benefits were incomprehensible to those they ruled. In the English West Indies in 1676, where English planters were busy exchanging useful knowledge in order to 'tame and improve nature', one of them reported (and perhaps embellished) the complaint of a slave in Barbados that 'the Devil was in the Englishman': 'He makes everything work; he makes the negro work, the horse work, the ass work, the wood work, the water work, and the wind work.'[21] There was no such cultural divide back home. Workers in Defoe's 'most lazy diligent nation in the world' were as conscious as their employers that improvement had given them some of the comforts of life, and as eager to acquire more.

The history of improvement is by definition a Whiggish story of progress, and I make no apology for the fact that this book is consequently a piece of Whig history. In 1848 the greatest of the Whig historians, Lord Macaulay, began his *History of England* from 1688 to his own day with the resounding declaration that 'the history of our country during the last hundred and sixty years is eminently the history of physical, of moral, and of intellectual improvement'.[22] Even at the time some of his readers found that assurance insufferably complacent, had diverse opinions about what moral improvement might consist in, and would have wanted to look more closely, as I have had tried to do, at improvement's contested history and its limitations. Material progress, the subject of this book, was plainly not sufficient for happiness, as its seventeenth-century critics eloquently complained. Writing in *The Theory of Moral Sentiments* about 'the pleasures of wealth and greatness' Adam Smith thought—as Locke had done half a century earlier—that it was a 'deception' to suppose that there was 'something grand and beautiful and noble' about them, since the quest for profit and power caused so much 'toil and anxiety'. Nevertheless, the deception was productive because it kept 'in continual motion the industry of mankind'. Mankind had been driven to 'cultivate the ground, to build houses, to found cities and commonwealths, and to invent and

---

[20] Bob Harris, 'The Enlightenment, Towns and Urban Society in Scotland, c.1760–1820', *EHR* 126 (2011), pp. 1104–5; Mark R. M. Towsey, *Reading the Scottish Enlightenment: Books and their Readers in Provincial Scotland, 1750–1820* (Leiden, 2010), p. 305 and *passim*; above, pp. 97–8, 161.

[21] Nuala Zahedieh, *The Capital and the Colonies: London and the Atlantic Economy, 1660–1700* (Cambridge, 2010), pp. 290–1; Jack P. Greene, 'Changing Identity in the British Caribbean: Barbados as a Case Study', in Nicholas Canny and Anthony Pagden, eds, *Colonial Identity in the Atlantic World 1500–1800* (Princeton, 1987), p. 222.

[22] Thomas Babington Macaulay, *The History of England*, ed. and abridged by Hugh Trevor-Roper (Harmondsworth, 1979), p. 52.

improve all the sciences and arts which ennoble and embellish human life' and all this had 'entirely changed the whole face of the globe'.[23]

Smith's cool and cosmopolitan Enlightenment vision is more acceptable now than Macaulay's insular and unqualified confidence in material progress. Still more acceptable nowadays, when we are more than ever conscious of finite natural resources, might be Robert Malthus's scepticism about whether 'the future improvement of society' could ever be absolutely assured.[24] But all three of them, even Malthus, had absorbed something from an improvement culture invented in England in the seventeenth century. Improvement had never been an intellectual straitjacket incapable of accommodating divergent interpretations of its implications, but neither was it a word which carried no meaning at all. It was an idiom of expression whose constant use encouraged some kinds of rhetoric and action while discouraging others. At its simplest it marked a mode of discourse which looked for inspiration to a better future and not to an ideal past, and it denoted an attitude of mind which supposed that people could improve their condition without damage to the public good and general happiness.

The language of improvement had made the English different from other people, in how they thought about themselves and their collective prospects, and how they acted to attain them. Daniel Defoe has been much cited already in this book, but there is a passage in his *Tour* which merits quotation before I conclude. It captures the essence of gradual and infinite improvement, and explains why any history of it must be incomplete, an interim statement, as this one has been. He refers to a bill going through parliament to make the river Nene navigable from Peterborough to Northampton. It was a major undertaking 'of infinite advantage to the country', had been embarked on earlier, and would take years to complete. But its supporters were 'content with performing it piecemeal, that is to say, some by some, that they may see how practicable it may be'. Defoe uses this as an illustration of the imperturbable determination of improvers and the boundless ambitions of improvement:

There will always be something new for those that come after... to describe... New foundations are always laying, new buildings always raising... new trades are every day erected, new projects enterpriz'd, new designs laid; so that as long as England is a trading, improving nation, no perfect description either of the place, the people, or the conditions and state of things can be given.[25]

Improvement never ended.

---

[23] Adam Smith, *The Theory of Moral Sentiments* (1759), pp. 348–9. Cf. above, pp. 206, 220.
[24] The subtitle of the first edition of Malthus's *Essay on the Principle of Population* (1798) referred to the effect of population on 'the future improvement of society'; it was replaced by a reference to population's 'past and present effects on human happiness' in the second edition (1803). In 1803, however, Malthus still thought a 'gradual and progressive improvement in human society' possible as well as desirable: *Essay* (1803), pp. 603–4.
[25] Daniel Defoe, *A Tour Through the Whole Island of Great Britain*, ed. G. D. H. Cole and D. C. Browning (2 vols, 1962), i. 252.

Bibliography

MANUSCRIPT SOURCES

British Library, London
Add. MSS 5540, 15642, 32523, 47131, 72850, 72866.
Cotton MSS Titus B.v, F.iv; Faustina, C.ii.
Lansdowne MSS 19, 55, 74, 238, 691.
Royal MS 18A.xxv.

Royal Society Library, London
Classified Papers, 10.iii/16 (Mellish, 'England's improvement').
MS 366/1/6 (Beale, 'From Utopia').

Beinecke Library, Yale University, New Haven
Osborn Shelves, b.153, b.395, fb.237.
OSB MSS 1 (Poley Papers), 2 (Blathwayt Papers), 6 (Danby Papers), 41 (Southwell Papers).

ELECTRONIC RESOURCE

The Hartlib Papers: A Complete Text and Image Database of the Papers of Samuel Hartlib (c.1600–1662) (2nd edn on CDROM, HROnline, University of Sheffield, 2002).

PRINTED PRIMARY SOURCES

(Place of publication is London unless otherwise stated.)

Abbot, George, *A Briefe Discription of the whole world* (1634 edn).
Adams, John, *Index Villaris* (1680).
Alleine, Joseph, *The True Way to Happiness* (1675).
Allen, Richard, *Insulae fortunatae: A discourse, shewing the happiness of these nations under our present governours & government* (1675).
Allestree, Richard, *The Art of Contentment* (Oxford, 1675).
Ananias and Saphira discover'd (1679).
Arbuthnot, John, *An Essay on the Usefulness of Mathematical Learning* (Oxford, 1701).
Astell, Mary, *The First English Feminist: Reflections upon Marriage and Other Writings*, ed. Bridget Hill (Aldershot, 1986).
Athenian Gazette (1691, 1693).
Aubin, Robert Arnold, ed., *London in Flames: London in Glory* (New Brunswick, NJ, 1943).
Aubrey, John, *Brief Lives*, ed. Oliver Lawson Dick (3rd edn, 1958).
[B., I.], *A Letter sent by I.B. gentleman* (1571).
Bacon, Francis, *The Works*, ed. James Spedding et al. (7 vols, 1859–64).
Bacon, Francis, *The Letters and Life*, ed. James Spedding (7 vols, 1861–74).
Bacon, Francis, *The Essayes or Counsels, Civill and Morall*, ed. Michael Kiernan (*OFB* xv, Oxford, 1985).

Bacon, Francis, *Francis Bacon: A Critical Edition of the Major Works*, ed. Brian Vickers (Oxford, 1996).
Bacon, Francis, *The Advancement of Learning*, ed. Michael Kiernan (*OFB* iv, Oxford, 2000).
Bacon, Francis, *The Instauratio magna Part II: Novum organum*, ed. Graham Rees (*OFB* xi, Oxford, 2004); Part III, ed. Rees (*OFB* xii, Oxford, 2007).
Bacon, Francis, *Early Writings 1584–1596*, ed. Alan Stewart with Harriet Knight (*OFB* i, Oxford, 2012).
Bacon, Francis, *The Historie of the raigne of King Henry the seventh*, ed. Michael Kiernan (*OFB* viii, Oxford, 2012).
[Barbon, Nicholas], *A Discourse Shewing the Great Advantages that New-Buildings and the Enlarging of Towns and Cities Do Bring to a Nation* (1678).
Barbon, Nicholas, *A Letter to a Gentleman in the Country giving an account of the two insurance-offices* (1684).
Barbon, Nicholas, *An Apology for the Builder* (1685).
Barbon, Nicholas, *A Discourse of Trade* (1690).
Barbon, Nicholas, *A Discourse Concerning Coining the New Money lighter* (1696).
Barlow, Roger, *A Brief Summe of Geographie*, ed. E. G. R. Taylor (Hakluyt Soc. 2nd ser. 69, 1932).
Barnes, Ambrose, *Memoirs of the Life of Mr. Ambrose Barnes*, ed. W. H. D. Longstaffe (Surtees Society 50, 1867).
Barry, Jonathan, and Morgan, Kenneth, eds, *Reformation and Revival in Eighteenth-Century Bristol* (Bristol Record Society 45, 1994).
Bateson, Mary, et al., eds, *Records of the Borough of Leicester* (7 vols, 1899–1974).
Baxter, Richard, *A Holy Commonwealth* (1659).
Baxter, Richard, *A Christian Directory* (1673).
Becher, Johann Joachim, *Magnalia naturae: or, The Truth of the Philosopher's Stone* (1680).
Bellers, John, *John Bellers: His Life, Times and Writings*, ed. George Clarke (1987).
Bentham, Joseph, *The Christian Conflict: A Treatise* (1635).
Besongne, Nicolas, *The Present State of France* (1671).
Best, Henry, *The Farming and Memoranda Books of Henry Best of Elmswell, 1642*, ed. Donald Woodward (British Academy, Records of Social and Economic History NS 8, Oxford, 1984).
Bethel, Slingsby, *The Present Interest of England Stated* (1671).
Bethel, Slingsby, *Observations on the Letter Written to Sir Thomas Osborn* (1673).
Bethel, Slingsby, *An Account of the French Usurpation upon the Trade of England* (1679).
Beverley, Robert, *The History and Present State of Virginia* (1705).
Blanch, John, *An Abstract of the Grievances of Trade which Oppress our Poor* (1694).
Blanch, John, *The Interest of England Considered* (1694).
Bland, John, *Trade Revived* (1659).
Blenerhasset, Thomas, *A Direction for the Plantation in Ulster* (1610).
Blith, Walter, *The English Improver* (1649).
Blith, Walter, *The English Improver Improved* (1652).
Blome, Richard, *A Geographicall Description of the four parts of the World* (1670).
Blome, Richard, *Britannia* (1673).
Blount, Henry, *Voyage into the Levant* (2nd edn, 1636).
Blundell, Nicholas, *The Great Diurnal of Nicholas Blundell of Little Crosby, Lancashire*, i: *1702–1711*, ed. J. J. Bagley (Lancashire and Cheshire Record Society 110, 1968).
Bodin, Jean, *The Six Bookes of a Common-weale* (1606).

Bohun, Edmund, *The Diary and Autobiography of Edmund Bohun Esq.*, ed. S. Wilton Rix (Beccles, 1853).
Boisguilbert, Pierre de, 'De la nécessité d'un traité de paix entre Paris et le reste du royaume', in *Pierre de Boisguilbert ou La Naissance de l'économie politique* (2 vols, Paris, 1966).
Bolton, Robert, *A Short and Private Discourse . . . concerning Usury* (1637).
Botero, Giovanni, *The Travellers Breviat, Or An historicall description of the most famous kingdomes in the World* (1601).
Botero, Giovanni, *A Treatise concerning the causes of the Magnificencie and greatness of Cities*, trans. Robert Peterson (1606).
Botero, Giovanni, *The Reason of State, and the Greatness of Cities*, ed. D. P. Waley (1956).
Bottarelli, Ferdinando, *The New Italian, English and French Pocket-Dictionary* (3 vols, 1777).
Boyle, Robert, *General Heads for the Natural History of a Country, Great or Small* (1692).
Boyle, Robert, *The Correspondence of Robert Boyle*, ed. Michael Hunter, Antonio Clericuzio, and Lawrence Principe (6 vols, 2001).
Brewster, Francis, *Essays on Trade and Navigation* (1695).
Britannia languens (1680), in McCulloch, ed., *Early English Tracts on Commerce*.
The British Merchant: or Commerce Preserv'd (1713, 1714).
Browne, Thomas, *Certain Miscellany Tracts* (1683).
Browning, Andrew, ed., *English Historical Documents, 1660–1714* (London, 1953).
Bruzen de la Martinière, Antoine Augustin, *Le Grand Dictionnaire géographique et critique* (The Hague, Amsterdam, Rotterdam, 9 vols, 1726–39).
Burton, Thomas, *Diary*, ed. John Towill Rutt (4 vols, 1828).
Butler, Joseph, *Fifteen Sermons Preached at the Rolls Chapel* (1726).
Camden, William, *Britain, or a chorographicall description of the most flourishing kingdoms, England, Scotland and Ireland, and the ilands adjoining*, trans. Philemon Holland (1610).
Cantillon, Philip, *The Analysis of Trade, Commerce, Coin, Bullion . . .* (1759).
Cantillon, Richard, *Essai sur la nature du commerce en général* (London [i.e. Paris?], 1755).
Cantillon, Richard, *Essai sur la nature du commerce en général*, ed. Henry Higgs (1931).
Carl, Ernst Ludwig, *Traité de la richesse des princes, et de leurs états* (3 vols, Paris, 1722–3).
Carpenter, Nathaniel, *Geography Delineated* (Oxford, 1625).
Carter, William, *England's Interest Asserted, in the Improvement of its Native Commodities* (1669).
Cary, John, *An Essay on the State of England, in Relation to its Trade* (Bristol, 1695).
Cary, John, *An Essay on the Coyn and Credit of England* (Bristol, 1696).
Cary, Walter, *The Present State of England* (1626).
Case, John, *Thesaurus oeconomiae* (Oxford, 1597).
Cawdrey, Robert, *A Table Alphabeticall* (3rd edn, 1613).
Certain Considerations Relating to the Royal African Company of England (1680).
Chamberlayne, Edward, *England's Wants: or several proposals probably beneficial for England* (1667).
Chamberlayne, Edward, *Angliae notitia: or, The Present State of England* (1st edn, 1669, numerous later edns).
Chamberlayne, Edward, *An Academy or Colledge, wherein young ladies and gentlewomen may . . . be duly instructed in the true Protestant religion* (1671).
Chamberlen, Hugh, *Dr Hugh Chamberlen's Proposal to make England Rich and Happy* (1690).
Chamberlen, Hugh, *Some Useful Reflections upon a Pamphlet called A Brief Account of The Intended Bank of England* (1694).

Chambers, Ephraim, *Cyclopaedia* (2 vols, 1728).
The Character and Qualifications of an Honest Loyal Merchant (1686).
Child, Josiah, *Brief Observations concerning Trade, and Interest of Money* (1668).
Child, Josiah, *A Discourse about Trade* (1690).
Child, Josiah, *A New Discourse of Trade* (1751 edn).
Child, Josiah, *Traités sur le commerce* (Amsterdam and Berlin, 1754).
Childrey, Joshua, *Britannia Baconica: or, the natural rarities of England, Scotland, & Wales* (1661).
Clement, Simon, *The Interest of England, as it Stands, with Relation to the Trade of Ireland* (1698).
Clifford, Anne, *The Diaries of Lady Anne Clifford*, ed. D. J. H. Clifford (Stroud, 1990).
Coke, Roger, *A Discourse of Trade* (1670).
Coke, Roger, *A Treatise Wherein is demonstrated, That the church and state of England, are in equal danger with the trade of it* (1671).
Coke, Roger, *England's Improvements* (1675).
Collins, John, *A Plea For the bringing in of Irish Cattel* (1680).
Considerations Requiring greater Care for Trade (1695).
Cooke, John, *Unum necessarium: or The Poore Mans Case* (1648).
Cooper, Anthony Ashley, 3rd Earl of Shaftesbury, *Characteristicks of Men, Manners, Opinions, Times*, ed. Philip Ayres (Oxford, 1999).
Cotton, John, *God's Promise to his Plantation* (1630).
Cradocke, Francis, *Wealth Discovered* (1661).
Crofton, Zachary, *Excise Anatomiz'd and Trade Epitomiz'd* (1659).
Crofts, Robert, *The Terrestriall Paradise, Or, Happinesse on Earth* (1639).
Crofts, Robert, *Paradise Within Us: Or, The happie Mind* (1640).
Crofts, Robert, *The Way to Happinesse on Earth: Concerning Riches, Honour, Conjugall Love, Eating, Drinking* (1641).
Culpeper, Cheney, 'The Letters of Sir Cheney Culpeper, 1641–1657', ed. M. J. Braddick and Mark Greengrass, in *Camden Miscellany XXXIII* (Camden Society 5th ser. 7, 1996).
Dasent, John Roche, et al., eds, *Acts of the Privy Council of England*, NS *1542–1631*, 46 vols (1890–1964).
Davenant, Charles, *The Political and Commercial Works*, ed. Sir Charles Whitworth (5 vols, 1771).
D'Avity, Pierre, *The Estates, Empires & Principalities of the World* (1615).
Dee, John, *The Elements of Geometrie of... Euclide* (1570).
Defoe, Daniel, *Taxes no Charge* (1690).
Defoe, Daniel, *An Essay upon Projects* (1697).
Defoe, Daniel, *An Argument shewing, that a Standing Army... Is not Inconsistent with a Free Government* (1698).
Defoe, Daniel, *The Poor Man's Plea* (1698).
Defoe, Daniel, *Reformation of Manners: A Satyr* (1702).
Defoe, Daniel, *Giving Alms no Charity* (1704).
Defoe, Daniel, *The Present State of the Parties* (1712).
Defoe, Daniel, *A General History of Trade: For the month of July* (1713).
Defoe, Daniel, *The Great Law of Subordination consider'd* (1724).
Defoe, Daniel, *A General History of Discoveries and Improvements* (in four monthly parts, 1725–6).
Defoe, Daniel, *Some Considerations upon Street-Walkers* (1726).
Defoe, Daniel, *The Complete English Tradesman* (1726).

Defoe, Daniel, *The Complete English Tradesman* (2 vols, 2nd edn, 1727).
Defoe, Daniel, *A Plan of the English Commerce* (1728).
Defoe, Daniel, *A Tour Through the Whole Island of Great Britain*, ed. G. D. H. Cole and D. C. Browning (2 vols, 1962).
De la Court, Pieter, *The True Interest and Political Maxims of the Republick of Holland and West-Friesland* (1702).
De la Court, Pieter and Johan, *Fables Moral and Political* (2 vols, 1703).
Derham, William, *Physico-Theology* (1714).
Desaguliers, John Theophilus, *A Course of Experimental Philosophy* (2 vols, 1734, 1763).
'De Souligné', *The Desolation of France Demonstrated* (1697).
'De Souligné', *A Comparison Between Old Rome in its Glory, as to the Extent and Populousness, and London as it is at present* (1706).
Diderot, Denis, and d'Alembert, Jean, eds, *Encyclopédie, ou Dictionnaire raisonné des sciences, des arts et des métiers* (17 vols, Paris, 1751–65).
Digges, Thomas, *An Arithmeticall Militare Treatise, named Stratioticos* (1579).
A Discourse of the Nature, Use, and Advantages of Trade (1694).
A Discourse of the Necessity of Encouraging Mechanick Industry (1690).
Donaldson, James, *Husbandry Anatomized, or An Enquiry into the Present Manner of teiling and manuring the ground in Scotland* (Edinburgh, 1697).
Donne, John, *A Sermon . . . To the Honourable Company of the Virginian Plantation* (1622).
Douglass, William, *A Summary, Historical and Political, Of the first Planting, progressive Improvements, and present State of the British Settlements in North-America* (2 vols, Boston, 1749–51).
Dyer, Alan, and Palliser, D. M., eds, *The Diocesan Population Returns for 1563 and 1603* (British Academy, Records of Social and Economic History ns 31, 2005).
Eburne, Richard, *Plaine Path-Way to Plantations* (1624).
Eliot, John, *The Survay or Topographical Description of France* (1592).
Englands Vanity; or, The voice of God against the monstrous sin of pride in dress and apparel (1683).
An Essay or Modest Proposal of a Way to encrease the Number of People (1693).
Evelyn, John, *The State of France* (1652).
Evelyn, John, *A Character of England* (1659).
Evelyn, John, *Fumifugium: or the Inconveniencie of the Aer and Smoak of London Dissipated* (1661).
Evelyn, John, *Navigation and Commerce, their Original and Progress* (1674).
Evelyn, John, *The Diary of John Evelyn*, ed. E. S. de Beer (6 vols, Oxford, reprint 2000).
Everett, George, *The Path-way to Peace and Profit* (1694).
Fenner, William, *A Treatise of the Affections* (1650).
Fenton, Roger, *A Treatise of Usurie* (1611).
Ferguson, Adam, *An Essay on the History of Civil Society* (Edinburgh, 1767).
[Ferguson, Robert], *The East-India-Trade: A Most Profitable Trade* (1677).
Fiennes, Celia, *The Journeys of Celia Fiennes*, ed. Christopher Morris (1947).
Fitzherbert, John, *The Boke of surveyeng and improumentes* (1523).
Fitzherbert, John, *The Boke of Husbandry* (?1533).
Fletcher, Andrew, *Political Works*, ed. John Robertson (Cambridge, 1997).
Florio, John, *A Worlde of Wordes* (1598).
Florio, John, *Vocabolario italiano & inglese* (1659).
Forset, Edward, *A Comparative Discourse of the Bodies Natural and Politique* (1606).
Fortrey, Samuel, *Englands Interest and Improvement* (Cambridge, 1663).

Fowler, Alastair, ed., *The New Oxford Book of Seventeenth-Century Verse* (Oxford, 1991).
Franklin, Benjamin, 'Observations on the Increase of Mankind', in *Political, Miscellaneous, and Philosophical Pieces* (1779).
Fuller, Thomas, *The History of the Worthies of England* (1662).
Furetière, Antoine, *Dictionnaire universel* (The Hague, 1690).
Gainsford, Thomas, *The Glory of England* (1618).
Gardiner, Samuel Rawson, ed., *The Constitutional Documents of the Puritan Revolution 1625–1660* (3rd edn, Oxford, 1906).
Gee, Joshua, *The Trade and Navigation of Great-Britain Considered* (1729).
Gibson, Edmund, *Camden's Britannia, Newly Translated into English with Large Additions and Improvements* (1695; 2nd edn, 1722).
Glanvill, Joseph, *The Way of Happiness* (1670).
Glasscock, Robin E., ed., *The Lay Subsidy of 1334* (British Academy, Records of Social and Economic History NS 2, Oxford, 1975).
Goodman, Godfrey, *The Fall of Man, or the corruption of nature proved by the light of our natural reason* (1616).
Gott, Samuel, *An Essay of the True Happines of Man* (1650).
Gough, Richard, *The History of Myddle*, ed. David Hey (Harmondsworth, 1981).
The Grand Concern of England Explained (1673).
Grew, Nehemiah, *Cosmologia sacra* (1701).
Grew, Nehemiah, *Nehemiah Grew and England's Economic Development: the means of a most ample increase of the wealth and strength of England, 1706–7*, ed. Julian Hoppit (British Academy Records of Social and Economic History NS 47, Oxford, 2012).
Grey, Anchitell, ed., *Debates of the House of Commons from the year 1667 to the year 1694* (10 vols, 1763).
Hagthorpe, John, *Englands-Exchequer. Or A Discourse Of The Sea and Navigation* (1625).
Haines, Richard, *Provision for the Poor* (1678).
Hale, Matthew, *The Primitive Origination of Mankind, considered and examined according to the light of nature* (1677).
Hale, Thomas, *An Account of Several New Inventions and Improvements now necessary for England* (1691).
Hall, Joseph, *The Discovery of A New World or A Description of the South Indies, Hetherto Unknowne* (1609).
Halley, Edmond, 'Some Further Considerations on the Breslaw Bills of Mortality', *Philosophical Transactions of the Royal Society*, 17.198 (1693), pp. 654–6.
Harrington, James, *The Political Works of James Harrington*, ed. J. G. A. Pocock (Cambridge, 1977).
Harriot, Thomas, *A Briefe and True Report of the New Found Land of Virginia*, ed. Theodore de Bry (Frankfurt, 1590).
Harris, James, *Three Treatises* (2nd edn, 1766).
Harris, John, *Lexicon technicum: or, An Universal English Dictionary of Arts and Sciences* (2 vols, 1704–10).
Harrison, William, *The Description of England*, ed. Georges Edelen (Ithaca, NY, 1968).
Harrold, Edmund, *The Diary of Edmund Harrold, Wigmaker of Manchester 1712–15*, ed. Craig Horner (Aldershot, 2008).
Hartley, T. E., ed., *Proceedings in the Parliaments of Elizabeth I* (3 vols, Leicester, 1995).
Hartlib, Samuel, *A Rare and New Discovery* (1652).
Hartlib, Samuel, *Samuel Hartlib his Legacie* (2nd edn, 1652).
Hartlib, Samuel, *A Discoverie for Division or Setting out of Land* (1653).

Hartlib, Samuel, *Samuel Hartlib and the Advancement of Learning*, ed. Charles Webster (Cambridge, 1970).
Hawkins, Richard, *A Discourse of the Nationall Excellencies of England* (1658).
Haynes, John, *Great Britain's Glory* (1715).
Herbert, Thomas, *A Relation of some yeares travaile, begunne Anno 1626* (1634).
Heylyn, Peter, *Cosmographie in Four Bookes* (1652).
Hobbes, Thomas, *Leviathan*, ed. Noel Malcolm (3 vols, Oxford, 2012).
Holden, Richard, *The Improvement of Navigation a great Cause of the Increase of Knowledge* (1680).
Hooke, Andrew, *An Essay on the National Debt, and National Capital* (1750).
Hooke, Robert, *The Diary of Robert Hooke 1672–80*, ed. Henry W. Robinson and Walter Adams (1935).
Hoppen, K. Theodore, ed., *Papers of the Dublin Philosophical Society, 1683–1709* (2 vols, Irish Manuscripts Commission, Dublin, 2008).
Houghton, John, *England's Great Happiness; or, A Dialogue between Content and Complaint* (1677).
Houghton, John, *A Proposal for Improvement of Husbandry and Trade* (1691).
Houghton, John, *An Account of the Acres & Houses with the Proportional Tax &c* (1693).
Houghton, John, *A Collection for the Improvement of Husbandry and Trade*, ed. Richard Bradley (4 vols, 1727–8).
Howell, James, *Instructions for Forreine Travell* (1642).
Howell, James, *Londinopolis; An Historical Discourse or Perlustration of the City of London* (1657).
Howell, James, *Lexicon tetraglotton: an English–French–Italian–Spanish dictionary* (1660).
Hughes, Paul L., and Larkin, James F., eds, *Tudor Royal Proclamations* (3 vols, New Haven, 1969).
Hulton, Mary H. M., ed., *Coventry and its People in the 1520s* (Dugdale Society 38, 1999).
Hume, David, *An Enquiry concerning the Principles of Morals* (1751).
Hume, David, *Political Discourses* (Edinburgh, 1752).
Hume, David, 'Of the Populousness of Antient Nations', in *Essays and Treatises on Several Subjects* (2nd edn, 1753).
Hume, David, *The History of England* (8 vols, 1763).
The Importance and Management of the British Fishery Consider'd (1720).
Interest of Money Mistaken, Or A treatise, proving, that the abatement of interest is the effect and not the cause of the riches of a nation (1668).
J., H., *A Letter from a Gentleman in the Country to his Friend in the City* (1691).
Jefferson, Thomas, *Notes on the State of Virginia* (2nd American edn, Philadelphia, 1794).
Johnson, Robert, *Nova Britannia* (1609).
Johnson, Thomas, *A Discourse Consisting of Motives for the Enlargement and Freedome of Trade* (1645).
Johnson, Thomas, *A Plea for Free-Mens Liberties* (1646).
Journal oeconomique, ou Mémoires, notes, et avis sur les arts, l'agriculture, le commerce (Paris, 1756).
Le Journal des sçavans pour l'année MDCLXVI (new edn, Paris, 1729).
Journals of the House of Commons.
Journals of the House of Lords.
Kayll, Robert, *The Trades Increase* (1615).

Kennett, White, *Parochial Antiquities attempted in the History of Ambrosden, Burcester and other adjacent parts...* (Oxford, 1695).
Keymer, John, *Original Papers regarding Trade in England and Abroad drawn up by John Keymer*, ed. M. F. Lloyd Prichard (New York, 1967).
King, Charles, ed., *The British Merchant* (3 vols, 1721).
King, Charles, ed., *Le Négotiant anglois*, trans. François Véron de Forbonnais (2 vols, Dresden, 1753).
King, Gregory, *A Scheme of the Rates and Duties...* (1695).
King, Gregory, *Two Tracts by Gregory King*, ed. G. E. Barnett (Baltimore, 1936).
King, Gregory, 'Natural and Political Observations and Conclusions upon the State and Condition of England, 1696', in Laslett, ed., *Earliest Classics*.
King, Gregory, 'Burns Journal', in Laslett, ed., *Earliest Classics*.
Langeren, Jacob van, *A Direction for the English Traviller* (1635).
Larkin, James F., and Hughes, Paul L., eds, *Stuart Royal Proclamations* (2 vols, Oxford, 1973, 1983).
Laslett, Peter, ed., *The Earliest Classics* (Farnborough, 1973).
Latham, Richard, *The Account Book of Richard Latham 1724–1767*, ed. Lorna Weatherill (British Academy, Records of Social and Economic History NS 15, Oxford, 1990).
Law, John, *Money and Trade considered, with a Proposal for supplying the Nation with Money* (Edinburgh, 1705).
Law, John, *Considerations sur le commerce et sur l'argent par Mr. Law, Controlleur Genéral des Finances* (The Hague, 1720).
Law, John, *Gedancken von Waaren- und Geld-Handel* (Leipzig, 1720).
Lee, Joseph, *Considerations Concerning Common Fields* (1654).
Lee, Joseph, *A Vindication of a Regulated Inclosure* (1656).
Leigh, Edward, *England Described* (1659).
Lemaître, Alexandre, *La Metropolitée ou De l'établissement des villes capitales* (Amsterdam, 1682).
A Letter to a Member of the Honourable House of Commons, In Answer to Three Queries (1697).
Lewis, Mark, *Proposals to increase Trade* (1677).
Lewis, Mark, *Proposals to the King and Parliament: Or A Large Model of a Bank* (1678).
Ligon, Richard, *A True and Exact History of the Island of Barbados* (1657).
Littleton, Edward, *The Groans of the Plantations* (1689).
Littré, Émile, *Dictionnaire de la langue française* (2 vols, Paris, 1863–73).
Locke, John, *Two Treatises of Government* (1690).
Locke, John, *An Essay concerning Human Understanding* (2 vols, 1715–16).
Locke, John, *Locke on Money*, ed. Patrick Hyde Kelly (2 vols, Oxford, 1991).
Locke, John, *Locke: Political Essays*, ed. Mark Goldie (Cambridge, 1997).
Lowndes, William, *A Report Containing an Essay for the Amendment of the Silver Coins* (1695).
Lowther, Sir John, *The Correspondence of Sir John Lowther of Whitehaven 1693–1698*, ed. D. R. Hainsworth (British Academy, Records of Social and Economic History NS 7, Oxford, 1983).
Luders, M., et al., eds, *Statutes of the Realm* (11 vols, 1820–8).
Ludovici, Christian, *A Dictionary English, German and French* (2nd edn, Leipzig and Frankfurt, 1736).
Luttrell, Narcissus, *The Parliamentary Diary of Narcissus Luttrell, 1691–3*, ed. Henry Horwitz (Oxford, 1972).

Macaulay, Thomas Babbington, Lord, *The History of England*, ed. and abridged by Hugh Trevor-Roper (Harmondsworth, 1979).
McCulloch, J. R., ed., *Early English Tracts on Commerce* (Cambridge, 1952).
Mackenzie, George, *The Moral History of Frugality with its opposite Vices* (1691).
Mackie, Erin, *The Commerce of Everyday Life: Selections from The Tatler and The Spectator* (1998).
[Mackworth, Humphrey], *England's Glory: or, The Great Improvement of Trade in general by a Royal Bank* (1694).
Maddison, Ralph, *Englands Looking In and Out* (1640).
Maddison, Ralph, *Great Britains Remembrancer Looking In and Out* (1655).
Maitland, Charles, *Mr Maitland's Account of Inoculating the Small Pox* (1722).
Malthus, Thomas Robert, *An Essay on the Principle of Population* (1798, 2nd edn, 1803).
Malynes, Gerard, *Saint George for England, Allegorically described* (1601).
Malynes, Gerard, *A Treatise of the Canker of Englands Common wealth* (1601).
Malynes, Gerard, *Englands View, in the Unmasking of Two Paradoxes* (1603).
Malynes, Gerard, *Consuetudo vel lex mercatoria, or The Ancient Law-Merchant* (1622).
Mandeville, Bernard, *The Fable of the Bees or Private Vices, Publick Benefits*, ed. F. B. Kaye (2 vols, Oxford, 1924).
Markham, Gervase, *Markhams farwell to Husbandry* (1620).
Marriage Promoted: In a Discourse of its Ancient and Modern Practice (1690).
Martin, Henry, *Considerations upon the East-India Trade* (1701).
Martindale, Adam, *The Life of Adam Martindale*, ed. Richard Parkinson (Chetham Society 4, 1845).
Massey, Edmund, *A Letter to Mr Maitland, in Vindication of the Sermon against Inoculation* (1722).
Massie, Joseph, *Calculations of Taxes for a Family of each Rank...* (1756).
Massie, Joseph, *A Plan For the Establishment of Charity-Houses* (1758).
Massie, Joseph, *Reasons Humbly Offered Against laying any farther Tax upon Malt or Beer* (1760).
Mather, Cotton, *Magnalia Christi Americana* (1702).
Mayerne, Louis Turquet de, *La Monarchie aristodémocratique* (Paris, 1611).
Melon, Jean-François, *Essai politique sur le commerce* (n.p., 1734; new edn, n.p., 1736).
Mercator: Or, Commerce Retrieved (1713, 1714).
Mexía, Pedro de, *The Treasurie of Aunciant and Moderne Times* (1613).
Miège, Guy, *A New Dictionary French and English* (1677).
Miège, Guy, *The Great French Dictionary. In Two Parts* (1688).
Miège, Guy, *The Present State of Great Britain* (2 vols, 1707).
Misselden, Edward, *Free Trade, or The Meanes to Make Trade Florish* (1622).
Misselden, Edward, *The Circle of Commerce: Or the Ballance of Trade* (1623).
Moffett, Thomas, *Healths Improvement* (1655).
Montesquieu, *Persian Letters*, trans. C. J. Betts (1977).
Morden, Robert, *The New Description and State of England* (1701).
More, Nicholas, *The Articles, Settlement and Offices of the Free Society of Traders in Pennsylvania...for the better Improvement and Government of Trade in that Province* (1682).
More, Thomas, *Utopia*, ed. Edward Surtz and Jack H. Hexter (New Haven, 1965).
Mun, Thomas, *A Discourse of Trade from England unto the East-Indies* (1621), in McCulloch, ed., *Early English Tracts on Commerce*.
Mun, Thomas, *England's Treasure by Forraign Trade* (1664).

Nalson, John, *The Present Interest of England; or, A Confutation of the Whiggish Conspiratours Anty-Monyan Principle* (1683).
Nelson, Robert, *A Companion for the Feastivals and Fasts of the Church of England* (3rd edn, 1705).
Nelson, Robert, *An Address to Persons of Quality and Estate* (1715).
Neve, Richard, *Arts Improvement* (1715).
A New and General Biographical Dictionary (11 vols, 1761–2).
Newton, A. P., ed., *Calendar of the Manuscripts of Lord Sackville*, i: *Cranfield Papers 1551–1612* (HMC 80, 1942).
Nicholls, William, *A Conference with a Theist* (1698).
Nicole, Pierre, *Moral Essays* (2 vols, 1677–80).
Nicolson, William, *A Seventeenth-Century Flora of Cumbria: William Nicolson's Catalogue of Plants, 1690*, ed. E. Jean Whittaker (Surtees Society 193, 1981).
Norden, John, *England: An Intended Guyde for English Travailers* (1625).
North, Dudley, 4th Baron, *Observations and Advices Oeconomical* (1669).
North, Dudley, *Discourses upon Trade* (1691).
North, Roger, *The Lives of the Norths* (3 vols, 1890).
Overall, W. H. and H. C., eds, *Analytical Index... to the Remembrancia... of the City of London* (1878).
Overbury, Thomas, *Sir Thomas Overbury, his Observations in his travailes* (1626).
Palmer, Thomas, *An Essay of the Meanes how to make our Trauailes, into forraine Countries, the more profitable and honourable* (1606).
Parker, Henry, *Of a Free Trade* (1648).
Paterson, William, *A Brief Account of the Intended Bank of England* (1694).
Paterson, William, *Proposals and Reasons for Constituting a Council of Trade* (Edinburgh, 1701).
Paterson, William, *The Occasion of Scotland's Decay in Trade* (n.p., 1705).
Paterson, William, *An Inquiry into the Reasonableness... of an Union with Scotland* (1706).
Paxton, Peter, *A Discourse Concerning the Nature, Advantage, and Improvement of Trade* (1704).
Peacham, Henry, *The Worth of a Peny* (1641).
Pepys, Samuel, *The Diary of Samuel Pepys*, ed. R. C. Latham and W. Matthews (11 vols, 1970–83).
Pett, Peter, *The Happy Future State of England* (1688).
Petty, William, *The Discourse... Concerning the Use of Duplicate proportion* (1674).
Petty, William, *Another Essay in Political Arithmetick* (1683).
Petty, William, *The Advice of W.P. to Mr Samuel Hartlib* (1648), in *Harleian Miscellany* (10 vols, 1808–13), vi. 141–58.
Petty, William, *The History of the Survey of Ireland, commonly called the Down Survey*, ed. T. A. Larcom (Dublin 1851).
Petty, William, *The Economic Writings of Sir William Petty, together with the Observations upon the Bills of Mortality*, ed. Charles Henry Hull (2 vols, Cambridge, 1899).
Petty, William, *The Petty Papers*, ed. Marquis of Lansdowne (2 vols, 1927).
Petty, William, *The Petty–Southwell Correspondence 1676–1687*, ed. Marquis of Lansdowne (1928).
Petty, William, *Petty Papers: Additional Manuscripts 72850–72908, Additional Charters 76966–76990* (British Library, 2000).
Petty, William, *William Petty on the Order of Nature: An Unpublished Manuscript Treatise*, ed. with an introduction by Rhodri Lewis (Tempe, Ariz., 2012).

Philalethes, *The Plain Man's Essay for England's Prosperity* (1698).
Philips, Erasmus, *An Appeal to Common Sense* (1720).
Philips, Erasmus, *The State of the Nation, In Respect to her Commerce, Debts, and Money* (1725).
Philo-Dicaeus, *The Standard of Equality in Subsidiary Taxes & Payments, or A Just and strong Preserver of Publique Liberty* (1647).
Philopatris, *A Treatise wherein is demonstrated... That the East-India trade is the most national of all foreign trades* (1681).
Plat, Hugh, *The Jewell House of Art and Nature* (1594).
Plattes, Gabriel, *A Discovery of Infinite Treasure, hidden since the worlds beginning* (1639).
[Plattes, Gabriel], *A Description of the Famous Kingdome of Macaria* (1641).
Plockhoy, Pieter Corneliszoon, *An Invitation to the aforementioned Society or Little Common-Weath* (1660).
Plot, Robert, *The Natural History of Oxford-shire* (2nd edn, Oxford, 1705).
Pollexfen, John, *A Discourse of Trade, Coyn, and Paper Credit* (1697).
Pollexfen, John, *England and East-India Inconsistent in their Manufactures* (1697).
Pollexfen, John, *A Vindication of Some Assertions relating to Coin and Trade* (1699).
Postlethwayt, Malachy, *A Dissertation on the Plan, Use, and Importance, of The Universal Dictionary of Trade and Commerce* (1749).
Potter, William, *The Key of Wealth, or, A new Way, for Improving of Trade* (1650).
Potter, William, *The Trades-Man's Jewel* (1650).
Povey, Charles, *The Unhappiness of England, as to its trade by sea and land* (1701).
Powell, Robert, *Depopulation Arraigned* (1636).
Powell, Thomas, *Humane Industrie: or, A History of most Manual Arts Deducing the Original, Progress and Improvement of them* (1661).
The Public Intelligencer, 24–31 Mar. 1656.
Puckle, James, *England's Path to Wealth and Honour* (2nd edn, 1700).
Pufendorf, Samuel, *An Introduction to the History of the Principal Kingdoms and States of Europe* (1695).
Purchas, Samuel, *Hakluytus Posthumus, or Purchas his Pilgrimes* (4 vols, 1625).
The Queen an Empress: And her Three Kingdoms one Empire (Dublin, 1706).
Raleigh, Walter, *The History of the World* (1614).
Raleigh, Walter, *Judicious and Select Essayes and Observations* (1650).
Raleigh, Walter, *Observations touching Trade and Commerce with the Hollander* (1653).
Reynell, Carew, *The True English Interest or An Account of the Chief National Improvements* (1674).
Richelet, Pierre, *Dictionnaire françois* (Geneva, 1680).
Risdon, Tristram, *A Chorographical Description or Survey of the County of Devon* (1811).
Roberts, Lewes, *The Merchants Mappe of Commerce* (1638).
Roberts, Lewes, *The Treasure of Traffike* (1641), in McCulloch, ed., *Early English Tracts on Commerce*.
Robinson, Henry, *Englands Safetie in Trades Increase* (1641).
Robinson, Henry, *Briefe Considerations, Concerning the advancement of Trade and Navigation* (1649).
Robinson, Henry, *The Office of Adresses and Encounters* (1650).
Robinson, Henry, *Certain Proposalls in order to the Peoples Freedome* (1652).
Rose, Philip, *An Essay on the Small-pox, Whether Natural or Inoculated* (1724).
[S., J.], *The Fourth Part of the Present State of England* (1683).
S., W., *The Golden Fleece* (1656).

Savery, Thomas, *Navigation Improv'd* (1698).
Savery, Thomas, *The Miners Friend* (1702).
Scott, William, *An Essay of Drapery, 1635*, ed. Sylvia L. Thrupp (Boston, 1953).
Seller, John, *Atlas minimus, or A Book of Geography* (1679).
Sergier, Richard, *The Present State of Spaine* (1594).
Serra, Antonio, *A Short Treatise on the Wealth and Poverty of Nations (1613)*, ed. Sophus A. Reinert, trans. Jonathan Hunt (2011).
Serres, Olivier de, *Le Théâtre d'agriculture et mesnage des champs* (Paris, 2nd edn, 1603).
Serres, Olivier de, *The Perfect Use of silk-wormes, and their benefit* (1607), trans. Nicholas Geffe.
Seyssel, Claude de, *La Monarchie de France et deux autres fragments politiques*, ed. Jacques Poujol (Paris, 1961).
Shaw, Wm A., ed., *Select Tracts and Documents Illustrative of English Monetary History 1626–1730* (1896).
Sheridan, Thomas, *A Discourse of the Rise and Power of Parliaments* (1677).
Short, Thomas, *New Observations on City, Town and Country Bills of Mortality* (1750).
Short, Thomas, *A Comparative History of the Increase and Decrease of Mankind in England, and Several Countries Abroad* (1767).
Simpson, William, *A Short Essay towards the History and Cure of Fevers* (1678).
Smart, John, *A Scheme of the Proportions the Several Counties in England paid to the land tax... compared with the Number of Members they send to Parliament* [1698].
Smith, Adam, *The Theory of Moral Sentiments* (1759).
Smith, Adam, *The Wealth of Nations*, ed. Andrew Skinner (2 vols, 1999).
Smith, John, *England's Improvement Reviv'd* (1670).
[Smith, Thomas], *A Discourse of the Common Weal*, ed. Elizabeth Lamond (Cambridge, 1954).
Smith, Thomas, *De republica Anglorum*, ed. Mary Dewar (Cambridge, 1982).
Smith, William, *The Particular Description of England, 1588*, ed. H. B. Wheatley and E. W. Ashbee (1879).
Some Observations upon the Bank of England (1695).
Somerset, Edward, Marquis of Worcester, *A Century of the Names and Scantlings of such Inventions, as at present I can call to mind to have tried and perfected* (1663).
Sorbière, Samuel, *Relation d'un voyage en Angleterre* (Cologne, 1666).
Sorbière, Samuel, *A Voyage to England... As also Observations on the same Voyage, by Dr Thomas Sprat* (1709).
Sprat, Thomas, *Observations on Monsieur de Sorbier's Voyage into England* (1668).
Sprat, Thomas, *History of the Royal Society*, ed. Jackson I. Cope and Harold Whitmore Jones (1959).
Stafford, J. Martin, ed., *Private Vices, Publick Benefits? The Contemporary Reception of Bernard Mandeville* (Solihull, 1997).
Steele, Richard, *The Husbandmans Calling* (1668).
Steele, Richard, *The Trades-man's Calling* (1684).
Stephens, Edward, *A Letter to a Lady, Concerning the due Improvement of her Advantages of Celibacie, Portion, and Maturing of Age and Judgment* (n.p., ?1695).
Stepney, George, *An Essay upon the Present Interest of England* (2nd edn, 1701).
Steuart, Sir James, *An Inquiry into the Principles of Political Oeconomy, being an Essay on the Science of Domestic Policy in Free Nations* (2 vols, 1767).
Stillingfleet, Edward, *Origines sacrae* (1662).

Stillingfleet, Edward, *A Sermon Preached on the Fast-Day, November 13. 1678* (5th edn, 1679).
Stout, William, *The Autobiography of William Stout of Lancaster 1665–1752*, ed. J. D. Marshall (Manchester, 1967).
Stow, John, *A Survey of London*, ed. Charles Lethbridge Kingsford (2 vols, Oxford, 1971).
[Streater, John], *Observations Historical, Political, and Philosophical, Upon Aristotle's first Book of Political Government* (1654).
Swift, Jonathan, *A Modest Proposal* (Dublin, 1729).
Sydenham, Thomas, *The Whole Works*, trans. John Pechey (1696).
Symson, Joseph, *'An Exact and Industrious Tradesman': The Letter Book of Joseph Symson of Kendal, 1711–1720*, ed. S. D. Smith (British Academy, Records of Social and Economic History ns 34, Oxford, 2002).
T., R., *The Art of Good Husbandry, Or, the Improvement of Time* (1675).
A Table of the cheiffest Citties, and Townes in England, as they ly from London (*c*.1600).
Tawney, R. H., and Power, Eileen, eds, *Tudor Economic Documents* (3 vols, 1924).
Taylor, Silvanus, *Common-good, or The Improvement of Commons, Forrests, and Chases, by Inclosure* (1652).
Temple, Richard, *An Essay upon Taxes* (1693).
Temple, William, *An Essay upon the Advancement of Trade in Ireland* (?Dublin, 1673).
Temple, William, *Remarques sur l'estat des Provinces Unies des Païs-bas* (The Hague, 1674).
Temple, William, *Miscellanea* (1680).
Temple, William, *Miscellanea: The Second Part* (1690).
Temple, William, *An Introduction to the History of England* (1695).
Temple, William, *Observations upon the United Provinces of the Netherlands*, ed. Sir George Clark (Oxford, 1972).
Templeman, Thomas, *A New Survey of the Globe* (1729).
Thirsk, Joan, and Cooper, J. P., eds, *Seventeenth-Century Economic Documents* (Oxford, 1972).
Thomas, Dalby, *An Historical Account of the Rise and Growth of the West-India Collonies* (1690).
[Thomas, Dalby], *Some Thoughts Concerning the Better Security of our Trade and Navigation* (1695).
Thoresby, Ralph, *Ducatus Leodiensis* (1715).
Trenchard, John, and Gordon, Thomas, *Cato's Letters* (4 vols, 1723–4).
Trevers, Joseph, *An Essay to the Restoring of our Decayed Trade* (1675).
Tryon, Thomas, *Some General Considerations Offered, Relating to our present Trade* (1698).
Tryon, Thomas, *England's Grandeur and the Way to Get Wealth* (1699).
[Tryon, Thomas], *A Brief History of Trade in England* (1702).
Tryon, Thomas, *Some Memoirs of the Life of Mr. Tho. Tryon* (1705).
Turner, Thomas, *The Diary of Thomas Turner, 1754–1765*, ed. David Vaisey (Oxford, 1985).
The Use and Abuses of Money, And the Improvements of it (1671).
Uztáriz, Gerónimo de, *The Theory and Practice of Commerce and Maritime Affairs*, trans. John Kippax (2 vols, 1751).
Uztáriz, Gerónimo de, *Theorica y practica de commercio y de marina* (Madrid, 1757).
Vanderlint, Jacob, *Money answers all Things* (1734).
Vauban, Sébastien Le Prestre de, *A Project for a Royal Tythe, or, General Tax* (1708).
Verney, Robert, *Englands Interest or the Great benefit to Trade by Banks or Offices of Credit* (1682).

Voltaire, *Letters Concerning the English Nation* (Dublin, 1733).
Voltaire, *Œuvres complètes de Voltaire*, ed. Louis Moland (Paris, 52 vols, 1877–85).
Voltaire, 'Observations sur MM Jean Law, Melon et Dutot sur le commerce, le luxe, les monnaies et les impôts', in *Œuvres complètes de Voltaire*, ed. Moland, vol. xxii.
Voltaire, *Essai sur les mœurs et l'esprit des nations*, ed. Bruno Bernard et al., vol. ii (*Les Œuvres complètes de Voltaire*, vol. xxii, Voltaire Foundation, Oxford, 2009).
Vossius, Isaac, *Variarum observationum liber* (1685).
Wallis, John, *The Correspondence of John Wallis*, vol. iii, ed. Philip Beeley and Christoph J. Scriba (Oxford, 2012).
Walwyn, William, *A Manifestation* (1649).
Walwyn, William, *The Writings of William Walwyn*, ed. Jack R. McMichael and Barbara Taft (Athens, Ga, 1989).
Wase, Christopher, *Dictionarium minus* (1662).
Watts, Isaac, 'Against Idleness and Mischief', in *Divine Songs . . . for the Use of Children* (2nd edn, 1716).
Weldon, Anthony, *A Perfect Description of the People and Country of Scotland* (1649).
Wheeler, John, *A Treatise of Commerce*, ed. George Burton Hotchkiss (New York, 1931).
Whiston, James, *England's State Distempers* (1704).
Whiston, James, *The Mismanagements in Trade Discovered* (1704).
Whiston, William, *A New Theory of the Earth from its Original to the Consummation of all Things* (1696).
Whitelocke, Bulstrode, *The Diary of Bulstrode Whitelocke 1605–1675*, ed. Ruth Spalding (British Academy Records of Social and Economic History ns 13, Oxford, 1990).
Widdrington, Sir Thomas, *Analecta Eboracensia: Some Remaynes of the Ancient City of York*, ed. Caesar Caine (1897).
Wilson, Thomas, *A Discourse upon Usury*, with an introduction by R. H. Tawney (1925).
Wilson, Thomas, 'The State of England Anno Dom. 1600', ed. F. J. Fisher, in *Camden Miscellany XVI* (Camden Society 3rd ser. 52, 1936).
Winstanley, Gerrard, *The Complete Works*, ed. Thomas N. Corns, Ann Hughes, and David Loewenstein (2 vols, Oxford, 2009).
Winthrop, John, *The Journal of John Winthrop, 1630–1649*, ed. Richard S. Dunn and Laetitia Yeandle (abridged edn, Cambridge, Mass., 1996).
Wood, William, *A Survey of Trade in Four Parts* (1718).
Woodward, John, *Brief Instructions For Making Observations in all Parts of the World* (1696).
Wotton, William, *Reflections upon Ancient and Modern Learning* (1694).
Wotton, William, *A Defense of the Reflections upon Ancient and Modern Learning* (1705).
Wright, Louis B., ed., *Advice to a Son: Precepts of Lord Burghley, Sir Walter Raleigh, and Francis Osborne* (Ithaca, NY, 1962).
Xenophons Treatise of Housholde (1544 edn).
Yarranton, Andrew, *The Improvement improved* (1663).
Yarranton, Andrew, *England's Improvement by Sea and Land* (2 vols, 1677, 1681).
[Yarranton, Andrew] *A Coffee-House Dialogue* (1679).
Yarranton, Andrew, *England's Improvements Justified* (1680).

SECONDARY WORKS

Allen, Robert C., *Enclosure and the Yeoman: The Agricultural Development of the South Midlands 1450–1850* (Oxford, 1992).

Allen, Robert C., 'The Great Divergence in European Wages and Prices from the Middle Ages to the First World War', *Explorations in Economic History*, 38 (2001), pp. 411–47.
Allen, Robert C., *The British Industrial Revolution in Global Perspective* (Cambridge, 2009).
Allen, Robert C., 'The British Industrial Revolution in Global Perspective', *Proceedings of the British Academy*, 167 (2011), pp. 199–224.
Allen, Robert C., *Global Economic History: A Very Short Introduction* (Oxford, 2011).
Allen, Robert C., 'Why the Industrial Revolution was British: Commerce, Induced Invention, and the Scientific Revolution', *EcHR* 64 (2011), pp. 357–84.
Allen, Robert C., Bengtsson, Tommy, and Dribe, Martin, eds, *Living Standards in the Past: New Perspectives on Well-Being in Asia and Europe* (Oxford, 2005).
Andrews, J. H., 'Land and People, c.1685', in T. W. Moody, F. X. Martin, and F. J. Byrne, eds, *A New History of Ireland*, iii: *Early Modern Ireland 1534–1691* (Oxford, 1976), pp. 454–77.
Andrews, J. H., *Shapes of Ireland: Maps and their Makers 1564–1839* (Dublin, 1997).
Andrews, J. H., 'How Many Acres? A Cartometric Exercise of 1642', *Irish Geography*, 34 (2001), pp. 1–10.
Appleby, Joyce Oldham, *Economic Thought and Ideology in Seventeenth-Century England* (Princeton, 1978).
Appleby, Joyce Oldham, 'Consumption in Early Modern Social Thought', in John Brewer and Roy Porter, eds, *Consumption and the World of Goods* (1993), pp. 162–74.
Archer, Ian W., 'The Nostalgia of John Stow', in David L. Smith, Richard Strier, and David Bevington, eds, *The Theatrical City: Culture, Theatre and Politics in London 1576–1649* (Cambridge, 1995), pp. 17–34.
Archer, Ian W., 'Material Londoners?', in Lena Cowen Orlin, ed., *Material London, ca. 1600* (Philadelphia, 2000), pp. 174–92.
Archer, Ian W., 'Social Order and Disorder', in Paulina Kewes, Ian Archer, and Felicity Heal, eds, *The Oxford Handbook of Holinshed's Chronicles* (Oxford, 2013), pp. 389–410.
Arkell, Tom, 'Illuminations and Distortions: Gregory King's Scheme Calculated for the Year 1688 and the Social Structure of Later Stuart England', *EcHR* 59 (2006), pp. 32–69.
Armitage, David, *The Ideological Origins of the British Empire* (Cambridge, 2000).
Armitage, David, *Foundations of Modern International Thought* (Cambridge, 2013).
Ash, Eric H., *Power, Knowledge, and Expertise in Elizabethan England* (Baltimore, 2004).
Ashton, Robert, 'Conflicts of Concessionary Interest in Early Stuart England', in Coleman and John, eds, *Trade, Government and Economy*, pp. 113–31.
Ashton, Robert, *The City and the Court 1603–1643* (Cambridge, 1979).
Aspromourgos, Tony, 'The Invention of the Concept of Social Surplus: Petty in the Hartlib Circle', *European Journal of the History of Economic Thought*, 12 (2005), pp. 1–24.
Baer, William C., 'Stuart London's Standard of Living: Re-examining the Settlement of Tithes of 1638 for Rents, Income, and Poverty', *EcHR* 63 (2010), pp. 612–37.
Baer, William C., 'Landlords and Tenants in London 1550–1700', *Urban History*, 38 (2011), pp. 234–55.
Barbour, Reid, *English Epicures and Stoics: Ancient Legacies in Early Stuart Culture* (Amherst, Mass., 1998).
Barley, M. W., 'Rural Building in England', in Thirsk, ed., *Agrarian History*, v.ii. 590–685.
Barley, M. W., 'Rural Housing in England', in Thirsk, ed., *Agrarian History*, iv. 696–766.
Barnard, Toby, 'The Hartlib Circle and the Cult and Culture of Improvement in Ireland', in Greengrass et al., eds, *Samuel Hartlib and Universal Reformation*, pp. 281–97.

Barnard, Toby, *Improving Ireland? Projectors, Prophets and Profiteers, 1641–1786* (Dublin, 2008).
Baron, Hans, 'The *Querelle* of the Ancients and Moderns as a Problem for Renaissance Scholarship', *Journal of the History of Ideas*, 20 (1959), pp. 3–22.
Barron, Caroline, *London in the Later Middle Ages: Government and People, 1200–1500* (Oxford, 2004).
Bastian, Frank, *Defoe's Early Life* (1981).
Bayly, C. A., *The Birth of the Modern World, 1780–1914* (Oxford, 2004).
Beckerman, Wilfred, *Economics as Applied Ethics: Value Judgements in Welfare Economics* (2010).
Beer, Anna R., *Sir Walter Ralegh and his Readers in the Seventeenth Century* (1997).
Ben-Amos, Ilana Krausman, *The Culture of Giving: Informal Support and Gift-Exchange in Early Modern England* (Cambridge, 2008).
Bennett, James A., Cooper, Michael, Hunter, Michael, and Jardine, Lisa, *London's Leonardo: The Life and Work of Robert Hooke* (Oxford, 2003).
Bennett, Robert J., *Local Business Voice: The History of Chambers of Commerce in Britain, Ireland and Revolutionary America 1760–2011* (Oxford, 2011).
Beresford, Maurice, 'Habitation versus Improvement: The Debate on Enclosure by Agreement', in F. J. Fisher, ed., *Essays in the Economic and Social History of Tudor and Stuart England in Honour of R. H. Tawney* (Cambridge, 1961), pp. 40–69.
Berg, Maxine, and Eger, Elizabeth, eds, *Luxury in the Eighteenth Century: Debates, Desires, and Delectable Goods* (Basingstoke, 2003).
Bernard, G. W., and Gunn, S. J., eds, *Authority and Consent in Tudor England: Essays Presented to C. S. L. Davies* (Aldershot, 2002).
Berry, Christopher J., *The Idea of Luxury: A Conceptual and Historical Investigation* (Cambridge, 1994).
Berry, Mary Elizabeth, *Japan in Print: Information and Nation in the Early Modern Period* (Berkeley, 2006).
Bindoff, S. T., 'The Making of the Statute of Artificers', in S. T. Bindoff, J. Hurstfield, and C. H. Williams, eds, *Elizabethan Government and Society: Essays Presented to Sir John Neale* (1961), pp. 56–94.
Blackstone, G. V., *A History of the British Fire Service* (1957).
Blom, Hans, 'Decay and the Political Gestalt of Decline in Bernard Mandeville and his Dutch Contemporaries', *History of European Ideas*, 36 (2010), pp. 153–66.
Bogart, Dan, 'Did the Glorious Revolution Contribute to the Transport Revolution? Evidence from Investment in Roads and Rivers', *EcHR* 64 (2011), pp. 1073–112.
Bohstedt, John, *The Politics of Provisions. Food Riots, Moral Economy, and Market Transition in England, c.1550–1850* (Farnham, 2010).
Bonney, Richard, 'Early Modern Theories of State Finance', in Richard Bonney, ed., *Economic Systems and State Finance* (Oxford, 1995), pp. 163–229.
Bonney, Richard, 'Towards the Comparative Fiscal History of Britain and France in the "Long" Eighteenth Century', in Prados de la Escosura, ed., *Exceptionalism and Industrialisation*, pp. 191–215.
Borsay, Peter, *The English Urban Renaissance: Culture and Society in the Provincial Town, 1660–1770* (Oxford, 1989).
Borsay, Peter, 'London, 1660–1800: A Distinctive Culture?', in Peter Clark and Raymond Gillespie, eds, *Two Capitals: London and Dublin 1500–1840* (Proceedings of the British Academy 107, Oxford, 2001), pp. 167–84.

Bottigheimer, Karl S., *English Money and Irish Land: The 'Adventurers' in the Cromwellian Settlement of Ireland* (Oxford, 1971).
Boulton, Jeremy, 'Clandestine Marriages in London: An Examination of the Neglected Urban Variable', *Urban History*, 20 (1993), pp. 191–210.
Boulton, Jeremy, 'Food Prices and the Standard of Living in London in the "Century of Revolution", 1580–1700', *EcHR* 53 (2000), pp. 455–92.
Boulton, Jeremy, 'London 1540–1700', in Clark, ed., *Urban History of Britain*, ii: *1540–1840*, pp. 315–46.
Bowie, Karin, 'New Perspectives on Pre-Union Scotland', in T. M. Devine and Jenny Wormald, eds, *The Oxford Handbook of Modern Scottish History* (Oxford, 2012), pp. 303–19.
Bowley, Marian, *Studies in the History of Economic Theory before 1870* (1973).
Braddick, Michael J., *State Formation in Early Modern England, c.1550–1700* (Cambridge, 2000).
Braddick, Michael J., *God's Fury, England's Fire: A New History of the English Civil Wars* (2008).
Brayshay, Mark, Harrison, Philip, and Chalkley, Brian, 'Knowledge, Nationhood and Governance: The Speed of the Royal Posts in Early Modern England', *Journal of Historical Geography*, 24 (1998), pp. 265–88.
Breen, T. H., 'Creative Adaptations: Peoples and Cultures', in Jack P. Greene and J. R. Pole, eds, *Colonial British America* (Baltimore, 1984), pp. 195–232.
Breen, T. H., *The Marketplace of Revolution: How Consumer Politics Shaped American Independence* (Oxford, 2004).
Brenner, Robert, *Merchants and Revolution: Commercial Change, Political Conflict, and London's Overseas Traders 1550–1653* (Princeton, 1993).
Brewer, John, *The Sinews of Power: War, Money and the English State, 1688–1783* (1989).
Broad, John, 'Cattle Plague in Eighteenth-Century England', *Agricultural History Review*, 31 (1983), pp. 104–15.
Broadberry, Stephen, and Gupta, Bishnupriya, 'The Early Modern Great Divergence: Wages, Prices and Economic Development in Europe and Asia, 1500–1800', *EcHR* 59 (2006), pp. 2–31.
Broadberry, Stephen, et al., 'British Economic Growth, 1270–1870', working paper.
Brockliss, Laurence, and Jones, Colin, *The Medical World of Early Modern France* (Oxford, 1997).
Brook, Timothy, *The Confusions of Pleasure: Commerce and Culture in Ming China* (Berkeley, 1998).
Brook, Timothy, *Vermeer's Hat: The Seventeenth Century and the Dawn of the Global World* (2008).
Brooks, Christopher W., *Law, Politics and Society in Early Modern England* (Cambridge, 2008).
Brooks, Colin, 'Projecting, Political Arithmetic and the Act of 1695', *EHR* 97 (1982), pp. 31–53.
Bruni, Luigino, and Porta, Pier Luigi, *Economics and Happiness: Framing the Analysis* (Oxford, 2006).
Buchwald, Jed Z., and Feingold, Mordechai, *Newton and the Origin of Civilization* (Princeton, 2013).
Burke, Peter, 'A Civil Tongue: Language and Politeness in Early Modern Europe', in Burke et al., eds, *Civil Histories*, pp. 31–48.

Burke, Peter, 'History, Myth, and Fiction: Doubts and Debates', in Rabasa et al., eds, *Oxford History of Historical Writing*, iii: *1400–1800*, pp. 261–81.

Burke, Peter, Harrison, Brian, and Slack, Paul, eds, *Civil Histories: Essays Presented to Sir Keith Thomas* (Oxford, 2000).

Burtt, Shelley, *Virtue Transformed: Political Argument in England 1688–1740* (Cambridge, 1992).

Canny, Nicholas, ed., *The Oxford History of the British Empire*, i: *The Origins of Empire* (Oxford, 1998).

Canny, Nicholas, *Making Ireland British, 1580–1650* (Oxford, 2001).

Canny, Nicholas, 'A Protestant or Catholic Atlantic World? Confessional Divisions and the Writing of Natural History', *Proceedings of the British Academy*, 181 (2012), pp. 83–121.

Capp, Bernard, *Astrology and the Popular Press: English Almanacs 1500–1800* (1979).

Carpenter, Kenneth H., *Dialogue in Political Economy: Translations from and into German in the 18th Century* (Boston, 1977).

Challis, C. E., *The Tudor Coinage* (Manchester, 1978).

Chandaman, C. D., *The English Public Revenue, 1660–1688* (Oxford, 1975).

Charlesworth, Lorie, *Welfare's Forgotten Past: A Socio-legal History of the Poor Law* (Abingdon, 2010).

Chartres, John A., 'The Marketing of Agricultural Produce', in Thirsk, ed., *Agrarian History*, v.ii. 406–502.

Chartres, John, 'No English Calvados? English Distillers and the Cider Industry in the Seventeenth and Eighteenth Centuries', in John Chartres and David Hey, eds, *English Rural Society 1500–1800: Essays in Honour of Joan Thirsk* (Cambridge, 1990), pp. 313–42.

Childs, Nick, *A Political Academy in Paris, 1724–1731: The Entresol and its Members* (Studies in Voltaire and the Eighteenth Century 10, Oxford, 2000).

Clark, Aidan, with Edwards, R. Dudley, 'Pacification, Plantation, and the Catholic Question, 1603–23', in T. W. Moody et al., eds, *A New History of Ireland*, iii: *Early Modern Ireland 1534–1691* (Oxford, 1976), pp. 187–232.

Clark, Gregory, 'The Condition of the Working Class in England 1209–2004', *Journal of Political Economy*, 113 (2005), pp. 1307–40.

Clark, Peter, *The English Alehouse: A Social History 1200–1830* (1983).

Clark, Peter, *British Clubs and Societies 1580–1800: The Origins of an Associational World* (Oxford, 2000).

Clark, Peter, ed., *The Cambridge Urban History of Britain*, ii: *1540–1840* (Cambridge, 2000).

Clark, Peter, 'The Multi-centred Metropolis: The Social and Cultural Landscapes of London, 1600–1840', in Peter Clark and Raymond Gillespie, eds, *Two Capitals: London and Dublin 1500–1840* (Proceedings of the British Academy 107, Oxford, 2001), pp. 239–56.

Clark, Peter, 'Improvement, Policy and Tudor Towns', in Bernard and Gunn, eds, *Authority and Consent*, pp. 233–47.

Clay, C. G. A., *Economic Expansion and Social Change: England 1500–1700* (2 vols, Cambridge, 1984).

Claydon, Tony, *William III and the Godly Revolution* (Cambridge, 1996).

Clunas, Craig, 'Things in Between: Splendour and Excess in Ming China', in Frank Trentmann, ed., *The Oxford Handbook of the History of Consumption* (Oxford, 2012), pp. 47–63.

Cockayne, Emily, *Hubbub: Filth, Noise and Stench in England 1600–1770* (New Haven, 2007).
Cole, Charles Woolsey, *Colbert and a Century of French Mercantilism* (2 vols, New York, 1939).
Coleman, D. C., 'Politics and Economics in the Age of Anne: The Case of the Anglo-French Trade Treaty of 1713', in Coleman and John, eds, *Trade, Government and Economy*, pp. 187–211.
Coleman, D. C., *The Economy of England 1450–1750* (Oxford, 1977).
Coleman, D. C., and John, A. H., eds, *Trade, Government and Economy in Pre-industrial England: Essays Presented to F. J. Fisher* (1976).
Coleman, Olive, 'What Figures? Some Thoughts on the Use of Information by Medieval Governments', in Coleman and John, eds, *Trade, Government and Economy*, pp. 96–112.
Collinson, Patrick, 'Christian Socialism in Elizabethan Suffolk: Thomas Carew and his *Caveat for Clothiers*', in Carole Rawcliffe, Roger Virgoe, and Richard Wilson, eds, *Counties and Communities: Essays on East Anglian History Presented to Hassell Smith* (Centre of East Anglian Studies, Norwich, 1996), pp. 161–78.
Collinson, Patrick, 'John Stow and Nostalgic Antiquarianism', in J. F. Merritt, ed., *Imagining Early Modern London: Perceptions and Portrayals of the City from Stow to Strype, 1598–1720* (Cambridge, 2001), pp. 27–51.
Colvin, Brenda, *Land and Landscape: Evolution, Design and Control* (1970).
Connolly, S. J., *Contested Island: Ireland 1460–1630* (Oxford, 2007).
Cook, Harold J., *Matters of Exchange: Commerce, Medicine, and Science in the Dutch Golden Age* (New Haven, 2007).
Cook, Harold J., 'Markets and Cultures: Medical Specifics and the Reconfiguration of the Body in Early Modern England', *TRHS* 6th ser. 21 (2011), pp. 123–45.
Cooper, Alix, '"The Possibilities of the Land": The Inventory of "Natural Riches" in Early Modern German Territories', *History of Political Economy*, 35 (2003), Supplement 1, pp. 129–53.
Cooper, Alix, *Inventing the Indigenous: Local Knowledge and Natural History in Early Modern Europe* (Cambridge, 2007).
Cooper, J. P., 'Economic Regulation and the Cloth Industry in Seventeenth-Century England', *TRHS* 5th ser. 20 (1970), pp. 73–99.
Cooper, J. P., 'Social and Economic Policies under the Commonwealth', in G. E. Aylmer, ed., *The Interregnum: The Quest for Settlement, 1646–1660* (1972), pp. 121–42.
Cooper, J. P., *Land, Men, and Beliefs: Studies in Early-Modern History*, ed. G. E. Aylmer and J. S. Morrill (1983).
Copenhaver, Brian P., 'The Historiography of Discovery in the Renaissance: The Sources and Composition of Polydore Vergil's De inventoribus rerum. I–III', *Journal of the Warburg and Courtauld Institutes*, 41 (1978), pp. 192–214.
Corfield, P. J., *The Impact of English Towns, 1700–1800* (Oxford, 1982).
Cox, Nancy, *The Complete Tradesman: A Study of Retailing, 1550–1820* (Aldershot, 2000).
Cramsie, John, *Kingship and Crown Finance under James VI and I, 1603–1625* (Woodbridge, 2002).
Crépel, P., 'Arithmétique politique et population dans les métamorphoses de l'*Encyclopédie*', in Martin, ed., *Arithmétique politique*, pp. 47–69.
Cressy, David, *Literacy and the Social Order: Reading and Writing in Tudor and Stuart England* (Cambridge, 1980).
Cressy, David, *Coming Over: Migration and Communication between England and New England in the Seventeenth Century* (Cambridge, 1987).

Cressy, David, *Saltpeter: The Mother of Gunpowder* (Oxford, 2013).
Crossley, Alan, ed., *VCH Oxfordshire*, iv: *The City of Oxford* (Oxford, 1979).
Crouzet, François, 'The Sources of England's Wealth: Some French Views in the Eighteenth Century', in P. L. Cottrell and Derek H. Aldcroft, eds, *Shipping, Trade and Commerce: Essays in Memory of Ralph Davis* (Leicester, 1981), pp. 61–79.
Crowley, John E., *The Invention of Comfort: Sensibilities and Design in Early Modern Britain and Early America* (Baltimore, 2001).
Cullen, Karen J., *Famine in Scotland: The 'Ill Years' of the 1690s* (Edinburgh, 2010).
Currie, C. R. J., and Lewis, C. P., eds, *A Guide to English County Histories* (Stroud, 1994).
Curtis, Mark H., 'The Alienated Intellectuals of Early Stuart England', *P&P* 23 (Nov. 1962), pp. 25–43.
Dabhoiwala, Faramerz, *The Origins of Sex: A History of the First Sexual Revolution* (2012).
Darby, H. C., *The Draining of the Fens* (2nd edn, Cambridge, 1956).
Darwin, John, 'Civility and Empire', in Burke et al., eds, *Civil Histories*, pp. 321–36.
Darwin, John, *After Tamerlane: The Rise and Fall of Global Empires, 1400–2000* (2008).
Daunton, Martin, ed., *Charity, Self-Interest and Welfare in the English Past* (1996).
Davis, J. C., *Utopia and the Ideal Society: A Study of English Utopian Writing 1516–1700* (Cambridge, 1981).
Davis, Ralph, 'English Foreign Trade, 1660–1700', *EcHR* 7 (1954), pp. 150–66.
Davis, Ralph, *English Overseas Trade 1500–1700* (1973).
Davison, Lee, Hitchcock, Tim, Keirn, Tim, and Shoemaker, Robert B., eds, *Stilling the Grumbling Hive: The Response to Social and Economic Problems in England, 1689–1750* (Stroud, 1992).
Dean, David M., *Law-Making and Society in Late Elizabethan England: The Parliament of England, 1584–1601* (Cambridge, 1996).
Dean, David M., 'Elizabeth's Lottery: Political Culture and State Formation in Early Modern England', *Journal of British Studies*, 50 (2011), pp. 587–611.
de Jongste, Jan A. F., 'The Restoration of the Orangist Regime in 1747: The Modernity of a "Glorious Revolution"', in Jacob and Mijnhardt, eds, *The Dutch Republic in the Eighteenth Century*, pp. 32–59.
De Roover, Raymond, *Gresham on Foreign Exchange* (Cambridge, Mass., 1949).
De Roover, Raymond, 'The Concept of the Just Price: Theory and Economic Policy', *Journal of Economic History*, 18 (1958), pp. 418–34.
Descendre, Romain, *L'État du monde: Giovanni Botero entre raison d'état et géopolitique* (Geneva, 2009).
de Vries, Jan, *The Industrious Revolution: Consumer Behaviour and the Household Economy, 1650 to the Present* (Cambridge, 2008).
de Vries, Jan, and van der Woude, Ad, *The First Modern Economy: Success, Failure, and Perseverance of the Dutch Economy, 1500–1815* (Cambridge, 1997).
Dickey, Laurence, 'Power, Commerce and Natural Law in Daniel Defoe's Political Writings, 1698–1707', in Robertson, ed., *Union for Empire*, pp. 63–96.
Dickinson, H. W., 'The Steam Engine to 1830', in Charles Singer and Richard Raper, eds, *A History of Technology* (7 vols, Oxford, 1954–78), iv. 168–98.
Dickson, P. G. M., *The Sun Insurance Office 1710–1960* (Oxford, 1960).
Dickson, P. G. M., *The Financial Revolution in England: A Study in the Development of Public Credit, 1688–1756* (1967).
Dobson, Mary J., 'The Last Hiccup of the Old Demographic Regime: Population Stagnation and Decline in Late Seventeenth and Early Eighteenth-Century South-East England', *Continuity and Change*, 4 (1989), pp. 395–428.

Dobson, Mary J., *Contours of Death and Disease in Early Modern England* (Cambridge, 1997).
Dodgson, John, 'Gregory King and the Economic Structure of Early Modern England: An Input–Output Table for 1688', *EcHR* 66 (2013), pp. 993–1016.
Donagan, Barbara, *War in England 1642–1649* (Oxford, 2008).
Doohwan, Ahn, 'The Anglo-French Treaty of Commerce of 1713: Tory Trade Politics and the Question of Dutch Decline', *History of European Ideas*, 36 (2010), pp. 167–80.
Dow, Frances, *Cromwellian Scotland, 1651–1660* (Edinburgh, 1979).
Drayton, Richard, *Nature's Government: Science, Imperial Britain, and the 'Improvement' of the World* (New Haven, 2000).
Dunn, Kevin, 'Milton among the Monopolists: *Areopagitica*, Intellectual Property and the Hartlib Circle', in Greengrass et al., eds, *Samuel Hartlib and Universal Reformation*, pp. 177–92.
Dupâquier, Jacques, 'Londres ou Paris? Un grand débat dans le petit monde des arithméticiens politiques (1662–1759)', *Population*, 53 (1998), pp. 311–25.
Dupâquier, Jacques and Michel, *Histoire de la démographie* (Paris, 1985).
Dyer, Christopher, *Standards of Living in the Later Middle Ages: Social Change in England c.1200–1520* (Cambridge, 1989).
Dyer, Christopher, *A Country Merchant, 1495–1520: Trading and Farming at the End of the Middle Ages* (Oxford, 2012).
Earle, Carville V., 'Environment, Disease and Mortality in Early Virginia', in Tate and Ammerman, eds, *Chesapeake in the Seventeenth Century*, pp. 96–125.
Earle, Peter, 'The Economics of Stability: The Views of Daniel Defoe', in Coleman and John, eds, *Trade, Government and Economy*, pp. 274–92.
Earle, Peter, *The Making of the English Middle Class: Business, Society and Family Life in London, 1660–1730* (1989).
Eastwood, David, *Government and Community in the English Provinces, 1700–1870* (Basingstoke, 1997).
Edie, Carolyn Andervont, 'New Buildings, New Taxes, and Old Interests: An Urban Problem of the 1670s', *Journal of British Studies*, 6 (1967), pp. 35–63.
Elliott, J. H., *The Old World and the New 1492–1650* (Cambridge, 1970).
Elliott, J. H., 'Self-Perception and Decline in Early Seventeenth-Century Spain', *P&P* 74 (Feb. 1977), pp. 43–57.
Elliott, J. H., 'Introduction', in Nicholas Canny and Anthony Pagden, eds, *Colonial Identity in the Atlantic World, 1500–1800* (Princeton, 1987), pp. 3–13.
Elliott, J. H., 'The Seizure of Overseas Territories by the European Powers', in David Armitage, ed., *Theories of Empire 1450–1800* (Aldershot, 1998), pp. 139–57.
Elliott, J. H., *Empires of the Atlantic World: Britain and Spain in America, 1492–1830* (New Haven, 2006).
Elmer, Peter, *The Miraculous Conformist: Valentine Greatrakes, the Body Politic, and the Politics of Healing in Restoration Britain* (Oxford, 2013).
Elton, G. R., *The Parliament of England, 1559–1581* (Cambridge, 1986).
Elvin, Mark, *The Pattern of the Chinese Past* (Stanford, Calif., 1973).
Endres, A. M., 'The Functions of Numerical Data in the Writings of Graunt, Petty and Davenant', *History of Political Economy*, 17 (1985), pp. 245–64.
Englard, Izhak, *Corrective and Distributive Justice: From Aristotle to Modern Times* (Oxford, 2009).
Estabrook, Carl B., *Urbane and Rustic England: Cultural Ties and Social Spheres in the Provinces, 1660–1780* (Manchester, 1998).

Evans, R. J. W., *The Making of the Habsburg Monarchy, 1550–1700* (Oxford, 1979).
Faccarello, Gilbert, *The Foundations of Laissez-faire: The Economics of Pierre de Boisguilbert* (1999).
Falkus, Malcolm, 'Lighting in the Dark Ages of English Economic History: Town Streets before the Industrial Revolution', in Coleman and John, eds, *Trade, Government and Economy in Pre-Industrial England*, pp. 248–73.
Farber, Lianna, *An Anatomy of Trade in Medieval Writing: Value, Consent, and Community* (Ithaca, NY, 2006).
Ferguson, Arthur B., *The Articulate Citizen and the English Renaissance* (Durham, NC, 1965).
Findlay, Ronald, and O'Rourke, Kevin H., *Power and Plenty: Trade, War, and the World Economy in the Second Millennium* (Princeton, 2007).
Finkelstein, Andrea, *Harmony and the Balance: An Intellectual History of Seventeenth-Century English Economic Thought* (Ann Arbor, 2000).
Finkelstein, Andrea, *The Grammar of Profit: The Price Revolution in Intellectual Context* (Leiden, 2006).
Finlay, Roger, *Population and Metropolis: The Demography of London 1580–1650* (Cambridge, 1981).
Fitzgerald, Timothy, *Discourse on Civility and Barbarity: A Critical History of Religion and Related Categories* (Oxford, 2007).
Fitzmaurice, Andrew, *Humanism and America: An Intellectual History of English Colonisation, 1500–1625* (Cambridge, 2003).
Fitzmaurice, Andrew, 'American Corruption', in John F. McDiarmid, ed., *The Monarchical Republic of Early Modern England* (Aldershot, 2007), pp. 217–31.
Foster, Stephen, *Their Solitary Way: The Puritan Social Ethic in the First Century of Settlement in New England* (New Haven, 1971).
Foster, Stephen, ed., *British North America in the Seventeenth and Eighteenth Centuries* (Oxford, 2013).
Foster, Stephen, and Haefeli, Evan, 'British North America in the Empire', in Foster, ed., *British North America*, pp. 18–66.
Fox, Adam, *Oral and Literate Culture in England, 1500–1700* (Oxford, 2000).
Fox, Adam, 'Sir William Petty, Ireland, and the Making of a Political Economist, 1653–87', *EcHR* 62 (2009), pp. 388–404.
Fox, Adam, 'Printed Questionnaires, Research Networks, and the Discovery of the British Isles 1650–1800', *Historical Journal*, 53 (2010), pp. 593–621.
Fox, Adam, 'Vernacular Culture and Popular Customs in Early Modern England: Evidence from Thomas Machell's Westmorland', *Cultural and Social History*, 9 (2012), pp. 329–47.
Fox, Adam, 'Food, Drink and Social Distinction in Early Modern England', in Hindle et al., eds, *Remaking English Society*, pp. 165–87.
Frace, Ryan K., 'Religious Toleration in the Wake of Revolution: Scotland on the Eve of the Enlightenment (1688–1710s)', *History*, 93 (2008), pp. 355–75.
French, H. R., *The Middle Sort of People in Provincial England, 1600–1750* (Oxford, 2007).
Friedel, Robert, *A Culture of Improvement: Technology and the Western Millennium* (Cambridge, Mass., 2007).
Friis, Astrid, *Alderman Cockayne's Project and the Cloth Trade* (Copenhagen and London, 1927).
Games, Alison, *The Web of Empire: English Cosmopolitans in an Age of Expansion, 1560–1660* (Oxford, 2008).

Gauci, Perry, *The Politics of Trade: The Overseas Merchant in State and Society, 1660–1720* (Oxford, 2001).
Gauci, Perry, ed., *Regulating the British Economy, 1660–1850* (Farnham, 2011).
Gaukroger, Stephen, *The Collapse of Mechanism and the Rise of Sensibility: Science and the Shaping of Modernity, 1680–1760* (Oxford, 2010).
Gay, E. F., 'The Midland Revolt and the Inquisitions of Depopulation of 1607', *TRHS* ns 18 (1904), pp. 195–244.
Gentles, Ian, 'The Management of the Crown Lands, 1649–60', *Agricultural History Review*, 19 (1971), pp. 25–41.
Geraghty, Anthony, *The Architectural Drawings of Sir Christopher Wren at All Souls College, Oxford* (Aldershot, 2007).
Gilman, Ernest B., *Plague Writing in Early Modern England* (Chicago, 2009).
Glaisyer, Natasha, *The Culture of Commerce in England, 1660–1720* (Woodbridge, 2006).
Glass, David V., *Numbering the People: The Eighteenth-Century Population Controversy and the Development of Census and Vital Statistics in Britain* (Farnborough, 1973).
Glennie, Paul, and Thrift, Nigel, *Shaping the Day: A History of Timekeeping in England and Wales 1300–1800* (Oxford, 2009).
Glennie, Paul, and Whyte, Ian, 'Towns in an Agrarian Economy 1540–1700', in Clark, ed., *Cambridge Urban History of Britain*, ii. 167–93.
Godfrey, Eleanor S., *The Development of English Glassmaking, 1560–1640* (Oxford, 1975).
Goldie, Mark, 'Sir Peter Pett, Sceptical Toryism and the Science of Toleration in the 1680s', in W. J. Sheils, ed., *Persecution and Toleration* (Studies in Church History 21, Oxford, 1984), pp. 247–73.
Goodare, Julian, *The Government of Scotland, 1560–1625* (Oxford, 2004).
Goose, Nigel, 'Immigrants and English Economic Development in the Sixteenth and Early Seventeenth Centuries', in Goose and Luu, eds, *Immigrants in Tudor and Early Stuart England*, pp. 136–60.
Goose, Nigel, and Luu, Lien, eds, *Immigrants in Tudor and Early Stuart England* (Brighton, 2005).
Gordon, Barry J., *Economic Analysis Before Adam Smith: Hesiod to Lessius* (1975).
Gould, J. D., *The Great Debasement: Currency and the Economy in Mid-Tudor England* (Oxford, 1970).
Grafe, Regina, 'Polycentric States: The Spanish Reigns and the "Failure" of Mercantilism', in Stern and Wennerlind, *Mercantilism Reimagined*, pp. 241–62.
Grafton, Anthony, *What Was History? The Art of History in Early Modern Europe* (Cambridge, 2007).
Graham, Aaron, 'Auditing Leviathan: Corruption and State Formation in Early Eighteenth-Century Britain', *EHR* 128 (2013), pp. 806–38.
Gramlich-Oka, Bettina, and Smits, Gregory, eds, *Economic Thought in Early Modern Japan* (Leiden, 2010).
Grassby, Richard B., 'Social Status and Commercial Enterprise under Louis XIV', *EcHR* 2nd ser. 13 (1960), pp. 19–38.
Grassby, Richard, *The English Gentleman in Trade: The Life and Works of Sir Dudley North 1641–1691* (Oxford, 1994).
Grassby, Richard, *The Business Community of Seventeenth-Century England* (Cambridge, 1995).
Grassby, Richard, *Kinship and Capitalism: Marriage, Family and Business in the English-Speaking World, 1580–1740* (Cambridge, 2001).
Greenblatt, Stephen, *The Swerve: How the Renaissance Began* (2011).

Greene, Jack P., 'Changing Identity in the British Caribbean: Barbados as a Case Study', in Nicholas Canny and Anthony Pagden, eds, *Colonial Identity in the Atlantic World 1500–1800* (Princeton, 1987), pp. 213–66.

Greengrass, Mark, Leslie, Michael, and Raylor, Timothy, eds, *Samuel Hartlib and Universal Reformation: Studies in Intellectual Communication* (Cambridge, 1994).

Grenier, Jean-Yves, *Histoire de la pensée économique et politique de la France d'ancien régime* (Paris, 2007).

Grice-Hutchinson, Marjorie, *The School of Salamanca: Readings in Spanish Monetary Theory, 1544–1605* (Oxford, 1952).

Grice-Hutchinson, Marjorie, *Economic Thought in Spain: Selected Essays*, ed. Laurence S. Moss and Christopher K. Ryan (Aldershot, 1993).

Griffin, Emma, 'A Conundrum Resolved? Rethinking Courtship, Marriage and Population Growth in Eighteenth-Century England', *P&P* 215 (May 2012), pp. 125–64.

Griffiths, Elizabeth, '"A Country Life": Sir Hamon Le Strange of Hunstanton in Norfolk, 1583–1654', in Hoyle, ed., *Custom, Improvement and Landscape*, pp. 203–34.

Griffiths, Paul, 'Local Arithmetic: Information Cultures in Early Modern England', in Hindle et al., eds, *Remaking English Society*, pp. 113–34.

Griffiths, Paul, and Jenner, Mark S. R., eds, *Londinopolis: Essays in the Cultural and Social History of Early Modern London* (Manchester, 2000).

Gurney, John, *Brave Community: The Digger Movement in the English Revolution* (Manchester, 2007).

Hacking, Ian, *The Emergence of Probability* (Cambridge, 1975).

Hadfield, Andrew, *Literature, Travel, and Colonial Writing in the English Renaissance, 1545–1625* (Oxford, 1998).

Hadfield, Andrew, *Amazons, Savages, and Machiavels: Travel and Colonial Writing in English, 1550–1630: An Anthology* (Oxford, 2001).

Hannah, Leslie, 'The Moral Economy of Business: A Historical Perspective on Business and Efficiency', in Burke et al., eds, *Civil Histories*, pp. 285–99.

Harding, Vanessa, *The Dead and the Living in Paris and London, 1500–1670* (Cambridge, 2002).

Harding, Vanessa, and Baker, Philip, *People in Place: Families, Households and Housing in Early Modern London* (Centre for Metropolitan History, 2008).

Harkness, Deborah, *The Jewel House: Elizabethan London and the Scientific Revolution* (New Haven, 2007).

Harley, David, 'Pious Physic for the Poor: The Lost Durham County Medical Scheme of 1655', *Medical History*, 37 (1993), pp. 148–66.

Harley, J. B., 'Meaning and Ambiguity in Tudor Cartography' In Sarah Tyacke, ed., *English Map-Making 1500–1650: Historical Essays* (1983), pp. 22–45.

Harris, Bob, 'The Enlightenment, Towns and Urban Society in Scotland, c.1760–1820', *EHR* 126 (2011), pp. 1097–136.

Harris, Frances, *Transformations of Love: The Friendship of John Evelyn and Margaret Godolphin* (Oxford, 2002).

Harris, Jonathan Gil, *Sick Economies: Drama, Mercantilism, and Disease in Shakespeare's England* (Philadelphia, 2004).

Harris, Tim, *Revolution: The Great Crisis of the British Monarchy, 1685–1720* (2007).

Harris, Victor, *All Coherence Gone: A Study of the Seventeenth Century Controversy over Disorder and Decay in the Universe* (1966).

Harrison, David, *The Bridges of Medieval England: Transport and Society 400–1800* (Oxford, 2004).

Harrison, Mark, *Disease and the Modern World: 1500 to the Present Day* (Cambridge, 2004).
Harrison, Peter, *The Fall of Man and the Foundations of Science* (Cambridge, 2007).
Harvey, Karen, *The Little Republic: Masculinity and Domestic Authority in Eighteenth-Century Britain* (Oxford, 2012).
Harvey, P. D. A., *Maps in Tudor England* (1993).
Hatcher, John, *The History of the British Coal Industry*, i: *Before 1700* (Oxford, 1993).
Hatcher, John, 'Labour, Leisure and Economic Thought before the Nineteenth Century', *P&P* 160 (Aug. 1998), pp. 64–115.
Hayton, David, 'Moral Reform and Country Politics in the Late Seventeenth-Century House of Commons', *P&P* 128 (Aug. 1990), pp. 48–91.
Headrick, Daniel R., *When Information Came of Age: Technologies of Knowledge in the Age of Reason and Revolution, 1700–1850* (Oxford, 2000).
Helgerson, Richard, *Forms of Nationhood: The Elizabethan Writing of England* (Chicago, 1992).
Heller, Agnes, *A Theory of Modernity* (Oxford, 1999).
Henning, Basil Duke, *The House of Commons 1660–1690* (3 vols, 1983).
Hey, David, *Packmen, Carriers and Packhorse Roads: Trade and Communications in North Derbyshire and South Yorkshire* (Leicester, 1980).
Higgs, Edward, *The Information State in England: The Central Collection of Information on Citizens since 1500* (Basingstoke, 2004).
Hill, Bridget, 'The Idea of a Protestant Nunnery', *P&P* 117 (Nov. 1987), pp. 107–30.
Hill, Christopher, *Puritanism and Revolution* (1958).
Hinde, Andrew, *England's Population: A History since the Domesday Survey* (2003).
Hindle, Steve, *On the Parish? The Micro-Politics of Poor Relief in Rural England c.1550–1750* (Oxford, 2004).
Hindle, Steve, 'Imagining Insurrection in Seventeenth-Century England: Representations of the Midland Rising of 1607', *History Workshop Journal*, 66 (2008), pp. 21–61.
Hindle, Steve, 'Dearth and the English Revolution: The Harvest Crisis of 1647–50', *EcHR* 61 (2008), Special Issue 1, pp. 64–98.
Hindle, Steve, 'Work, Reward and Labour Discipline in Late Seventeenth-Century England', in Hindle et al., eds, *Remaking English Society*, pp. 255–79.
Hindle, Steve, Shepard, Alexandra, and Walter, John, eds, *Remaking English Society: Social Relations and Social Change in Early Modern England* (Woodbridge, 2013).
Hingley, Richard, *The Recovery of Roman Britain 1586–1906* (Oxford, 2008).
Hirsch, Jean-Pierre, *Les Deux rêves du commerce: Entreprise et institution dans la région lilloise (1780–1860)* (Paris, 1991).
Hirschman, Albert O., *The Passions and the Interests: Political Arguments for Capitalism before its Triumph* (Princeton, 1976).
Hirst, Derek, *Dominion: England and its Island Neighbours 1500–1707* (Oxford, 2012).
Holleran, Claire, *Shopping in Ancient Rome: The Retail Trade in the Late Republic and the Principate* (Oxford, 2012).
Hollingsworth, T. H., *The Demography of the British Peerage* (1964).
Holmes, G. A., 'The "Libel of English Policy"', *EHR* 76 (1961), pp. 193–216.
Hont, Istvan, *Jealousy of Trade: International Competition and the Nation-State in Historical Perspective* (Cambridge, Mass., 2005).
Hont, Istvan, 'The Early Enlightenment Debate on Commerce and Luxury', in Mark Goldie and Robert Wokler, eds, *The Cambridge History of Eighteenth-Century Political Thought* (Cambridge, 2006), pp. 377–418.

Hoppen, K. Theodore, *The Common Scientist in the Seventeenth Century: A Study of the Dublin Philosophical Society, 1683–1708* (1970).
Hoppit, Julian, 'Political Arithmetic in Eighteenth-Century England', *EcHR* 49 (1996), pp. 516–40.
Hoppit, Julian, 'Patterns of Parliamentary Legislation, 1600–1800', *Historical Journal*, 39 (1996), pp. 109–31.
Hoppit, Julian, *A Land of Liberty? England 1689–1727* (Oxford, 2000).
Hoppit, Julian, 'The Contexts and Contours of British Economic Literature, 1660–1760', *Historical Journal*, 49 (2006), pp. 79–110.
Hoppit, Julian, 'The Nation, the State, and the First Industrial Revolution', *Journal of British Studies*, 50 (2011), pp. 307–31.
Hoppit, Julian, 'Bounties, the Economy and the State in Britain, 1689–1800', in Gauci, ed., *Regulating the British Economy*, pp. 139–60.
Horn, Jeff, *The Path not Taken: French Industrialization in the Age of Revolution, 1750–1830* (Cambridge, Mass., 2006).
Horne, Thomas A., *The Social Thought of Bernard Mandeville: Virtue and Commerce in Early Eighteenth-Century England* (1978).
Horsefield, J. Keith, *British Monetary Experiments, 1650–1710* (1960).
Hoskins, W. G., *Provincial England: Essays in Social and Economic History* (1963).
Houlbrooke, Ralph, *Death, Religion and the Family in England 1480–1750* (Oxford, 1998).
Houlbrooke, Ralph, 'England', in Paulina Kewes et al., eds, *The Oxford Handbook of Holinshed's Chronicles* (Oxford, 2013), pp. 629–45.
Houston, Alan, *Benjamin Franklin and the Politics of Improvement* (New Haven, 2008).
Houston, Alan, and Pincus, Steve, eds, *A Nation Transformed: England after the Restoration* (Cambridge, 2001).
Houston, Rab, 'Custom in Context: Medieval and Early Modern Scotland and England', *P&P* 211 (May 2011), pp. 35–76.
Hoyle, Richard W., ed., *The Estates of the English Crown 1558–1640* (Cambridge, 1992).
Hoyle, Richard W., 'Disafforestation and Drainage: The Crown as Entrepreneur', in Hoyle, ed., *The Estates of the English Crown*, pp. 353–88.
Hoyle, Richard W., ed., *Custom, Improvement and the Landscape in Early Modern Britain* (Farnham, 2011).
Hoyle, Richard W., 'Introduction: Custom, Improvement and Anti-Improvement', in Hoyle, ed., *Custom, Improvement and the Landscape*, pp. 1–38.
Hsia, R. Po-chia, *A Jesuit in the Forbidden City: Matteo Ricci 1552–1610* (Oxford, 2010).
Hughes, Ann, *Politics, Society and Civil War in Warwickshire, 1620–1660* (Cambridge, 1987).
Hundert, E. G., *The Enlightenment's Fable: Bernard Mandeville and the Discovery of Society* (Cambridge, 1994).
Hunter, Michael, *John Aubrey and the Realm of Learning* (1975).
Hunter, Michael, *Science and Society in Restoration England* (Cambridge, 1981).
Hunter, Michael, *Boyle: Between God and Science* (New Haven, 2009).
Hutchison, Terence W., *Before Adam Smith: The Emergence of Political Economy, 1662–1776* (Oxford, 1988).
Iliffe, Rob, 'Material Doubts: Hooke, Artisan Culture and the Exchange of Information in 1670s London', *British Journal for the History of Science*, 28 (1995), pp. 285–318.
Innes, Joanna, 'The "Mixed Economy of Welfare" in Early Modern England: Assessment of the Options from Hale to Malthus (*c.*1683–1803)', in Daunton, ed., *Charity, Self-Interest and Welfare*, pp. 139–80.

Innes, Joanna, *Inferior Politics: Social Problems and Social Policies in Eighteenth-Century Britain* (Oxford, 2009).
Innes, Stephen, *Creating the Commonwealth: The Economic Culture of Puritan New England* (New York, 1995).
Israel, Jonathan I., *The Dutch Republic: Its Rise, Greatness and Fall, 1477–1806* (Oxford, 1995).
Israel, Jonathan I., *Radical Enlightenment: Philosophy and the Making of Modernity 1650–1750* (Oxford, 2001).
Israel, Jonathan, *A Revolution of the Mind: Radical Enlightenment and the Intellectual Origins of Modern Democracy* (Princeton, 2010).
Jacob, James R., 'The Political Economy of Science in Seventeenth-Century England', *Social Research*, 59 (1992), pp. 505–32.
Jacob, Margaret C., *The Cultural Meaning of the Scientific Revolution* (Philadelphia, 1988).
Jacob, Margaret C., and Mijnhardt, Wijnand W., eds, *The Dutch Republic in the Eighteenth Century: Decline, Enlightenment, and Revolution* (Ithaca, NY, 1992).
Jacobsen, Gertrude A., *William Blathwayt: A Late Seventeenth Century English Administrator* (New Haven, 1932).
Jardine, Lisa, *The Curious Life of Robert Hooke: The Man who Measured London* (2003).
Jenner, Mark, '"Another *Epocha*"? Hartlib, John Lanyon and the Improvement of London in the 1650s', in Greengrass et al., eds, *Samuel Hartlib and Universal Reformation*, pp. 343–56.
Jenner, Mark, 'The Politics of London Air: John Evelyn's *Fumifugium* and the Restoration', *Historical Journal*, 38 (1995), pp. 535–51.
Jenner, Mark S. R., 'From Conduit Community to Commercial Network? Water in London 1500–1725', in Griffiths and Jenner, eds, *Londinopolis*, pp. 250–72.
Johnson, Matthew, *Housing Culture: Traditional Architecture in an English Landscape* (1993).
Johnston, Warren, *Revelation Restored: The Apocalypse in Later Seventeenth-Century England* (Woodbridge, 2011).
Jones, D. W., *War and Economy in the Age of William III and Marlborough* (Oxford, 1988).
Jones, D. W., 'The Workings and Measurement of Pre-Industrial "Organic" Economies: Conjectures on English Agrarian Growth, 1660–1820', *Journal of European Economic History*, 35 (2006), pp. 177–212.
Jones, E. L., *The European Miracle: Environments, Economies and Geopolitics in the History of Europe and Asia* (3rd edn, Cambridge, 2003).
Jones, E. L., and Falkus, M. E., 'Urban Improvement and the English Economy in the Seventeenth and Eighteenth Centuries', in Peter Borsay, ed., *The Eighteenth-Century Town: A Reader in English Urban History 1688–1820* (1990), pp. 116–58.
Jones, Gareth, *History of the Law of Charity, 1532–1827* (Cambridge, 1969).
Jones, Norman, *God and the Moneylenders: Usury and Law in Early Modern England* (Oxford, 1989).
Jordan, W. K., *Philanthropy in England 1480–1660: A Study of the Changing Pattern of English Social Aspirations* (1959).
Jordan, W. K., *The Charities of London, 1480–1660* (1960).
Jordan, W. K., *The Forming of the Charitable Institutions of the West of England* (Transactions of the American Philosophical Society NS 50, Part 8, Philadelphia, 1960).
Kadane, Matthew, *The Watchful Clothier: The Life of an Eighteenth-Century Protestant Capitalist* (New Haven, 2013).

Kain, Roger J. P., and Baigent, Elizabeth, *The Cadastral Map in the Service of the State: A History of Property Mapping* (Chicago, 1992).

Kelly, Morgan, and Ó Gráda, Cormac, 'The Waning of the Little Ice Age: Climate Change in Early Modern Europe', *Journal of Interdisciplinary History*, 44.3 (2014), pp. 301–25.

Klein, Bernhard, *Maps and the Writing of Space in Early Modern England and Ireland* (Basingstoke, 2001).

Knights, Mark, *Politics and Opinion in Crisis, 1678–81* (Cambridge, 1994).

Knights, Mark, *Representation and Misrepresentation in Later Stuart Britain* (Oxford, 2005).

Knights, Mark, 'Towards a Social and Cultural History of Keywords and Concepts by the Early Modern Research Group', *History of Political Thought*, 31 (2010), pp. 427–48.

Knights, Mark, *The Devil in Disguise: Deception, Delusion, and Fanaticism in the Early English Enlightenment* (Oxford, 2011).

Knights, Mark, 'Regulation and Rival Interests in the 1690s', in Gauci, ed., *Regulating the British Economy*, pp. 63–81.

Knights, Mark, and the Early Modern Research Group, 'Commonwealth: The Social, Cultural, and Conceptual Contexts of an Early Modern Keyword', *Historical Journal*, 54 (2011), pp. 659–87.

Konvitz, Josef W., *Cartography in France 1660–1848: Science, Engineering and Statecraft* (Chicago, 1987).

Kreager, Philip, 'New Light on Graunt', *Population Studies*, 42 (1988), pp. 129–40.

Kreager, Philip, 'John Graunt', in K. Kempf-Leonard, ed., *Encyclopedia of Social Measurement* (2005), ii. 161–6.

Kuchta, David, *The Three-Piece Suit and Modern Masculinity: England, 1550–1850* (Berkeley, 2002).

Landers, John, *Death and the Metropolis: Studies in the Demographic History of London 1670–1830* (Cambridge, 1993).

Landreth, David, *The Face of Mammon: The Matter of Money in English Renaissance Literature* (Oxford, 2012).

Langford, Paul, *Public Life and the Propertied Englishman, 1689–1798* (Oxford, 1991).

Langford, Paul, *Englishness Identified: Manners and Character, 1650–1850* (Oxford, 2000).

Lass, Roger, ed., *The Cambridge History of the English Language*, iii: *1476–1776* (Cambridge, 1999).

Layard, Richard, *Happiness: Lessons from a New Science* (2006).

Lemire, Beverly, *Fashion's Favourite: The Cotton Trade and the Consumer in Britain, 1660–1800* (Oxford, 1991).

Leng, Thomas, 'Commercial Conflict and Regulation in the Discourse of Trade in Seventeenth-Century England', *Historical Journal*, 48 (2005), pp. 933–54.

Leng, Thomas, *Benjamin Worsley (1618–1677): Trade, Interest and the Spirit in Revolutionary England* (Woodbridge, 2008).

Leng, Thomas, '"A Potent Plantation Well Armed and Policeed": Huguenots, the Hartlib Circle, and British Colonization in the 1640s', *William and Mary Quarterly*, 3rd ser. 66 (2009), pp. 173–94.

Leng, Thomas, 'Epistemology: Expertise and Knowledge in the World of Commerce', in Stern and Wennerlind, eds, *Mercantilism Reimagined*, pp. 97–116.

Leslie, Michael, 'The Spiritual Husbandry of John Beale', in Leslie and Raylor, eds, *Culture and Cultivation*, pp. 151–72.

Leslie, Michael, and Raylor, Timothy, eds, *Culture and Cultivation in Early Modern England: Writing and the Land* (Leicester, 1992).

Letwin, William, *The Origins of Scientific Economics* (New York, 1964).
Levin, Jennifer, *The Charter Controversy in the City of London, 1660–88, and its Consequences* (1969).
Levine, Joseph M., *The Autonomy of History: Truth and Method from Erasmus to Gibbon* (Chicago, 1999).
Levy, Barry, *Town Born: The Political Economy of New England from its Founding to the Revolution* (Philadelphia, 2009).
Lindert, Peter H., and Williamson, Jeffrey G., 'Revising England's Social Tables 1688–1812', *Explorations in Economic History*, 19 (1982), pp. 385–408.
Lindley, Keith, *Fenland Riots and the English Revolution* (1982).
Lloyd, Sarah, *Charity and Poverty in England, c.1680–1820: Wild and Visionary Schemes* (Manchester, 2009).
Loft, Philip, 'Political Arithmetic and the English Land Tax in the Reign of William III', *Historical Journal*, 56 (2013), pp. 321–43.
Loveman, Kate, 'Samuel Pepys and "Discourses touching Religion" under James II', *EHR* 127 (2012), pp. 46–82.
Low, Anthony, *The Georgic Revolution* (Princeton, 1985).
McCloskey, Deirdre N., *The Bourgeois Virtues: Ethics for an Age of Commerce* (Chicago, 2006).
McCloskey, Deirdre N., *Bourgeois Dignity: Why Economics Can't Explain the Modern World* (Chicago, 2010).
McCormick, Ted, *William Petty and the Ambitions of Political Arithmetic* (Oxford, 2009).
MacCulloch, Diarmaid, *Suffolk and the Tudors: Politics and Religion in an English County 1500–1600* (Oxford, 1986).
Macfarlane, Alan, *The Savage Wars of Peace: England, Japan and the Malthusian Trap* (Oxford, 1997).
Machin, R., 'The Great Rebuilding: A Reassessment', *P&P* 77 (Nov. 1977), pp. 33–56.
Macinnes, Allan I., *The British Revolution, 1629–1660* (2005).
McIntosh, Marjorie Keniston, *Poor Relief in England 1350–1600* (Cambridge, 2012).
McKellar, Elizabeth, *The Birth of Modern London: The Development and Design of the City, 1660–1720* (Manchester, 1999).
Maclean, Ian, *Scholarship, Commerce, Religion: The Learned Book in the Age of Confessions, 1560–1630* (Cambridge, Mass., 2012).
MacLeod, Christine, *Inventing the Industrial Revolution: The English Patent System, 1660–1800* (Cambridge, 1988).
MacLeod, Christine, 'The European Origins of British Technological Predominance', in Prados de la Escosura, ed., *Exceptionalism and Industrialisation*, pp. 111–26.
McMahon, Darrin M., *The Pursuit of Happiness: A History from the Greeks to the Present* (2006).
McRae, Andrew, *God Speed the Plough: The Representation of Agrarian England, 1500–1660* (Cambridge, 1996).
McRae, Andrew, *Literature and Domestic Travel in Early Modern England* (Cambridge, 2009).
Maddison, Angus, *Contours of the World Economy, 1–2030AD: Essays in Macro-Economic History* (Oxford, 2007).
Maginn, Christopher, *William Cecil, Ireland, and the Tudor State* (Oxford, 2012).
Malcolm, Noel, *Aspects of Hobbes* (Oxford, 2002).
Malcolm, Noel, *Reason of State, Propaganda, and the Thirty Years' War: An Unknown Translation by Thomas Hobbes* (Oxford, 2007).

Manning, Robert B., *Village Revolts: Social Protest and Popular Disturbances in England 1509–1640* (Oxford, 1988).
Marshall, P. J., *Remaking the British Atlantic: The United States and the British Empire after American Independence* (Oxford, 2012).
Marshall, Tristan, *Theatre and Empire: Great Britain on the London Stages under James VI and I* (Manchester, 2000).
Martin, John E., *Feudalism to Capitalism: Peasant and Landlord in English Agrarian Development* (1983).
Martin, Thierry, ed., *Arithmétique politique dans la France du XVIIIe siècle* (Paris, 2003).
Mathias, Peter, *The Transformation of England: Essays in the Economic and Social History of England in the Eighteenth Century* (1979).
Mayhew, Nicholas, *Sterling: The History of a Currency* (1999).
Mayhew, Nicholas, 'Prices in England, 1170–1750', *P&P* 219 (May 2013), pp. 3–39.
Mayhew, Robert, *Enlightenment Geography: The Political Languages of British Geography, 1650–1850* (Basingstoke, 2000).
Melton, James Van Horn, *The Rise of the Public in Enlightenment Europe* (Cambridge, 2001).
Mendelson, Sara, and Crawford, Patricia, *Women in Early Modern England, 1550–1720* (Oxford, 1998).
Mendyk, Stan A. E., *'Speculum Britanniae': Regional Study, Antiquarianism, and Science in Britain to 1700* (Toronto, 1989).
Metzler, Mark, and Smits, Gregory, 'Introduction: The Autonomy of Market Activity and the Emergence of *Keizai* Thought', in Gramlich-Oka and Smits, eds, *Economic Thought in Early Modern Japan*, pp. 1–19.
Meusnier, N., 'Vauban: Arithmétique politique, Ragot et autre *cochonnerie*', in Martin, ed., *Arithmétique politique dans la France du XVIIIe siècle*, pp. 91–132.
Mijnhardt, Wijnand, 'The Dutch Enlightenment: Humanism, Nationalism, and Decline', in Jacob and Mijnhardt, eds, *Dutch Republic in the Eighteenth Century*, pp. 197–223.
Mitter, Rana, *Modern China: A Very Short Introduction* (Oxford, 2008).
Mokyr, Joel, *The Gifts of Athena: Historical Origins of the Knowledge Economy* (Princeton, 2002).
Mokyr, Joel, 'Knowledge, Enlightenment and the Industrial Revolution: Reflections on *The Gifts of Athena*', *History of Science*, 45 (2007), pp. 185–96.
Mokyr, Joel, *The Enlightened Economy: An Economic History of Britain, 1700–1850* (New Haven, 2009).
Morera, Raphaël, *L'Assèchement des marais en France au XVIIe siècle* (Rennes, 2011).
Morgan, Prys, 'Wild Wales: Civilizing the Welsh from the Sixteenth to the Nineteenth Centuries', in Burke et al., eds, *Civil Histories*, pp. 265–83.
Morgan, Victor, 'The Cartographical Image of "The Country"', *TRHS* 5th ser. 29 (1979), pp. 129–54.
Mortimer, Ian, 'The Triumph of the Doctors: Medical Assistance to the Dying, c.1570–1720', *TRHS* 6th ser. 15 (2005), pp. 97–116.
Mortimer, Ian, *The Dying and the Doctors: The Medical Revolution in Seventeenth-Century England* (Woodbridge, 2009).
Mukerjii, Chandra, 'Demonstration and Verification in Engineering: Ascertaining Truth and Telling Fictions along the Canal du Midi', in Roberts et al., eds, *Mindful Hand*, pp. 169–86.
Muldrew, Craig, 'Interpreting the Market: The Ethics of Credit and Community Relations in Early Modern England', *Social History*, 18 (1993), pp. 163–83.

Muldrew, Craig, *The Economy of Obligation: The Culture of Credit and Social Relations in Early Modern England* (Basingstoke, 1998).

Muldrew, Craig, *Food, Energy and the Creation of Industriousness: Work and Material Culture in Agrarian England, 1550–1780* (Cambridge, 2011).

Muldrew, Craig, 'From Credit to Savings? An Examination of Debt and Credit in Relation to Increasing Consumption in England [c.1650–1770]', *Quaderni storici*, 137 (Aug. 2011), pp. 1–24.

Muldrew, Craig, ' "Th'ancient Distaff" and "Whirling Spindle": Measuring the Contribution of Spinning to Household Earnings and the National Economy in England, 1550–1770', *EcHR* 65 (2012), pp. 498–526.

Muldrew, Craig, 'From Commonwealth to Public Opulence: The Redefinition of Wealth and Government in Early Modern Britain', in Hindle et al., eds, *Remaking English Society*, pp. 317–39.

Murphy, Anne L., *The Origins of English Financial Markets: Investment and Speculation before the South Sea Bubble* (Cambridge, 2009).

Murphy, Anne L., 'Demanding "Credible Commitment": Public Reactions to the Failures of the Early Financial Revolution', *EcHR* 66 (2013), pp. 178–97.

Murphy, Antoin E., *Richard Cantillon: Entrepreneur and Economist* (Oxford, 1989).

Murphy, Antoin E., *The Genesis of Macroeconomics: New Ideas from Sir William Petty to Henry Thornton* (Oxford, 2009).

Nash, Gary B., *The Urban Crucible: Social Change, Political Consciousness, and the Origins of the American Revolution* (Cambridge, Mass., 1979).

Nayar, Pramod K., *English Writing and India, 1600–1920: Colonizing Aesthetics* (2008).

Neeson, J. M., *Commoners: Common Right, Enclosure and Social Change in England 1700–1820* (Cambridge, 1993).

Nef, John U., *The Rise of the British Coal Industry* (2 vols, 1932).

Nelson, Eric, *The Greek Tradition in Republican Thought* (Cambridge, 2004).

Newton, Gill, 'Infant Mortality Variations, Feeding Practices and Social Status in London between 1550 and 1750', *Social History of Medicine*, 24 (2011), pp. 260–80.

Noonan, John T., *The Scholastic Analysis of Usury* (Cambridge, Mass., 1957).

North, Douglass C., *Institutions, Institutional Change and Economic Performance* (Cambridge, 1990).

North, Douglass C., and Weingast, Barry R., 'Constitutions and Commitment: The Evolution of Institutions Governing Public Choice in Seventeenth-Century England', *Journal of Economic History*, 49 (1989), pp. 803–32.

Novak, Maximilian E., *Daniel Defoe: Master of Fictions* (Oxford, 2001).

O'Brien, Patrick K., 'The Political Economy of British Taxation, 1660–1815', *EcHR* 41 (1988), pp. 1–32.

O'Brien, Patrick K., *Power with Profit: The State and the Economy, 1688–1815* (Inaugural Lecture, University of London, 1991).

O'Brien, Patrick K., 'The Nature and Historical Evolution of an Exceptional Fiscal State and its Possible Significance for the Precocious Commercialization and Industrialization of the British Economy from Cromwell to Nelson', *EcHR* 64 (2011), pp. 408–46.

Offer, Avner, *The Challenge of Affluence: Self-Control and Well-Being in the United States and Britain since 1950* (Oxford, 2006).

Ogborn, Miles, *Indian Ink: Script and Print in the Making of the English East India Company* (Chicago 2007).

Ogilvie, Sheilagh, 'Consumption, Social Capital, and the "Industrious Revolution" in Early Modern Germany', *Journal of Economic History*, 70 (2010), pp. 287–325.

Orlin, Lena Cowen, ed., *Material London, ca. 1600* (Philadelphia, 2000).
Ormrod, David, *The Rise of Commercial Empires: England and the Netherlands in the Age of Mercantilism, 1650–1770* (Cambridge, 2003).
Ormrod, W. M., 'The English Crown and the Customs, 1349–63', *EcHR* 2nd ser. 40 (1987), pp. 27–40.
Overton, Mark, *Agricultural Revolution in England: The Transformation of the Agrarian Economy, 1500–1850* (Cambridge, 1996).
Overton, Mark, Whittle, Jane, Dean, Darron, and Hann, Andrew, *Production and Consumption in English Households, 1600–1750* (2004).
Pagden, Anthony, *Lords of All the World: Ideologies of Empire in Spain, Britain and France c.1500–c.1800* (New Haven, 1995).
Palliser, D. M., *The Age of Elizabeth: England under the Later Tudors, 1547–1603* (1983).
Palmer, Stanley H., *Economic Arithmetic: A Guide to the Statistical Sources for English Commerce, Industry, and Finance, 1700–1850* (New York, 1977).
Parker, Geoffrey, *Global Crisis: War, Climate Change and Catastrophe in the Seventeenth Century* (New Haven, 2013).
Parkin, Jon, *Science, Religion and Politics in Restoration England: Richard Cumberland's De legibus naturae* (1999).
Parry, Graham, *The Trophies of Time: English Antiquarians of the Seventeenth Century* (Oxford, 1995).
Pastorino, Cesare, 'The Mine and the Furnace: Francis Bacon, Thomas Russell, and Early Stuart Mining Culture', *Early Science and Medicine*, 14 (2009), pp. 630–60.
Pearl, Valerie, 'Puritans and Poor Relief: The London Workhouse 1649–1660', in Donald Pennington and Keith Thomas, eds, *Puritans and Revolutionaries: Essays in Seventeenth-Century History Presented to Christopher Hill* (Oxford, 1978), pp. 206–32.
Peltonen, Markku, *Classical Humanism and Republicanism in English Political Thought, 1570–1640* (Cambridge, 1995).
Pennell, Sara, 'The Material Culture of Food in Early Modern England, c.1650–1750', in Sarah Tarlow and Susie West, eds, *Familiar Pasts? Archaeologies of Later Historical Britain 1550–1860* (1999), pp. 35–50.
Pennell, Sara, '"Great quantities of gooseberry pye and baked clod of beef": Victualling and Eating out in Early Modern London', in Griffiths and Jenner, eds, *Londinopolis*, pp. 228–49.
Pennington, David, 'Beyond the Moral Economy: Economic Change, Ideology and the 1621 House of Commons', *Parliamentary History*, 25 (2006), pp. 214–31.
Penovich, Katherine R., 'From "Revolution Principles" to Union: Daniel Defoe's Intervention in the Scottish Debate', in Robertson, ed., *Union for Empire*, pp. 228–42.
Perrot, Jean-Claude, *Une histoire intellectuelle de l'économie politique: XVIIe–XVIIIe siècle* (Paris, 1992).
Pettegree, Andrew, 'Centre and Periphery in the European Book World', *TRHS* 18 (2008), pp. 101–28.
Pettigrew, William, 'Regulatory Inertia and National Economic Growth: An African Trade Case Study, 1660–1714', in Gauci, ed., *Regulating the British Economy*, pp. 25–40.
Pickstone, John V., 'Dearth, Dirt and Fever Epidemics: Rewriting the History of British "Public Health", 1780–1850', in Terence Ranger and Paul Slack, eds, *Epidemics and Ideas* (Cambridge, 1992), pp. 125–48.
Pincus, Steve, 'Popery, Trade and Universal Monarchy: The Ideological Context of the Outbreak of the Anglo-Dutch War', *EHR* 107 (1992), pp. 1–29.

Pincus, Steve, 'From Butterboxes to Wooden Shoes: The Shift in English Sentiment from Anti-Dutch to Anti-French in the 1670s', *Historical Journal*, 38 (1995), pp. 333–61.
Pincus, Steve, *Protestantism and Patriotism: Ideologies and the Making of English Foreign Policy, 1650–1668* (Cambridge, 1996).
Pincus, Steve, 'Neither Machiavellian Moment nor Possessive Individualism: Commercial Society and the Defenders of the English Commonwealth', *American Historical Review*, 103 (1998), pp. 705–36.
Pincus, Steve, 'From Holy Cause to Economic Interest: The Study of Population and the Invention of the State', in Houston and Pincus, eds, *A Nation Transformed*, pp. 272–98.
Pincus, Steve, 'John Evelyn: Revolutionary', in Frances Harris and Michael Hunter, eds, *John Evelyn and his Milieu* (2003), pp. 185–219.
Pincus, Steve, *1688: The First Modern Revolution* (New Haven, 2009).
Pincus, Steve, 'Rethinking Mercantilism: Political Economy, the British Empire, and the Atlantic World in the Seventeenth and Eighteenth Centuries', *William and Mary Quarterly*, 69 (2012), pp. 3–34.
Pincus, Steve, and Wolfram, Alice, 'A Proactive State? The Land Bank, Investment and Party Politics in the 1690s', in Gauci, ed., *Regulating the British Economy*, pp. 41–62.
Pocock, J. G. A., *The Machiavellian Moment: Florentine Political Thought and the Atlantic Republican Tradition* (Princeton, 1975).
Pocock, J. G. A., *Barbarism and Religion*, ii: *Narratives of Civil Government* (Cambridge, 1999).
Pocock, J. G. A., *Barbarism and Religion*, iv: *Barbarians, Savages and Empires* (Cambridge, 2005).
Pomeranz, Kenneth, *The Great Divergence: China, Europe, and the Making of the Modern World Economy* (Princeton, 2000).
Poole, Robert, *Time's Alteration: Calendar Reform in Early Modern England* (1998).
Poole, William, *The World Makers: Scientists of the Restoration and the Search for the Origins of the Earth* (Oxford, 2010).
Poovey, Mary, *A History of the Modern Fact: Problems of Knowledge in the Sciences of Wealth and Society* (Chicago, 1998).
Porter, Stephen, *The Great Fire of London* (Stroud, 1996).
Powell, Martyn J., *The Politics of Consumption in Eighteenth-Century Ireland* (Basingstoke, 2005).
Prados de la Escosura, Leandro, ed., *Exceptionalism and Industrialisation: Britain and its European Rivals, 1688–1815* (Cambridge, 2004).
Prestwich, Menna, *Cranfield: Politics and Profits under the Early Stuarts* (Oxford, 1966).
Price, William Hyde, *The English Patents of Monopoly* (Cambridge, Mass., 1906).
Quintrell, B. W., 'The Making of Charles I's Book of Orders', *EHR* 95 (1980), pp. 553–72.
Rabasa, José, et al., eds, *The Oxford History of Historical Writing*, iii: *1400–1800* (Oxford, 2012).
Rabb, Theodore K., *Enterprise and Empire: Merchant and Gentry Investment in the Expansion of England, 1575–1630* (Cambridge, Mass., 1967).
Rabb, Theodore K., *Jacobean Gentleman: Sir Edwin Sandys, 1561–1629* (Princeton, 1998).
Ramsay, G. D., *The City of London in International Politics at the Accession of Elizabeth Tudor* (Manchester, 1975).
Ramsay, G. D., 'Industrial Discontent in Early Elizabethan London: Clothworkers and Merchants Adventurers in Conflict', *London Journal*, 1 (1975), pp. 227–39.
Ramsay, G. D., *The Politics of a Tudor Merchant Adventurer: A Letter to the Earls of Friesland* (Manchester, 1979).

Ravenhill, William, 'John Adams, his Map of England, its Projection, and his *Index Villaris* of 1680', *Geographical Journal*, 144 (1978), pp. 424–37.

Raylor, Timothy, 'Samuel Hartlib and the Commonwealth of Bees', in Leslie and Raylor, eds, *Culture and Cultivation*, pp. 91–129.

Raymond, Joad, *Pamphlets and Pamphleteering in Early Modern Britain* (Cambridge, 2003).

Razzell, Peter, *The Conquest of Smallpox* (Firle, 1977).

Reed, Michael, *The Age of Exuberance 1500–1700* (1987).

Reinart, Erik S., 'Emulating Success: Contemporary Views of the Dutch Economy before 1800', in Oscar Gelderblom, ed., *The Political Economy of the Dutch Republic* (Farnham, 2009), pp. 19–39.

Reungoat, Sabine, *William Petty: Observateur des Îles Britanniques* (Paris, 2004).

Richeson, A. W., *English Land Measuring to 1800: Instruments and Practices* (Cambridge, Mass., 1966).

Riley, James C., *Population Thought in the Age of the Demographic Revolution* (Durham, NC, 1985).

Riley, James C., *The Eighteenth Century Campaign to Avoid Disease* (Basingstoke, 1987).

Roberts, Lissa, Schaffer, Simon, and Dear, Peter, eds, *The Mindful Hand: Inquiry and Invention from the Late Renaissance to Early Industrialisation* (Amsterdam, 2007).

Robertson, J. C., 'Reckoning with London: Interpreting the Bills of Mortality before John Graunt', *Urban History*, 23 (1996), pp. 325–50.

Robertson, John, ed., *A Union for Empire: Political Thought and the British Union of 1707* (Cambridge, 1995).

Robertson, John, 'An Elusive Sovereignty: The Course of the Union Debate in Scotland 1698–1707', in Robertson, ed., *Union for Empire*, pp. 198–227.

Robertson, John, *The Case for the Enlightenment: Scotland and Naples 1680–1760* (Cambridge, 2005).

Roche, Daniel, *A History of Everyday Things: The Birth of Consumption in France, 1600–1800* (Cambridge, 2000).

Rodger, N. A. M., *The Safeguard of the Sea: A Naval History of Britain 660–1649* (1997).

Rodger, N. A. M., *The Command of the Ocean: A Naval History of Britain, 1649–1815* (2004).

Rollison, David, *The Local Origins of Modern Society, Gloucestershire 1500–1800* (1992).

Rosen, G., 'Cameralism and the Concept of Medical Police', *Bulletin of the History of Medicine*, 27 (1953), pp. 21–42.

Rossi, Paolo, *Francis Bacon: From Magic to Science* (1968).

Røstvig, Maren-Sofie, *The Happy Man: Studies in the Metamorphoses of a Classical Ideal* (2 vols, Oslo and Oxford, 1954–8).

Rothkrug, Lionel, *Opposition to Louis XIV: The Political and Social Origins of the French Enlightenment* (Princeton, 1965).

Rowlands, Guy, *The Financial Decline of a Great Power: War, Influence, and Money in Louis XIV's France* (Oxford, 2012).

Roy, Ian, 'England Turned Germany? The Aftermath of the Civil War in its European Context', *TRHS* 5th ser. 28 (1978), pp. 127–44.

Rundle, David, 'Was There a Renaissance Style of Politics in Fifteenth-Century England?' in Bernard and Gunn, eds, *Authority and Consent*, pp. 15–32.

Rusnock, Andrea A., *Vital Accounts: Quantifying Health and Population in Eighteenth-Century England and France* (Cambridge, 2002).

Sacks, David Harris, 'The Greed of Judas: Avarice, Monopoly, and the Moral Economy in England, ca.1350–ca.1600', *Journal of Medieval and Early Modern Studies*, 28 (1998), pp. 263–308.
Schabas, Margaret, *The Natural Origins of Economics* (Chicago, 2005).
Schabas, Margaret, and De Marchi, Neil, 'Oeconomies in the Age of Newton', *History of Political Economy*, 35 (2003), Supplement 1, pp. 1–13.
Schaeper, Thomas J., *The French Council of Commerce, 1700–1715* (Columbus, Oh., 1983).
Schaffer, Simon, 'The Earth's Fertility as a Social Fact in Early Modern Britain', in Mikuláš Teich, Roy Porter, and Bo Gustafsson, eds, *Nature and Society in Historical Context* (Cambridge, 1997), pp. 124–47.
Schaffer, Simon, 'Introduction', in Roberts et al., eds, *Mindful Hand*, pp. 85–93.
Schiffman, Zachary Sayre, *The Birth of the Past* (Baltimore, 2011).
Schneewind, Jerome B., *The Invention of Autonomy: A History of Modern Moral Philosophy* (Cambridge, 1998).
Scott, Jonathan, *Commonwealth Principles: Republican Writing of the English Revolution* (Cambridge, 2004).
Scott, Jonathan, *When the Waves Ruled Britannia: Geography and Political Identities, 1500–1800* (Cambridge, 2011).
Serjeantson, R. W., 'Proof and Persuasion', in Katharine Park and Lorraine Daston, eds, *The Cambridge History of Science*, iii: *Early Modern Science* (Cambridge, 2006), pp. 132–75.
Sewell, William H., Jr, 'The Empire of Fashion and the Rise of Capitalism in Eighteenth-Century France', *P&P* 206 (Feb. 2010), pp. 81–120.
Shammas, Carole, 'English-Born and Creole Elites in Turn-of-the-Century Virginia', in Tate and Ammerman, eds, *Chesapeake in the Seventeenth Century*, pp. 274–96.
Shammas, Carole, *The Pre-Industrial Consumer in England and America* (Oxford, 1990).
Shannon, Bill, 'Approvement and Improvement in the Lowland Wastes of Early Modern Lancashire', in Hoyle, ed., *Custom, Improvement and the Landscape in Early Modern Britain*, pp. 175–202.
Shapiro, Barbara J., *A Culture of Fact: England, 1550–1720* (Ithaca, NY, 2002).
Sharpe, Kevin, *Sir Robert Cotton 1586–1631: History and Politics in Early Modern England* (Oxford, 1979).
Sharpe, Kevin, 'Sir Thomas Witherings and the Reform of the Foreign Posts, 1632–40', *BIHR* 57 (1984), pp. 149–64.
Sharpe, Kevin, *Rebranding Rule: The Restoration and Revolution Monarchy, 1660–1714* (2013).
Sharpe, Pamela, *Population and Society in an East Devon Parish: Reproducing Colyton 1540–1840* (Exeter, 2002).
Shepard, Alexandra, and Spicksley, Judith, 'Worth, Age, and Social Status in Early Modern England', *EcHR* 64 (2011), pp. 493–530.
Shovlin, John, *The Political Economy of Virtue: Luxury, Patriotism, and the Origins of the French Revolution* (Ithaca, NY, 2006).
Simpson, James, 'European Farmers and the British "Agricultural Revolution"', in Prados de la Escosura, ed., *Exceptionalism and Industrialisation*, pp. 69–85.
Skinner, Quentin, *Visions of Politics* (3 vols, Cambridge, 2002).
Slack, Paul, 'Books of Orders: The Making of English Social Policy, 1577–1631', *TRHS* 5th ser. 30 (1980), pp. 1–22.
Slack, Paul, *The Impact of Plague in Tudor and Stuart England* (1985).
Slack, Paul, *Poverty and Policy in Tudor and Stuart England* (1988).

Slack, Paul, *The English Poor Law 1531–1782* (Basingstoke, 1990).
Slack, Paul, *From Reformation to Improvement: Public Welfare in Early Modern England* (Oxford, 1999).
Slack, Paul, 'Perceptions of the Metropolis in Seventeenth-Century England', in Burke et al., eds, *Civil Histories*, pp. 161–80.
Slack, Paul, 'Great and Good Towns 1549–1700', in Clark, ed., *Cambridge Urban History of Britain*, ii. 347–76.
Slack, Paul, 'Measuring the National Wealth in Seventeenth-Century England', *EcHR* 57 (2004), pp. 607–35.
Slack, Paul, 'Government and Information in Seventeenth-Century England', *P&P* 184 (Aug. 2004), pp. 33–68.
Slack, Paul, 'The Politics of Consumption and England's Happiness in the Later Seventeenth Century', *EHR* 122 (2007), pp. 609–31.
Slack, Paul, 'Material Progress and the Challenge of Affluence in Seventeenth-Century England', *EcHR* 62 (2009), pp. 576–603.
Slack, Paul, *'Plenty of People': Perceptions of Population in Early Modern England* (Stenton Lecture 2010, University of Reading, 2011).
Smith, A. Hassell, *County and Court: Government and Politics in Norfolk, 1558–1603* (Oxford, 1974).
Smith, J. R., *The Speckled Monster: Smallpox in England 1670–1970, with Particular Reference to Essex* (Chelmsford, 1987).
Smith, Pamela H., *The Business of Alchemy: Science and Culture in the Holy Roman Empire* (Princeton, 1994).
Smith, Richard M., 'Charity, Self-Interest and Welfare: Reflections from Demographic and Family History', in Daunton, ed., *Charity, Self-Interest and Welfare*, pp. 23–49.
Smith, Richard M., 'Plagues and Peoples: The Long Demographic Cycle, 1250–1670', in Paul Slack and Ryk Ward, eds, *The Peopling of Britain: The Shaping of a Human Landscape* (Oxford, 2002), pp. 177–210.
Smith, Richard M., 'Periods, Structures and Regimes in Early Modern Demographic History', *History Workshop Journal*, 63 (2007), pp. 202–18.
Smout, T. C., *Nature Contested: Environmental History in Scotland and Northern England since 1600* (Edinburgh, 2000).
Smyth, William J., *Map-Making, Landscapes and Memory: A Geography of Colonial and Early Modern Ireland, c.1530–1750* (Cork, 2006).
Soll, Jacob, *The Information Master: Jean Baptiste Colbert's Secret State Intelligence System* (Ann Arbor, 2009).
Solomon, Howard M., *Public Welfare, Science, and Propaganda in Seventeenth Century France: The Innovations of Théophraste Renaudot* (Princeton, 1972).
Solomon, Julie Robin, *Objectivity in the Making: Francis Bacon and the Politics of Inquiry* (Baltimore, 1998).
Sonenscher, Michael, 'The Emergence of Fashion and the Rise of Capitalism in Eighteenth-Century France', *P&P* 216 (Aug. 2012), pp. 247–58.
Sowerby, Scott, *Making Toleration: The Repealers and the Glorious Revolution* (Cambridge, Mass., 2013).
Spadafora, David, *The Idea of Progress in Eighteenth-Century Britain* (New Haven, 1990).
Spengler, J. J., *French Predecessors of Malthus: A Study of Eighteenth-Century Wage and Population Theory* (Durham, NC, 1942).
Spicksley, Judith, 'Usury Legislation, Cash, and Credit: The Development of the Female Investor in the Late Tudor and Stuart Periods', *EcHR* 61 (2008), pp. 277–301.

Spufford, Margaret, *The Great Reclothing of Rural England: Petty Chapmen and their Wares in the Seventeenth Century* (1984).
Spufford, Margaret, 'The Cost of Apparel in Seventeenth-Century England and the Accuracy of Gregory King', *EcHR* 53 (2000), pp. 677–705.
Statt, Daniel, *Foreigners and Englishmen: The Controversy over Immigration and Population, 1660–1760* (Newark, Del., 1995).
Stern, Philip J., *The Company-State: Corporate Sovereignty and the Early Modern Foundation of the British Empire in India* (Oxford, 2011).
Stern, Philip J., and Wennerlind, Carl, eds, *Mercantilism Reimagined: Political Economy in Early Modern Britain and its Empire* (Oxford, 2014).
Stevenson, Laura Caroline, *Praise and Paradox: Merchants and Craftsmen in Elizabethan Popular Literature* (Cambridge, 1984).
Stewart, Larry, *The Rise of Public Science: Rhetoric, Technology, and Natural Philosophy in Newtonian Britain, 1660–1750* (Cambridge, 1992).
Stone, Lawrence, 'Elizabethan Overseas Trade', *EcHR* 2nd ser. 2 (1949–50), pp. 30–58.
Stone, Lawrence, 'The Educational Revolution in England, 1560–1640', *P&P* 28 (July 1964), pp. 41–80.
Stone, Lawrence, *The Crisis of the Aristocracy, 1558–1641* (Oxford, 1965).
Stone, Lawrence, *The Family, Sex and Marriage in England, 1500–1800* (1977).
Stone, Richard, *Some British Empiricists in the Social Sciences, 1650–1900* (Cambridge, 1997).
Studenski, Paul, *The Income of Nations: Theory, Measurement, and Analysis* (New York, 1958).
Styles, John, *The Dress of the People: Everyday Fashion in Eighteenth-Century England* (New Haven, 2007).
Supple, B. E., *Commercial Crisis and Change in England, 1600–1642* (Cambridge, 1964).
Suranyi, Anna, *The Genius of the English Nation: Travel Writing and National Identity in Early Modern England* (Newark, Del., 2008).
Tarlow, Sarah, *The Archaeology of Improvement in Britain, 1750–1850* (Cambridge, 2007).
Tate, Thad W., and Ammerman, David L., eds, *The Chesapeake in the Seventeenth Century: Essays on Anglo-American Society and Politics* (New York 1979).
Taylor, John A., *British Empiricism and Early Political Economy: Gregory King's 1696 Estimates of National Wealth and Population* (Westport, Conn., 2005).
Temin, Peter, and Voth, Hans-Joachim, *Prometheus Shackled: Goldsmith Banks and England's Financial Revolution after 1700* (Oxford, 2013).
Théré, C., 'Economic Publishing and Authors, 1566–1789', in Gilbert Faccarello, ed., *Studies in the History of French Political Economy: From Bodin to Walras* (1998), pp. 1–56.
Thirsk, Joan, ed., *The Agrarian History of England and Wales*, iv: *1500–1640* (Cambridge, 1967); v: *1640–1750* (2 vols, Cambridge, 1984–5).
Thirsk, Joan, 'Enclosing and Engrossing', in *Agrarian History*, iv. 200–55.
Thirsk, Joan, 'Agricultural Policy: Public Debate and Legislation', in *Agrarian History*, v. ii. 298–388.
Thirsk, Joan, 'Agricultural Innovations and their Diffusion', in *Agrarian History*, v.ii. 533–89.
Thirsk, Joan, *Economic Policy and Projects: The Development of a Consumer Society in Early Modern England* (Oxford, 1978).
Thirsk, Joan, 'The Crown as Projector on its Own Estates, from Elizabeth I to Charles I', in Hoyle, ed., *Estates of the English Crown*, pp. 297–352.

Thirsk, Joan, 'Making a Fresh Start: Sixteenth-Century Agriculture and the Classical Inspiration', in Leslie and Raylor, eds, *Culture and Cultivation*, pp. 15–34.
Thirsk, Joan, *Alternative Agriculture: A History from the Black Death to the Present Day* (Oxford, 1997).
Thomas, Keith, *Religion and the Decline of Magic* (1971).
Thomas, Keith, *Man and the Natural World: Changing Attitudes in England 1500–1800* (1983).
Thomas, Keith, *The Perception of the Past in Early Modern England* (Creighton Trust Lecture, University of London, 1983).
Thomas, Keith, 'The Meaning of Literacy in Early Modern England', in G. Baumann, ed., *The Written Word: Literacy in Transition* (Wolfson College Lectures 1985, Oxford, 1986), pp. 97–131.
Thomas, Keith, 'Cleanliness and Godliness in Early Modern England', in Anthony Fletcher and Peter Roberts, eds, *Religion, Culture and Society in Early Modern Britain: Essays in Honour of Patrick Collinson* (Cambridge, 1994), pp. 56–83.
Thomas, Keith, *The Ends of Life: Roads to Fulfilment in Early Modern England* (Oxford, 2009).
Thompson, Ann, *The Art of Suffering and the Impact of Seventeenth-Century Anti-Providential Thought* (Aldershot, 2003).
Thompson, E. P., 'The Moral Economy of the English Crowd in the Eighteenth Century', *P&P* 50 (Feb. 1971), pp. 76–136.
Thompson, S. J., 'Parliamentary Enclosure, Property, Population, and the Decline of Classical Republicanism in Eighteenth-Century Britain', *Historical Journal*, 51 (2008), pp. 621–42.
Thompson, S. J., 'Census-Taking, Political Economy and State Formation in Britain, c.1790–1840', Ph.D. thesis, University of Cambridge, 2010.
Thompson, S. J., 'The First Income Tax, Political Arithmetic, and the Measurement of Economic Growth', *EcHR* 66 (2013), pp. 873–94.
Tilley, Morris Palmer, *A Dictionary of the Proverbs in England in the Sixteenth and Seventeenth Centuries* (Ann Arbor, 1950).
Tilmouth, Christopher, *Passion's Triumph over Reason: A History of the Moral Imagination from Spenser to Rochester* (Oxford, 2007).
Tittler, Robert, *Nicholas Bacon: The Making of a Tudor Statesman* (1976).
Todd, Barbara J., 'Demographic Determinism and Female Agency: The Remarrying Widow Reconsidered . . . Again', *Continuity and Change*, 9 (1994), pp. 421–50.
Tomlins, Christopher L., *Freedom Bound: Law, Labor, and Civic Identity in Colonizing English America, 1580–1865* (New York, 2010).
Towsey, Mark R. M., *Reading the Scottish Enlightenment: Books and their Readers in Provincial Scotland, 1750–1820* (Leiden, 2010).
Trevor-Roper, Hugh, *Religion, the Reformation and Social Change* (1967).
Trevor-Roper, Hugh, *Europe's Physician: The Various Life of Sir Theodore de Mayerne* (2006).
Tribe, Keith, *Governing Economy: The Reformation of German Economic Discourse, 1750–1840* (Cambridge, 1988).
Tuck, Richard, *Philosophy and Government, 1572–1651* (Cambridge, 1993).
Turnbull, G. H., *Hartlib, Dury and Comenius: Gleanings from Hartlib's Papers* (Liverpool, 1947).
Tuttle, Leslie, *Conceiving the Old Regime: Pronatalism and the Politics of Reproduction in Early Modern France* (Oxford, 2010).

Underdown, David, *Revel, Riot and Rebellion: Popular Politics and Culture in England, 1603–1660* (Oxford, 1985).
Van Zanden, Jan Luiten, 'Taking the Measure of the Early Modern Economy: Historical National Accounts for Holland in 1510/14', *European Review of Economic History*, 6 (2002), pp. 161–93.
Vardi, Liana, *The Physiocrats and the World of the Enlightenment* (Cambridge, 2012).
Velema, Wyger R. E., 'Ancient and Modern Virtue Compared: De Beaufort and Van Effen on Republican Citizenship', *Eighteenth-Century Studies*, 30 (1997), pp. 437–43.
Vine, Angus, *In Defiance of Time: Antiquarian Writing in Early Modern England* (Oxford, 2010).
Vivo, Filippo de, *Information and Communication in Venice: Rethinking Early Modern Politics* (Oxford, 2007).
Voigtländer, Nico, and Voth, Hans-Joachim, 'Malthusian Dynamics and the Rise of Europe: Make War, Not Love', *American Economic Review*, 99 (2009), pp. 248–54.
Waddell, Brodie, *God, Duty and Community in English Economic Life, 1660–1720* (Woodbridge, 2012).
Wakefield, Andre, *The Disordered Police State: German Cameralism as Science and Practice* (Chicago, 2009).
Walker, Mack, *German Home Towns: Community, State and General Estate, 1648–1871* (Ithaca, NY, 1971).
Walsham, Alexandra, *Charitable Hatred: Tolerance and Intolerance in England, 1500–1700* (Manchester, 2006).
Walsham, Alexandra, *The Reformation of the Landscape: Religion, Identity, and Memory in Early Modern Britain and Ireland* (Oxford, 2011).
Walter, John, 'Grain Riots and Popular Attitudes to the Law: Maldon and the Crisis of 1629', in John Brewer and John Styles, eds, *An Ungovernable People: The English and their Law in the Seventeenth and Eighteenth Centuries* (1980), pp. 47–84.
Warde, Paul, *Ecology, Economy and State Formation in Early Modern Germany* (Cambridge, 2005).
Warde, Paul, 'The Idea of Improvement c.1520–1700', in Hoyle, ed., *Custom, Improvement and the Landscape in Early Modern Britain*, pp. 127–48.
Warner, Jessica, 'Faith in Numbers: Quantifying Gin and Sin in Eighteenth-Century England', *Journal of British Studies*, 50 (2011), pp. 76–99.
Watts, John, '*The Policie in Christen Remes*: Bishop Russell's Parliamentary Sermons of 1483–84', in Bernard and Gunn, eds, *Authority and Consent*, pp. 33–59.
Watts, John, *The Making of Polities: Europe, 1300–1500* (Cambridge, 2009).
Weatherill, Lorna, *Consumer Behaviour and Material Culture in Britain 1660–1760* (1988).
Webster, Charles, *The Great Instauration: Science, Medicine and Reform, 1626–1660* (1975).
Webster, Charles, *Utopian Planning and the Puritan Revolution: Gabriel Plattes, Samuel Hartlib and MACARIA* (Wellcome Unit for the History of Medicine, Oxford, 1979).
Webster, Charles, 'Benjamin Worsley: Engineering for Universal Reform from the Invisible College to the Navigation Act', in Greengrass et al., eds, *Samuel Hartlib and Universal Reformation*, pp. 213–35.
Wells, Robert V., *The Population of the British Colonies in America before 1776: A Survey of Census Data* (Princeton, 1975).
Wennerlind, Carl, 'Credit-Money as the Philosopher's Stone', *History of Political Economy*, 35 (2003), Supplement 1, pp. 234–61.

Wennerlind, Carl, *Casualties of Credit: The English Financial Revolution, 1620–1720* (Cambridge, Mass., 2011).
Weststeijn, Arthur, *Commercial Republicanism in the Dutch Golden Age: The Political Thought of Johan and Pieter de la Court* (Leiden, 2012).
Whaley, Joachim, *Germany and the Holy Roman Empire* (2 vols, Oxford, 2012).
Whatley, Christopher A., *The Scots and the Union* (Edinburgh, 2006).
White, Stephen D., *Sir Edward Coke and the Grievances of the Commonwealth* (Manchester, 1979).
Whittle, Jane, *The Development of Agrarian Capitalism: Land and Labour in Norfolk, 1440–1580* (Oxford, 2000).
Whittle, Jane, and Griffiths, Elizabeth, *Consumption and Gender in the Early Seventeenth-Century Household: The World of Alice Le Strange* (Oxford, 2012).
Wiles, Richard C., 'The Theory of Wages in Later English Mercantilism', *EcHR* 2nd ser. 21 (1968), pp. 113–26.
Willan, T. S., *The Inland Trade* (Manchester, 1976).
Willmoth, Frances, *Sir Jonas Moore: Practical Mathematics and Restoration Science* (Woodbridge, 1993).
Wilson, Catherine, *Epicureanism at the Origins of Modernity* (Oxford, 2008).
Winter, Anne, and Lambrecht, Thijs, 'Migration, Poor Relief and Local Autonomy: Settlement Policies in England and the Low Countries in the Eighteenth Century', *P&P* 218 (Feb. 2013), pp. 91–126.
Withers, Charles W. J., 'How Scotland Came to Know Itself: Geography, National Identity and the Making of a Nation 1680–1790', *Journal of Historical Geography*, 21 (1995), pp. 371–97.
Withers, Charles W. J., 'Geography, Science and National Identity in Early Modern Britain: The Case of Scotland and the Work of Sir Robert Sibbald (1641–1722)', *Annals of Science*, 53 (1996), pp. 29–73.
Withington, Phil, *Society in Early Modern England: The Vernacular Origins of Some Powerful Ideas* (Cambridge, 2010).
Withington, Phil, 'Intoxicants and the Early Modern City', in Hindle et al., eds, *Remaking English Society*, pp. 135–63.
Wong, R. Bin, *China Transformed: Historical Change and the Limits of European Experience* (Ithaca, NY, 1997).
Wood, Diana, *Medieval Economic Thought* (Cambridge, 2002).
Wood, Gordon S., *Empire of Liberty: A History of the Early Republic, 1789–1815* (Oxford, 2009).
Wood, Neal, *Foundations of Political Economy: Some Early Tudor Views on State and Society* (Berkeley, 1994).
Woodward, Donald, 'The Background to the Statute of Artificers: The Genesis of Labour Policy, 1558–63', *EcHR* 2nd ser. 33 (1980), pp. 32–44.
Woodward, Donald, *Men at Work: Labourers and Building Craftsmen in the Towns of Northern England, 1450–1750* (Cambridge, 1995).
Woodward, Walter W., *Prospero's America: John Winthrop Jr., Alchemy and the Creation of New England Culture, 1606–1676* (Chapel Hill, NC, 2010).
Woolf, Daniel R., *The Idea of History in Early Stuart England* (Toronto, 1990).
Woolf, Daniel R., *Reading History in Early Modern England* (Cambridge, 2000).
Woolf, Daniel R., *The Social Circulation of the Past: English Historical Culture, 1500–1730* (Oxford, 2003).

Woolf, Daniel, R., 'Historical Writing in Britain from the Late Middle Ages to the Eve of the Enlightenment', in Rabasa et al., eds, *Oxford History of Historical Writing*, iii: *1400–1800*, pp. 473–96.

Worden, Blair, 'The Question of Secularization', in Houston and Pincus, eds, *A Nation Transformed*, pp. 20–40.

Worden, Blair, *The English Civil Wars 1640–1660* (2009).

Wrigley, E. A., *People, Cities and Wealth: The Transformation of Traditional Society* (Oxford, 1987).

Wrigley, E. A., 'A Simple Model of London's Importance in Changing English Society and Economy, 1650–1750', in Wrigley, *People, Cities and Wealth*, pp. 133–56.

Wrigley, E. A., 'The Transition to an Advanced Organic Economy: Half a Millennium of English Agriculture', *EcHR* 59 (2006), pp. 435–80.

Wrigley, E. A., *Energy and the English Industrial Revolution* (Cambridge, 2010).

Wrigley, E. A., Davies, R. S., Oeppen, J. E., and Schofield, R. S., eds, *English Population History from Family Reconstitution, 1580–1837* (Cambridge, 1997).

Wrigley, E. A., and Schofield, R. S., *The Population History of England 1541–1871: A Reconstruction* (2nd edn, Cambridge, 1989).

Wu, Jiang, *Enlightenment in Dispute: The Reinvention of Chan Buddhism in Seventeenth-Century China* (Oxford, 2008).

Yamamoto, Koji, 'Piety, Profit and Public Service in the Financial Revolution', *EHR* 126 (2011), pp. 806–34.

Yamamoto, Koji, 'Reformation and the Distrust of the Projector in the Hartlib Circle', *Historical Journal*, 55 (2012), pp. 375–97.

Yeh, Sarah, 'Colonial Identity and Revolutionary Loyalty: The Case of the West Indies', in Foster, ed., *British North America*, pp. 195–226.

Yokota, Kariann Akemi, *Unbecoming British: How Revolutionary America became a Postcolonial Nation* (Oxford, 2011).

Yolton, Jean S., *John Locke as Translator* (Studies on Voltaire and the Eighteenth Century, Oxford, 2000).

Zahedieh, Nuala, *The Capital and the Colonies: London and the Atlantic Economy, 1660–1700* (Cambridge, 2010).

Zahedieh, Nuala, 'Colonies, Copper, and the Market for Inventive Activity in England and Wales, 1680–1730', *EcHR* 66 (2013), pp. 805–25.

Zell, Michael, 'Walter Morrell and the New Draperies Project, c.1603–1631', *Historical Journal*, 44 (2001), pp. 651–75.

Index

English place-names are identified by reference to the historic counties as they were before 1974.

Abbot, George, abp of Canterbury 33 n. 94
acres:
  measurement of 26–8, 49, 119
  per person 197
Act of Union (1707) 187, 189–90
Adams, John 16, 28
Addison, Joseph 213, 251
advanced organic economy 11, 199
affluence 130, 170, 202
Africa 35, 70, 187
agrarian improvement 4–5, 11, 30, 58–60,
    231–5, 258–9
  advocated 18–19, 40, 59–60, 94, 106–8,
    133, 154, 167, 193, 232–5
  arguments against 58, 108, 234–5
  books on 106–7, 230–1, 234
  in continental Europe 250–1, 252
  in fens and marshes 30, 60, 61, 134, 224,
    232, 234, 245
  in forests and wastes 55, 58, 60, 234
  in Ireland 66–7, 98, 161–2, 188–9, 228, 232
  by manuring 36, 58, 66, 93, 94 n. 15,
    220 n. 29
  and new crops 11, 30, 232
  in North America 68–70, 255, 258–9
  on open fields 235
  in Scotland 97, 161–2, 189, 231–4
  by soldiers 97, 106–7, 231–2
  see also agricultural productivity; enclosure
agricultural depression 132–3, 136, 138
agricultural productivity 154, 232–6
air pumps 228, 237
alchemy 77, 100, 109, 124, 151
alehouses 60, 157
Allen, Richard 143 n. 72
Allestree, Richard 113, 150–1
almanacs 37–8, 167
Alsted, Johann Heinrich 99
America:
  improvement in 67–70, 109, 186–7,
    254–5, 258–9
  natives of 31–2, 35–6, 38
  plantations in 22, 31–2, 96, 116, 125, 258–9
  see also empire; Spain; Virginia
amour propre, *see* self-love
Amsterdam 10, 33, 74, 148
Ancient Britons 36, 38
Ancients *versus* Moderns 39–41, 194–5,
    197, 199
Andrewes, Thomas 99
Anglo-Saxons 96–7, 194

antiquaries 16–17, 20–1, 37
Antwerp 55–6, 74, 244
anxiety 166, 206, 220–2, 263
apocalyptic expectations, *see* millennium
apparel, *see* dress
appetites, *see* passions
apprentices 64, 143, 160
Approvement (Improvement) Act (1549) 5, 233
Arbuthnot, John 223
arcana imperii 48, 104, 119–20
Aristotle 76
  on justice 79–80
  on nature 102
  on population 45, 78
army 93, 97, 221
  and improvement 97, 106–7, 231–2
Artificers Act (1563) 56
Athenian Gazette 230
Aubrey, John 21, 41, 108, 117, 157
avarice 17–18, 54, 69, 75, 112, 152, 205, 209;
  see also corruption; covetousness; profit

Bacon, Sir Francis:
  on agrarian improvement 5, 18–19, 59, 235
  on balance of trade 84
  on common weal 61–2
  on greatness of states 46–7
  on history 40–1, 193
  on improvement of knowledge 2–3, 5
  on inventions 52, 74–5, 87
  on justice and proportion 80
  on London 61, 75
  on population 46, 59, 78
  on schools 64–5
  on sedition 50–1
  on travel 32
  on Ulster 67
  on usury 87
  influence 98, 107–8, 127, 134–5, 167, 201
  works:
    History of Henry VII 235
    Instauratio magna 99
    New Atlantis 100
    Novum Organum 40–1, 101–2
Bacon, Sir Nathaniel 234
bad harvests 44, 81, 82; *see also* Books of Orders;
    famine
balance of trade:
  early use of concept 45, 56, 83–6
  exploited in argument 96, 123, 131, 134,
    150, 179–80, 184, 192

balance of trade: (*cont.*)
  with France 131, 184–5
  after 1700 179, 201
Banbury (Oxon.) 20
Bank of England 179, 182–3, 196–7
banks, projects for:
  in England 93, 95, 109, 115, 141, 142, 173, 244–5, 248
  in Europe 243, 245
  see also Bank of England
Barbados 174–5, 263
barbarism 19, 32, 35–6, 55, 66, 68, 200, 216–17, 235; *see also* civility
Barbon, Nicholas 129, 172, 210, 217, 242, 246, 247, 261
  debts 165, 211
  dress 146
  insurance companies 173
  on cities 143–7, 153
  on consumption 143–7, 153
  on empire 186
  on prodigality 145–6
  on recoinage 183–4
Barbon, Praise-God 143, 165
Barlow, Roger 31 n. 81
Barnes, Ambrose 135
Bath (Som.) 177
battle of the books, *see* Ancients *versus* Moderns
Baxter, Richard 112, 150
Bayle, Pierre 209
Beale, John 108, 114, 126–7
  on improvement 153–4, 161
  on prodigality 149, 153–4
  on Utopia 101, 153–4
Becher, Johann Joachim 151, 246
bees 128, 251
  as metaphor and model 113–14, 135, 141, 207–9, 212, 221, 255
  see also Mandeville
Beijing 10 n. 40
Bell, Henry 177, 178
Bellers, John 195
Berkeley, George, bp of Cloyne 211
Bermuda 36
Berwick (Northumb.) 22
Bethel, Slingsby 135, 138, 139
bigamy 137–8
bills of mortality, *see* London
Birmingham (Warks.) 228
Black Death 11, 12, 260; *see also* plague
Blanch, John 192
Bland, John 120 n. 142
Blathwayt, William 163
Blewitt, George 210–11
Blith, Walter 7, 106–7, 114
Blome, Richard 16, 35, 36
Blount, Henry 35
Board of Trade (1696-) 176, 184, 186, 225, 231; *see also* Commission on Trade;

Council for Foreign Plantations; councils of trade
Boate, Gerard 98
Bodin, Jean:
  on history 37, 39–40
  on reason of state 45, 242, 243
  on resources of France 49, 246
Bodley, Sir Thomas 52, 74
Bohemia 99
Bohun, Edmund 194, 212
Boisguilbert, Pierre Le Pesant de 151, 192, 246–7, 253
Bombay 188
Books of Orders 62–3, 81–2, 87; *see also* moral economy
Boroughs, John 96
Boston, Mass. 70, 254
Botero, Giovanni:
  on economic growth 45, 77
  on population 46–7
  on reason of state 33, 242, 243
bourgeois virtues 72 n. 88, 240, 260 n. 7
Bourne, Henry 21
Boyle, Robert 32, 114, 134, 149, 254
Boyle family 98
Braddon, Laurence 227
Bradley, Humphrey 230
Breslau 196
Brewster, Francis 204
Bristol 175, 176–7, 207
Britannia languens (1680) 64 n. 53, 150, 159, 167
British Merchant 185, 250
Brown, Lancelot ('Capability') vii–viii
Brutus 37
Bufton, Joseph 206–7, 222
Bunyan, John 23
Burnet, Thomas 42
Bush, Rice 93
Butler, Joseph, bp of Durham 210
Byzantium 38; *see also* Constantinople

calendar reform 130, 225
Camden, William:
  Britannia 16–17, 19, 22, 30, 37
  on new discoveries 40
Cameralism 151, 242, 246, 247–8, 251, 253
Canal du Midi (France) 245–6
Cannon, John 221
Canterbury 56
Cantillon, Philip 250
Cantillon, Richard 247, 249–50, 252
Carew, Richard 17
Carl, Ernst Ludwig 247–8, 253
Carpenter, Nathaniel 35, 37 n. 117
carpets 159
cartography, *see* maps
'Cato' 216, 224 n. 56

Cary, John 175–6, 177, 183, 194, 204, 207, 250
Cecil, Sir Robert 44, 48–9, 75
Cecil, William, 1st lord Burghley:
  and cartography 22
  economic attitudes 55–7, 82, 87
  and economic projects 54–7, 60, 230
  on foreign travel 32
  and information 44
  on usury 56
censuses:
  partial 44
  proposed 171, 225–6, 243
  in Spain 251
  see also Compton Census
Chamberlayne, Edward 28–9, 33, 43, 130, 160, 237
Chamberlen, Hugh 173, 182
Chamberlen, Peter 101
chambers of commerce 228 n. 70, 247
Chapman, George 18
Charitable Corporation 208
Charitable Uses Acts 64
charity, concept of 88–9; *see also* philanthropy
charity schools 105, 177, 208; *see also* education
Charlemont (Ulster) 67
Charles I, king of England 61, 113
Charles II, king of England:
  as improver 115, 131
  on Petty and Royal Society 116
Charterhouse 64
Cheshire 165
Chester 228
Chichester (Sussex) 103
Child, Sir Josiah 133, 164, 183, 184, 201, 246
  criticised 192–3
  on Dutch West Indies 259
  on interest rates 132–3, 135
  on Irish improvement 135, 232
  on trade 172
  translated 7, 250
Child, Robert 19
children 105–6, 144, 156, 159, 160–1, 212, 213, 227, 230, 239; *see also* education
Childrey, Joshua 20
China:
  English knowledge of 33, 34, 35–6, 38, 199
  in 16th-17th century 8–9, 35 n. 108, 45
chocolate 143
chorography 3–4, 20, 28–9
chronicles 37
circulation, as economic metaphor 25, 71, 73, 78, 133, 214, 263
civic virtue 53, 89, 94
civility 19, 200
  in Britain 17, 30, 60, 97
  in continental Europe and Asia 34–6
  in plantations 22, 32

civil society 65
Clarke, Edward 177
cleanliness 219–20
climate 11, 20–1
clocks and watches 158, 231, 237
clockwork 228, 237
cloth industry 55–6, 71, 82, 138, 231
  new draperies 56, 57, 61, 87, 230, 244
  see also Cockayne project; cottons; linen; Merchant Adventurers Company
clubs and societies 21, 190, 222, 227–8, 239 n. 122, 251; *see also* Royal Society
coaches 25, 139–40, 158, 192, 205
coal 11, 30, 37, 134, 154–5, 236, 259
coal trade 61, 81, 204
Cockayne project 71, 82
coffee 158
coffee-houses 134, 168, 173, 174, 227–8
coinage:
  quantity of 45, 183
  scarcity of 82, 132, 183
  see also recoinage
Coke, Sir Edward 71 n. 86, 81
Coke, Roger 80 n. 133, 130–1, 150
Colbert, Jean-Baptiste 136, 151, 213, 224, 244, 245–6
Colchester (Essex) 30, 56, 177
Collins, John 139
Comenius, Jan 99, 103
comfort:
  and conveniences of life 205–7, 226, 248
  domestic 17, 155–9, 201, 215–17, 219–20,
  English invention 215–16, 248
  as material goal 93, 113, 123, 142, 152, 158–9, 166, 198, 207–8, 212, 248
  see also consumption; ease; housing
commerce, *see* balance of trade; trade
Commission on Trade (1622), 82–4, 87–8;
  see also Board of Trade; Council for Foreign Plantations; councils of trade
common weal:
  in 15th century 65, 260–1
  in Germany 243, 246, 253
  as ideal 53–5, 57–61, 77–8, 79, 88–9
  see also Discourse of the Common Weal
commonwealths, as political entities 34, 45, 65–6, 69, 71–2, 73, 80, 86, 88–90
  in Interregnum 93, 95–6, 97, 101, 105, 111, 114
  see also common weal
communication 23–5; *see also* information; post office; transport; travel
companies 70–1, 93, 168, 173; *see also* East India Company; Eastland Company; Levant Company; Merchant Adventurers Company; Royal African Company; Russia Company; Virginia Company
Compton Census (1676) 140, 163, 173
Constantinople 12, 147; *see also* Byzantium

consumer revolutions 3, 8; see also consumption; luxury
consumption:
  as economic stimulus 123, 141–2, 144–6, 154
  high living 131, 144–5, 154, 155
  moderation in 86, 112, 131, 240–1
  patterns of 155–9
  see also comfort; consumer revolutions; dress; emulation; luxury; prodigality; sumptuary laws
conveniences of life, see comfort
Cook, John 95
copper industry 238
Coram, Thomas 227
corn bounties 186, 233, 244
Cornwall 17, 21 n. 37, 158, 181
corruption:
  of civic virtue 53, 94, 255
  in colonising 38, 67–8
  of England 244, 254
  from extravagance 210–11
  in improvement 59–60
  in India 36
  of manners 130
  in philanthropy 64, 208
  from private enterprise 53, 206
Cotton, John 68
Cotton, Sir Robert 37, 83, 201
  and enclosure 50–2, 53
  and London 147
cottons 9, 156, 214, 231, 238
Council for Foreign Plantations 116
Council of Commerce (French) 247
councils of trade 73, 86, 93, 95–6, 116, 132; see also Board of Trade; Commission on Trade
counterfactual questions 258
Coventry (Warks.) 44
covetousness 54, 62, 146, 209; see also avarice; corruption; profit
Cradocke, Francis 115
Cradocke, Samuel 115
Cranbrook (Kent) 119
Cranfield, Sir Lionel 84
crape 162
Creation:
  date of 37, 42, 126, 197
  Scale of 127–8, 200
Crofts, Robert 115
Cromwell, Oliver, lord protector 92, 97, 98, 105, 134
Cromwell, Thomas, earl of Essex 44
Culpeper, Cheney 93, 100–1, 103, 104, 109
Culpeper, Thomas, the elder 132
Culpeper, Sir Thomas, the younger 132–3
Cumberland 231
Curaçao 259

curtains 158
customs accounts 43, 84, 86, 138, 154, 172; see also taxation

Danby, Sir Thomas Osborne, 1st earl of 135, 163, 167
Daniel, prophecies of 38, 40, 127, 135
Darien scheme 187, 189
Davenant, Charles 183–4, 185, 217, 218, 219, 232, 244
  on agrarian improvement 30
  on costs of war 180–2
  on English prosperity 187, 191–3, 207, 216
  on London 148–9
  on luxury 193, 206
  and political arithmetic 47, 116–7, 124, 163, 225
  on population 195
  on Scotland and Ireland 189
  on taxation 180–2
Declaration of Indulgence (1687) 137
Dee, John 74
Defoe, Daniel 185, 187
  Complete English Tradesman 165
  on costs of war 191
  on history 218–19
  on industry and idleness 221
  on invention and improvement 29
  on London 148
  on luxury 153, 206, 211–12
  on marriage 240
  on projects 173
  on Scotland 97, 189, 231–2
  on taxation 181
  Tour 30–1, 264
De la Court, Johan and Pieter 244, 246
Delhi 10 n. 40
demography 118–19, 251; see also population
Derby 228, 231, 236
Derbyshire 134
Derham, William 200
Desaguliers, John Theophilus 237
Devon 37 n. 117, 51
dialect surveys 21
diaries 221, 222
dictionaries 5–6, 215–16, 226
diet, see food and drink
Diggers 94
Digges, Sir Dudley 72
Digges, Thomas 74
Discourse of the Common Weal 51, 54, 59
disease, see dysentery; malaria; plague; smallpox
divorce 137
Dockwra Copper Company 238
Domesday Book 43
Donaldson, James 233–4, 248
Donne, John 34, 68
Dover harbour 57

Drayton, Michael 18, 25, 245
dress:
  expenditure on 203–5
  extravagance in 131, 145–6, 149–50, 153, 155, 165
  see also sumptuary laws; three-piece suit
Dublin:
  improvement society 228
  Philosophical Society 152, 161, 163
  Trinity College 105 n. 70
Dugdale, Sir William 234–5
Dunton, John 230
Durham 105
Dury, John 99, 103, 104
Dutch Republic, *see* United Provinces
Duty Act (1695) 181, 213
dyeing industry 238
dysentery 224

ease 146, 150, 152, 158, 179, 193, 207–8, 210, 212, 216–17, 220, 240–1; *see also* comfort
East India Company 70–2, 100, 168, 184, 186, 188
Eastland Company 88
Eburne, Richard 78
economic growth:
  in Asia 8–9
  and collective morality 240–1, 252–3, 262
  and culture 4, 13–14, 229–30
  and institutions 229, 257
  measurement of 12–13, 82–3, 163–4, 180, 191
  and political structures 80, 89–90, 244–5, 260
  preconditions for 229
  and religion 126, 134–5, 167
  and social capital 229–30
  and towns 120, 143–9
  see also advanced organic economy; national income; population; trade
economic thought:
  in Asia 8–9
  in continental Europe 2, 242–54
  in economic crises 82, 95
  historical and comparative 43–4, 134–6, 193–4, 248–9
  see also balance of trade; circulation; economy; mercantilism; moral economy; political economy
economy:
  mechanical models of 78, 145
  organic models of 45, 77–8, 108–9
  as separate thing 240, 261
  use of term 76–7, 123, 131, 245
Edgar, king of England 96–7
Edinburgh 97, 161, 203, 228, 248
Edo (Japan) 9, 147

education 64–5, 93, 101, 103, 105–6, 160, 177, 208, 227, 248; *see also* children; literacy
Edward III, king of England 43, 45, 134
effeminacy 38, 150, 170, 202, 205, 209, 211, 239
Egypt 38, 39
Elizabeth I, queen of England 47
Elizabetha, proposed town in Ulster 67
Elliott, John 259
emigration, *see* migration
empire, English and British 4, 69–70, 238, 258–9
  disputed character of 185–6
  early concepts of 31–2, 67, 96–7, 101, 115–16, 145, 179
  problems of 186–9
employment exchange 103
emulation:
  between consumers 142, 145–6, 150–3, 155, 158, 166, 198, 204–6, 209–10, 254
  between countries 244, 250
  between towns 178, 219
enclosure:
  debated 50–2, 58–9, 80–1, 132, 235
  efforts to control 44, 51–2, 59–60
  in practice 58, 93–4, 232–3
  see also agrarian improvement
English, reputed to be
  backward in invention 29, 172–3
  brave 29
  comfortable 215–16, 248
  idle 29, 134, 221
  industrious 29, 93, 174, 186, 221
  prone to dissension 244
  see also improvement
enlightened economy 241, 250, 252
Enlightenment 38–9, 102, 180, 241, 250–2
Entresol club 247
Epicureanism 152, 208–9, 242–3, 246, 261
equity, *see* justice
eschatology 126–7; *see also* Last Judgement; millennium
Essex 17, 51, 156
Estienne, Claude 234
ethnography 21, 32
Evelyn, John:
  on France 34
  on London 115, 148
  on multitudes 136, 140–1
  on navigation and commerce 136
  on Petty 128
  on post office 173
  on 1688 revolution 170–1
excess 36, 86, 112, 130–1, 141, 150, 155, 206, 210, 221, 240–1; *see also* luxury; prodigality
excise, *see* taxation
Exclusion Crisis 139

Exeter 177
Experiment (Petty's boat) 116
experiments 74–5, 99–100, 228, 230

Fall of Man 39
famine 11, 140, 189, 219; *see also* bad harvests; Books of Orders; moral economy
felicity, *see* happiness
Fénelon, François de Salignac de La Mothe 208, 247, 253
fens, *see* agrarian improvement
Ferguson, Adam 217
fertility, of land 18, 55
fertility, of populations 63, 118, 136–7, 140, 159–60, 195–6
  policies to increase 137–8, 181
  preventive checks to 11, 199, 238–40
  see also marriage; population
Fiennes, Celia 29–30, 156
fire brigade 95
fire engines 219, 236
fire insurance 165, 173, 219
fires 178; *see also* London: Great Fire
Fish, Simon 47
fish-days 57
fisheries 57, 83, 96, 115, 216
Fitzherbert, John 5, 26–7, 74
Fletcher, Andrew, of Saltoun 190, 212–13
Florence 10
Florida 67
Florio, John 6
Foley, Paul 195
food and drink 13, 48, 153, 155, 157–9, 204, 219–20; *see also* alehouses
foreign exchange 79, 82–5, 95
forests 44, 93, 227; *see also* agrarian improvement
Forset, Edward 77, 113
Fortrey, Samuel 115, 131, 141, 150, 185
France 243
  comparisons with England 33–4, 49–50, 72, 122, 163, 191, 244, 247
  concerns about population 119 n 136, 136, 137, 253
  economic projects 244–5
  economic thinking in 2, 76, 245–50, 252–4
  hostility to 138, 144
  information about 33–4, 49
  maps of 25
  trade with 175, 184, 185, 201
  war with 179–83, 191
  see also Colbert
Franklin, Benjamin 255
freedom of conscience 126, 137, 168–9
free ports 95, 120
free-rider problem 241
free trade 71, 87–9, 94, 111; *see also* laissez-faire
frugality 132–3, 139, 141–2, 165–6, 205–6, 209, 210; *see also* United Provinces
Fuller, Thomas 21

Gainsford, Thomas 33, 147
Gassendi, Pierre 101
Gay, John 227
gentry:
  and status distinctions 18, 158
  and trade 69, 135, 168, 259–60
  wealth of 58
Georgic literature 18–19
Germany 98–9, 242, 243, 251; *see also* Cameralism
Gibson, Edmund, bp of London 16, 30
glass-making 57, 115, 172–3, 228, 231; *see also* windows
Gloucester 177
Gloucestershire 18, 174
Godolphin, Sidney, 1st earl of 244–5
Goodman, Godfrey, bp of Gloucester 41 n. 139
Gordon, Thomas *see* 'Cato'
Gott, Samuel 112–13
Gough, Richard 26, 166
Graunt, John:
  career 118, 164–5
  on date of Creation 116
  as demographer 118–20
  on global population 125–6
  on London fertility 140
  on London population 119, 147, 148
  on polygamy 126
  on population of England 119, 154, 195
  on public knowledge 119–20, 175
  reputation 226
  on trade 138
great divergence 8–9
Greeks, ancient 38, 39, 40, 96 n. 30, 135, 193, 202
Gresham, Sir Thomas 79
Grew, Nehemiah 200–1
Gross Domestic Product (GDP), *see* national income
Guy, Thomas 209

Hagthorpe, John 72–3
Haines, Richard 161
Hakewill, George 41
Hakluyt, Richard 31–2
Hale, Sir Matthew 127, 140, 149 n. 107
Hales, John 58
Hall, Joseph, bp of Norwich 35
Halle (Germany) 247
Halley, Edmond:
  on acreage of England 28
  on fertility and marriage 196, 239
Hanbury, William 227
Hanway, Jonas 227
happiness:
  common aspiration 33, 77, 214, 216, 248, 252–3, 255, 264 n. 24
  debated 3, 142–3, 149, 151–3
  evolution of concept 101–2, 104, 111–13, 150–1

happiest place to live 134, 168
happiness and welfare 112
happy man 112, 206–7
  not achieved 166–7, 220–1, 263
  see also anxiety
Hardwicke's Marriage Act (1753) 225–6
Harrington, James:
  on cities 111
  on population 137 n. 44
  on property distribution 135, 168
  Rota Club 117
Harris, James 220–1
Harrison, William 15, 17, 156
Hartlib, Samuel:
  on agrarian improvement 106–8
  background 98–9
  and Bacon 99, 101–2
  and bee-hives 114
  on communication 23, 103
  correspondence 91
  on happiness 111–12
  on improvement of children 220 n. 33
  and Malynes 109
  and millenarian expectations 101
  and parliament 99, 104–5, 117
  and Petty 101, 117
  and projects 104, 106
  on reformation 92
  reputation 143, 154
  on schools and workhouses 103, 105–6
  as State Agent for Universal Learning 104–5
  on universal improvement 92, 99
Hartlib circle 92, 98–102, 242
  on agrarian improvement 89, 106–8
  on moderation in consumption 112
  and Office of Address 102–5
  and technology 19, 236
Harvey, William 78
health 220, 222–3, 224, 251; *see also* disease; medical services; mortality
hearth tax, *see* taxation
Hebrews 39
Henry IV, king of France 245
Herbert, Sir Thomas 33, 34
Herefordshire 134
Heresbach, Conrad 234
Heritage, John 58
Hesiod 18
Heveningham, Sir Arthur 60
Heylyn, Peter 33
histories:
  of husbandry 114
  of population 125–6, 140, 197–8
  of trades 114, 218
  of world 39, 218–19
  see also history
history:
  Biblical 38, 42
  in China 8

  perceptions of 35–42:
    apocalyptic 38–9, 40
    cyclical 38, 41, 55, 145, 193–4
    linear 38–42, 255
    providential 31–2, 38–9, 125–6, 197, 218–19
  Scottish school of 217, 241
  Whig 263
  see also almanacs; antiquaries; chorography; economic thought; histories; myths and legends; natural history
Hobbes, Thomas:
  on economy 76
  on felicity 152
  on justice 80
  and Petty 101
  on population 197 n. 135
Holland, *see* United Provinces
Hooke, Andrew 225
Hooke, Robert 142
  and urban planning 148, 178
  on utility of science 236–7
Horneck, Anthony 205
hospitality 17, 58, 139
hospitals 64, 93, 101, 103, 130, 148, 209, 227, 248
  Foundling Hospital 227
  Westminster Hospital 227
  see also medical services
Houghton, John:
  and Beale 149, 153
  on coffee-houses 174
  Collection for the Improvement of Husbandry and Trade 20, 144–5, 172–3, 174
  on consumption 119, 142–6, 149–50, 153
  on distribution of taxation 181
  on East India trade 184
  on economic growth 145, 194, 196
  England's Great Happiness 142–6, 149–50
  on improvement 28, 108, 143–4
  on Ireland 188
  on London 144, 210
  on plantations 187 n. 77
  on prodigality 144–5, 149, 153
  as shopkeeper 143, 164–5, 173
housing 17, 75, 148, 155–6, 159, 201, 217, 219–20
Howell, James 6, 29, 147
Hull, 177
Hume, David:
  on economic growth 199, 260
  History of England 218
  on improvement 241, 250
  on jealousy of trade 180, 213
  on luxury 217
Hunstanton (Norf.) 233
husbandmen 18, 29, 56, 58, 64, 74, 94, 147, 160–1, 179
husbandry, *see* agrarian improvement

314   Index

Hutcheson, Francis 211
Huygens, Christiaan 237
Hyde, Edward, 1st earl of Clarendon 61, 97
hypotheses, use of 117, 123, 124, 198, 246, 262

immigration, *see* migration
improvement:
   English reputation for 29, 172–3, 230
   etymology and usage of word 1, 4–6
   evolution as culture 75–6, 212–13, 220, 227–8
   finite 109–10, 120
   infinite 104, 108–11, 122, 127–8, 138, 145–6, 183, 199, 200, 204, 252–3
   lack of alternative culture 235
   piecemeal 75, 108, 264
   in titles of books 7, 26–7, 106–8, 110, 115, 161–2
   in translations 5–6, 248
   see also agrarian improvement; America; Hartlib; health; housing; invention; Ireland; London; manufactures; Scotland; self-improvement; time; transport; urban improvement; women
improvement commissions 178
improvement societies 228
India 9, 34, 35 n. 108, 36, 195; *see also* cottons; East India Company
industriousness 29, 64, 72, 93, 134, 154, 159, 166, 231–2; *see also* industrious revolution
industrious revolution 3, 8, 13
industry 115, 134, 154–5, 236, 238, 254;
   see also cloth industry; coal; copper industry; patents
inflation 55, 78
information:
   collection 43–4, 83, 245–6
   exchange 103–4, 173
   private or public 119–20, 175
   uses of 43–52, 83, 261–2
   see also knowledge; Office of Address
information state 262
inoculation 223–4, 247, 254
insurance, *see* fire insurance
interest rates 56, 78, 86–7, 95, 132–3, 135, 201; *see also* usury
invention 2–3, 7, 52, 102, 103, 104, 200
   and improvement 87, 92–3, 100, 229–30, 255, 262
   labour-saving 187–8, 238
Invention I, Invention II (Petty's boats) 116
inventories, probate 157–9
Inverness 97
Ireland:
   Child on 135, 231–2
   commercial relations with England 187, 188
   English improvement in 66–7, 97–8, 154, 161–3, 188–90, 228, 231–2
   and Hartlib circle 98
   luxury issue in 123, 211
   maps and surveys 22, 26, 48, 98
   Petty on 122, 124–5
   plantations in 56, 66–7
   Temple on 135–6
   see also Dublin
Irish, barbarous 35, 66
Irish Woollens Act (1699) 188 n. 86, 231 n. 83
iron age (17th century as) 138, 204
Italy 2, 10, 135, 242

Jackman, Henry 59
Jamaica 97, 228, 259
James I, king of England 19, 245
James II, king of England 137, 163, 168–9
Jansenists, *see* Nicole
Japan 8–9, 34
Jefferson, Thomas 255–6
Jenner, Edward 224
Johnson, Robert 38
Johnson, Samuel 48, 128
Johnson, Thomas 92–3
Julius Caesar 36
Jurin, James 223
justice:
   distributive and commutative 79–81, 82, 88, 94, 254
   natural 81–2
   just price 2, 80

Kendal (Westm.) 151 n. 113
Kent 51, 158
Keymer, John 73
King, Gregory 47, 163
   on costs of war 180
   and Duty Act (1695) 181
   on national wealth 124, 163–4
   on patterns of expenditure 48
   on Petty 124
   on population 195
   against publication 175
   social table 164, 225, 226
King's Lynn (Norf.) 177
knowledge:
   certain 3, 26–8, 37, 43, 47
   clear 119, 175–6
   new 1, 43
   public 119–20, 175
   see also information; mathematics; numbers
knowledge economy 229–31, 234, 236–7, 240–1, 252
Kyoto (Japan) 9

labour, value of 120–3, 128
labourers 3, 12–13, 58, 121, 156, 158–60, 226–7
labour-saving, *see* invention
laissez-faire 89, 246–7; *see also* free trade

Lambarde, William 15, 16–17
Lancashire 151, 154–5, 165
landscape 18–19, 22, 30, 167, 232, 234
land and trade, interests of 121, 139, 182, 240–1, 259–60
Last Judgement 38–9, 42, 197
Lautern (Germany) 251
Law, John 244–5, 247–8, 249
Lee, William 238
Leeds 29–30, 228
Leicester 133–4
Leiden 143, 161, 208
leisure preferences 159, 221
Leland, John 16
Leroy, Louis 39–40
Le Strange, Sir Hamon 233–4
Le Strange, Sir Nicholas 233
Levant Company 70, 206
Levellers 94
Lewis, Mark 141
Lhuyd, Edward 20, 30
Licensing Act (1693), lapse of 173, 176
linen 144, 156–7, 167, 171 n. 6, 186–7, 188;
  see also cottons
literacy 65, 230
Liverpool 177
Lloyd, John 30
Locke, John 124, 151, 183, 261
  on emulation 205–6
  on Hakluyt and Purchas 31
  on happiness 152, 220
  on labour-saving inventions 188
  on London and Paris 147
  on prodigality 131, 205–6
  on self-improvement 220
Lombe, John 231, 236
London:
  attempts to regulate 61, 75, 139–40, 144, 147
  bills of mortality 44–5, 47, 51, 118–19, 200, 226
  city politics 69, 70, 99
  companies 67, 70, 71, 73–4, 97–8
  compared with Paris 34, 147
  compared with imperial Rome 194
  contemporary perceptions of 17, 29, 129, 134, 138–40, 143, 145–9, 192–3, 209
  Corporation for the Poor 105–6, 176–7
  economic importance 10, 13, 25, 75
  fertility in 140
  Great Fire (1666) 7, 147–8, 165
  images of 23–4, 147
  improvement in 29, 75, 105–6, 115, 147–8
  as information centre 21, 73–5, 103, 134–5, 172, 231, 258
  London Bridge 37
  mortality and morbidity in 140, 148, 222–3, 224
  New Exchange 75

New River 75, 81, 118, 219
Old Exchange 103
plague (1625) 165; (1665) 126, 136, 140, 143–4
population 11–12, 73, 119, 146, 198
Reformation Societies 203
schools 64, 75, 105
standard of living 10, 155–6, 258
streets and districts:
  Fulham 228
  St James's Park 115
  Strand 148
  Threadneedle St 103, 105
  Vauxhall 105, 237
  West End 75
Londonderry 67
longitude 25–6, 218, 238
lottery 57
Louis XIV, king of France 25, 191, 208
Lowndes, William 183
Lowther family 161, 177, 233, 236
  Sir John, of Lowther, 1st bart 234
  Sir John, of Whitehaven, 2nd bart 231
Lowther New Town (Westm.) 177
luxury:
  in China 8
  debated 3–4, 130–1, 139, 149–53, 167, 170, 193
  demoralized 146, 202–14
  and imports 71, 86, 144
  in North America 255
  old and new 153, 158, 193
  termed refinement 217
  see also consumption; emulation

Macaulay, Thomas Babington, 1st lord 263
Macclesfield (Ches.) 219
Machiavelli, Niccolò 47
Machiavellian moment 242
Mackenzie, Sir George 205, 210
Mackworth, Sir Humphrey:
  on Bank of England 183, 196–7, 207–8
  mining projects 174 n. 22, 205
Maddison, Sir Ralph 83, 85, 94–5
Madras 188
Madrid 147
Maidstone (Kent) 228
malaria 224
Mallet, Michael 137–8
Malthus, Thomas Robert 264
Malynes, Gerard:
  on absolute government 89
  compares England and France 49–50
  on foreign exchange 82–3, 84–5, 88
  on justice 79–80, 89
  on minerals 77, 109
  posthumous influence 174, 242
  on riches 77
  on usury 87

Manchester 30, 228
Mandeville, Bernard:
　background 208–9, 261
　on English as improvers 230
　Fable of the Bees 3–4, 208–10, 212, 217
　and Fénelon 208
　and Franklin 255
　impact in Britain 210–12
　impact in continental Europe 243, 251, 252
　on stages of civilization 217
manufactures 83, 120, 122, 133, 167, 236–8;
　　see also cloth industry; cottons; industry; linen
maps 16, 20, 21–3, 25–6, 34, 98, 161
Markham, Gervase 19, 40
marriage:
　concern about 126, 137–8, 150, 195–6, 201, 238–40
　determinants of 63, 159–60, 239–40
　N. W. European pattern 11, 199
　see also fertility; population
marriage market 104
Marseilles 223
Martin, Henry 185, 187–8, 196, 217
Martindale, Adam 165, 166
Massachusetts 68
Massie, Joseph 225, 226, 253
mathematics 74, 80, 103, 114, 117–18, 130;
　see also numbers; political arithmetic
Mather, Cotton 70, 254–5
Mede, Joseph 99
medical police 251
medical services 105, 130, 222–3; *see also* hospitals
Mellish, Reason 141
Melon, Jean-François 247, 248–9, 252
mercantilism 86, 87–8, 185–6, 187
Mercator 185
Merchant Adventurers Company 55–6, 70, 72, 73
　monopoly attacked 71, 87, 92, 185–6
merchants:
　status 71–3, 75, 88–9, 94, 144, 245, 247
　too few 87, 132, 209
　too many 17, 56
　values of 65–6, 75, 204
　see also land and trade
Mexía, Pedro 31
Middle Claydon (Bucks.) 137
middlemen 87, 155
middle sort 94, 207
Midland Rising (1607) 50, 59
Miège, Guy 6, 43, 207, 215–16
migration:
　emigration 69, 120, 136–7
　　encouraged 24, 56, 66, 78
　　opposed 137, 195, 201, 231
　immigration encouraged 56–7, 137, 141, 230–1

internal migration 63, 119, 140
　see also naturalization; population
military-fiscal state 43, 190, 212, 257
millennium, expectations of 39, 99, 100–1, 114, 126–7; *see also* Last Judgement
Milton, John 92
minerals 34, 77, 109
mines 61, 134, 159, 205, 231, 236, 259; *see also* coal
mirrors 158, 159
Misselden, Edward 82
　on balance of trade 82, 85, 86
　on commerce and politics 78, 88, 89
　on usury 87
model villages 177
modernity 1–2; *see also* Ancients *versus* Moderns
modern system of politics 180, 190
Mokyr, Joel 229, 241
Molina, Luis 85 n. 160
Molyneux, William 161, 163, 188
money supply, *see* Bank of England; banks; coinage; recoinage
monopolies, *see* companies; Merchant Adventurers Company; patents
Montagu, Henry, 1st earl of Manchester 83
Montchrétien, Antoine de 76, 242, 245
Montesquieu, Charles Louis de Secondat, baron de 186, 247
moral economy 81–2, 88–9
Morden, Robert 28
More, Henry 104
Morgan, William 147, 161
mortality 140, 148, 159, 195, 198–200, 221–4;
　see also Black Death; health; London; plague; population
Mortlake (Surrey) 228
Moyle, Walter 217
Mun, Thomas
　on balance of trade 82–3, 84–6, 88–9
　on consumption 86, 112
　defence of East India Company 72, 86
　posthumous influence 131, 150, 151, 174, 242, 246
　on usury 87
Munster (Ireland) 66–7
muster rolls 46, 51
Myddle (Salop) 26, 166
myths and legends 36–7, 39–40

Nanjing (China) 9 n. 32
Naples 250–1
national accounts 121
National Debt 179, 182, 191, 201
national income, of England:
　compared with France and Dutch Republic 122, 163, 191
　measured 12–13, 120–1, 163–4, 225
　per head 12, 121–2
　predicted increase 122–3, 200

unequally distributed 71, 139, 164, 168, 192–3, 204–5, 226–7
  see also economic growth; standard of living
national wealth, see national income; riches
natural history 19–21, 98, 200
naturalization 137, 168–9
natural theology 42, 149 n. 107, 197, 200
naval power 11–12, 46, 96–7, 136, 145, 180, 186; see also navy
navigation:
  and commerce 72, 88, 200, 218
  encouraged 7, 31, 93, 95, 115, 189
  knowledge of 40, 74, 135, 165, 168, 171, 258
  see also naval power
Navigation Acts 88, 95, 130, 257
navy 30, 93, 96–7, 115–16, 135, 168, 171, 229, 257; see also naval power; navigation
Neale, Thomas 195
Needham, Marchamont 97
Nelson, Robert 205, 206
Nene, river 264
Netherlands, see United Provinces
Neve, Richard 228
Newcastle-upon-Tyne 30, 134, 154–5, 228
Newcomen, Thomas 237
New Haven, Conn. 69 n. 74
Newsham, Richard 219
newspapers 172–3
Newton, Sir Isaac 183
Nicholls, William 197
Nicole, Pierre 152–3, 205, 242, 246, 255
Nicolson, William 30
Norden, John 21–2, 60
Norman Conquest 37, 194 n. 114
Norman Yoke 94
North, Dudley, 4th lord 131, 152–3, 166
North, Sir Dudley 172
North, Roger 146
Northampton 178, 264
Northamptonshire 50, 51–2
Norwich 56, 177, 231
Nourse, Timothy 177
Nowell, Laurence 22
numbers:
  appetite for 26–8, 163, 261–2
  number, weight, and measure 26, 47, 80, 117, 188, 213
  see also mathematics
numeracy 230

occupations, distribution of 10, 154–5, 226, 239
oeconomy, see economy: use of term
Office of Address 102–6, 173
Ogilby, John 20, 23, 163
Osaka (Japan) 9
Ottoman Empire 35, 63, 135; see also Turkey
Overbury, Sir Thomas 34

Oxford 25, 105
  Bodleian Library 205
Oxfordshire 19, 20

Palmer, Thomas 32, 35
pansophia 99
papermaking 231
Papin, Denis 237
Paris 12, 33, 103
  bills of mortality 119
  compared with London 75, 127
  French perceptions of 151
  urban improvement in 151, 245
parishes:
  number of 47, 49–50
  officers of 137, 239
parish registers 44, 118
Parker, Henry 94, 111
parliament:
  distribution of seats 48, 181
  opinion in 22, 27–8, 59, 71, 82, 133, 137–8, 195, 225–6
  powerful 178, 229, 245
  reactive 89, 93, 104–6, 175–8
  see also political parties
passions 166, 203, 206, 208–9, 246
  and interests 3, 151–2, 242–3
  see also anxiety; happiness; self-love
patents 57, 60, 230, 236
Paterson, William:
  on Bank of England 182, 197
  on Duty Act (1695) 181
  on Ireland 188
  on Scotland 189, 203, 248
pawnshops 103, 130, 208
peace and plenty 86, 101, 123, 145, 216
peace and prosperity 145
Peacham, Henry 166 n. 180
pedlars 155
Penn, William 161, 254
Pennsylvania 254
Pepys, Samuel 30, 138, 160, 163
perfection:
  of agriculture 232, 248
  of France 33
  of inventions 29, 238
  of knowledge 41, 100–1, 127, 198
  of manufactures 231, 248
  of USA 255
Persia 33
Peru, 232–3
Peter I (the Great), of Russia 224
Peter, Hugh 95
Peterborough (Northants.) 264
petitioning 175–6, 185
Pett, Sir Peter:
  on Danby 163, 167
  Happy Future State 166–7, 171
  on improvement 167

Pett, Sir Peter: (*cont.*)
  and Petty 124, 195
  on population 167
Petty, Elizabeth, Lady 146
Petty, Sir William 32, 101, 116–18, 133, 151, 152, 242
  accused of atheism 126
  on cities 120, 127
  coat of arms 114
  on consumption and emulation 123, 142, 198
  double-bottomed boat 116, 123, 236
  Down Survey 26, 98
  on economic growth 121–3, 125, 198–9
  on empire 187
  on end of world 126–7, 197–8
  and Graunt 118–20, 123
  and Hartlib circle 91, 101–2
  and hearth tax 120, 140
  on improvement 120, 125, 127–8, 148, 198
  influence 123–4, 128, 143, 154, 168, 246
  and Ireland 26, 98, 122, 140
  on London 140, 146, 147–8, 198
  and mathematics 103, 114, 117
  methodology 116–18, 120–5, 197–8
  motto 102, 117
  on national income 120–3
  on plague 121–2, 140, 198
  and political arithmetic 3, 47–8, 89, 116–17, 123–4
  and political economy 76–7, 116, 120–1, 261
  on population 120, 140, 253
    marriage and fertility 159, 239
    projected increase 125, 127, 146, 197–8
    transplantation of 124–5
  religious opinions 125–7
  reputation 180–1, 185, 213, 225–6, 249
  on value of labour 120–3
  works:
    Advice of W.P. 103–4, 127
    Another Essay 146, 197–8
    Political Anatomy of Ireland 122–4, 171
    Political Arithmetick 117, 122–6, 129–30, 138, 142, 146, 171
    'Scale of Creatures' 127–8
    'treatise of naval philosophy' 171
    Treatise of Taxes 120, 121, 123
    Use of Duplicate proportion 118 n. 128
    Verbum sapienti 121–4, 127, 128, 171–2
Petyt, William, see *Britannia languens*
pewter 17, 159, 201
philanthropy 64–5, 227; *see also* charity schools
Philips, Erasmus 201–2
physico-theology, *see* natural theology
Physiocrats 252, 253
Picts 36
pin-making, pins 57, 149, 238

plague 44–5, 62–3, 140, 223; *see also* Black Death; London
plantations, *see* America; empire; Ireland
Plat, Sir Hugh 74
Plattes, Gabriel:
  on bee-hives 113–14
  on cities 148
  on inventions and improvements 99–100, 127, 231
  projects 104, 106
Plockhoy, Peter Cornelius 112
Plot, Robert 20
ploughmen 18–19, 103, 144, 209
Poland 98
political arithmetic 3, 47–8, 114, 143, 168, 175, 205
  applications 117–19, 124–5, 163–4, 172, 175, 188, 191–2, 195, 197–8
  in 18th century 123, 213, 217–18, 225–6
  in France 246, 249
  satirized 163, 213
  see also Davenant; Graunt; King; Petty
political economy 3, 52, 116, 120–3, 261
  early use of term 76–7, 123
  in France 76–7, 245, 247–50, 252
  in Holland 251
  number of publications on 174, 250
  in Scotland 189, 250
  see also economic thought
political parties 138–9, 145, 150, 166, 171, 177, 182, 184
Pollexfen, John 184, 192–3
polygamy 35, 126, 210
Pontefract (Yorks.) 30
poor relief:
  corporations for 105–6, 176–7
  cost of 63, 176, 193
  impact of 63–4, 137, 221, 227
  legislation for 62, 63–4
  proposals for 93, 105, 171, 227
Popish Plot 149
population 11–13, 45, 58, 154, 225, 241
  and economic productivity 111, 120–5, 133, 136, 140–1, 154, 199, 238
  measurement of 28–9, 47, 49–50, 119, 122, 140, 194–5, 201, 225, 246
  multiplication of mankind 111, 125, 140, 197
  17th-century end to growth 136–8, 195, 198–9, 225, 258
  thought too large 31, 50–1, 56, 59, 66, 78, 120
  thought too small 54, 111, 126, 136–8, 140, 167, 181, 187 n. 77
  see also fertility; London; marriage; migration; mortality
Postlethwayt, Malachy 227, 250
post office, posts 23, 25, 173

Potter, Thomas 225–6
Potter, William 93, 104, 109–11, 123
Povey, Charles 173, 174, 204, 236
Powell, Robert 80–1
Powell, Thomas 115
Pre-Adamites 42, 126
printing, discovery of 35, 38
Privy Council 50, 52, 71, 82–4
prodigality 86, 131, 144–6, 205;
  see also excess; luxury
profit, private 3, 53, 54–5, 129
  from land 17–18, 51, 76, 234
  from plantations 66, 69, 254
  from projects 58, 81, 104, 112, 174
  from trade 71, 73, 88
  see also avarice; covetousness
progress 3, 5, 35–6, 41, 252, 253
  use of term 93, 136, 152 n. 120, 197, 201, 212, 217–18, 249
projects 54–5, 60–1, 104, 173–5
proportion 80, 120, 122, 180
proverbs 29, 172–3, 230, 233–4
providences 39, 102–3, 139, 165, 198; *see also* history
public good 53–4, 62, 66, 72, 92, 171
Pufendorf, Samuel 31, 244
Purchas, Samuel 31–2, 39
Pym, John 99

quarantine 63, 140, 223–4
Quarantine Act (1721) 223, 224
Quesnay, François 253, 261
questionnaires 20, 114, 163

Radcliffe, John 209
Ralegh, Sir Walter 133, 136
  on enclosure 59
  on history 39, 218
  on trade and navigation 73, 96
Ray, John 200
reason of state 33, 45–6, 50, 88, 94, 242
rebellions 54; *see also* Midland Rising
recoinage:
  of 1560 55
  of 1696 174, 179, 183–4, 191
Recorde, Robert 26
referenda 117
refinement 217, 255
reformation 1, 75, 83, 114, 116
  proposals for 53–5, 60–1, 66, 91–2
  Protestant 17, 44, 65, 167, 171
  see also reformation of manners
reformation of manners 169, 171, 211
  societies for 203, 206
Renaudot, Théophraste 103, 104, 108
rents 83, 88, 129, 132–3, 136, 139
revolution:
  as cyclical event 38, 40–1, 55, 91, 202–3
  as radical change 1, 91–2, 93, 167, 171

Reynell, Carew 138, 183
Rich, Robert, 2nd earl of Warwick 99
riches, definition of 76–8, 123, 192, 217; *see also* national income
roads, *see* transport
Roberts, Lewes:
  on commerce 72, 74
  on land 77, 89
  on luxury 86
Robinson, Henry:
  on economic growth and happiness 109, 111, 113
  and Hartlib circle 100, 103–4
  on invention and improvement 93, 100
  on sovereignty of seas 96
Roe, Sir Thomas 71
Romans, in Britain 36, 37
Rome:
  imperial 38, 40, 134, 193, 194, 202
  in 17th century 147
Romsey (Hants.) 118, 119
Rota Club, *see* Harrington
Rotterdam 95
Rouen 147
Royal African Company 168
Royal Society 2–3, 41, 91, 98, 105, 118, 141, 151, 172, 196, 246
  activities 20, 23, 116, 134, 223
  aspirations 42, 114, 126–7, 149
  Georgical committee 114
  reputation 128, 143, 242
Russia Company 185
Ryder, Joseph 221

St Michael's Mount (Corn.) 22
Salisbury 178
Sandys, Sir Edwin 69, 71
Saussure, César de 230
Savery, Thomas 237
saving:
  by labourers 159
  in national accounts 122–3, 124
Saxton, Christopher 15, 21–2
Say, Jean-Baptiste 216
schools, *see* education; London
Schröder, Wilhelm von 246
science:
  balance of trade as 85
  demography as 226
  and industry 236–7
  political arithmetic as 117
Scotland 46
  improvement in 97, 154, 161, 187–90, 228, 233–4, 248
  migration from 67, 97
  Petty on 125
  poverty in 135, 233–4
  and Union 189–90
Scots:

Scots: (*cont.*)
  barbarous 32
  dirty 220
Scott, William, 72
Selden, John 97
self-improvement 32, 65, 69, 165–6, 220; *see also* education; literacy
self-love 60, 131, 152, 205, 208–9, 210
Seller, John 146 n. 89, 161
Serres, Olivier de 245
servants 69, 144, 156, 160, 165, 204
Shaftesbury, Anthony Ashley Cooper, 3rd earl of 209, 221
Shelburne, William Petty, 2nd earl of 185
Sheridan, Thomas:
  on improvement 141
  on population 137
  on wealth and trade 123–4, 128, 134, 139
Sibbald, Sir Robert 161
silk, silkworms 35, 70, 111, 231, 245
Silk Act (1700) 184
slaves 70, 263
smallpox 68–9, 200, 223–4, 228, 247, 254
Smart, John 182 n. 52
Smith, Adam 149, 218
  on improvement 263–4
  on political arithmetic 123
  on stationary state 191
Smith, John, promoter of Virginia 69
Smith, John, writer on trade 133
Smith, Sir Thomas 66, 78; *see also* Discourse of the Common Weal
Smyth, John, of Nibley 18
Society for Promoting Christian Knowledge 205
Socinianism 118
Somerset, Edward, 2nd marquis of Worcester 105, 237
Somerset 51
Sorbière, Samuel 134, 154
Sorocold, George 219, 236
South Sea Bubble 208
Southwell, Sir Robert 127, 160, 168, 198
Spain 168
  in America 68, 232–3, 258–9
  census 251
  decline of 33, 135, 193–4
  economic thinking in 2, 243
  trade of 72–3, 217
Speed, John 21–2
Sprat, Thomas 114, 134, 138, 154
Staffordshire 20
standard of living 10–13, 155–6, 160, 238–40, 260; *see also* national income
state secrets, see *arcana imperii*
stationary state, of economy 11, 191, 253
steam power 237–8
Steele, Richard 166
Stepney, George 192
Stillingfleet, Edward, dean of St Paul's 149
stocking-making 57, 133–4, 228, 238
stock market 173, 181

Stow, John 17
Streater, John 111, 120
Suffolk 30
sugar 70, 98, 157, 228, 238
Sully, Maximilien de Béthune 245
sumptuary laws 86, 131, 142, 203, 206; *see also* luxury
surveying 26
Süssmilch, Johann Peter 251
Sutton, Robert 223
Sutton, Thomas 64
Swift, Jonathan 213
Sydenham, Thomas 222, 224

Taunton (Som.) 177
taxation 80, 121, 180–2
  customs 131, 185–6
  excises 93, 94, 120, 135, 163, 175, 180–1
  hearth tax 28, 47, 120, 134, 140, 163, 181, 195
  land tax 28, 93, 94, 181–2
  subsidies 44, 45, 48, 51
  see also customs accounts; Duty Act
tea 158
technology-transfer 57, 230
telescopes 230, 237
Temple, Sir William 134, 152, 163, 185, 210
  on history 38, 41, 194
  on Ireland 135–6
  on luxury 141–2, 150
  on United Provinces 6–7, 135–6
Templeman, Thomas 194 n. 118
Terra Australis 35
terra nullius 31–2, 67–8
Thomas, Sir Dalby 204–5, 210
Thompson, E.P. 81
Thoresby, Ralph 29
three-piece suit 131, 150, 169
thrift 18–19, 72, 75, 166, 244
time, improvement of 7, 228
Tiverton (Devon) 119
tobacco 37, 69, 70, 71, 134, 143, 157, 172
Tobago 259
Tocqueville, Alexis de 221
Tone, river 177
Tories, see political parties
trade:
  growth of:
    internal 23, 111, 155
    foreign 11–12, 70, 86, 129, 138, 154
  and international politics 72–3, 95, 179–80
  opinion about 80, 89, 135–6, 168, 212–13
  as zero-sum game 109, 120, 253
  see also companies; gentry; histories; land and trade; merchants
transport 22–5
  improvements in 23–5, 177–8, 264
travel 22–3, 29–31, 32–3, 100
Trenchard, John, *see* 'Cato'
Tryon, Thomas 193, 200, 210, 255
  career 174–5

on consumption 175, 204
on French trade 175, 185
on improvement 175
on marriage 196
on population 195–6
Turkey 168; *see also* Constantinople; Ottoman Empire
turnips 11, 30, 232
Turnpike Acts 25, 178
Tusser, Thomas 18

Ulster 26, 66–7, 97
United Provinces of the Netherlands 100, 169, 208–9
  Child on 259
  compared with England and France 122, 163, 191
  economic thinking in 244, 251
  economy of 10–11, 136, 199, 243
  English perceptions of:
    as competitor 72–3, 95–6, 130, 133–4, 138, 153
    as model 93, 94–5, 135, 141, 151, 168, 246
    frugality in 34, 131, 209, 244, 251
    as information exchange 242
    Mandeville on 209
    Petty on 122, 126, 127
    Temple on 135–6, 141–2
urban improvement 95, 177–8, 219–20, 260
usury 56, 80, 86–7; *see also* interest rates
Usury Acts (1571, 1624) 56, 86–7
Utopias 54, 92, 101, 112, 153, 167–8, 227
  Macaria 100, 111
  New Atlantis 100
Utrecht, Peace of (1713) 170, 180, 185
Uztáriz, Gerónimo de 7 n. 24

vaccination 224
Vanderlint, Jacob 226–7
Vauban, Sébastien Le Prestre, maréchal de 192, 246
Venice 133, 136
Vermuyden, Sir Cornelius 230
Verney, Robert 142
Vienna 10, 151
Virginia:
  improvement in 36, 67, 70, 96, 255
  natives of 36, 38
Virginia Company 68, 71, 83
Voltaire, François Marie Arouet 247, 250
  on China 8
  on comfort 216 n. 3, 248 n. 163
  on English agriculture 232–3
  on invention 236
  on luxury 211

Waad, Armagil 55, 58
wages 10, 12–13, 56, 188

Wales 30, 35–6
Waller, Edmund 115
Walwyn, William 100, 112–13
war (1689–1713), cost of 179–80, 182, 190–1
Ward, John 37–8
Warwick 178
waterworks 177, 219, 236, 238, 249; *see also* London: New River
Watts, Isaac 212
wealth, *see* national income; riches
weather, *see* climate
Weber, Max 126
weights and measures 130, 172, 176, 177
West Indies 98, 168, 187, 238, 259, 263; *see also* Barbados; Jamaica
Westmorland 177, 179 n. 41
Whigs, *see* political parties
Whiston, James 202–3, 204
Whiston, William 197, 238
Whitehaven (Cumb.) 231
Whitelocke, Sir Bulstrode 22
Widdrington, Sir Thomas 29
widows, widowers 64, 160, 181
wig-making 167, 222
William III, king of England 169, 202
Wilson, Sir Thomas 18, 46, 58
Wiltshire 21
windows 37, 148, 156, 201
Winstanley, Gerrard 94
Winthrop, John, junior 109
Winthrop, John, senior 68–9
Witney (Oxon.) 20
Wolsey, Cardinal Thomas 44
women, improvement of 230, 239
Wood, William, writer on America 68
Wood, William, writer on trade 201
Woodward, John 32
Worcester, marquis of, *see* Somerset, Edward
Worcester 177
workhouses 64, 93, 101, 176–7, 207, 221
  literary 103, 105
Worsley, Benjamin 98, 132, 242, 261 n. 15
  on consumption 96, 142
  on empire 96, 116, 186
  on improvement 96, 133
  and millennium 100–1
  saltpetre project 104
Wotton, William 41
Wren, Sir Christopher 29 n. 71, 134, 148, 183

Xenophon 18, 76

Yarranton, Andrew 115, 161–2
yeomen 18, 26, 58, 158, 235
  wealth of 58, 160
York 29
Yorkshire (West Riding) 29, 154–5
Young, Arthur 252

The manufacturer's authorised representative in the EU for product safety is
Oxford University Press España S.A. of el Parque Empresarial San Fernando de
Henares, Avenida de Castilla, 2 – 28830 Madrid (www.oup.es/en or product.
safety@oup.com). OUP España S.A. also acts as importer into Spain of products
made by the manufacturer.

www.ingramcontent.com/pod-product-compliance
Ingram Content Group UK Ltd.
Pitfield, Milton Keynes, MK11 3LW, UK
UKHW021250180426
11946UKWH00003B/59